Lecture Notes in Electrical Engineering

Volume 857

The book series *Lecture Notes in Electrical Engineering* (LNEE) publishes the latest developments in Electrical Engineering - quickly, informally and in high quality. While original research reported in proceedings and monographs has traditionally formed the core of LNEE, we also encourage authors to submit books devoted to supporting student education and professional training in the various fields and applications areas of electrical engineering. The series cover classical and emerging topics concerning:

- Communication Engineering, Information Theory and Networks
- Electronics Engineering and Microelectronics
- Signal, Image and Speech Processing
- Wireless and Mobile Communication
- Circuits and Systems
- Energy Systems, Power Electronics and Electrical Machines
- Electro-optical Engineering
- Instrumentation Engineering
- Avionics Engineering
- Control Systems
- Internet-of-Things and Cybersecurity
- Biomedical Devices, MEMS and NEMS

For general information about this book series, comments or suggestions, please contact leontina.dicecco@springer.com.

To submit a proposal or request further information, please contact the Publishing Editor in your country:

China

Jasmine Dou, Editor (jasmine.dou@springer.com)

India, Japan, Rest of Asia

Swati Meherishi, Editorial Director (Swati.Meherishi@springer.com)

Southeast Asia, Australia, New Zealand

Ramesh Nath Premnath, Editor (ramesh.premnath@springernature.com)

USA, Canada:

Michael Luby, Senior Editor (michael.luby@springer.com)

All other Countries:

Leontina Di Cecco, Senior Editor (leontina.dicecco@springer.com)

**** This series is indexed by EI Compendex and Scopus databases. ****

More information about this series at https://link.springer.com/bookseries/7818

Andrey A. Radionov · Vadim R. Gasiyarov
Editors

Advances in Automation III

Proceedings of the International Russian Automation Conference, RusAutoCon2021, September 5–11, 2021, Sochi, Russia

Springer

Editors
Andrey A. Radionov
South Ural State University
Chelyabinsk, Russia

Vadim R. Gasiyarov
South Ural State University
Chelyabinsk, Russia

ISSN 1876-1100 ISSN 1876-1119 (electronic)
Lecture Notes in Electrical Engineering
ISBN 978-3-030-94201-4 ISBN 978-3-030-94202-1 (eBook)
https://doi.org/10.1007/978-3-030-94202-1

This Springer imprint is published by the registered company Springer Nature Switzerland AG
The registered company address is: Gewerbestrasse 11, 6330 Cham, Switzerland

Preface

International Russian Automation Conference (RusAutoCon) took place during September 5–11, 2021, in Sochi, Russian Federation. The conference was organized by South Ural State University (national research university). The international program committee has selected 230 reports.

The conference was divided into eight sections, including

1. Process Automation;
2. Modeling and Simulation;
3. Control Theory;
4. Machine Learning, Big Data, Internet of Things;
5. Industrial Robotics and Mechatronic Systems;
6. Computer Vision;
7. Industrial Automation Systems Cybersecurity;
8. Diagnostics and Reliability of Automatic Control Systems;

The international program committee has selected totally 48 papers for publishing in Lecture Notes in Electrical Engineering (Springer International Publishing AG).

The organizing committee would like to express our sincere appreciation to everybody who has contributed to the conference. Heartfelt thanks are due to authors, reviewers, participants and all the team of organizers for their support and enthusiasm which granted success to the conference.

Andrey A. Radionov
Conference Chair

Contents

Machine Learning, Big Data, Internet of Things

Computer Vision

Industrial Automation Systems Cybersecurity

Process Automation

Development of an Automatic Process Control System for Biological Wastewater Treatment

Yu. V. Bebikhov, Yu. A. Podkamenniy, and A. S. Semenov(✉)

Polytechnic Institute (Branch), Mirny Ammosov North-Eastern Federal University, bldg. 1, 5, st. Tikhonov, Mirny 678175, Russia

Abstract. This paper describes the upgrade of the existing wastewater treatment facilities in the town of Mirny, which consists in adopting an automatic process control system (APCS) for the second stage of biological wastewater treatment. Upgrading the existing biological treatment process was necessary due to the qualitative and quantitative changes in households' wastewater, as well as due to the need to adopt an automatic process control system using state-of-the-art wastewater treatment technology. The existing wastewater treatment facility design documentation has undergone major revisions regarding solutions, electricity delivery, and comprehensive automation. The paper describes the structure and computerization of the system; it also shows the automation diagrams and explains which solutions were chosen for the upgrade and why. The paper further dwells upon the development of the operator GUI. The paper also highlights the positive outcomes in the form of using less manual labor. The system is highly reliable and error-proof, as it contains an alarm system, is capable of issuing alerts, can lock the actuators in order to protect its integrity, self-diagnose the malfunctions, process data, generate and archive reports. In order to complete the implementation of the biological wastewater treatment APCS, there are plans to compose a blue paper on the tertiary treatment process that keeps suspended matter on the inner surface of self-washing disc filters. This reduces the amount of filtered particulate matter as well as chemical and biological consumption of oxygen and phosphates; as a result, wastewater fees become lower.

Keywords: Automatic control system · Process · Biological treatment · Wastewater · Software · Controller · Operator · Energy efficiency · Reliability

1 Introduction

Most of the wastewater systems used today in Russia are undergoing reconstruction as of writing this paper [1–3]. The reasons are many: stations have better performance, but the equipment is often outdated and worn-out, while wastewater contains ever more pollutants as people increase their usage of household chemicals [4–6]. Household habits have changed as well, and wastewater now often contains large-sized contaminants such as cotton pins, wet wipes, etc. [7–10].

A. A. Radionov and V. R. Gasiyarov (Eds.): RusAutoCon 2021, LNEE 857, pp. 3–13, 2022.
https://doi.org/10.1007/978-3-030-94202-1_1

Wastewater treatment facilities in the town of Mirny receive wastewater from pipelines that deliver it to the intake tank of the first pumping station, where it is pre-filtered using automated grates; then such wastewater goes into sand traps that remove sand to further transport it for disposal. Once the sand traps are done, wastewater is gravity-drained into the tank of the second pumping station that delivers it to the biological treatment unit, which consists of several treatment tanks in three parallel lines that connect a primary sedimentation tank, a double-passage, four-section aerated lagoon, a secondary sedimentation tank, and a purified water tank (listed in the order of placement). Primary and secondary sedimentation tanks are squared tanks that have conical bottoms and a central pipe; submersible pumps remove sediments from the cones and transport it to the dehydration unit; dried sediments go to the landfill. Bio-treated wastewater is further disinfected by UV radiation.

Since the recent tightening of the regulations on treated wastewater to be discharged into water bodies, the conventional wastewater treatment process developed per SNiP 2.04.03-85 [11, 12] is no longer capable of regulation-required performance, which is why it was necessary to find a new solution.

2 Goals and Objectives

The goal hereof was to develop an automatic process control system for biological wastewater treatment, as it was necessary to upgrade the existing treatment facilities due to qualitative and quantitative changes in households' wastewater. To that end, the researchers addressed the following objectives: propose an advanced wastewater treatment technology, describe the structure and computerization, develop the automation diagram and choose the hardware, and create an operator GUI integrating the station information system.

3 Research Results

3.1 Solutions in Use

Since biogenic contaminants are mainly removed by biological treatment facilities, it is recommendable to split the nitrogen and phosphorus removal procedures into biological simultaneous nitrification–denitrification (SNdN) in aerated lagoons and chemical deposition of phosphorus in secondary sedimentation tanks [13–15]. To continue the upgrade effort, it was decided to revise and update the existing design documentation in sections covering the solutions, electricity and mechanics, and comprehensive automation.

When making changes to the documentation on Mirny's wastewater treatment facilities capable of treating up to 19,000 m^3 a day, the authors opted for enhanced biological SNdN and phosphorus removal plus prefermentation in primary sedimentation tanks; this is an advanced wastewater treatment technology that has some fundamental differences from the conventional process. For SNdN, polymer loads that had been used as biocenosis support were removed from the existing aerated lagoons and replaced with fine-bubble disc diffusers with variable-frequency drive-equipped air blowers to provide surface-wide oxygenation.

Nitrification zones of aerated lagoons contain six dissolved oxygen control systems provide cascaded control over the air blowers. The implemented SNdN process consists in creating an anoxic environment in the denitrification zone and supplying wastewater to it. Figure 1 shows how the selected SNdN technology removes organic nitrogen.

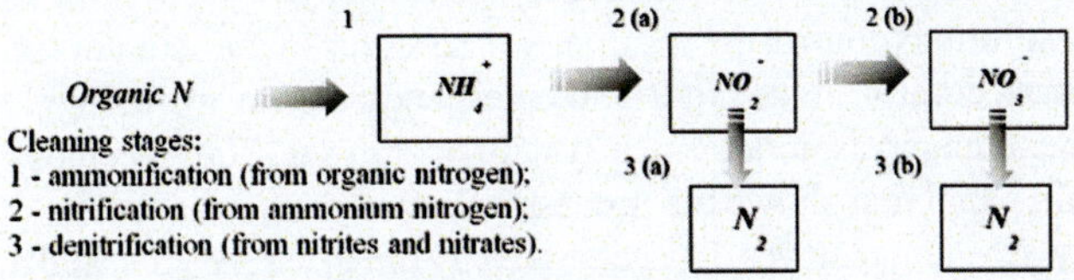

Fig. 1. SNdN process.

During denitrification, nitrites are reduced biologically (Stage 3) to molecular nitrogen, and so are nitrates (Stage 3b); this nitrogen is blown from wastewater into the atmosphere. Denitrification is provided by heterotrophic facultative anaerobes that feed on the organic substrate and oxygen contained in the nitrates. The presence of dissolved oxygen switches them to aerobic metabolism; therefore, in order to oxidize the organics, denitrifying bacteria use oxygen not from nitrites or nitrates, but the dissolved oxygen, i.e., the process is similar to aerobic biological treatment. The environment that contains chemically bonded oxygen rather than dissolved oxygen is referred to as anoxic environment. Anoxic environment is a result of using mechanical mixing instead of aeration in order to keep activated sludge suspended.

The conventional biological wastewater treatment process is not capable of effective phosphorus removal, which is why it was decided to use chemical phosphorus deposition in secondary sedimentation tanks in order to reduce the concentration of phosphates in purified water [16–18]. In order to prevent possible release of suspended matter from secondary sedimentation tanks, the design should provide for appropriate placement state-of-the-art disc mesh filters.

Once the design documentation was revised, the engineering team purchased and installed the following APCS equipment: 2 automated workstations, 8 automation cabinets, 1 coagulant dosing unit, 4 automatic samplers, 27 electrically actuated valves, 6 oxygen meters, 6 sludge/sediment level sensors, 6 excess sludge flow meters (Fig. 2).

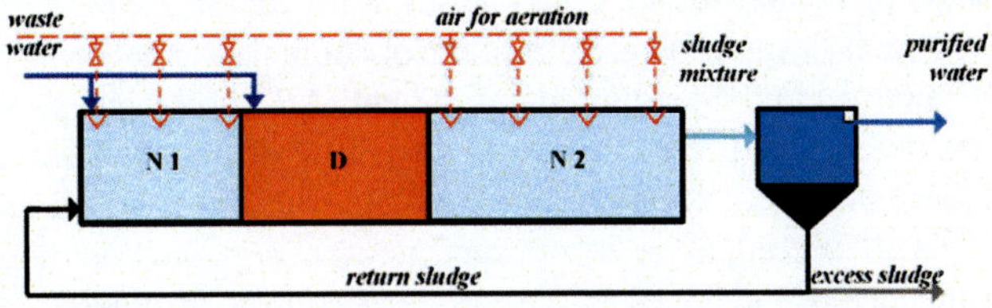

Fig. 2. Novel biological treatment technology: N is the nitrification zone (aerobc conditions); D is the denitrification zone (anoxic conditions).

3.2 Structure and Computerization

The design provided for a hierarchical, two-tiered control system. The lower tier comprises local control systems based on programmable controllers and provides automatic local control in real time as well as data collection and processing according to the signal lists; data is then communicated to the upper tier [19, 20].

Lower-tier solutions control the parameters and status of the actuators on site; provide automatic process control; issue alerts and emergency alarms while logging the events; enable equipment diagnosis; monitor the system operations. Staff receives process status data on a mimic panel that shows the process parameters in dynamics.

The facility's system receives data generated by the PLCs, AWS's, and other software and hardware; data is composed of discrete and analog signals from sensors in the process zone; parameters of analog signal software processing; current values of parameters, process messages, and reports; patterns and history of the process (shown on screens or printed); signaling of parameter deviations (alerts and emergency alarms).

The upper tier communicates data according to the functional hierarchy; data consists of: core process parameters; data exchanged with the adjacent automation systems; readings and data reportable to the management.

Archiving and archival data collection is a function that logs events and reports them to control-room personnel and other staff, including process history, automation events, and operator's actions. The archive is filled when changes occur. The archives contain such data as events (including alerts and alarms), reports on customer-approved emergency and contingency algorithms, as well as hardware and software package (HSP) logs. Archival data can be rendered as tables, graphs, logs, whether on monitors or in printed format. This data can further be used for calculations and other tasks.

Data is mainly shown to the control-room personnel on color displays as mimic panels, histograms, tables, and text messages. Mimic panels are callable library fragments. The content of each fragment must be specified by technologists when developing software applications. This content comprises static and dynamic data such as measurement results, status change, actuator events, flows, etc. Dynamic data is rendered as change in numbers or color, or as flashing. Data follows a general-to-particular hierarchy. The workscreen shows the core data that provides an overview of the situation. Should any parameters deviate from the normative values, or should any status change, the mimic panel must change color or start flashing to be noticed by staff. In that case, staff should be able to 'zoom in' the mimic panel. Each fragment shows its name, current time, as well as equipment, valve, and stream designations according to the process diagram.

The alarm system is designed to alert the control-room personnel on any process disruptions or malfunctions in the diagnosed equipment. All the signals are visualized on monitors. Alarms include alerts on any process parameter deviations beyond the acceptable limits as well as emergency alarms should such deviation constitute an emergency. Any and all alarms are accompanied by appropriate light signaling on monitors.

Event logging includes changes in the status of controlled objects, the timing of parameter deviations beyond the acceptable limits, issued alarms, malfunctions, actions on the part of relay protection and automations, as well as switching of equipment and automations by operational controls. The onset and completion of an event should be logged chronologically. The system must be able to display and/or print time-specific event logs upon the technologist's request. Beside event logs, technologists must be able to request time-specific status logs for controlled objects. Operating parameters logs must be printable or displayable on the control-room personnel's request. Such requests (queries) specify the 'before' and 'after' timepoints.

Each table line contains the following data: parameter ID and short name, unit of measurement, current value in physical units (as digits), and the parameter-assigned timestamp.

Data is recorded in printed forms. These forms are designed in special software to match the existing reporting format and then added to the form library. The user can make a call to show or print a form.

The upper tier collects, receives, and processes analog and discrete data in HSP, generates and runs discrete control actions, makes decisions on whether to trigger protections or initiate emergency shutdowns, and runs various control functions (laws). The upper tier controls the units of the system in a variety of operating conditions, monitors the parameters, and handles emergencies. It consists of a control cabinet (CC) and a power cabinet (PC). CC receives and measures temperature and current signals, controls a single facility, and generates commands to switch different units on and off. PC receives commands from the CC to switch on and off the units within a single facility. The upper tier is Logix-based. It is powered from the biological treatment unit control cabinet (BCC), which has automatic circuit breakers as well as pulsed Class 2 surge protections (SP). The workstation and the monitor of each AWS are powered via an uninterrupted power supply (UPS) that keeps the AWS up and running for at least 10 min after a power outage. UPS also stabilizes and filters the power input to provide an extra layer of surge protection.

Workstations, UPS's, power sockets, and data outlets are mounted in a server rack under the desktops. Upon desktops stand workstation monitors, peripherals, and a printer. The monitors are fitted with basic speakers to issue alerts or emergency alarms.

Each AWS can be used as an engineering station with the provided USB key. Backup power sockets and data outlets are provided to set up a mobile engineering station for the commissioning crew.

The operator remains in full control of the process should upper-tier equipment fail as long as no system-wide failure occurs; thus, an AWS failure, a UPS failure, a lower-tier network switch failure, or a triggered automatic circuit breaker are tolerable when isolated.

The lower tier and the field tier of the system remain functional even if two AWS's fail. The printer is connected to the network to print MS Excel reports or screenshots of the visuals from each AWS.

The BTU cabinet contains backup network switch ports to expand the system and connect to a higher-level network.

The upper tier of the APCS uses servers, AWS's, and engineering stations. Figure 3 shows the APCS flowchart.

Mid-tier runs automatic controls, startups and shutdowns, provides logic and command-based controls, initiates emergency shutdowns and protections. It uses PLCs.

The lower tier (the field tier) collects data on the process parameters and equipment status; it also effects the control actions. Core lower-tier hardware and subunits

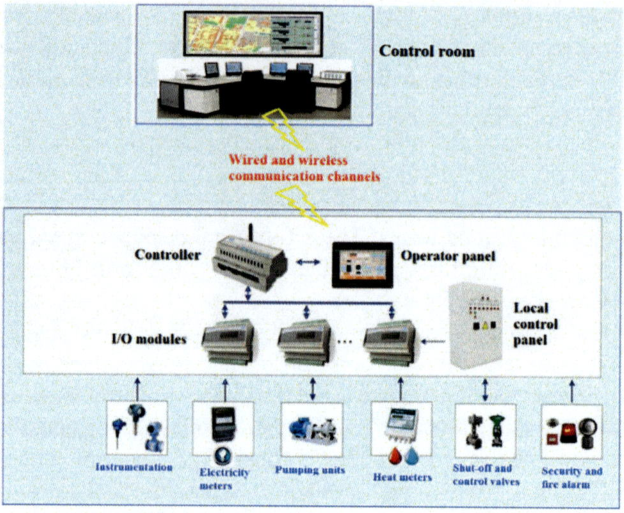

Fig. 3. Flowchart of the automatic process control system

are: sensors and actuators, distributed input/output stations, starters, limit switches, and frequency converters.

The core idea behind developing this system was to replace the existing obsolete process controls and to make a more reliable system that has more extensive and advanced automated controls, enhances process control, and is fail-safer. The APCS stabilizes core process parameters, provides timely and accurate reporting to the staff, and protects the equipment in case of a looming emergency.

The upper tier of the core system enables technologists and engineers to interact with the controlled equipment, arranges all system operations, and prepares data arrays for use by administrative and engineering staff other than control room personnel. Besides, the upper tier enables the APCS engineer to interact with the serviced HSP. It consists of the technologist' AWS and the APCS engineer's AWS.

Upper-tier computers and controllers are connected to a doubled Ethernet network. Logix Allen-Bradley provides a single integrated architecture for discrete control, drive control, servo drive control, and continuous process control; it uses Rockwell Automation controllers that must have at least 20% redundant inputs/outputs to potentially expand the system's functionality [21, 22].

For this particular case, the recommended option is to use ControlLogix controllers with the following minimum requirements: at least two backup slots on the bus to install additional modules; 1756-IF16 4…20 mA analog input modules equipped with 20-pin RTB 1756-TBNH screw clamps; 1756-IV32 discrete input modules equipped with 36-pin RTB 1756-TBCH screw clamps; 1756-IV32 discrete output modules equipped with 1756-OB32 36-pin RTB 1756-TBCH screw clamps; 1756-OF6CI analog modules equipped with 20-pin RTB 1756-TBNH screw clamps. The following connectivity modules are mandatory: 1756-DNB (DeviceNet), 1756-CNB (ControlNet), 1756-ENBT (Ethernet 10/100Мбит/с), MVI56-MCM (Modbus-RTU, ProSoft). The main controller must be equipped with two connectivity modules: CNB and ENBT. The developer estimates the required controller power source capacity and selects a controller processor module with a minimum of 1.5 Mb of memory. Use packaged Series 1492 terminals with fuses, including connection cables, to connect inputs and outputs in controller cabinets.

For electric drives, the system has actuator locking dependencies to prevent jamming in case of an unscheduled shutdown or another emergency. HSP user develops the logic behind the locking dependency. The complete control system also has additional inputs and outputs to execute the corresponding locks in the designed hardware circuit.

3.3 Automation Diagram and Choice of Hardware

The automation diagram is the core design document that defines the structure and the extent of process automation at the facility, as well as which automations and hardware, including computers, the facility is to be equipped with.

When designing an automation system, the automation diagram (AD) becomes the main document defining the functional layout and the extent of automation; it also describes the functions of the automation system and how they pertain to the facility.

AD design effort follows this algorithm: define and outline the automation problem, research the facility (object) to automate, develop an automation system, choose the hardware, choose the auxiliaries, install automation on boards and in cabinets, and prepare the documentation.

The automation diagram shows the process and engineering equipment, the piping (pipelines, gas lines, and air ducts), automation hardware or control circuits, communication lines between separate automation hardware pieces or circuits.

Analysis of the process as an object of automation produced the functional diagram and the production mimic panel.

When outlining the optimization problems, constraints should be specified on all the process parameters and variables, as well as on process-affecting changes. Unit-specific functions enable the system to work as intended. Computer hardware is tasked to control equipment startups and shutdowns, monitor its status and prevent overloads, keep it within a certain range of parameters, stabilize some specific process parameters, and optimize the qualitative and quantitative readings from some specific units.

The recommendation is to consider the following in order to choose specific automations: (i) for easier purchasing, configuring, maintenance, and repair, use identical automations to control identical process parameters; (ii) prefer mass-produced automations; (iii) when controlling multiple identical parameters, use multipoint and centralized controls; (iv) to automate complex processes, use computers and control units; (v) accuracy class must match the process requirements; (vi) for local control, use simple and reliable instrumentation, as it will often have to perform in an aggressive environment; (vii) to automate the hardware operating in aggressive environments, provide special instrumentation or protect the instruments if they are not designed for heavy-duty use.

Nearly all the lower-tier instruments have 4–20 mA output and are installed in open-air; they are therefore designed to work at −40 °C to +90 °C and are watertight.

To cable sensors to secondary hardware and to the automation system cabinet, run shielded cables (with earthed shields) on the cabinet side.

To control the process, use sensors that provide standard 4…20 mA output via ModBus RTU interfaces in order to connect the controller to smart devices: heat meters, electrical grid controls, as well as DeviceNet/ControlNet-connected devices.

The selection and quantity of sensors must provide complete control over the process equipment status and flows.

Electricity delivered from external grids must have a low voltage of ~380/220 V ±15% at 50 Hz (solid-earthed neutral, TN-S earthing).

For startup protections, use TeSys magnetic starters from Schneider Electric.

DANFOSS main pump has variable frequency drives (VFDs) connected to the control system via Ethernet/IP interfaces in order to control the process. VFD power electronics contains an automatic input circuit breaker (uses special fast-acting fuses), a magnetic starter, an input filter, and a converter.

Electrical power is measured by the following solutions: voltage and electric power quality measurements with LCD-equipped switchboard indicators; current measurement on each smart starters; controllers calculating the power of each electrical unit and metering electricity for each consumer.

3.4 Operator GUI Development

PLCs are programmed in an integrated production control system using RSLogix 5000 software.

Purified wastewater level stabilization loop controls wastewater flow to the station. Control systems based on this principle are simple and least inertial. Such a system controls an LT level sensor and a PLC that sets a value to the wastewater feed controller. This PID controller is highly accurate in steady state. Figure 4 shows the automatic purified wastewater level control system (LCS) coded in FBD.

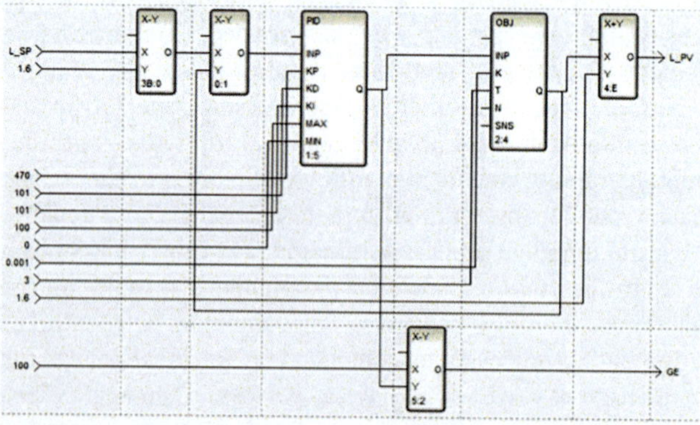

Fig. 4. Automatic purified wastewater level control system.

Operator's AWS is based on FACTORYTALK VIEW, an integrated industrial process control system.

FACTORYTALK VIEW provides a single line of programming, i.e., a single tool to develop all the APCS modules. Single-line programming enables developing human-machine interfaces, resource accounting systems, industrial controller programs, and WEB interfaces within a single project. To that end, FACTORYTALK VIEW has a set of specialized editors.

The main mimic panel is Biological Treatment Unit and Release of Structures, see Fig. 5.

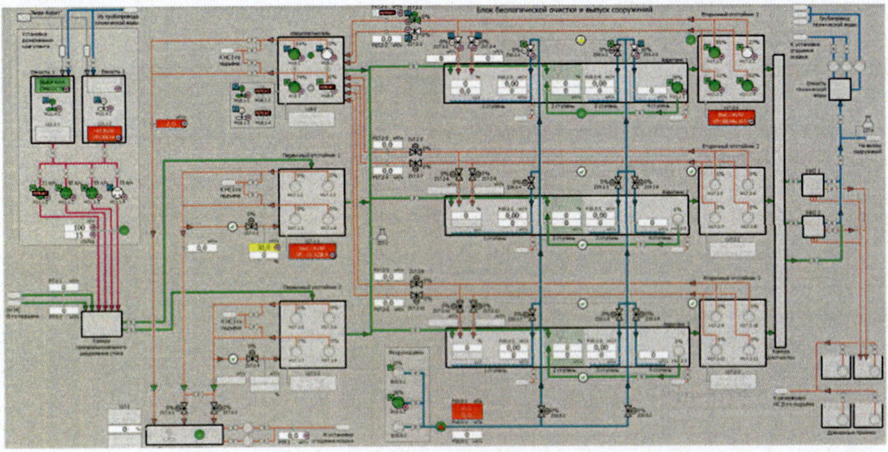

Fig. 5. Biological treatment unit and release of structures.

To further implement the APCS, the authors plan to research and compose a blue paper on tertiary treatment; develop the mathematical methods, models, algorithms, and software; and perform the startup and commissioning of the tertiary treatment process. This effort will include the purchase and installation of self-washing disc filters according to the design. They will retain suspended matter on the inner disc surface. Fine particulate matter removal rates will go up as the filters become clogged. Once preconfigured maximum pressure difference is reached, suspended matter will be removed automatically from the filtering surface. This also reduces the amount of filtered particulate matter as well as chemical and biological consumption of oxygen and phosphates; as a result, wastewater fees become lower.

4 Conclusions

Controlling such a complex facility in practice is difficult, as the problem is multidimensional. This is why the automation problem is usually split into several subproblems to create a tiered control system referred to as an integrated APCS. In tiered systems material flow control subsystems serve as the lower-tier high-frequency cascade. Drainage systems in large cities or at large industrial production sites often contain dozens or even hundreds of such subsystems. The quality of control over material flows in such systems has substantial impact on the accuracy of running the control actions in the APCS, and therefore the upper-tier control over the processes. Besides, flow instability worsens the operating parameters of many units and structures, resulting in non-productive use (waste) of reagents and energy. Treated water, sludge, or sediments are strongly aggressive to equipment that runs measurements or control actions. In this light, the reliability, durability, and energy capacity of flow control systems largely determine the operating performance of the entire integrated APCS. However, to date, wastewater treatment facility control effectiveness is yet to be researched as a function of subsystem structure and material flow control methods; therefore, evidence-based synthesis of such subsystems has not yet been developed.

Therefore, research and development of material flow algorithms and control systems remains relevant for improving the efficiency and performance of wastewater treatment APCS.

The system presented herein can serve as a baseline for further research seeking to comprehensively automate water supply and drainage systems.

Acknowledgment. The authors thank the Russian Science Foundation, grant no. 21-79-10303.

References

1. Suchkova, M.V.: Current state of the water and sanitation services of Russia in the regions. Top. Issues Rational Use Nat. Resour. **2**, 752–758 (2020). https://doi.org/10.1201/9781003014638-36
2. Polyakova, O.S., Semyonov, S.: Reconstruction experience of the wastewater treatment plant (Kargasok village, Tomsk region, Russia) using "constructed wetlands" technology. IOP Conf. Ser. Earth Environ. Sci. **400**(1), 012020 (2019). https://doi.org/10.1088/1755-1315/400/1/012020
3. Dmitrienko, V., Merenkova, N., Zanina, I., Dmitrienko, N.: Use of low potencial wastewater heat. E3S Web Conf. **104**, 01007 (2019). https://doi.org/10.1051/e3sconf/201910401007
4. Korotkova, T.G., Altukhova, D.V., Istoshina, N.Y., Demin, V.I.: Technology of wastewater treatment production of vegetable oils and fats and evaluation of aeration tank efficiency on the basis of microanalysis of activated sludge. J. Ecol. Eng. **20**(7), 70–78 (2019). https://doi.org/10.12911/22998993/109878
5. Iyare, P.U., Ouki, S.K., Bond, T.: Microplastics removal in wastewater treatment plants: a critical review. Environ. Sci. Water Res. Technol. **6**(10), 2664–2675 (2020). https://doi.org/10.1039/d0ew00397b
6. Aksenov, V., Tsarev, N., Maksimova, E.: Implementation of zero liquid discharge systems in Russian industry. IOP Conf. Ser. Mater. Sci. Eng. **972**(1), 012033 (2020). https://doi.org/10.1088/1757-899X/972/1/012033
7. Vialkova, E., Maksimova, S., Zemlyanova, M., Maksimov, L., Vorotnikova, A.: Integrated design approach to small sewage systems in the Arctic climate. Environ. Process. **7**(2), 673–690 (2020). https://doi.org/10.1007/s40710-020-00427-6
8. Primin, O.: Clean water of Russia: problems and solutions. IOP Conf. Ser. Mater. Sci. Eng. **365**(2), 022064 (2018). https://doi.org/10.1088/1757-899X/365/2/022064
9. Smirnova, E., Alexeev, M.: The problem of dephosphorization using waste recycling. Environ. Sci. Pollut. Res. **24**(14), 12835–12846 (2017). https://doi.org/10.1007/s11356-017-8857-0
10. Pariy, A.V., Lysov, A.V.: Creating a national benchmarking system for the utilities of water supply and wastewater sector in Russia. Water Sci. Technol. Water Supply **14**(3), 438–443 (2014). https://doi.org/10.2166/ws.2013.220

11. SNiP 2.04.03-85: Sewerage. External networks and facilities, Moscow (2008)
12. Code of rules SP 32.13330.2012 Sewerage: External networks and facilities. Updated edition of SNiP 2.04.03-85, Moscow (2012)
13. Xu, X., Ma, S., Jiang, H., Yang, F.: Start-up of the anaerobic hydrolysis acidification (ANHA)-simultaneous partial nitrification, anammox and denitrification (SNAD)/enhanced biological phosphorus removal (EBPR) process for simultaneous nitrogen and phosphorus removal for domestic sewage treatment. Chemosphere **275**, 130094 (2021). https://doi.org/10.1016/j.chemosphere.2021.130094
14. Iannacone, F., Di Capua, F., Granata, F., Gargano, R., Esposito, G.: Shortcut nitrification-denitrification and biological phosphorus removal in acetate- and ethanol-fed moving bed biofilm reactors under microaerobic/aerobic conditions. Biores. Technol. **330**, 124958 (2021). https://doi.org/10.1016/j.biortech.2021.124958
15. Chen, S., Chen, Z., Dougherty, M., Zuo, X., He, J.: The role of clogging in intermittent sand filter (ISF) performance in treating rural wastewater retention pond effluent. J. Clean. Prod. **294**, 126309 (2021). https://doi.org/10.1016/j.jclepro.2021.126309
16. Pestryak, I.V., Morozov, V.V.: Flotation of copper-molybdenum ores with household wastewater recycling. Obogashchenie Rud **2020**(4), 3–9 (2020). https://doi.org/10.17580/or.2020.04.01
17. Pestryak, I.V., Morozov, V.V.: Modeling and investigation of the effect of calcium ions on the floatability of molybdenite. Obogashchenie Rud **2019**(3), 22–29 (2019). https://doi.org/10.17580/or.2019.03.04
18. Jargalsaikhan, E., Morozov, V.V.: Optimization of reagent modes in copper–molybdenum ore flotation using economy-oriented criteria. Min. Inf. Anal. Bull. **2019**(3), 210–220 (2019). https://doi.org/10.25018/0236-1493-2019-03-0-210-220
19. Semenov, A., Bebikhov, Y., Yakushev, I.: On the characteristic features of implementing the NET Linx open network architecture in the control logix system. In: Radionov, A.A., Gasiyarov, V.R. (eds.) RusAutoCon 2020. LNEE, vol. 729, pp. 32–41. Springer, Cham (2021). https://doi.org/10.1007/978-3-030-71119-1_4
20. Bebikhov, Y.V., Podkamenniy, Y.A., Golikov, V.V., Spiridonov, V.M.: On the issue of complex automation of mining operations in the diamond mining industry. In: 2020 International Multi-Conference on Industrial Engineering and Modern Technologies (FarEastCon 2020), no. 9271623 (2020). https://doi.org/10.1109/FarEastCon50210.2020.9271623
21. Nikolaev, A.V., Kamakin, A.N., Vasilkov, Yu.V.: Improving the productivity of equipment by modernizing the process control system. Math. Methods Eng. Technol. MMTT **5**, 11–13 (2017)
22. Churkin, G.M., Shilovsky, V.I., Razvin, A.A.: Formation of alternative configurations of structures for local automation of process control systems. Math. Methods Eng. Technol. MMTT **6**, 9–13 (2017)

Automation the Processes of Wood Processing by Drilling, Due to the Development a Mathematical Apparatus for Accounting the Interrelated Feed Drive and Drive of Cutting

V. P. Lapshin(✉), I. A. Turkin, and V. U. Omelechko

Don State Technical University, 1, Gagarin Sq., Rostov-on-Don 344000, Russia

Abstract. The article is devoted the issues of mathematical modeling of cutting processes, which today represent a rather complex and nontrivial scientific problem that requires the use the newest methods for synthesizing models based on a synergistic concept and reflecting the interconnection parameters and coordinates state of processing processes, as well as the use of modern software packages to assess the adequacy. By the synergetic concept modeling process, we mean the construction of such models that would reflect the formation reactions to the shaping movements the tool in a woodworking machine in the coordinates of the cutting process, such as processing speed and feed rate. Our goal of modeling in the work has been achieved, a model of the dynamics the cutting system has been developed in two versions, the first option takes into account only the moments and forces associated with the formation of the cutting force in the processing zone, and the second model includes an additional moment of resistance associated with the accumulation of chips in the chip channel drill. The paper presents the results of modeling the obtained models in the MATLAB environment, these results confirm the adequacy of the developed models.

Keywords: Drilling · Cutting force · Dynamics of the processing process

1 Introduction

The modern economy today needs new approaches to solving old problems, which can increase the efficiency of solving old problems. In the case of woodworking, we can talk about the introduction of new management principles, ensuring a higher quality of processing on old woodworking machines and centers. However, the introduction of such methods and principles is not possible without the development of mathematical models of the cutting process, including wood drilling processes.

Not looking that the technologists describing the process of wood processing by drilling [1–14] have formed in sufficient detail a mathematical model that forms a system of moments, there is currently no general model of the dynamics cutting process, taking into account the drive part of cutting. At the same time, when forming the control cutting process on a CNC machine, it is required to synthesize the feed and cutting

A. A. Radionov and V. R. Gasiyarov (Eds.): RusAutoCon 2021, LNEE 857, pp. 14–24, 2022.
https://doi.org/10.1007/978-3-030-94202-1_2

control, taking into account the relationship between the drives, a general, consistent mathematical model cutting process is provided… With this such problem has already been solved, that is, the models have already been formed and modeling these models has been carried out, the results of the work, in sufficient detail, this is all described in [3–9]. In this regard, the purpose of our work is to synthesize a mathematical model that reflects the process of wood drilling by drilling. As the tasks of studying the formation of the model itself, as well as a qualitative study of this model, numerical modeling in the MATLAB/Simulink environment, the purpose of numerical modeling was to assess the degree adequacy developed mathematical model to the real of wood drilling processes.

2 Research Methodology

At the heart woodworking machine is a modern electric drive. The electric drive itself is a device that converts electrical energy into mechanical, the model of such a system based on a DC electric drive with a collector is given in the expression (1) [3–5]:

$$\begin{cases} \frac{di}{dt} = \frac{1}{L}U - \frac{1}{L}c_e\omega - \frac{1}{L}Ri \\ \frac{d\omega}{dt} = \frac{1}{J}c_m i - \frac{1}{J}f(\omega) \end{cases} \quad (1)$$

where U is the voltage supplied to the motor collector, i is the current consumed by the motor, R, L are the parameters of the electrical part of the motor, J is the parameter characterizing the inertial properties of the motor rotor, the reduced inertial moment of all rotating masses, ω is the rotational speed of the motor rotor, $f(\omega)$ - external, applied moment resistance, cm, ce - mechanical and electrical constants of the motor. For modeling, we denote the variable ω as x_1, and the variable i as x_2, in addition to this, the following notation for the constants is introduced: $a_{11} = \frac{1}{J}$, $a_{12} = \frac{c_m}{J}$, $a_{21} = \frac{c_e}{L}$, $a_{22} = \frac{R}{L}$, $b = \frac{1}{L}$ after that from the systems (1), we obtain the following system:

$$\begin{cases} \frac{dx_1}{dt} = -a_{11}f(x_1) + a_{12}x_2 \\ \frac{dx_2}{dt} = -a_{21}x_1 - a_{22}x_2 + bU \end{cases} \quad (2)$$

The static equations for system (2) will take the form:

$$\begin{cases} 0 = -a_{11}f(x_{01}) + a_{12}x_{02} \\ 0 = -a_{21}x_{01} - a_{22}x_{02} + bU \end{cases} \quad (3)$$

where x_{01}, x_{02} - the values coordinates of the state system at the equilibrium point.

In Eqs. (2) and (3), f(x_1) the functional dependence, the values the coordinates state system at the equilibrium point reflecting the connection of resistance from the cutting side, the shaping movement the tool, in our case we talking about drilling holes. When describing this functional dependence, we must start from two processes that perform when the drill penetrates into the part, the first process is associated with the drilling process itself, when the leading cutting edges drill cut off part the material due to rotation, and the second process is associated with the fact that the cut material must be removed from the cutting area. The removal of such material occurs due to its movement along the chip channel drill, while the more the drill penetrates into the workpiece, the deeper cutting is in this channel, therefore, this additional moment increases as the drill sinks into the workpiece.

Taking into account the given structure of the construction of spiral (screw) drills, we will develop a mathematical model structure of the moment of resistance to cutting, relying on this system of construction, known in [1, 2].

According to [2], the circumferential cutting force is found as $F = 1000P/Vs$, where P is the cutting power (kW), Vs is the average cutting speed (m/s). The final expression for calculating the circumferential cutting force $F = 500F_{cf}D\frac{V_s}{n}$. Or, after conversion in mm/s and rad/s $F \approx 10\pi F_{cf}D\frac{V}{\omega}$, V is the feed rate in mm/s, F_{cf}- specific value the cutting force (MPa), and ω is the speed of rotation spindle with the drill fixed in it in rad/s. Axial force on the drill, find how $F_{oc} = (0.25 + 0.07D)F$.

Moment of resistance to the main movement $M = RF$, where R is the radius of the twist drill (mm).

Based on these considerations, the general model the cutting system, taking into account the influence on the processing process of two drives, the drive providing the main movement (cutting) and the feed drive, will take the following form (4) where x_1 is the rotor speed driving motion drive in the machine,

$$\begin{cases} \frac{dx_1}{dt} = -a_{11}f_1(x_1, x_3) + a_{12}x_2 \\ \frac{dx_2}{dt} = -a_{21}x_1 - a_{22}x_2 + b_1U_1 \\ \frac{dx_3}{dt} = -a_{31}f_3(x_1, x_3) + a_{32}x_4 \\ \frac{dx_4}{dt} = -a_{41}x_3 - a_{42}x_4 + b_2U_2 \\ f_1(x_1, x_3) = 10\pi F_{cf}DRK_v\frac{x_3}{x_1} \\ f_3(x_1, x_3) = (0.25 + 0.07D)10\pi F_{cf}DK_v\frac{x_3}{x_1} \end{cases} \tag{4}$$

x_3 is the rotor speed drive providing the feed in the machine, x_2 and x_4 are the currents consumed by the drives that provide cutting and feed, respectively, K_v is the transmission ratio of the gearbox and screw pair in the feed drive.

Based on the above model the processing process, the process itself is a kind of interconnected and mutually supportive system of drives, combined processes of forming a force reaction from the side the processed material, shaping movements of the tool.

It should be noted here that the mathematical model given in system (4) reflects only the case of processing when the effect of chips in the drill's chip channels is not so significant, that is, it does not have a noticeable effect on the dynamics of the processing process. In a real machine, when carrying out deep drilling operations, the accumulation of chips in the chip channel of the drill has a significant effect on the cutting process [2]. So, in the work of Glebova I.T. it is indicated: "Difficulties in drilling are associated with the removal of chips from the hole being machined. At a several depths of drilling, the volume of the forming chips begins poor the removed from the holes. The chips are compacted, briquettes are formed in the chip flutes drill, and the drill is jammed in the holes" [2, 15–17].

When drilling, both in metals and in wood, the calculated drilling value is considered associated with the hole diameter. The formula for such a relationship usually looks like this $m = t/d_0$, where t is the drilling depth (mm), and d_0 is the hole diameter (mm).

In the case of metalworking, it is customary to consider values of no more than 3, 3.5, and in the case of drilling in wood, m should be less than 10 [6, 18, 19]. Based on these considerations, from a certain value the influence the chips accumulated in the chip channels on the cutting dynamics becomes critical.

To take into account the friction the chips in the chip channels, an additional integral operator is introduced into the system of Eqs. (4), taking into account the effect of drilling on the formed cutting resistance moment and the formed axial force in the direction of feed:

$$\begin{cases} \frac{dx_1}{dt} = -a_{11}f_1(x_1, x_3) + a_{12}x_2 \\ \frac{dx_2}{dt} = -a_{21}x_1 - a_{22}x_2 + b_1U_1 \\ \frac{dx_3}{dt} = -a_{31}f_3(x_1, x_3) + a_{32}x_4 \\ \frac{dx_4}{dt} = -a_{41}x_3 - a_{42}x_4 + b_2U_2 \\ f_1(x_1, x_3) = 10\pi F_{cf} DRK_v \frac{x_3}{x_1} + \frac{K_m}{d_0} \int_0^t x_3 dt \\ f_3(x_1, x_3) = (0.25 + 0.07D)10\pi F_{cf} DK_v \\ \cdot \frac{x_3}{x_1} + \frac{K_m}{d_0} \int_0^t x_3 dt \end{cases} \tag{5}$$

where K_m is the coefficient characterizing the influence of the accumulation of chips in the drill's chip channel on the dynamics cutting process. It should be noted here that the coefficient can be both linear and non-linear.

In this work, we will consider a linear case in modeling, then the K_m coefficient is such that at m = 9 the value of the component describing the increase in cutting resistance becomes comparable or even greater than the moment of cutting resistance. In this case, the amplification factor can be provided by the effect of compaction the wood in the chip channels (briquetting).

Taking into account the above material, the general mathematical system can be represented by the system of equations (3) for the case of shallow drilling and the system (4) for the case of drilling deep holes.

3 Simulation Results

Before modeling the cutting control system, we will conduct a preliminary analysis the static analysis the system of differential equations (4) with the determination of a possible equilibrium cutting mode [4]. In this case, with static, we mean a mode that will be established in the system after the end of transient processes, that is, such a mode in which all derivatives the state is equal to zero. In this case, the equations of system (5) will take the following form:

$$\begin{cases} 0 = -a_{11}f_1(x_1, x_3) + a_{12}x_2 \\ 0 = -a_{21}x_1 - a_{22}x_2 + b_1U_1 \\ 0 = -a_{31}f_3(x_1, x_3) + a_{32}x_4 \\ 0 = -a_{41}x_3 - a_{42}x_4 + b_2U_2 \\ f_1(x_1, x_3) = K_{f1}\frac{x_3}{x_1} \\ f_3(x_1, x_3) = K_{f2}\frac{x_3}{x_1} \end{cases} \tag{6}$$

where K_{f1} and K_{f2} the corresponding constants.

After reduction and transformation, system (6) can be represented as:

$$\begin{cases} b_1U_1 = a_{21}x_1 + \dfrac{a_{22}a_{11}}{a_{12}}K_{f1}\dfrac{x_3}{x_1} \\ b_2U_2 = a_{41}x_3 + \dfrac{a_{42}a_{31}}{a_{32}}K_{f2}\dfrac{x_3}{x_1} \end{cases} \tag{7}$$

Or, after highlighting x3 in the first equation

$$x_3 = x_1(\frac{b_1a_{12}}{a_{22}a_{11}K_{f1}}U_1 + \frac{a_{21}a_{12}}{a_{22}a_{11}K_{f1}}x_1)$$

After substitution into the second equation of system (7), we get:

$$\begin{aligned} &\frac{a_{41}a_{12}a_{21}}{a_{22}a_{11}K_{f1}}x_1^2 + (\frac{a_{42}a_{12}b_1}{a_{22}a_{11}K_{f1}}U_1 + \frac{a_{42}a_{31}a_{12}a_{21}K_{f2}}{a_{32}a_{22}a_{11}K_{f1}})x_1 \\ &+(\frac{a_{42}a_{31}a_{12}b_1K_{f2}}{a_{32}a_{22}a_{11}K_{f1}}U_1 - b_2U_2) = 0 \end{aligned} \tag{8}$$

Equation (8) has a solution for:

$$(\frac{a_{42}a_{12}b_1}{a_{22}a_{11}K_{f1}}U_1 + \frac{a_{42}a_{31}a_{12}a_{21}K_{f2}}{a_{32}a_{22}a_{11}K_{f1}})^2 - 4(\frac{a_{42}a_{31}a_{12}b_1K_{f2}}{a_{32}a_{22}a_{11}K_{f1}}U_1 - b_2U_2) \cdot (\frac{a_{41}a_{12}a_{21}}{a_{22}a_{11}K_{f1}}) \succ 0 \tag{9}$$

For the convenience of further reasoning, consider the case of cutting with two identical motors providing cutting and feed. Let as such an electric motor be a DC motor with the following design parameters: $a_{11} = a_{31} = 6.5$, $a_{12} = a_{32} = 4.8$, $a_{21} = a_{41} = 12$, $a_{22} = a_{42} = 9.3$, $b_1 = b_2 = 14$, $Y_1 = Y_2 = 220$ (taken from [17, 18]).

For the case of drilling with a drill with a diameter of D = 10 mm, with the gear ratio of the reducer and the screw pair in $K_v = 0.01$ and, accordingly, with the feed Sz = 1 mm/rev, when drilling pine Fud ≈ 0.47 N/mm^2 (taken from [1]). Based on this, K_{f1} = 7.38 N/mm, and K_{f2} = 7 N/mm. Then Eq. (9) will look like this numerically:

$$1.55x_1^2 + 319.4x_1 - 159 = 0 \tag{10}$$

Let's check the correctness of our reasoning by directly simulating the system of Eqs. (4), for this development the system research program in the MATLAB/Simulink environment, the simulation results, for the case the previously given parameters the drives and drilling elements, are presented in Fig. 1, 2.

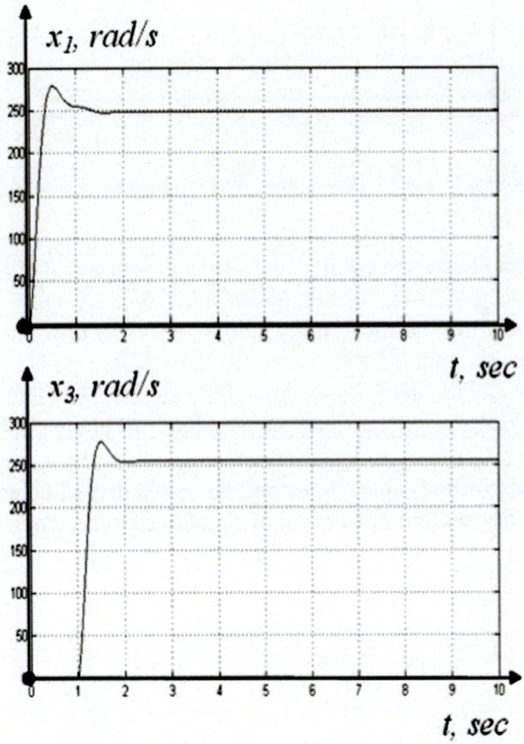

Fig. 1. Results of modeling the system of equations; transient characteristics of angular frequencies of rotation of rotors of electric drives, providing cutting (x1) and feed (x3), respectively.

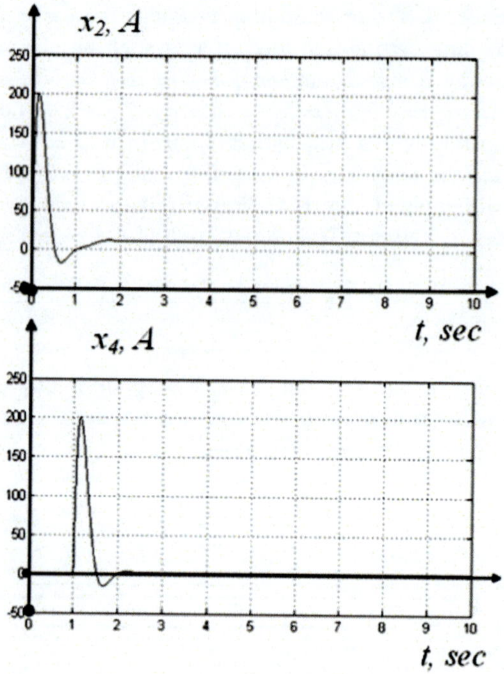

Fig. 2. Currents consumed by electric drives, providing cutting (x2) and supply (x4), respectively.

In the case of simulation, the cutting control system shown in Fig. 1, 2, the electric drive providing the feed will turn on after turning on and reaching the set cutting speed, the electric drive ensuring the rotation the spindle with the drill fixed in it. The feed rate itself, in connection with the transformation the angular motion into translational, through the gearbox and the helical pair, will have a significantly lower value in mm/s.

To clarify the processing process and the possibility of automatic stopping the process, we modernize the system of Eqs. (5) by adding a logical condition to it that stops the feed drive when m values close to 9 dimensions the drilled hole are reached. The results of modeling the modernized system of equations are shown in Fig. 3, 4.

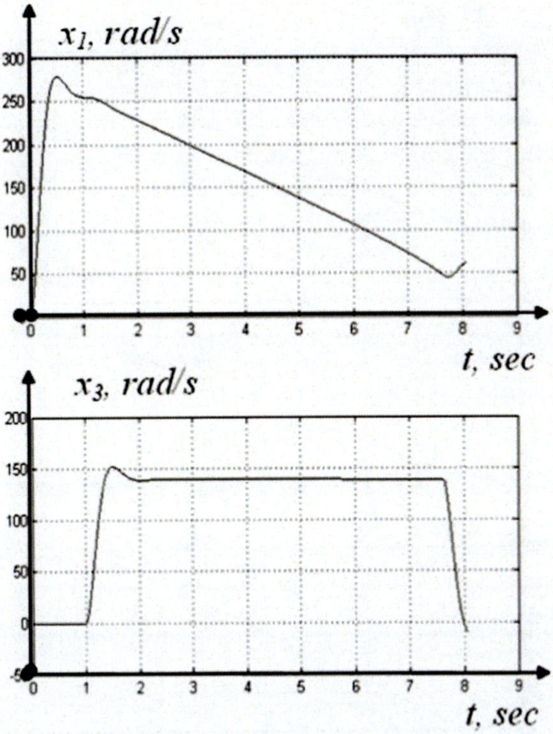

Fig. 3. The results of modeling the system for the case of a system modernized in terms of supply; a) transient characteristics angular frequencies rotation of rotors, electric drives, providing cutting (x1) and feed (x3), respectively.

Figure 3, 4 the results of modeling the system for the case of a system modernized in terms of supply; a) transient characteristics angular frequencies rotation of rotors, electric drives, providing cutting (x1) and feed (x3), respectively, b) moment of resistance on the drive, providing cutting, axial force from the process side, on the drive providing feed, graph of real feed in providing mm, respectively.

As can be seen from Fig. 3 and 4, this is valid when the feed reaches approximately 90 mm. At the same time, the moment of resistance to cutting stops its growth, and the axial force stabilizes. In a real cutting system, after this, the operation of removing the drill from the workpiece begins, after which the chips accumulated in the chip channels are removed and the drill is cleaned.

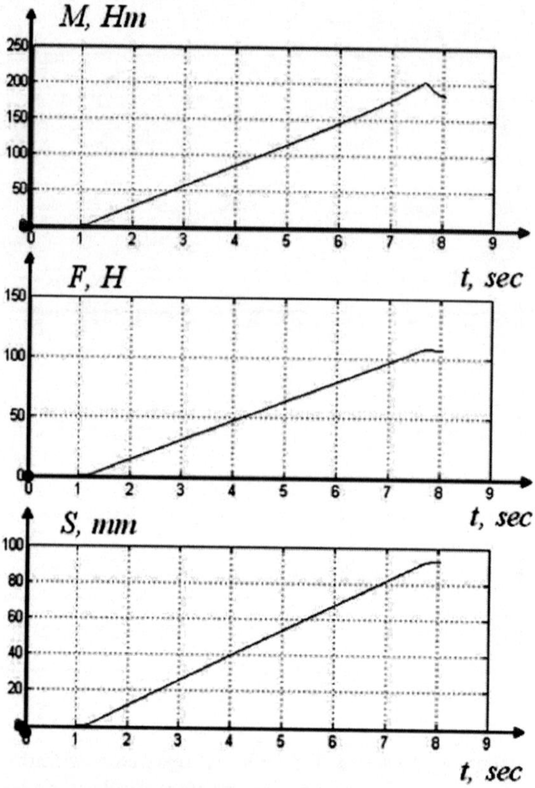

Fig. 4. Moment of resistance on the drive, providing cutting, axial force from the process side, on the drive providing feed, graph of real feed in providing mm, respectively.

4 Conclusion and Discussion of the Results

Equations The article presents an in-depth analysis of two systems, offering two types of mathematical model the drilling system. The section presents two models developed in the Simulink environment the MATLAB system that solve these problems. In the case the first model, both the statics and dynamics the cutting process were assessed, the main regularities were identified and it was determined that the force the interfering machining process (we are talking about the axial component the cutting force) is not large enough, and in this regard, it is small affects the dynamics the drilling process. The analysis the second equation the dynamics machining process shows that the process of accumulation of chips in the chip-leading channels the channel has a decisive influence on the dynamics the machining process. It is this process that has a significant effect on the stability the chips in excess of a certain one, the system loses its stability and the option of losing both the drill and the entire spindle assembly is possible.

In general, it can be stated that the research task, the achievement, the developed mathematical model the drilling control system, the presented system (5), set in the introduction, is quite adequate for the process of processing wood by drilling on woodworking equipment.

Acknowledgments. Authors wishing to acknowledge assistance or encouragement from colleagues, special work by technical staff or financial support from organizations should do so in an unnumbered.

References

1. Glebov, I., Martinon, A.: Permissible depth of wood drilling. Woodworking: Technologies, Equipment, Management of the XXI Century 1:124–129 (2017)
2. Glebov, I., Martinon, A.: Transportation of shavings when drilling wood. Woodworking: Technologies, Equipment, Management of the XXI Century 1:129–135 (2017)
3. Lapshin, V., Turkin, I.: Modeling the dynamics of shaping movements when drilling deep holes of small diameter. Bull. Adyghe State Univ. Ser. 4 Nat.-Math. Tech. Sci. **4**(110), 226–233 (2012)
4. Ivanovsky, E.: Wood Cutting. Lesnaya promyshlennost, Moscow (1975)
5. Lapshin, V., Turkin, I., Bocharov, K.: To the issue of stability of an electromechanical system synthesized by the ACAR method with a nonlinear load. In: Dynamics of Technical Systems "DTS-2017", pp. 8–11 (2017)
6. Jiang, X.: Study of phase shift control in high-speed ultrasonic vibration cutting. IEEE Trans. Industr. Electron. **65**(3), 2467–2474 (2017)
7. Qian, P.: Research on cooling technology of shredder cutting tool with ultrasonic vibration-assisted cutting. IEEE Access **7**, 140513–140523 (2019)
8. Li, X., Zhang, D.: The mechanism of influence on the surface roughness with ultrasonic elliptical vibration cutting. In: 2010 International Conference on Mechanic Automation and Control Engineering, pp. 6172–6175 (2010)
9. Lapshin, P.: Influence of the temperature in the tool-workpiece contact zone on the deformational dynamics in turning. Russ. Eng. Res. **40**, 259–265 (2020)
10. Kozochkin, P.: Vibration features in hard cutting of harden steel workpiece. In: Quality Management, Transport and Information Security, Information Technologies, pp. 458–461 (2018)
11. Sabirov, F.S., Vainer, L.G., Rivkin, A.V.: Vibroacoustic diagnostics of bidirectional end milling. Russ. Eng. Res. **35**(6), 458–461 (2015). https://doi.org/10.3103/S1068798X15060179
12. Porvatov, A.: About possibility of vibroacoustic diagnostics of electrical discharge machining and characterization of defects. Mech. Ind. **16**(7), 707 (2015)
13. Jozwik, J.: Measurement and analysis of vibration in the milling process of sintered carbide workpiece. In: 2019 IEEE 5th International Workshop on Metrology for AeroSpace, pp. 376–380 (2019)
14. Zakovorotny, V., Nguyen, D., Fam, D.: Management of evolutionary processes during processing on metal-cutting machines. In: Proceedings of the VIII International Scientific and Technical Conference on the Dynamics of Technological Systems, pp. 142–145 (2007)
15. Jáuregui, J.: Frequency and time-frequency analysis of cutting force and vibration signals for tool condition monitoring. IEEE Access **6**, 6400–6410 (2018)

16. Puchkin, V., Ryzhkin, A., Turkin, I.: Influence of carbides and alloying elements on intercrystallite corrosion in cutting-ceramic inserts. Russ. Eng. Res. **39**(1), 79–85 (2019)
17. Kruk, M., Jegorowa, A., Kurek, J., Osowski, S., Gorski, J.: Automatic recognition of drill condition on the basis of images of drilled holes. In: 2016 17th International Conference on Computational Problems of Electrical Engineering, pp. 1–4 (2016)
18. Jemielniak, K., Urbanski, T., Kossakowska, J., Bombiński, S.: Tool condition monitoring based on numerous signal features. Int. J. Adv. Manuf. Technol. **59**, 73–81 (2012)
19. Hong, J., Zhou, J.H., Chan, H.L., Zhang, C., Xu, H., Hong, G.S.: Tool condition monitoring in deep hole gun drilling: a data-driven approach. In: 2017 IEEE International Conference on Industrial Engineering and Engineering Management, pp. 2148–2152 (2017)

Analytical Review of Electronic Devices of Modern Supercomputing Systems

M. Sidorova, L. Gorbushin(✉), and N. Koneva

Moscow Polytechnic University (Moscow Polytech), 38, Bolshaya Semyonovskaya Street, Moscow 107023, Russia

Abstract. This study analyzes modern hardware solutions for high-performance data processing. High performance of computing systems is achieved by using appropriate architectural solutions and modern electronic devices, which are constantly being improved and modernized. The tables provided in this study describe the latest high-performance computing and storage systems. Modern hardware solutions have been analyzed that break the exaFLOP barrier. Problems associated with achieving super high performance have been formulated.

Keywords: Electronic devices · Computing · Supercomputing systems

1 Introduction

The volumes of transmitted and processed data have been growing rapidly in recent years, inevitably requiring advances in computing technology in terms of performance and reliability. Consequently, this leads to changes in the structure and scale of data processing devices, as well as to the emergence of data processing centers (DPC) [1–3].

Handling large amounts of data requires high computational performance of processing devices and a sufficient level of reliability.

High performance of computing systems is achieved by using appropriate architectural solutions and modern electronic devices, which are constantly being improved and updated.

Computing performance is typically measured in FLOPS (floating-point operations per second), as well as its derivatives. Currently, systems with a computing performance exceeding 10 teraFLOPS are considered supercomputers ($10 * 10^{12}$ or ten trillion FLOPS; for comparison, an average modern desktop computer has a performance of about 0.1 teraFLOPS) [3].

Many benchmarks exist for the evaluation of supercomputinig systems. Performing a number of these benchmarks allows adequately assessing the supercomputer's characteristics in various application conditions.

The most commonly used benchmark is LINPAK, which is based on solving a system of linear algebraic equations using the Gauss method. This is the standard benchmark for evaluating supercomputer performance.

A. A. Radionov and V. R. Gasiyarov (Eds.): RusAutoCon 2021, LNEE 857, pp. 25–33, 2022.
https://doi.org/10.1007/978-3-030-94202-1_3

2 Problem Statement

2.1 Modern Supercomputers

The LINPAK benchmark is used to rank the world's top supercomputing systems (Table 1) [4].

China and the United States are leading by the number of systems included in the TOP-500 in 2020. A serious supercomputer race has unfolded between these countries—212 and 113 supercomputers were included in the November ranking, respectively.

In 2020, almost all the countries on the list of the TOP-500 most powerful computing systems in the world have increased their supercomputer capacities. Japan made an extreme leap [5].

According to the LINPAK benchmark, the most powerful computing system is the Japanese Supercomputer Fugaku—its performance exceeds four hundred petaFLOPS ($400 * 10^{15}$ FLOPS).

A supercomputing system is a computing system with a performance many times higher compared to mainstream computers due to parallel data processing. Parallel data processing implies simultaneous operation of many independent CPUs [2, 3].

Such powerful computational abilities of modern supercomputing systems are due to the cluster architecture that has been used for more than 15 years.

Cluster architecture means that the supercomputers are built by combining serial computing components using high-speed local area networks controlled by special software.

The performance of cluster solutions is improved by various accelerators, most often—GPUs. The utilization of GPUs in conjunction with a standard processor architecture significantly not only increases the graphical performance, but allows solving a wide range of problems in various fields.

The most advanced solutions in this area are AMD Instinct MI100 and Nvidia A100 GPUs.

AMD Instinct MI100 is the world's fastest GPU for high-performance computing. AMD CDNA-based MI100 GPUs is a significant leap in processing power and connectivity offering up to 3.5x higher performance for high-performance computing (FP32 Matrix) and up to 7x higher performance for artificial intelligence (FP16) tasks compared to the previous generation of AMD GPUs.

This GPU provides theoretical peak computational performance for double-precision numbers (FP64) up to 11.5 teraFLOPS (Fig. 1) and an increased computation performance for double-precision numbers up to 74% for high-performance computing applications [6].

The developer, AMD, plans to launch a supercomputing system "El Capitan Supercomputer" in 2023 using these GPUs, as well as certain modern CPU solutions. The computing power of the designed system will be at least 2 exaFLOPS [7].

However, supercomputing systems are not limited to cluster solutions based on x86-64 processors in combination with CPUs.

One of the most promising approaches is the use of ARMv8-a CPUs with scalable vector extension technology (SVE). Such CPUs have proven well in the Supercomputer Fugaku system bringing it to the top of the TOP-500 supercomputers rating for 2020

Table 1. Ten top best computing systems from the TOP-500.

No.	Rmax Rpeak (PFLOPS)	Name	Architecture, processor type, network	Manufacturer	Location country, year	Operating system
1	415.530 513.855	Supercomputer Fugaku	A64FX 48C, Tofu interconnect D	Fujitsu	Kobe Research Institute, Japan, 2020	Linux (RHEL)
2	148.600 200.795	Summit	Power System AC922 POWER9, Tesla V100, Infiniband EDR	IBM	Oak Ridge National Laboratory, USA, 2018	Linux (RHEL)
3	94.640 125.712	Sierra	Power System S922LC POWER9, Tesla V100, Infiniband EDR	IBM	Livermore National Laboratory, USA, 2018	Linux (RHEL)
4	93.015 125.436	Sunway Taihu Light	Sunway MPP SW26010, Sunway	NRCPC	National Supercomputer Center in Wuxi, China, 2016	Linux (Raise)
5	63.460 79.215	Selene	Nvidia DGX A100, AMD EPYC 7742 64C 2.25 GHz, Nvidia A100, Mellanox HDR Infiniband	Nvidia	Nvidia Corporation, USA, 2020	Linux (Ubuntu 20.04.1 LTS)
6	61.445 100.679	Tianhe-2	TH-IVB-FEP Xeon E5-2692, Matrix-2000, TH Express-2	National University of Defense Technology (NUDT)	National Supercomputer Center in Guangzhou, China, 2013	Linux (Kylin)
7	44.120 70.980	JUWELS Booster Module	Bull Sequana XH2000, AMD EPYC 7402 24C 2.8 GHz, NVIDIA A100, Mellanox HDR InfiniBand/ParTec ParaStation ClusterSuite	Atos	Julich Research Center, Germany, 2020	Linux (CentOS)

(*continued*)

Table 1. (*continued*)

No.	Rmax Rpeak (PFLOPS)	Name	Architecture, processor type, network	Manufacturer	Location country, year	Operating system
8	35.450 51.721	HPC5	PowerEdge C4140, Xeon Gold 6252 24C 2.1 GHz, NVIDIA Tesla V100, Mellanox HDR Infiniband	Dell EMC	Eni, Italy, 2020	Linux (CentOS)
9	23.516 38.746	Frontera	Dell C6420, Xeon Platinum 8280 28C 2.7 GHz, Mellanox InfiniBand HDR	Dell EMC	University of Texas at Austin, USA, 2019	Linux (CentOS)
10	61.445 100.679	Dammam-7	Cray CS-Storm, Xeon Gold 6248 20C 2.5 GHz, NVIDIA Tesla V100 SXM2, InfiniBand HDR 100	HPE	Saudi Arabian Oil Company (ARAMCO), Saudi Arabia, 2020	Linux (RHEL)

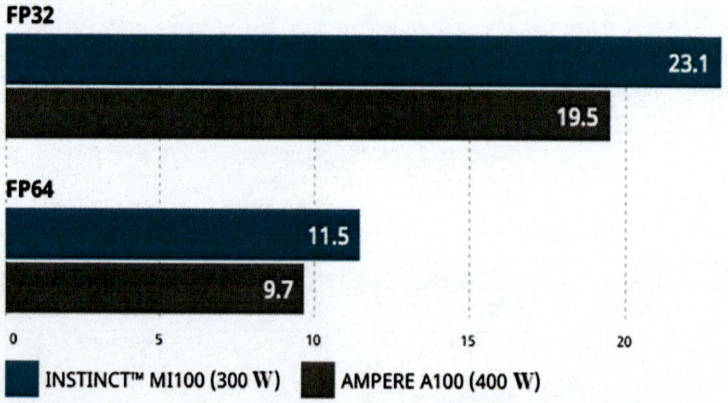

Fig. 1. Performance comparison for AMD MI100 and Nvidia A100.

[4]. This system is based on 152 thousand A64FX CPUs. Each processor has 48 cores, which gives the supercomputer 7,299,072 cores. The peak performance of the system is 514 teraFLOPS, which is 2.56 times faster than the system that took the second place in the list. Furthermore, the Fujitsu supercomputer has the highest energy consumption in the top ten consuming about 28 MW [8].

This CPU type is also considered for the Tianhe-3 computing system developed in China (NUDT). The expected performance is 1.29 exaFLOPS [9].

In 2019, representatives of the U.S. Department of Energy (U.S. The Department of Energy), which is the customer, have officially confirmed that Intel and Cray work together on the Aurora supercomputer "capable of sustaining 1 exaFLOPS performance", which will start its operation at Argonne National Laboratory in the end of 2021.

Aurora's 1 exaFLOPS performance, meaning 10^{18} (quintillion) floating-point operations per second, will be provided by the new, classified next-generation Intel Xe architecture, which characteristics are still a rumor.

The design of the Aurora supercomputer is based on two hundred unified Cray Shasta cluster systems, united by the Cray Slingshot Interconnect and the Shasta software stack.

Each Shasta system is based on new generation Intel Xeon Scalable CPUs, Intel Xe computing architecture, new generation Intel Optane Data Center Persistent Memory, as well as the Intel One API software stack [10].

However, a rumor appeared in 2020 that Intel's one-year delay with the transition to the 7-nm technology, which postpones the release of the Intel Xe Ponte Vecchio GPU, leaves in a doubt the possibility of launching the exaFLOPS-level Aurora supercomputer on schedule [11].

HPE unveils high-performance Cray systems—HPE Cray supercomputers represent one of the most important technological advances in decades. It opens up fundamentally new opportunities for finding solutions to questions arising in the modern world. HPE Cray supercomputers open a new era of supercomputer development, which in turn will become the beginning of a new stage in scientific progress, giving impetus to new discoveries and achievements.

The developers have completely redesigned and created an entirely new solution to meet various requirements. Hardware and software innovations are addressing the challenges that arise with the growth of the number of cores, computing node architectures, and the expansion of workflows due to the widespread adoption of AI technologies.

Due to rapidly changing tasks, it is crucial to be able to choose the right architecture for their execution. With HPE Cray supercomputers, multiple CPU options can be deployed, as well as a mix of different CPUs with a unified management and application development infrastructure. Flexible choices appear of single or multi-socket nodes, GPUs, FPGAs, and other new processing components such as dedicated GPUs for AI.

HPE Slingshot is a newly designed, high-performance switch for supercomputers. It supports Ethernet connectivity and has advanced adaptive routing, unique congestion control, and granular QoS. Support for both IP routing and remote memory operations expands the range of applications beyond traditional modeling and simulation.

3 Discussion

3.1 Supercomputers in the Russian Federation

Russian supercomputers are regularly included in the rating of the world's top supercomputers. At the end of 2020, the Russian computer Christofari is ranked 40th in the TOP-500. At the same time, the computing systems of Rosatom (the most powerful according to some sources) are not included in this list.

At the end of November 2020, one of the top experts in the field of supercomputers in the Russian Federation, the head of the Institute of Software Systems (ISS) of the Russian Academy of Sciences, revealed that the Russian Federation is 12.5 years behind USA and 9.5 years behind China in terms of available real computing power.

In the summer of 2020, the rumor appeared about the Rosatom's intention to make its own "child", the Russian Federal Nuclear Center—All-Russian Scientific Research Institute of Experimental Physics (RFNC-VNIIEF), the only supplier of supercomputer modeling systems for departments and state companies for 2020–2024 [5].

The 33rd edition of the TOP-50 of the most powerful supercomputers in the CIS is presented in Table 2.

The total peak performance among all rating systems was 29.9 PFLOPS. In just six months, two new supercomputing systems appeared on the list and three more systems were updated.

The leader of the list is the Christofari supercomputer developed by SberCloud and NVIDIA, and installed in Sberbank, demonstrating a peak performance of 8.8 PFLOPS. The second position was retained by the Lomonosov-2 sypercomputing system developed by T-Platforms and installed at Moscow State University named after M.V. Lomonosov with the peak performance of 4.9 PFLOPS. A supercomputer manufactured by T-Platforms and Cray with a peak performance of 1.3 PFLOPS takes third place in the rating and is installed in the main data center of the Federal Service for Hydrometeorology and Environmental Monitoring.

The newcomers to the rating were supercomputers installed at the Ioffe Physical-Technical Institute (peak performance of 92.2 teraFLOPS) and the Higher College of Informatics of Novosibirsk State University (196.7 teraFLOPS). Both supercomputers are used to solve scientific and research problems.

49 out of 50 systems in the updated ranking are based on Intel processors. The number of hybrid architectures using GPUs for computing has decreased from 27 to 26. The number of supercomputers using the InfiniBand network communication protocol decreased from 33 to 32, while the number of supercomputers using only the Gigabit Ethernet network communication protocol for node interaction remained equal to 8. The number of systems on the list based on Intel Omni-Path technology has increased from 5 to 6 [12].

Table 2. Leaders of TOP-50 CIS supercomputing systems.

No.	Name Installation location	Nodes CPU GPU	Architecture: number of nodes: node configuration network: computing/service/transport	Rmax Rpeak (teraFLOPS)	Developer Application area
1	Cristofari SberCloud (Oblachnye tekhnologii LLC) SberBank, Moscow	75 150 1200	75: NVIDIA DGX-2 CPU: 2x Intel Xeon Platinum 8168 24C 2.7 GHz, 1536 GB RAM GPU: 16x NVIDIA Tesla V100 EDR Infiniband/100 Gigabit Ethernet/10 Gigabit Ethernet	6669.0 8789.76	SberCloud (Oblachnye tekhnologii LLC) NVIDIA Cloud provider
2	Lomonosov-2 Moscow State University named after M.V. Lomonosov, Moscow	1696 1696 1856	160: CPU: 1x Intel Xeon Gold 6126, 96 GB RAM GPU:2x NVIDIA Tesla P100 1536: CPU: 1x Intel Xeon E5-2697v3, 64 GB RAM GPU: 1x NVIDIA Tesla K40M	2478.0 4946.79	T-Platforms Science and education
3	FSBI Main Computer Center of Roshydromet, Moscow	976 1952 n/a	976: CPU: 2x Intel Xeon E5-2697v4, 128 GB RAM Aries/Aries + Gigabit Ethernet/Aries + Infiniband	1200.35 1293.0	T-Platforms Cray Research
4	Politekhnik - RSK Tornado Supercomputer Center, St. Petersburg Polytechnic University, St. Petersburg	784 1568 128	2: CPU: 2x Intel Xeon Platinum 8268, 768 GB RAM GPU: 8x NVIDIA Tesla V100 3: CPU: 2x Intel Xeon E5-2697v3, 128 GB RAM 64: CPU: 2x Intel Xeon Platinum 8268, 192 GB RAM 36: CPU: 2x Intel Xeon E5-2697v3, 128 GB RAM 56: CPU: 2x Intel Xeon E5-2697v3, 64 GB RAM GPU: 2x NVIDIA Tesla K40 623: CPU: 2x Intel Xeon E5-2697v3, 64 GB RAM FDR Infiniband/Gigabit Ethernet/Gigabit Ethernet	910.31 1309.0	RSK Group Of Companies Science and education

4 Conclusion

A number of challenges arise on the way to exoFLOP performance levels:

1. Power usage. If the current trends in technology development persist, the exaFLOP-level system based on computational elements will have a power consumption of about 100 MW in the foreseeable future. The current energy efficiency of supercomputing systems is approximately 1 GFLOPS/W, while values of 40–75 GFLOPS/W are required for exascale systems.
2. RAM and storage systems. In many aspects, modern technologies do not yet meet the requirements of exaFLOP-level computations in terms of data storage volumes and access speeds, which is true both for RAM and long-term data storage systems. Exascale systems need about 128 PB of RAM, combined into a single address space and having a multi-level architecture.
3. Parallelism. Due to the slowdown in the growth of the clock frequency, the only way to increase the peak performance of computing systems is to use parallel computing. Currently, no computing system architecture exists capable to efficiently handle such a number of cores. Modern application development technologies and their current implementations (OpenMP, TBB, Cilk+, MPI, CUDA, OpenCL, etc.) are also not ready for the efficient use of such numbers of computing resources.
4. Reliability. The fast increase in the number of components leads to a decrease in the mean time between failures (MTBF). Currently, 20% of the processing power of supercomputer systems is lost due to failures and recovery from them, with typical recovery times ranging from 8 h to 15 days. For exascale systems, this time shall be between a few minutes and 1 h.
5. Heterogeneity. The architecture heterogeneity allows using the computing capabilities of high-performance and energy-efficient specialized devices, such as GPUs or FPGAs, which replace VLSIs, while maintaining the functionality of a general-purpose system. However, such a heterogeneity can be used effectively only with a software model providing a unified approach to the development of software for various hardware configurations.

Increasing the computational performance and achieving an exaFLOPS level is one of the most urgent directions in the development of science and technology, which will allow solving the most complex problems of modeling, design and development.

References

1. Gorbushin, L.N., Koneva, N.Ye.: High-performance systems in data processing centers. Sci. Tech. Bull. Volga Reg. 4Ж135–140 (2019)
2. Yablonsky, S.V., Koneva, N.Ye., Konev, F.B.: Development of high-performance data processing systems. – Natural and mathematical sciences in the modern world. In: Collection of Articles Based on the Proceeding of the XIX International Scientific and Practical Conference, vol. 6, no. 18, pp. 49–56 (2014)
3. Yablonskiy, S.V., Koneva, N.Ye.: High-performance computing systems. – Natural and mathematical sciences in the modern world. In: Collection of Articles of the 30th International Conference “Development of Science in the XXI Century” Part 1, Kharkov, pp. 30–36 (2017)

4. https://www.top500.org/lists/top500/list/2020/11/
5. https://www.tadviser.ru/index.php/Статья:Суперкомпьютеры_(рынок_России)
6. https://www.amd.com/ru/products/server-accelerators/instinct-mi100
7. https://www.amd.com/ru/press-releases/2020-03-04-next-generation-amd-epyc-cpus-and-radeon-instinct-gpus-enable-el-capitan1989
8. https://www.ferra.ru/news/computers/nazvana-strana-s-samym-moshnym-superkompyuterom-23-06-2020.htm
9. https://servernews.ru/987026
10. https://www.cnews.ru/news/top/2019-03-19_cray_stroit_pervyj_ekzaflopsnyj_superkompyuter
11. https://servernews.ru/1022102
12. http://www.sib-science.info/ru/news/opublikovana-novaya-redaktsiya-21092020

The Reiterated Neural Network Parametric Identification of Nonlinear Dynamic Models of Objects

A. V. Volkov[1] (✉), A. D. Semenov[2], and B. A. Staroverov[3]

[1] Ogarev Mordovia State University, 68, Bolshevistskaya Street, Saransk 430005, Russia
[2] Penza State University, 40, Krasnaya Street, Penza 440026, Russia
[3] Kostroma State University, 17, Dzerginskogo Street, Kostroma 156005, Russia

Abstract. We posed a problem and developed an algorithm for the neural network parametric identification of nonlinear dynamic models of objects with a computational experiment, the formation of training samples on its basis and the subsequent sequential training of two neural networks. Neural networks perform bijective mapping of object parameters into the original model to the output variables of the second neural network. Sequential training or sequential bijective identification of a neural network consists in preliminary training of the first neural network based on experimental data and using its synaptic coefficients to train the second neural network. An example of parametric identification of a nonlinear dynamic model will be a 600 W high pressure sodium lamp. The computational experiment was carried out in the MATLAB environment. The computational experiment technique and the results of experimental studies are presented in the article. Taking into account the good approximating ability of neural networks, the proposed algorithm can be considered as an effective method for parametric identification of nonlinear models.

Keywords: Parametric identification · Nonlinear object · Neural networks · Bijective mapping

1 Introduction

Quite a lot of approaches and methods have been developed for the identification of nonlinear objects [1, 2]. However, they do not allow us to directly evaluate the parameters of a nonlinear dynamic model. Modern heuristic methods for minimizing the discrepancy between the calculated and experimental data on the parameters of the model [3] make it possible to quickly obtain results of acceptable quality but do not guarantee the finding of an unambiguous solution to the identification problem. The use of neural networks for this purpose [4–19] shows that parametric identification is in principle possible. However, in the case of nonlinear objects, an "individual" approach is required in choosing the type, structure, composition, and algorithm for training the network which in general is a rather complex problem.

A. A. Radionov and V. R. Gasiyarov (Eds.): RusAutoCon 2021, LNEE 857, pp. 34–42, 2022.
https://doi.org/10.1007/978-3-030-94202-1_4

2 Statement of the Identification Problem

A mathematical model of a dynamic nonlinear system is given:

$$F(x, \dot{x}, y, u, \theta) = 0, \tag{1}$$

where x, (x)˙ are vectors of state variables and their derivatives, y, u are vectors of output and input variables; θ is the vector of model parameters.

Experimental waveforms of system variables are given (1)

$$x_e = x(t);\ y_e = y(t);\ u_e = u(t). \tag{2}$$

Using the model of a dynamic nonlinear object or system (1) and experimental data (2), it is necessary to find the mapping

$$\left[x_e, y_e, u_e\right] \Rightarrow \hat{\theta}, \tag{3}$$

where $\hat{\theta}$ is an estimate of the model parameters that minimizes the discrepancy E_m parameters θ taken relative to their nominal values θ_n: $\theta_o = \theta/\theta_n$

$$E_m = \min\left|\theta_o - \hat{\theta}_o\right| \tag{4}$$

3 Two-Dimensional Neural Network Approximation of the Inverse Operator of Parametric Identification

The parametric identification algorithm consists in mapping a set of model parameters θ into an estimate of these parameters $\hat{\theta}$ through a particular identification procedure. Traditionally, this procedure is carried out in three stages [20, 21]. At the first stage, an array of input variables of the U model is formed which has statistical characteristics close to the characteristics of white noise. This condition is necessary to improve the convergence of the model estimates during identification. At the second stage, the initial parameters of the model are mapped into a set of model state variables X and output variables Y using a physical (on the object) or computational (on the model) experiment. In the third stage, the sets of input U and output X, Y variables are mapped using the identification procedure to the estimate of the set of model parameters θ.

Without reducing the generality of reasoning we will assume that the dynamic model under study is completely observable, otherwise an array of state variables X should be used as the set of output variables.

In this formulation, the identification problem is the inverse of the problem of modeling the dynamic model under study. The primal problem is to find an array of output variables Y from the given arrays of input variables U and parameters θ of the model. The inverse problem, the identification problem on the contrary is to find an array of model parameter estimates ($\hat{\theta}$) from the arrays of its input U and output Y variables that can be represented by the following structures (Fig. 1).

Obviously, when solving any inverse problem questions arise about the existence, correctness, and physical realizability of the inverse operator that provides an array of

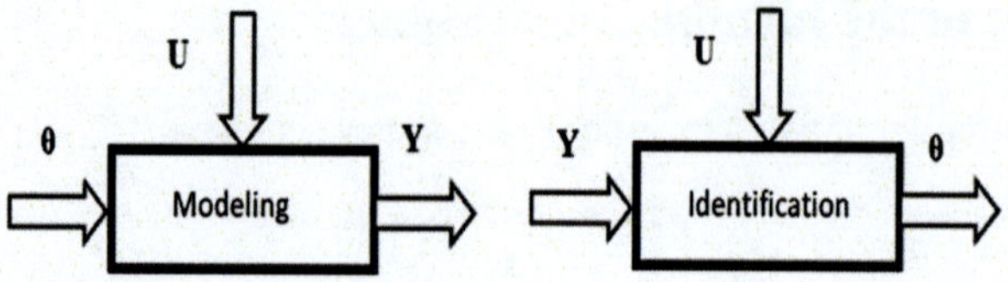

Fig. 1. Block diagrams of modeling and identification of dynamic models.

estimates of the parameters of the model θ. In solving such problems, the method of regularization of A. N. Tikhonov is widely used.

In this case, to improve the accuracy and convergence of estimates it is proposed to use the method of two-dimensional neural network approximation of the inverse operator that implements the identification procedure.

The first neural network based on a training sample of output **U**, output **Y** variables, and state **X** variables grouped into a training sample.

$$\begin{aligned} P_r &= [X, Y, U]; \\ T_r &= Y; \\ X &= X(t);\ Y = Y(t);\ U = U(t), \end{aligned} \quad (5)$$

where **X, Y, U** are arrays of calculated data including all the model variables obtained in each of the N experiments of the experiment plan. The resulting training arrays $\mathbf{P_r}$ $\mathbf{T_r}$ are mapped to the array of synaptic and weight coefficients of the first neural network $\mathbf{W_k}$.

$$[P_r, T_r] \overset{Training}{\longrightarrow} W_k \quad (6)$$

The second neural network based on a training sample consisting of an array of synaptic coefficients and an array of model parameters θ obtained as a result of a computational experiment (6) calculates an estimate of the model parameters $\hat{\Theta}$.

$$\begin{aligned} P_c &= W_k; \\ T_c &= \Theta, \end{aligned} \quad (7)$$

Taking into account the powerful approximating capabilities of neural networks, we should expect high accuracy of parametric identification.

4 Algorithm of Parametric Identification

The following algorithm is proposed for obtaining a mapping (3). On the mathematical model (1) by varying the parameters of θ is conducted an experiment (e.g. full factorial experiment 2N where N is dimension of vector θ) and a sample of the parameters of the Θ model is formed and samples with an array of inputs Pr and an array of outputs Tr s formed to build simulated models:

$$P_c = [X, Y, U];$$

$$T_c = Y;$$
$$X = X(t);\ Y = Y(t);\ U = U(t), \tag{8}$$

where **X, Y, U** are arrays of calculated data that include all the model variables obtained in each of the *N* experiments of the experiment plan.

On the sample $\mathbf{P_r}$, $\mathbf{T_r}$ the simulated model is constructed and the matrix of its parameters $\mathbf{W_r}$ is formed. As a simulated model a neural network is used in which the synaptic coefficients $\mathbf{W_r}$ play the role of identifiable parameters although any dynamic model can be used.

A new training sample is formed for training a static neural network:

$$\begin{aligned} P_c &= W_k; \\ T_c &= \Theta, \end{aligned} \tag{9}$$

1. The second static network is being trained and the estimation of the vector of parameters θ model (1) is calculated in each of the N experiments.
2. Experimental waveforms (2) hat did not participate in the training experiments are fed to the input of the obtained simulated model (the first trained neural network) and the vector of parameters of the simulated model (synaptic coefficients of the dynamic network) $\mathbf{W_r}$ for the current state of the object is calculated.
3. The obtained coefficients are fed to the input of a static neural network and the estimation of the vectors of model parameters for the current state of the real object $\hat{\Theta}$ is being calculated.

5 Forming a Model of a Nonlinear Object

As an example, the parametric identification of a high-pressure sodium lamp of the type high-pressure sodium arc lamp with a power of 600 W is considered [8, 9, 12, 13, 15–17].

$$\begin{cases} \frac{dx_1}{dt} = \frac{1}{L}\left[U_s - \left(\frac{1}{x_2x_3} + R\right)x_1\right], \\ \frac{dx_2}{dt} = A_l U_0^2 x_2^2 \frac{\left(\frac{x_1}{U_0x_2x_3}\right)^2 - 1}{1 + k_1\left(\frac{|x_1|}{U_0x_2x_3} - 1\right)}, \\ \frac{dx_3}{dt} = \left[k_2 + k_3\left(\frac{|x_1|}{U_0x_2x_3}\right)^{k_4}\right]\left[1 + k_1\left(\frac{|x_1|}{U_0x_2x_3} - 1\right) - x_3\right], \end{cases} \tag{10}$$

where x_1 is the lamp current; x_2 is reduced conductance of lamp that takes into account the average electron concentration; x_3 is a dimensionless quantity that takes into account the electron mobility; L, R are respectively the inductance and active resistance of the limiting choke; U_s, U_0 are respectively the supply voltage and the rated voltage on the lamp; A_l is a factor determined by the design of the lamp; k_1–k_4 are electrical coefficients determined for a particular type of lamp.

The lamp parameters are given in Table 3, the choke parameters R = 14 Ω; L = 0.062 H [8, 10–12].

The solution of Eqs. (9) was carried out in the MATLAB environment. The calculated and experimental waveforms of the lamp voltage and current are shown in Fig. 2.

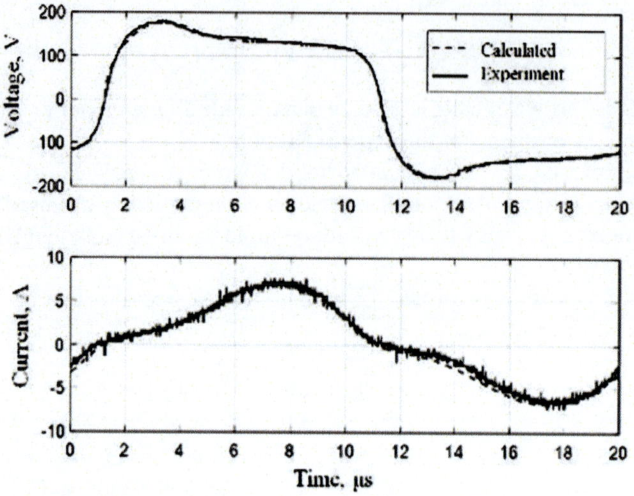

Fig. 2. The calculated and experimental waveforms of the lamp voltage and current.

A close agreement between the calculated and experimental data confirms an adequate description of the object by the system of Eqs. (9). Therefore, to demonstrate the effectiveness of the proposed notification method, we will later use experimental waveforms (Fig. 2) obtained by numerical solution of the system of Eqs. (9).

6 Parametric Identification of a Nonlinear Object Model

Parametric identification was carried out in accordance with the proposed algorithm:

An experiment was conducted in which the parameters of the model (7) were randomly changed in a limited range (Table 1).

A numerical solution of the system (7) was carried out in each experiment and a training sample (5) was formed based on the calculated voltage U_l and current I_l of the lamp, which in this example has the form:

$$P_r = [U_l, I_l];$$
$$T_r = I_l; \quad (11)$$

where U_l, I_l are arrays of waveforms of currents and voltages on the lamp obtained as a result of solving system (9) in each of the experiments specified in Table 1.

Sample (10) was used to train a two-layer forward neural network with 2 neurons in a hidden layer and with linear activation functions in each layer. Thus, the first neural network has 2 synatic connections in the hidden layer, 1 in the output layer, and 2 shift coefficients. The network was trained using the Levenberg-Markwart method with Bayesian regularization. The learning error in each experiment was almost zero. The maximum error value did not exceed 2×10^{-10} A.

A new training sample is formed for training the static neural network from the obtained synaptic coefficients of the dynamic neural network. Table 2 shows the synaptic

Table 1. An experiment plan.

Experience number/variable	U0	Al	K1	K2 × 104	K3 × 104	K4
1	136,2374	5,5335	0,6680	1,5546	3,2885	1,5107
2	135,7986	5,5887	0,6029	1,6960	3,0672	1,5979
3	147,5063	5,5083	0,6133	1,5694	3,1564	1,5831
4	145,7572	5,5269	0,6423	1,7191	3,3771	1,5418
5	148,1105	5,5645	0,6376	1,5764	3,1713	1,5482
6	131,4503	5,5425	0,6313	1,5646	3,0715	1,5423
7	139,4691	5,5091	0,6266	1,5615	3,1124	1,5440
8	131,1343	5,5602	0,6387	1,8664	3,0005	1,5462
9	136,7500	5,5091	0,6576	1,7733	3,2186	1,5426
10	143,2692	5,5582	0,6541	1,8480	3,1059	1,5318

coefficients of the network obtained as a result of training a dynamic neural network according to Table 1 and the subsequent numerical solution (7).

Table 2 was used for training a static neural network which was chosen as a three-layer direct transmission network with linear activation functions in each layer. The network was also trained similarly using the Levenberg-Markwart method with Bayesian regularization. The relative learning errors in each experiment are shown in Fig. 3.

Table 2. Training sample of a static neural network (synaptic coefficients of a dynamic neural network).

Experience number/coefficient	1	2	3	4	5	6	7	8	9	10	Ex.
iwIl	0,1766	0,6514	0,4177	0,1359	–0,2045	0,0802	–0,0223	–0,3555	0,5468	–0,6127	–0,0652
	1,6867	–0,6473	–0,8001	1,7077	–1,5911	–1,2301	2,6223	–2,3779	0,7177	1,5596	1,0806
	0,0735	–0,0824	–0,2155	0,1276	–0,2188	0,2594	–0,1059	0,3422	1,3720	–0,0132	0,1374
	–0,3327	–0,3907	0,3008	–0,3238	0,0329	–0,5883	0,0777	–0,6251	0,5716	0,1266	0,2285
	0,0716	–0,6689	0,3082	–0,0062	0,0953	–0,0249	0,0138	–0,0822	–0,2940	0,0153	0,6529
	0,3149	0,7860	0,5773	–0,4153	0,5312	–0,0028	–0,0305	–0,0315	0,0006	0,4952	–0,5880
iwIl	–0,1726	0,1492	–0,8230	0,3497	–0,2942	0,1310	0,1631	0,1926	–0,2865	0,8928	–,0615
	1,8399	1,7404	–0,4187	–1,2906	0,6604	0,4250	–0,4964	–1,0430	1,0078	0,3248	–1,3381
	–0,0719	–0,0189	0,4246	0,3284	–0,3147	0,4235	0,7739	–0,1854	–0,7189	0,0193	0,1297
	0,3252	–0,0895	–0,5926	–0,8332	0,0473	–0,9608	–0,5675	0,3386	–0,2995	–0,1845	0,2156
	–0,0699	–0,1532	–0,6072	–0,0159	0,1370	–0,0406	–0,1005	0,0445	0,1540	–0,0224	0,6162
	–0,3077	0,1800	–0,1375	–1,0687	0,7640	–0,0046	0,2226	0,0171	–0,0003	–0,7217	–0,5549
Iw2	0,7162	–0,1006	–0,3167	0,7157	–0,9204	0,8455	–0,8275	0,4559	0,4969	–0,8803	0,4002
	0,7329	0,4394	–0,6683	–0,2781	0,6400	–0,5178	–0,1132	0,8415	0,9483	–0,6041	0,4241
b1	–0,0740	–0,7109	–0,5874	0,1046	–0,2845	0,7231	–0,8304	0,3004	0,3473	–0,7435	–0,3563
	0,7870	0,3743	–0,1682	0,7948	0,1287	0,4130	–0,4933	–0,6826	0,4390	–0,7484	–0,5076
b2	0,3909	0,3720	–0,8988	–0,3189	0,7151	–0,2607	0,6461	0,9997	–0,5413	–0,1565	–0,0727

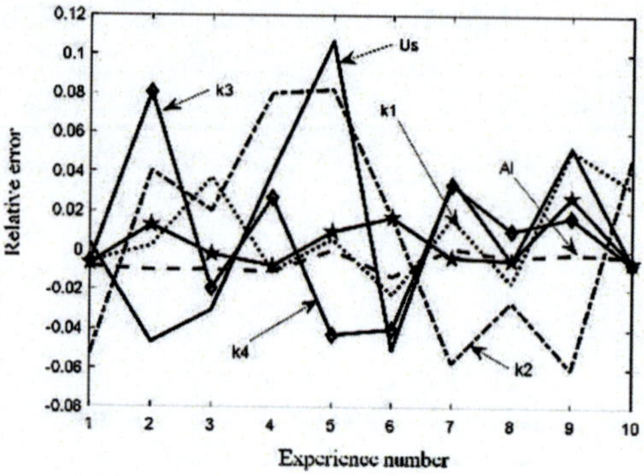

Fig. 3. The errors in calculating the parameters of the analytical model (7) obtained after training.

The experimental values of the lamp current and voltage were fed to the input of the first neural network and the vector of its parameters was calculated (the last column of Table 2).

Calculated vector of parameters of the dynamic neural network was fed to the input of the trained neural static, and calculated estimation of the model parameter vector $\hat{\theta}$ of the real object (Table 3) [20–22].

Table 3. Estimates of the nonlinear model parameters obtained as a result of identification.

Characteristics	U0	Al	k1	k2	k3	k4	Rms deviation	U0
Unit	V	1/J	–	1/s	1/s	–	V	A
Calculated data	4.38	5.5	0.6	1.5 × 104	3 × 104	1.5	10.44	0.697
Identification	126.72	5.48	0.53	1.5907 × 104	2.3676 × 104	1.42	4.38	0.659

We can note a good agreement between the calculated data of the model (7) and the data obtained as a result of identification. In addition, the root-mean-square deviation from the experimental values of current and voltage is less during identification.

7 Summary

1. The problem of parametric identification of nonlinear models of an object is posed, which consists in obtaining the mapping of experimental data of an object in the parameters of its model in its parameters using neural networks.
2. The algorithm of parametric identification is developed. It consists of carrying a computational experiment on a given nonlinear model, forming training samples based on the results of the experiment, then training dynamic and static neural networks, and calculating estimates of the parameters of the nonlinear model based on experimental data using trained networks.

3. A combination of two neural networks is proposed in which, during training and subsequent work, the synaptic coefficients of the first neural network are fed to the input of the second neural network.
4. In such a network the experimental data is displayed in the model parameters sequentially. The experimental data is first displayed in the synaptic coefficients of the first neural network. Then the synaptic coefficients of this network are fed to the inputs of the second neural network the output of which is the required parameter of the nonlinear model.
5. Experimental validation of the proposed method of neural network parametric identification on the example of a high-pressure sodium lamp model showed that the root-mean-square deviation of current and voltage from the nominal values does not exceed 4% for voltage and 10% for current.
6. Taking into account the good approximating ability of neural networks the proposed algorithm and neural networks can be considered as an effective identification method.

References

1. Tsibizova, T.Y.: Identification methods for nonlinear control systems (Metody identifikacii nelinejnyh sistem upravlenija Sovremennye problemy nauki i obrazovanija). Modern Probl. Sci. Educ. **2**(14), 3070–3074 (2015)
2. Pupkov, K.A., Kapalin, V.I., Yushchenko, A.S.: Functional Series in the Theory of Nonlinear Systems (Funkcional'nye rjady v teorii nelinejnyh system). Nauka, Moscow (1976)
3. Khodashinsky, I.A.: Identification of fuzzy systems: methods and algorithms. Manag. Issues **9**, 15–23 (2009)
4. Benderskaya, E.N., Nikitin, K.V.: Recurrent neural network as dynamical system and approaches to its training. Comput. Telecommun. Contr. **4**(176), 29–39 (2013)
5. Schrauwen, B., Verstraeten, D., Campenhout, J.V.: An overview of reservoir computing theory, applications and implementations. In: Proceedings of the 15th European Symposium on Artificial Neural Networks, pp. 471–482 (2007)
6. Fedorov, M.M.: The neural network methods for solving the problems of objects identification (Ispol"zovanie nejrosetevyh metodov dlja reshenija zadach identifikacii objektov). Modern Sci. Res. Innov. **9** (2013). http://web.snauka.ru/issues/2013/09/26285. Accessed 14 Sep 2021
7. Shumikhin, A.G., Boyarshinova, A.S.: The use of neural network dynamic models in the problem of parametric identification of a technological object as part of a control system. Vestnik PNRPU. Chem. Technol. Biotechnol. **3**, 21–38 (2015)
8. Kharchenko, V.F., Yagup, V.G., Yakunin, A.A.: The development of high-pressure discharge lamp computer model. Light Electr. **2**, 52–57 (2013)
9. Semenov, A.D., Volkov, A.V., Shchipakina, N.I.: Operational control algorithm of parameters of high-pressure sodium lamps based on a statistical time series model. IOP Conf. Series Mater. Sci. Eng. **971**(3), 0320842020 (2020). https://iopscience.iop.org/article/10.1088/1757-899X/971/3/032084. Accessed 14 Sep 2021
10. Hecht-Nielsen, R.: Kolmogorov's mapping neural network existence theorem. In: IEEE First Annual International Conference on Neural Networks, vol. 3, pp. 11–13 (1987)
11. Shumikhin, A.G.: The use of neural network dynamic models in the problem of parametric identification of a technological object as part of a control system. Vestnik PNRPU. Chem. Technol. Biotechnol. **3**, 21–38 (2015)

12. Herrick, P.R.: Mathematical models for High Intensity Discharge lamps. IEEE Trans. Power Electron. **5**(16), 648–654 (1980)
13. Volkov, A.V., Shikov, S.A., Temaeva, O.O.: Statistical methods of data analysis of the control system of high-pressure sodium lamps. Sci. Tech. Bull. Volga Reg. **1**, 107–109 (2020)
14. Semenov, A.D., Artamonov, D.V., Brjuhachev, A.V.: Identification of Control Objects. PGU, Penza (2005)
15. Goljandina, N.Je.: Caterpillar-SSA Method: Time Series Forecasting: A Tutorial. St. Petersburg State University Publishing House, St. Petersburg (2004)
16. Hooker, J.D.: The low-pressure sodium lamp. Plasma Science. IEEE Conference Record. Abstracts, p. 289 (1997)
17. Hopfield, J., Tank, D.: Computing with neural circuits: a model. Science **4764**(233), 625–633 (1986)
18. Koprnicky, J.: Electric conductivity model of discharge lamps. Selfreport of the Ph.D. thesis. Liberec (2007)
19. Ajzenberg, J.: Lighting Engineering Reference Book. Energoatomiizdat, Moscow (1983)
20. Lionel, S.: Influence de l'alimentation par commutateur de courant sur le rayonnement des lampes sodium haute-pression. Thèse doctor at engénie électrique en université Paul Sabatier, Toulouse, France (1995)
21. Grossberg, S. (ed.): Neural Networks and Natural Intelligence, p. 637. MIT Press, Cambridge (1988)
22. Risteski, I., Trencevski, K.G.: Principal values and principal subspaces of two subspaces of vector spaces with inner product. Beitragezur Algebra und Geometrie, Contrib. Algebra Geom. **42**, 289–300 (2001)

Application of the Theory of Statistical Hypotheses in Tasks of Automating Technological Processes

S. I. Polyakov[1] (✉), V. I. Akimov[2], and A. V. Polukazakov[2]

[1] Voronezh State University of Forestry and Technologies named after G.F. Morozov, 8, Timiryazeva Street, Voronezh 394613, Russia

[2] Voronezh State Technical University, 84, 20th Anniversary of October Street, Voronezh 394006, Russia

Abstract. This paper substantiates methods of applying the theory of statistical hypotheses in tasks of automating technological processes. Automated system operations are considered to be multi-alternative in terms of the theory of statistical hypotheses: two hypotheses of the correct mode and two hypotheses of the erroneous operation of the system. This paper provides general expressions for calculating the probabilities of these modes for two statistical models of the controlled parameter of the technological process and standard models of the metrological spread of the measuring channel. Herewith the accuracy class of the measuring channel should be expressed as a reduced error. A program has been developed for the statistical evaluation of the operating modes of the automation system for symmetric and asymmetric statistical models of the controlled parameter spread and several models of the metrological evaluation spread of this technological parameter.

Keywords: Statistical hypothesis · Automation · Probabilistic model · Technological process · Decision making · Criterion · Modeling

1 Introduction

If we consider that the functioning of a technological process automation system is the management of a complex technical system, then the theory of statistical hypotheses is the most acceptable one [1–4]. The main advantages of this method are its versatility and simplicity of the mathematical apparatus. From the point of view of operating the automated technological process, two alternative hypotheses can be distinguished: H0 - the system operates in a normal mode (the controlled technological parameter does not go beyond the permissible limits and the measuring channel evaluates this) and hypothesis H1 - the system operates in an emergency mode (the controlled technological parameter goes beyond limits and the measuring channel evaluates this). Let us elaborate on this in greater detail. We introduce the following parameters:

- Controlled parameter A_N;

A. A. Radionov and V. R. Gasiyarov (Eds.): RusAutoCon 2021, LNEE 857, pp. 43–56, 2022.
https://doi.org/10.1007/978-3-030-94202-1_5

- Allowable technological spread A_N: $\pm\,\theta$;
- Density of distribution of the technological parameter: ω (A);
- Assessment of the controlled parameter: $\hat{A}$
- Metrological scatter (absolute error): $\pm\,\Delta$;
- Conditional distribution density: $q\,(A\,/\,A_N) = q\,(\hat{A})$.

Figure 1 shows the conventional designations of the modes of operation of the automation system in the form of zones relative to the controlled technological parameter A_N.

The boundaries of the zones are determined by the parameters θ and Δ. Let us have a closer look at the qualitative differences in automation system behaviour in various zones. Hypothesis H0 is applicable in zone 1. Zones 2 and 3 implement hypothesis H_1. Zones 4 and 5 are characterized by the uncertainty of accepting any of the hypotheses. Let us calculate the prior probabilities of finding the operating modes of the automation system in each zone. These probabilities will be determined by the joint probability density distribution ω (A) and $q\,(A\,/\,A_N)$.

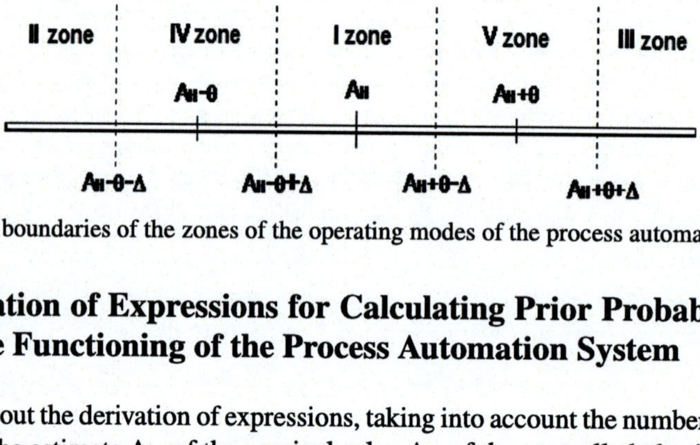

Fig. 1. The boundaries of the zones of the operating modes of the process automation system.

2 Derivation of Expressions for Calculating Prior Probabilities for the Functioning of the Process Automation System

Let us carry out the derivation of expressions, taking into account the number of the zone into which the estimate A_N of the nominal value A_N of the controlled physical quantity of the technological process can fall. Figure 1 shows the numbers of zones considering the type of distribution density of the controlled physical quantity $A_N \to \omega\,(A)$ and the conditional probability density of assessing the controlled physical quantity $\hat{A} \to q\,(A\,/\,A_N) = q\,(\hat{A})$ by the measuring channel of the process automation system. This operation is valid for the metrological characteristics of the measuring channel that is considered by the system as a form of accuracy class within the reduced error [5–8].

In accordance with the figure, we write down the boundaries of the zones.

Zone 1: from $(A_N - \theta - \Delta)$ to $(A_N - \theta + \Delta)$, the width of the first zone: $2\,(\theta - \Delta)$.

Zone 2: $-\infty$ to $(A_N - \theta - \Delta)$. Zone 3: from $(A_N + \theta + \Delta)$ to $+\infty$.

Zone 4: from $(A_N - \theta - \Delta)$ to $(A_N - \theta + \Delta)$.

Zone 5: from $(A_N - \theta - \Delta)$ to $(A_N - \theta - \Delta)$, the width of the fourth and fifth zones is 2Δ.

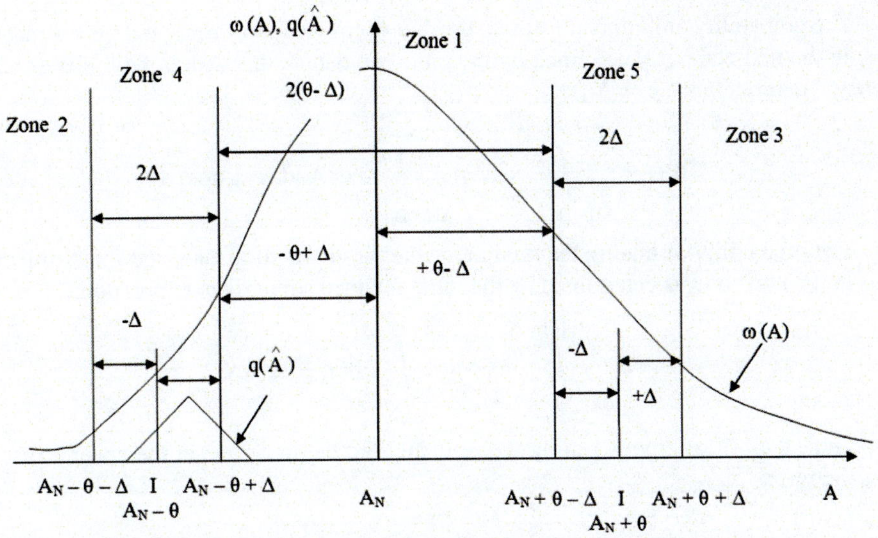

Fig. 2. The boundaries of the zones of the operating modes of the process automation system.

2.1 Calculation of the Prior Probability of Hitting the Estimate A_N of the Controlled Values A_N into Zone 1

The sought value is determined by combining the probabilities of events: the controlled value of the technological process A_N is within the boundaries of the first zone, the true value of A_N is in the fourth zone, and the estimate A_N is in the first zone, and, finally, the last event is associated with the location of A_N in the fifth zone, and the estimate A_N is in the first zone. Let us move on to calculating the probabilities of the listed events. The probability of the first event is determined by the type of distribution ω(A) [9–12].

$$P_{11} = \int_{AH-\Theta+\Delta}^{AH+\Theta-\Delta} \omega(A)dA. \quad (1)$$

To calculate the conditional probability of the second event, we will use Fig. 3.

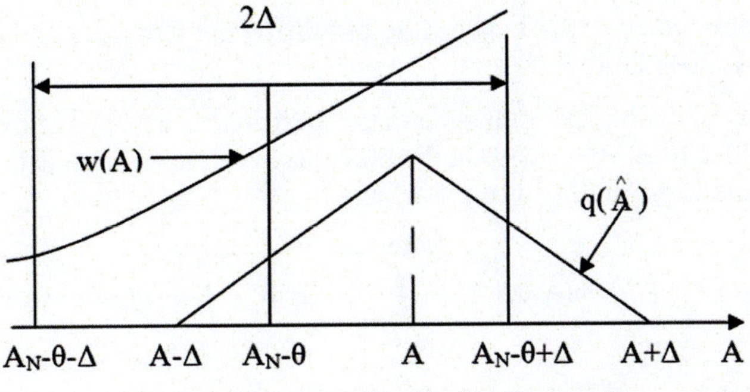

Fig. 3. Alignment of distributions in zone 2.

The probability of finding the nominal value A_N in the fourth zone, and the estimate A_N in the first zone, is determined by the ratio (we denote this within the limits of all integral expressions $A_N = A_H$):

$$P_{12} = \int_{AH-\Theta-\Delta}^{AH-\Theta+\Delta} \omega(A)\left[\int_{AH-\Theta+\Delta}^{A+\Delta} q(\hat{A})d\hat{A}\right]dA. \quad (2)$$

The probability of finding the nominal value A_N in the fifth zone, and the estimate A_N in the first zone, is determined by the ratio obtained similarly to expression (2):

$$P_{13} = \int_{AH+\Theta-\Delta}^{AH+\Theta+\Delta} \omega(A)\left[\int_{A-\Delta}^{A+\Theta-\Delta} q(\hat{A})d\hat{A}\right]dA. \quad (3)$$

Thus, the probability of getting the estimate into the first zone in the general case has the form:

$$P_1 = \int_{AH-\Theta+\Delta}^{AH+\Theta-\Delta} \omega(A)dA + \int_{AH-\Theta-\Delta}^{AH-\Theta+\Delta} \omega(A)\left[\int_{AH-\Theta+\Delta}^{A+\Delta} q(\hat{A})d\hat{A}\right]dA$$
$$+ \int_{AH+\Theta-\Delta}^{AH+\Theta+\Delta} \omega(A)\left[\int_{A-\Delta}^{A+\Theta-\Delta} q(\hat{A})d\hat{A}\right]dA \quad (4)$$

2.2 Calculation of the Prior Probability of Hitting the Estimate of the Controlled Values A_N into Zone 2

The probability of the estimate falling into the second zone is determined by two events: P_{21} is in the second zone, P_{22} - A_N is in the fourth zone, and the estimate is in zone 2. The sought values of the probabilities of these events can be found using Fig. 1 and the method by which expression (3) was obtained:

$$P_2 = \int_{-\infty}^{AH-\Theta-\Delta} \omega(A)dA + \int_{AH-\Theta-\Delta}^{AH-\Theta+\Delta} \omega(A)\left[\int_{A-\Delta}^{AH-\Theta-\Delta} q(\hat{A})d\hat{A}\right]dA \quad (5)$$

2.3 Calculation of the Prior Probability of Hitting the Estimate of the Verified Values A_N into Zone 3

The probability of getting the estimate into the third zone is determined by two events: P_{31} is in the third zone, P_{32} - A_N is in the fifth zone, and the estimate is in zone 3. The sought value of the probability of these events:

$$P_3 = \int_{AH+\Theta+\Delta}^{+\infty} \omega(A)dA + \int_{AH+\Theta-\Delta}^{AH+\Theta+\Delta} \omega(A)\left[\int_{AH+\Theta+\Delta}^{A+\Delta} q(\hat{A})d\hat{A}\right]dA \quad (6)$$

2.4 Calculation of the Prior Probability of Hitting the Estimate of the Controlled Values A_N into Zone 4

The sought value of the probability of getting the estimate into the fourth zone is determined by the sum of the probabilities of three events: the nominal value of A_N is within the boundaries of the fourth zone, the value of A_N is in the first zone, and the estimate is in the fourth zone. And finally, the last event is associated with the presence of A_N in the second zone, and the estimate in the fourth zone. Let us move on to calculating these probabilities. According to Fig. 1, we have:

$$P_{41} = \int_{AH-\Theta-\Delta}^{AH+\Theta+\Delta} \omega(A)dA = \int_{0}^{2\Delta} \omega(A)dA \tag{7}$$

$$P_{42} = \int_{AH-\Theta+\Delta}^{A+\Delta} \omega(A)[\int_{A-\Delta}^{AH-\Theta+\Delta} q(\hat{A})d\hat{A}]dA. \tag{8}$$

$$P_{43} = \int_{A-\Delta}^{AH-\Theta-\Delta} \omega(A)[\int_{AH-\Theta-\Delta}^{A+\Delta} q(\hat{A})d\hat{A}]dA. \tag{9}$$

Result:

$$P_4 = P_{41} + P_{42} + P_{43} = \int_{0}^{2\Delta} \omega(A)dA + \int_{AH-\Theta+\Delta}^{A+\Delta} \omega(A)[\int_{A-\Delta}^{AH-\Theta+\Delta} q(\hat{A})d\hat{A}]dA$$
$$+ \int_{A-\Delta}^{AH-\Theta-\Delta} \omega(A)[\int_{AH-\Theta-\Delta}^{A+\Delta} q(\hat{A})d\hat{A}]dA \tag{10}$$

2.5 Calculation of the Prior Probability of Hitting the Estimate of the Verified Values A_N into Zone 5

The sought value of the probability of getting the estimate into the fifth zone is determined by the sum of the probabilities of three events: the value of A_N is within the boundaries of the fifth zone, the controlled value of A_N is in the first zone, and the estimate is in the fifth zone, and, finally, the last event is associated with the presence of A_N in the third zone, and the estimate are in the fifth zone. The calculation of these probabilities, according to Fig. 2, is as follows:

$$P_{51} = \int_{AH+\Theta-\Delta}^{AH+\Theta+\Delta} \omega(A)dA = \int_{0}^{2\Delta} \omega(A)dA \tag{11}$$

$$P_{52} = \int_{A-\Delta}^{AH+\Theta-\Delta} \omega(A)\left[\int_{AH+\Theta-\Delta}^{A+\Delta} q(\hat{A})d\hat{A}\right]dA \tag{12}$$

$$P_{53} = \int_{AH+\Theta+\Delta}^{A+\Delta} \omega(A)\left[\int_{A-\Delta}^{AH+\Theta+\Delta} q(\hat{A})d\hat{A}\right]dA \tag{13}$$

The total probability P_5 according to expressions (11–13) is in the following form:

$$P_5 = \int_0^{2\Delta} \omega(A)dA + \int_{A-\Delta}^{AH+\Theta-\Delta} \omega(A)\left[\int_{AH+\Theta-\Delta}^{A+\Delta} q(\hat{A})d\hat{A}\right]dA + \int_{AH+\Theta+\Delta}^{A+\Delta} \omega(A)\left[\int_{A-\Delta}^{AH+\Theta+\Delta} q(\hat{A})d\hat{A}\right]dA \tag{14}$$

The technique used and the obtained general expressions (4), (5), (6), (10) and (14) allow us to calculate the prior probabilities of the functioning of the automation system of the technological process within the framework of the theory of statistical decisions [13]: the probabilities of the correct automation system operation, in which the decision made coincide with the actual operating mode of the technological equipment, where both the specified operating limits of the controlled parameter change according to the measuring channel estimates, and the operational malfunction when these parameters go beyond the permissible technological tolerances 2θ and the probabilities of the priori probabilities of standard distributions, often used in metrological practice, within 2Δ. To find the desired values of the probabilities, one can use the results presented in [14] or apply the convolution method [15].

3 Modeling

Based on the value of the expressions (4), (5), (6), (11) and (14), a program was developed and the operating modes of the process automation system were modeled within the framework of the theory of statistical hypotheses. Below are the simulation results for the normal probability density of the statistical model of the controlled value, uniform and triangular statistical models for estimating the controlled parameter [16, 17] (Fig. 4).

On the control panel, you can set the number of tested devices (input from 20 to 1000), the nominal value A (input from 0 to 1000), the error value of the working tool (input from 1 to 50), the distribution law of the error values of the working tool (normal, Rayleigh), ratios $\frac{\theta}{\sigma}$ and $\frac{\theta}{\Delta}$ (input from 1.1 to 10), as well as the distribution law of the error of an exemplary tool (uniform, Simpson, antimodal 1, antimodal 2) (Figs. 5, 6 and 7).

For the program to work, the.NET Framework version 4.0 or a later version must be installed on the computer.

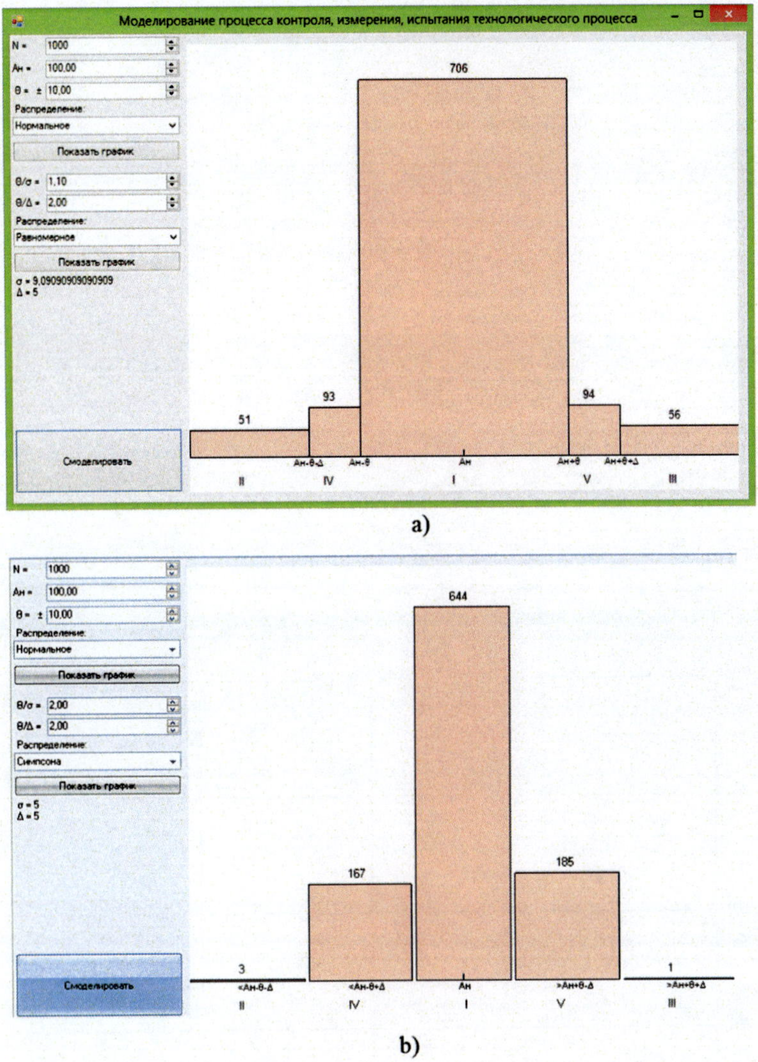

Fig. 4. Modeling results: a) normal and uniform; b) normal and triangular (Simpson) distributions.

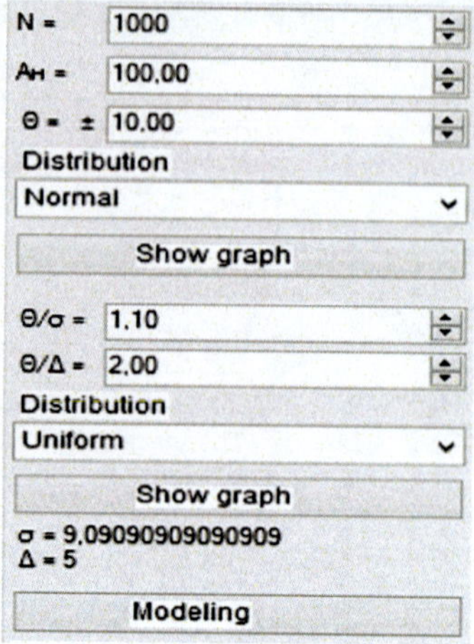

Fig. 5. Control panel

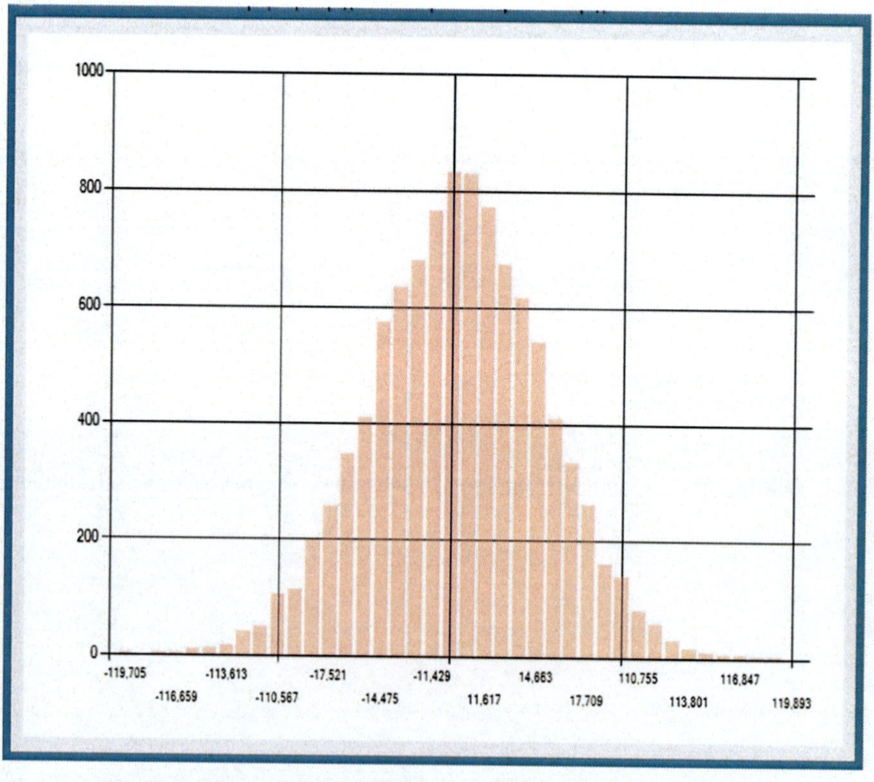

Fig. 6. Histogram of the distribution model of the controlled parameter

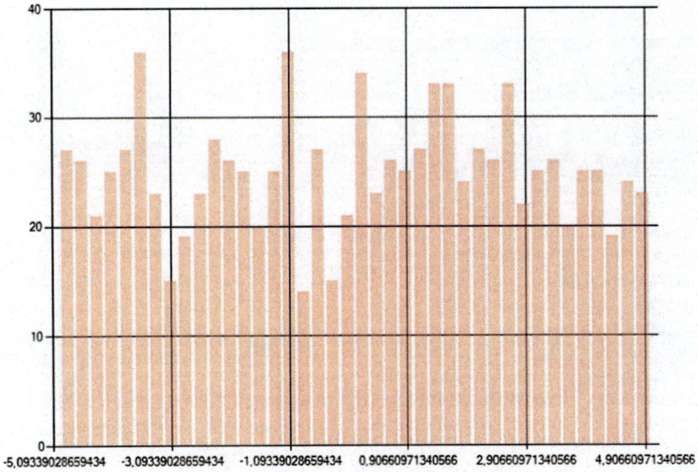

a)

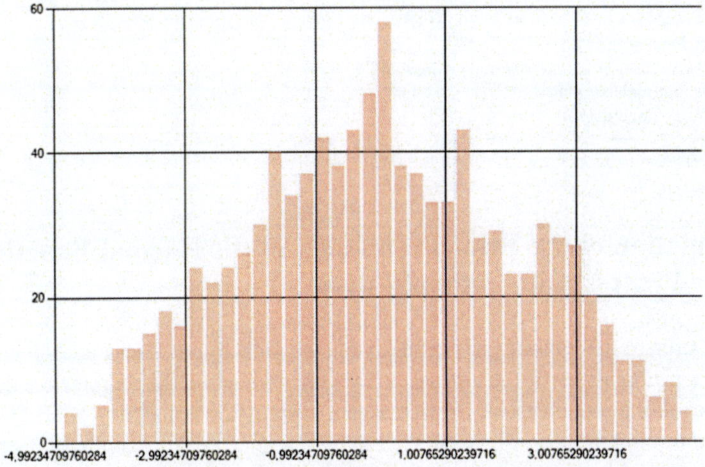

b)

Fig. 7. Histogram of models of distribution of metrological characteristics of the measuring channel a) equal; b) triangular (Simpson) distribution.

Fragment of the program

```
Microsoft Visual Studio Solution File, Format Version 12.00
# Visual Studio 2013
VisualStudioVersion = 12.0.21005.1
MinimumVisualStudioVersion = 10.0.40219.1
Project("{FAE04EC0-301F-11D3-BF4B-00C04F79EFBC}") = "Metr", "Metr\Metr.csproj", "{2DB30441-AAEC-4060-8F58-308AFB38A51C}"
EndProject
Global
GlobalSection(SolutionConfigurationPlatforms) = preSolution
    Debug|Any CPU = Debug|Any CPU
    Debug|x86 = Debug|x86
    Release|Any CPU = Release|Any CPU
    Release|x86 = Release|x86
    EndGlobalSection
    GlobalSection(ProjectConfigurationPlatforms) = postSolution
    {2DB30441-AAEC-4060-8F58-308AFB38A51C}.Debug|Any CPU.ActiveCfg = Debug|Any CPU
    {2DB30441-AAEC-4060-8F58-308AFB38A51C}.Debug|Any CPU.Build.0 = Debug|Any CPU
    {2DB30441-AAEC-4060-8F58-308AFB38A51C}.Debug|x86.ActiveCfg = Debug|x86
    {2DB30441-AAEC-4060-8F58-308AFB38A51C}.Debug|x86.Build.0 = Debug|x86
    {2DB30441-AAEC-4060-8F58-308AFB38A51C}.Release|Any CPU.ActiveCfg = Release|Any CPU
    {2DB30441-AAEC-4060-8F58-308AFB38A51C}.Release|Any CPU.Build.0 = Release|Any CPU
    {2DB30441-AAEC-4060-8F58-308AFB38A51C}.Release|x86.ActiveCfg = Release|x86
    {2DB30441-AAEC-4060-8F58-308AFB38A51C}.Release|x86.Build.0 = Release|x86
    EndGlobalSection
    GlobalSection(SolutionProperties) = preSolution
    HideSolutionNode = FALSE
    EndGlobalSection
EndGlobal
```

4 Calculation of the Prior Probability of the Normal Functioning of the Process Automation System

The application of the methods of the theory of statistical hypotheses makes it possible to estimate the probabilities of the normal (it does not go beyond the permissible deviations of the controlled parameter and the assessment system confirms this - zone 1). Another correct functioning of the automation system is characterized by leaving the permissible limits of the controlled parameter (and the evaluation system confirms this - zones 2 and 3). Particular attention should be paid to the modes of operation in zones 4 and 5, since it is possible to combine the modes of the first, second and third zones in them. Let us calculate the probabilities of these events.

In accordance with past findings, we present the desired value of the probability as the sum of the probabilities of several events. In our case, these events are as follows: the nominal value of the monitored parameter A_N should be in the range from $(A_N - \theta - \Delta)$ to $(A_N - \theta + \Delta)$, i.e., in the first zone. The probability of this event is determined by expression (1), which suggests that it does not depend on the distribution density of the exemplary tool q $(\hat{A})$, but is determined only by the ratio θ/Δ and the distribution density of the device being verified ω (A).

Unlike the probability P_1, which consists of three events, the probability has two additional events associated with the presence of A_N not in zones 2 and 3, but only in a part of these zones, enclosed in the intervals from $(A_N - \theta)$ to $(A_N - \theta + \Delta)$ and from ($A_N + \theta - \Delta$) to ($A_N + \theta$). Thus, the second event determines the requirement to find the monitored parameter A_N in the interval from ($A_N - \theta$) to ($A_N - \theta + \Delta$) and the probability of its estimation by the measuring channel $\hat{A}_N$ in the same interval. The third event is associated with the condition of finding the desired values in the range from ($A_N + \theta - \Delta$) to ($A_N + \theta$). Let us write down the probabilities of these events by analogy with expressions (2) and (3).

$$P_{2N\hat{N}} = \int_{AH-\Theta}^{AH-\Theta+\Delta} \omega(A)\Big[\int_{A-\Delta}^{AH-\Theta+\Delta} q(\hat{A})d\hat{A}\Big]dA. \tag{15}$$

$$P_{3N\hat{N}} = \int_{AH+\Theta-\Delta}^{AH+\Theta} \omega(A)\Big[\int_{AH+\Theta-\Delta}^{A+\Delta} q(\hat{A})d\hat{A}\Big]dA. \tag{16}$$

The total $P_{N\hat{N}}$ probability is determined as the sum of the probability of events (1), (15) and (16).

$$\begin{aligned} P_{N\hat{N}} = & \int_{AH-\Theta+\Delta}^{AH+\Theta-\Delta} \omega(A)dA + \int_{AH-\Theta}^{AH-\Theta+\Delta} \omega(A)\Big[\int_{A-\Delta}^{AH-\Theta+\Delta} q(\hat{A})d\hat{A}\Big]dA \\ & + \int_{AH+\Theta-\Delta}^{AH+\Theta} \omega(A)\Big[\int_{AH+\Theta-\Delta}^{A+\Delta} q(\hat{A})d\hat{A}\Big]dA. \end{aligned} \tag{17}$$

For symmetric distributions ω (A) and q ($\hat{A}$), expression (17) can be simplified and represented in a form that coincides with the analogous from [2]

$$P_{N\hat{N}} = 2\int_{AH}^{AH+\Theta} \omega(A)\Big[\int_{A-\Delta}^{AH+\Theta} q(\hat{A})d\hat{A}\Big]dA. \tag{18}$$

Calculation of the priori probability of the controlled parameter exit from the nominally permissible with fixing this mode of the automation system $P_{A\hat{A}}$.

This probability is determined by the sum of the probabilities of four events:

A_N is in the range from $-\infty$ to ($A_N - \theta - \Delta$), in the range from ($A_N + \theta + \Delta$) to $+\infty$. In these intervals, the probabilities of events are determined only by the ratio θ/Δ and the distribution density of the device being verified ω (A). The sought values of these events are equal to the first term in expressions (7) and (8). The third event determines the location of A_N and $\hat{A}$ in the zone from ($A_N - \theta - \Delta$) to ($A_N - \theta$). Finding A_N and $\hat{A}$ in the zone from ($A_N + \theta$) to ($A_N + \theta + \Delta$) constitutes the fourth event.

Thus, the desired value of the probability is calculated based on expression (19).

$$P_{A\hat{A}} = \int_{-\infty}^{AH-\Theta-\Delta} \omega(A)dA + \int_{AH+\Theta+\Delta}^{+\infty} \omega(A)dA + \int_{AH-\Theta-\Delta}^{AH-\Theta} \omega(A)[\int_{A-\Delta}^{AH-\Theta} q(\hat{A})d\hat{A}]dA$$
$$+ \int_{AH+\Theta}^{AH+\Theta+\Delta} \omega(A)[\int_{AH+\Theta}^{A+\Delta} q(\hat{A})d\hat{A}]dA = 2\{\int_{AH+\Theta}^{+\infty} \omega(A)[\int_{AH+\Theta}^{A+\Delta} q(\hat{A})d\hat{A}]dA\}. \quad (19)$$

Calculation of the priori probability of erroneous operation of the automation system (the monitored parameter does not go beyond the permissible limits, and the measuring channel fixes the exit of this parameter outside the permissible limits) $P_{N\hat{A}}$ [18–21].

This probability includes two events: A_N is in the range from $(A_N - \theta)$ to $(A_N - \theta + \Delta)$, and $\hat{A}$ in the first case falls into the interval from $(A_N - \theta - \Delta)$ to $(A_N - \theta)$, and in the second case - from $(A_N + \theta)$ to $(A_N + \theta + \Delta)$, while A_N should be in the range from $(A_N + \theta - \Delta)$ to $(A_N + \theta)$.

$$P_{N\hat{A}} = \int_{AH-\Theta}^{AH-\Theta+\Delta} \omega(A)[\int_{AH-\Theta-\Delta}^{A+\Delta} q(\hat{A})d\hat{A}]dA + \int_{AH+\Theta-\Delta}^{AH+\Theta} \omega(A)[\int_{AH+\Theta}^{A+\Delta} q(\hat{A})d\hat{A}]dA$$
$$= \int_{AH-\Theta}^{AH+\Theta} \omega(A)[\int_{A-\Delta}^{AH-\Theta} q(\hat{A})d\hat{A} + \int_{AH+\Theta}^{A+\Delta} q(\hat{A})d\hat{A}]dA = 2\{\int_{AH}^{AH+\Theta} \omega(A)[\int_{AH+\Theta}^{A+\Delta} q(\hat{A})d\hat{A}]dA\}. \quad (20)$$

Calculation of the priori probability of erroneous operation of the automation system (the monitored parameter goes beyond the permissible limits, and the measuring channel fixes the normal value of this parameter outside the permissible limits) $P_{A\hat{N}}$.

This probability includes two events: A_N is in the interval from $(A_N - \theta - \Delta)$ to $(A_N - \theta)$, and $\hat{A}$ in the first case falls into the interval from $(A_N - \theta)$ to $(A_N - \theta + \Delta)$, and in the second case - from $(A_N + \theta - \Delta)$ to $(A_N + \theta)$, while A_N should be in the range from $(A_N + \theta)$ to $(A_N + \theta + \Delta)$.

$$P_{A\hat{N}} = \int_{AH-\Theta-\Delta}^{AH-\Theta} \omega(A)[\int_{AH-\Theta}^{A+\Delta} q(\hat{A})d\hat{A}]dA + \int_{AH+\Theta}^{AH+\Theta+\Delta} \omega(A)dA[\int_{AH+\Theta-\Delta}^{A+\Delta} q(\hat{A})d\hat{A}]dA$$
$$= \int_{-\infty}^{AH-\Theta} \omega(A)[\int_{AH-\Theta}^{A+\Delta} q(\hat{A})d\hat{A}]dA + \int_{AH+\Theta}^{\infty} \omega(A)dA[\int_{A-\Delta}^{AH+\Theta} q(\hat{A})d\hat{A}]dA = 2\{\int_{AH+\Theta}^{\infty} \omega(A)[\int_{A-\Delta}^{AH+\Theta} q(\hat{A})d\hat{A}]dA\}. \quad (21)$$

5 Conclusion

1. General expressions have been obtained for priori assessment of the operating modes of the process automation system.
2. In zones 4 and 5, the choice of statistical hypotheses (H_0 and H_1) requires additional research.
3. A program for finding the posterior probability modes of the automation system operation by the method of statistical modeling has been developed.

References

1. Lehman, E.: Testing Statistical Hypotheses, 2nd edn., p. 408. Nauka, Moscow (1979). Trans. from English
2. Naumenko, A.P., Kudryavtseva, I.S., Odinets, A.I.: Probabilistic and Statistical Methods of Decision-Making: Theory, Priery, Tasks: Textbook. Allowance, p. 108. Ministry of Education and Science of Russia, OmSTU, Publishing House of OmSTU, Omsk (2018)
3. Budnikov, I.K.: Theory and Practice of a Scientific Experiment: Textbook, p. 132. State Energ. Un-t, Kazan (2014)
4. De Groot, M.: Optimal Statistical Solutions, p. 492. Mir, Moscow (1974)
5. Zachs, S.: Theory of Statistical Inferences, p. 776. Mir, Moscow (1975)
6. Gren, E.: Statistical Games and Their Application, p. 176. Statistics, Moscow (1975)
7. Vorobiev, N.N.: Game Theory for Cybernetic Economists, p. 271. Nauka, Moscow (1985)
8. Kremlev, A.G.: Basic Concepts of Game Theory: Textbook; Scientific Editor A. M. Tarasyev; Ministry of Education and Science of the Russian Federation, Ural Federal University, p. 144. Ural University Publishing House, Yekaterinburg (2016)
9. Ferguson, T.S.: Mathematical Statistics: A Decision Theoretic Approach. Academic Press, New York (1987)
10. Akimov, V.I., Polyakov, S.I., Polukazakov, A.V.: Software life management systems for "smart" residential houses. In: 2020 International Russian Automation Conference (RusAutoCon), Sochi, Russia, pp. 267–272 (2020). 10.1109/RusAutoCon49822.2020.9208215
11. Polyakov, S.I.: Automation and Automation of Production Processes [Text]: Textbook. Allowance, p. 373; Fed. Education Agency, GOU VPO "VGLTA", Voronezh (2007)
12. Polyakov, S.I., Akimov, V.I., Polukazakov, A.V.: Simulation of a heating control system for a "smart" residential building. Scientific. J. Model. Syst. Process. **13**(1), 68–76 (2020)
13. Akimov, V.I., Desyatirikova, E.N., Polukazakov, A.V., Polyakov, S.I., Mager, V.E.: Development and research of a "smart home" heating control system. In: 2020 IEEE Conference of Russian Young Researchers in Electrical and Electronic Engineering (EIConRus), St. Petersburg and Moscow, Russia, pp. 574–580 (2020). https://doi.org/10.1109/EIConRus49466.2020.9039541
14. Volkov, V.D., Shashkin, A.I., Smolyaninov, A.V., Desyatirikova, E.N.: Theory of Automatic Control: Textbook, p. 745. Scientific Book, Voronezh (2015)
15. Akimov, V.I.: Fundamentals of Metrology and Radio Measurement: Textbook. Allowance, p. 130. IN AND. Voronezh. Polytechnic in-t, Voronezh (1992)
16. Akimov, V.I., Polukazakov, A.V., Sitnikov, N.V.: Selecting criteria for optimizing parameters of ADC for digital signal processing. In: International Russian Automation Conference (RusAutoCon) (2019). https://doi.org/10.1109/RUSAUTOCON.2019.8867651(Scopus)

17. Volkova, V.N., Loginova, A.V., Desyatirikova, E.N.: Simulation modeling of a technological breakthrough in the economy. In: Proceedings of 2018 IEEE Conference of Russian Young Researchers in Electrical and Electronic Engineering, ElConRus, pp. 1293–1297 (2018). https://doi.org/10.1109/EIConRus.2018.8317332
18. Akimov, V.I., Polyakov, S.I., Polukazakov, A.V.: Design and development of cascade heating control for a "smart" residential housing. In: 2020 International Russian Automation Conference (RusAutoCon), Sochi, Russia, pp. 42–48 (2020). https://doi.org/10.1109/RusAutoCon49822.2020.9208225
19. Polyakov, S.I., Zuikin, N.P.: Design of Control Systems. Textbook. Allowance, p. 133. Voronezh. State forestry engineering. Acad, Voronezh (2001)
20. Lysenko, E.V.: Designing of Automated Control Systems for Technological Processes, p. 129. Radio and Communication, Moscow (2007)
21. Akimov, V.I.: The Solution of Metrological Problems by the Method of Statistical Modeling: Textbook. Allowance, p. 117. IN AND. Akimov; Voronezh. State Technical University, Voronezh (1996)
22. Smorkalova, V.M.: Tasks of Testing Statistical Hypotheses: Study Guide, p. 23. Nizhny Novgorod State University, Nizhny Novgorod (2015)

The Automated Method of Metrological Inspection of Parts Manufactured According to Additive Technology Using the 3D Scanning Method

L. O. Fedosova, A. M. Mukletsov, and A. V. Zolotov(✉)

Nizhniy Novgorod State Technical University n.a. R. E. Alekseev, b.24, Minina Street, Nizhniy Novgorod 603950, Russia

Abstract. The article provides an overview of the existing methods of metrology use in the control of the accuracy of manufacturing processes. The choice of using the 3D-scanning procedure for metrological control of parts manufactured using additive technologies is justified. The article provides an example of using the 3D scanning method for metrological control of a part. A structured-light 3D scanner was used as a measuring tool. The article describes the method of using a 3D scanner to calibrate a 3D printer to compensate for the shrinkage of various materials: PLA, ABS, PETG, Nylon. For metrological control, a part with a complex geometry was selected, namely, a centrifugal turbine impeller printed by an FDM printer. A comparative analysis using measured material shrinkage coefficient was carried out on a centrifugal impeller. The article also describes the effect of supports on the curvature of the impeller blades. The study of the influence of the printing speed and the layer height on the accuracy of the part manufacturing was carried out.

Keywords: Metrological inspection · Additive technology · 3D printing · 3D scanning · Reverse engineering · Measurement automation

1 Introduction

Additive technologies are widely used in various industries. As a rule, 3D printing is recommended to be used with parts of complex geometry that are difficult or impossible to manufacture by conventional methods. In additive manufacturing, there are many 3D printing technologies, a wide variety of kinematics of the printers themselves. In addition, for each task, there is a certain type of material being used: wax, rubber, plastic, metal, and even biomaterial (for printing human organs). In this regard, there is a need for metrological control, since in 3D printing, the material can shrink, sagging of unsupported spans can occur, and the quality of the part can be affected by the printing parameters (speed, layer height, extrusion temperature, etc.).

A. A. Radionov and V. R. Gasiyarov (Eds.): RusAutoCon 2021, LNEE 857, pp. 57–67, 2022.
https://doi.org/10.1007/978-3-030-94202-1_6

2 Technology Review

To date, several measuring methods and tools are used to control the geometry of parts. All methods differ in both cost and measurement accuracy. Below is a brief description of such tools.

The first method is connected with the use of conventional measuring instruments [1]. Typical contact measurement tools include: micrometer, caliper, passameter/snap gauge, etc. This method is widely used to control the external and internal dimensions, depths of grooves and holes. The main advantage of the method is its low cost; however, there are some disadvantages. The use of such measuring instruments is labor-intensive, time-consuming and requires the use of a great number of extra tooling for measuring items of various sizes. With standard tools, it is very difficult to obtain all the necessary information to assess the accuracy of part manufacturing, and it is impossible to measure parts with complex geometries.

The second measurement method also applies to the contact method. The coordinate measuring machine is a well-proven metrology solution featuring the highest accuracy. Coordinate measuring machines are used when it is necessary to conduct high-precision measurements. Inspection of parts carried out by coordinate measuring machines allows ensuring high performance, accuracy and reliability of measurements, with low labor intensity. In addition to the advantages, there are also some disadvantages, namely, when operating a coordinate measuring machine, there are some strict requirements for the workspace, as well as for the operators, that shall be met [2].

Currently, a promising method for assessing the accuracy of parts manufacturing is 3D scanning, which refers to the non-contact method of measurement [3–5]. This technology has already become widespread. The authors of the article [6] in their work conducted a study of the geometry deviations of a 3D printed ship model caused by shrinkage-induced deformations for subsequent tests of its hydro-mechanical properties in the pilot pool. The analysis of geometry deviations was carried out using 3D scanning with the development of a technological scheme for describing this process.

In the article [7], the authors attempted to evaluate the serviceability of a large-sized cast bar using 3D scanning. As an example, a large-sized workpiece of the "body" type, produced by casting and having a rather complex geometry, was chosen. Based on the results of the measurements, the suitability of the part was assessed, in accordance with the reference CAD-model.

The advantages of the 3D scanning method, in comparison with the use of conventional tools or coordinate measuring machines, include the following: better price/quality ratio, high measurement speed, as well as ease of equipment operation. 3D scanning gives an opportunity to evaluate not only the accuracy of manufacturing, but also the quality of the surface. In this regard, let us take a closer look at the 3D scanning technology.

3 Theory

Mainly 3D scanners are divided into optical and laser. According to the installation method, there are stationary scanners and manual (mobile) scanners.

A stationary laser 3D scanner is mostly used for scanning large-sized objects, featuring very high accuracy and scanning speed. A mobile 3D scanner is widely used in cases when it is impossible to place a stationary scanner. It is less accurate and more expensive.

The most appropriate 3D scanning technology for geometry control in FDM printing would be an optical structured-light 3D scanner [8]. The gist of the technology is that various template images are projected onto the scanned object using a conventional LED-DLP projector. A camera installed nearby captures the resulting distorted images. After that, special software for these distortions estimates the shape and depth of the object and converts all this data into a three-dimensional image. The disadvantages of an optical 3D scanner are special surface requirements to be met, for example, the surface shall not be transparent, mirrored, black or glossy.

The process of evaluating the accuracy of part manufacturing using 3D scanning can be divided into several sequential stages (Fig. 1).

The first step is to set the parameters for printing the part in Polygon X. The printing parameters are the most important stage in reaching the quality of the part manufacturing [9–11]. The program allows selecting parameters such as: filling percentage, layer height, number of perimeters, and scale of the model. At this stage, the supporting structures are also placed, the printing speed and the extrusion temperature of the material are specified. Since each material differs in its shrinkage when 3D printing, it is necessary to calibrate the 3D printer for each material. The shrinkage factor is determined by printing calibration cubes [12]. After printing, the calibration cube is scanned, after that its geometry is compared with the reference CAD model. The deviations found will represent a shrinkage. Compensation for material shrinkage will be conducted by adjusting the scale of the part.

3D scanning starts with the calibration of the 3D scanner itself [13]. In RangeVision Spectrum, there are three calibration fields for scanning objects of various sizes. Due to the small dimensions of printed parts, choose the smallest calibration field for scanning objects of no more than $130 \times 100 \times 100$ mm. The scanning accuracy for the selected area is 0.04 mm. Calibration, scanning and further scanner operation take place in the ScanCenter NG program.

The next step is to prepare the part for 3D scanning. The optical 3D scanner cannot be used with black and glossy objects. The first problem is solved easily by choosing the material of a different color; however, after printing it often appears that the surface of the part is glossy. To eliminate the glossy surface, the part shall be covered with a light matting spray.

Following the above-mentioned steps, the 3D scanning stage begins. Rangevision 3D scanner supports operation in three scanning modes, which differ primarily in the method of combining fragments:

- scanning without using markers;
- scanning using markers;
- scanning on the turntable.

The next stage is the processing of the received scans. When scanning, unnecessary objects fall into the field of view of the 3D scanner, which must be removed during processing. Since the scanner used is optical, digital glitches appear in the obtained

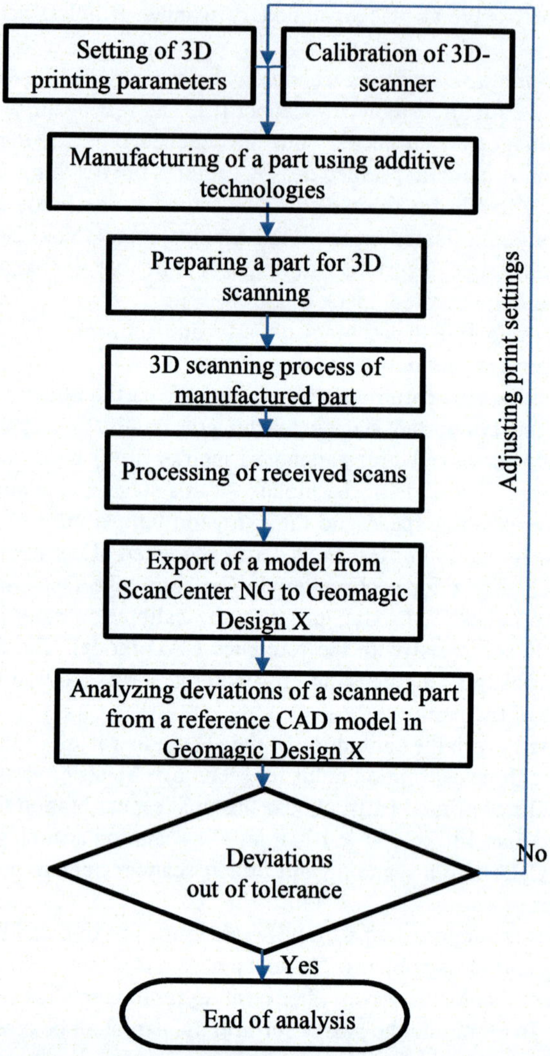

Fig. 1. Part manufacturing accuracy evaluation algorithm using 3D scanning technology.

point clouds, which directly affect the accuracy, so that when processing the scan, it is necessary to eliminate them.

When scanning a single fragment, a three-dimensional model of a section of the object's surface is obtained. To create a full-scale model of an object, it is necessary to scan it from different angles. The resulting fragments shall be aligned, automatically or manually, as per the surface geometry.

To continue working with the resulting model, it shall be exported from ScanCenter NG. The most popular file format for working with 3D models obtained by 3D scanning is stl. In this file format, data on the object is stored as a list of triangular faces that define the surface of the model, and their normals. When exporting models, it is often necessary to reduce the number of triangular faces, because a great number of them on large-sized models can negatively affect the size of the exported file and complicate further work with it.

Geometry deviation analysis is conducted using Geomagic Design X software [14]. To find deviations, the scanned model shall be compared with the reference CAD model. In the case of deviations that do not meet the specified tolerance limits, the process of determining material shrinkage or adjusting the print parameters is repeated.

To evaluate the accuracy of manufacturing using additive technology, a part with a complex geometry was selected - a centrifugal turbine impeller. The impeller is designed to allow a fluid to pass through it [15]. Using 3D scanning, it is possible to determine the angle of installation of the blade relative to the root frame and separate sections, the deviation of the surface of the blade airfoil, as well as the profile thickness. Since the top blade portions will be printed on the supports, the method would help to find deviations from the reference CAD model.

4 Practical Implementation

Initially, experiments were carried out to determine the shrinkage factor of the material. For this purpose, calibration cubes were printed from a selected range of materials: PLA, ABS, PETG, Nylon [16].

As a result, a graph of the material shrinkage in XYZ directions was built (Fig. 2). To find the average deviation, 5 points are selected from each side of the cube, 4 closer to the corners of the cube and 1 in the center of the side. When generating a deviation map, the minimum and maximum values are set to 0.4 mm, and the tolerance is set to 0.05 mm.

The average deviation for each of the cube axes is calculated using the formula:

$$\sigma = (b_1 + \cdots + b_5)/5 + (c_1 + \cdots + c_5)/5 \tag{1}$$

where b is the value of the calibration cube geometry deviation after 3D printing from the reference CAD model at a specific point, c - from the opposite side.

As we can see from the graph, "Bestfilament" PLA materials, "eSUN" and "Bestfilament" PETG materials have the minimum shrinkage, which corresponds to the required 3D printing accuracy shown by the green line on the graph. Nylon and ABS materials do not meet the required accuracy limits, so it is necessary to repeat the printing procedure, taking into account the determined shrinkage along the XYZ axes. Moreover, in the case of all four cubes, the shrinkage factor in the Y-axis is less than in the X-axis by an average of 47 microns, which is more than the claimed positioning accuracy of the printhead of 11 microns. This may be due to two options. The first option is associated with the incorrect generation of the G-code command by the Polygon X slicer. It can be checked by generating a new G-code for the calibration cube using the same settings. The second option may be associated with a slight stretching of the drive belts.

To compensate for shrinkage of Nylon material, the scale of the calibration cube model was increased by 1.24% in the X-direction, 1.05% in the Y-direction, and 0.4% in the Z-direction, for ABS material by 0.88% in the X-direction, 0.8% in the Y-direction, and 0.65% in the Z-direction.

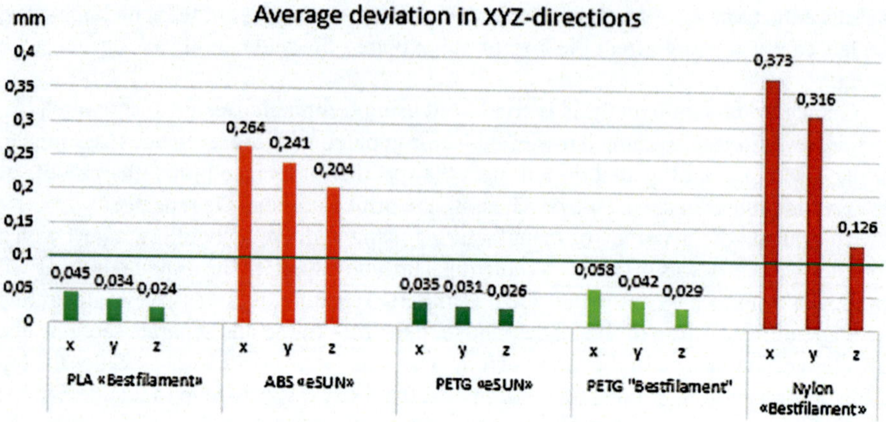

Fig. 2. XYZ average deviation graph.

Following the application of the shrinkage factor, the calibration cubes were re-3D printed and 3D scanned. Based on the result of comparing the cube scans with the reference CAD model, a graph was built (Fig. 3). The graph shows that the geometry deviations are very close to the positioning accuracy of the printerhead (11 microns in the XY direction and 1.25 microns in the Z direction), so trying to further increase the accuracy seems to be pointless.

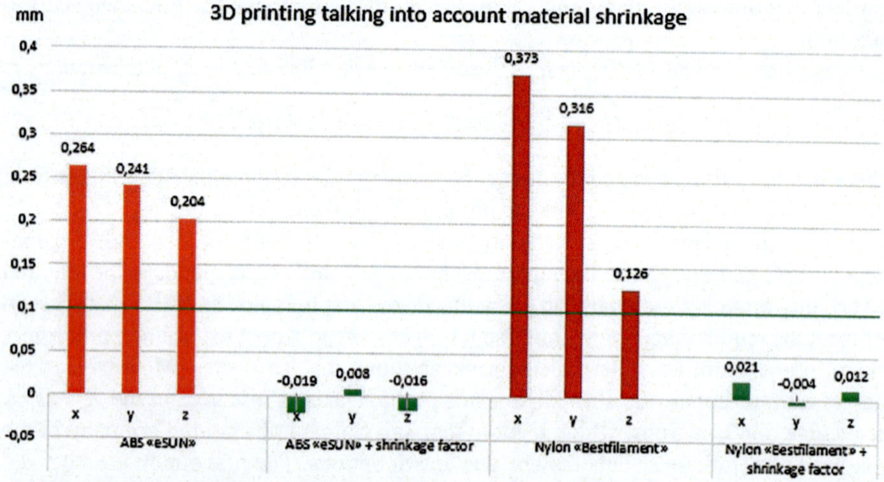

Fig. 3. Result of applying shrinkage factor to calibration cubes made of Nylon and ABS.

For metrological inspection, a centrifugal turbine impeller was selected as an item with complex geometry. The manufacturing process of a centrifugal impeller from a CAD model to a physical model is shown in Fig. 4. The three-dimensional model of the part was prepared in the Autodesk Inventor environment [17]. Next, the model of the impeller was prepared for printing in the Polygon X software.

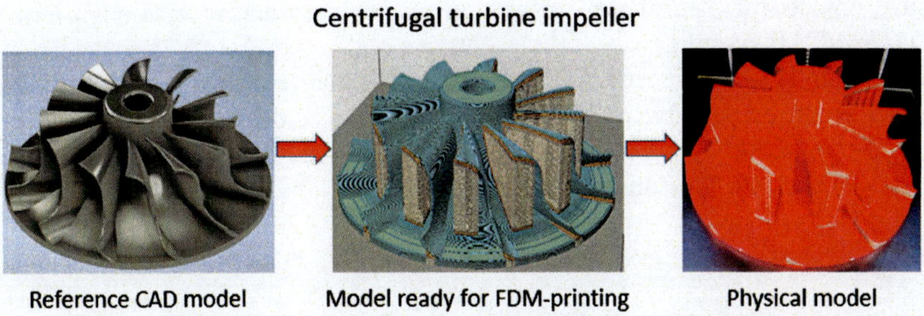

Fig. 4. The manufacturing process of turbine impeller using 3D printing.

ABS-plastic from the company "Bestfilament" was chosen as the first material for 3D printing. Since the item has unsupported parts-blades, it was necessary to place supports for their construction. For their easy separation, a gap was set between the model and the supports: in the Z direction - 0.15 mm, and in the XY directions - equal to the width of the extrusion - 0.39 mm.

3D scanning of the impeller took place on a turntable, which allowed speeding up the process of obtaining a 3D model. The item was also covered with a matting spray, as it had a glossy surface. To create a full-scale model of the object, all the fragments of the surfaces had to be scanned, for this purpose, four angles were selected. In one complete rotation of the table, 12 scans are made, which are combined into a group. During the scanning process, the scans from the group are automatically combined with one another.

The process of combining is complicated by the symmetry of the part, so it has to be optimized. For the correct stitching of scans in a group and groups with one another, markers were applied to the surface of the impeller, in the form of small pieces of plasticine. The scanning took place in four positions: in the first one, the impeller blades were scanned, in the second-the lower part, in the third-the upper part, in the fourth-the lower and upper parts simultaneously. Before combining the groups, the noise and the support for the impeller placement were removed (Fig. 5).

Fig. 5. 3D model acquisition steps.

The shrinkage factor will be determined by aligning the scanned model with the reference CAD model. To work with data obtained by 3D scanning, there is used software called Geomagic Design X. The program allows designing virtual 3D models of physical objects for metrological inspection of geometry and reverse engineering in CAD/CAM/CAE systems.

To download the reference CAD model of the impeller, export it in stp. file format in the Autodesk Inventor software, which will contain data on the three-dimensional image of the part. In addition, in Design X, import the previously prepared stl. model of the impeller obtained by 3D scanning (Fig. 6).

Fig. 6. The sequence of deviation analysis.

To determine the deviations of the printed impeller, it is necessary to combine the imported models. The alignment takes place in two stages. At the first stage, the scan is roughly compared with the surface of the CAD model, and at the second stage, the program aligns the geometry of the two models as accurately as possible. Then the geometry of the scanned impeller is analyzed. A deviation map is generated, where the deviation value is shown in color. In the settings, it is possible to set the minimum and maximum deviation, as well as the tolerance limits.

"Bestfilament" ABS-plastic was chosen as the material for printing the impeller. Since the item has unsupported parts-blades, supporting structures were placed for their construction. For their easy separation, a gap was set between the model and the supports: in the Z direction - 0.15 mm, and in the XY directions - equal to the width of the extrusion - 0.39 mm.

As we can see from the generated deviation map (Fig. 7a), there is a significant shrinkage in the Z-direction on the impeller. The lower parts of all the blades are deformed. This may have been caused by deflection of the blade, as it is very thin. In addition, the deformation could be caused by overheating of the material. The upper parts of the blades have sagging, in comparison with the reference CAD model. The reason may be the installed supports with a gap of 0.15 mm in the Z-axis direction.

After adjusting the scale of the impeller model, taking into account the determined shrinkage factor, for the ABS material, a printing procedure was repeated. To eliminate the sagging of the upper part of the blades, a gap of 0.1 mm was set between the impeller model and the supports. The deviation map shows (Fig. 7b) that some of the sagging was eliminated, but to eliminate it completely, it is necessary to print the supports without a gap in the Z direction with the use of another material, that will dissolve after printing.

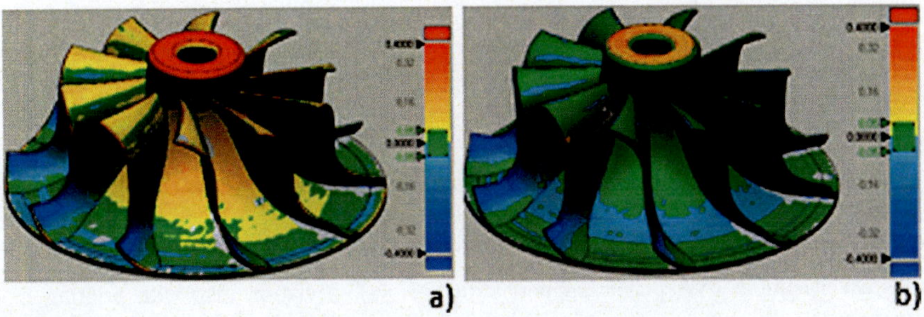

Fig. 7. ABS-plastic impeller geometry deviations: before the application of shrinkage factor (a) and after (b).

The lower part of the blades still has deviations; the application of the shrinkage factor did not bring a desirable result. In addition, due to the layer height of 0.2 mm, there are deviations in the form of “steps” on the lower part of the impeller.

To smooth out the upper section of the blades and eliminate the “stepped surface” in the lower part of the impeller, the layer height was changed from 0.2 mm to 0.1 mm in the printing parameters [18]. Reducing the height of the print layer also had a positive effect on the sagging of the blades. However, the digital map (Fig. 8) still shows deviations on the upper surface of the impeller. The lower parts of the blades are still deformed; in addition, all the blades have enlarged faces.

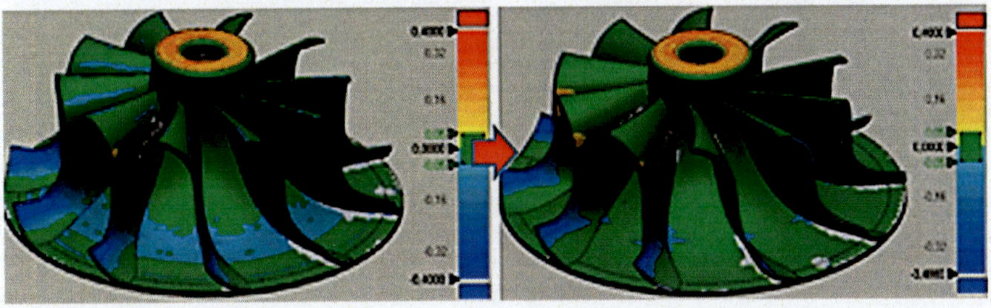

Fig. 8. Impeller geometry deviations after adjusting the height of the print layer.

The next step to eliminate geometry deviations in the impeller blades was the use of dual-extrusion printing [19]. As previously stated, to eliminate the sagging of the upper part of the blade completely, it is necessary to use a soluble material, so the supporting structures were printed using a water-soluble PVA plastic (Fig. 9a). In this case, it was possible to eliminate the Z-axis gap between the blade and the support. To eliminate the deformation of the lower part of the blades, the extrusion temperature was reduced from 245 °C to 230 °C, in addition, the air circulation in the 3D printer chamber was completely turned off, to eliminate the through the flow of air.

After 3D printing of the impeller with the adjustment of the parameters, a perfect state of the blade surface was reached, with the only exception of blade edges (Fig. 9b). The problem is the inertia of the print head of the printer, it cannot stop immediately,

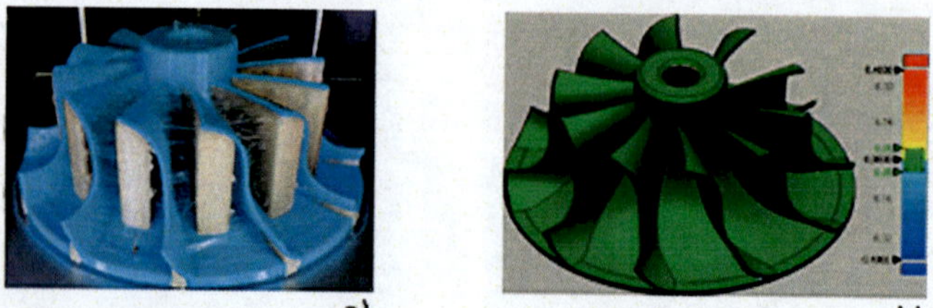

a) b)

Fig. 9. Water-soluble PVA plastic support structures (a), deviation map with printing mode adjusted (b).

which is why there is an extra influx of plastic on the sharp edges. To eliminate this defect, the printing speed for the perimeter was reduced by half, from 60 mm/s to 30 mm/s [20].

Figure 10 shows the result of using 3D scanning technology as a method for calibrating the 3D printing parameters of a centrifugal turbine impeller. On the digital map, there are no deviations in the geometry of the scanned impeller from the reference CAD model, which confirms the optimal choice of 3D printing parameters for this item.

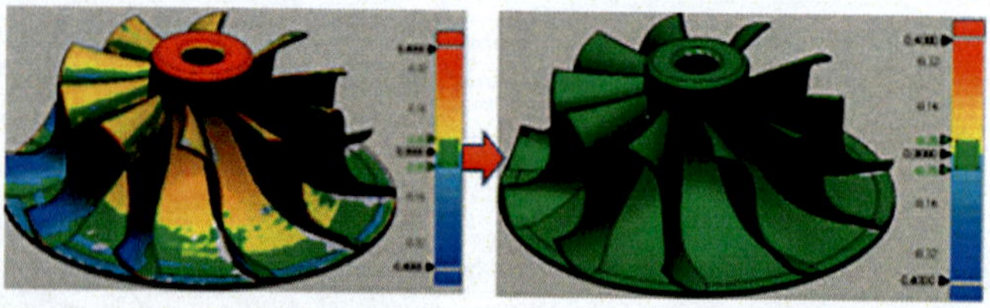

Fig. 10. Adjustment of 3D printing settings using 3D scanning technology.

5 Conclusion

In general, 3D scanning method is used in the absence of a three-dimensional model to solve the issue of reverse engineering. However, this method is also used for metrological analysis. The use of the 3D scanning method allows conducting the inspection of complex geometry in additive manufacturing. According to the generated map of geometry deviations for the part, the user can adjust the printing parameters and calibrate the 3D printer for different types of materials, which will allow producing the part with a given accuracy. The use of 3D scanning technology for metrological inspection of parts, when the use of conventional measurement methods is difficult, is particularly relevant.

References

1. Sekatsky, V.S., Pikalov, Y., Merzlikina, N.V.: Methods and Means of Measurement and Control: Textbook, p. 316. Siberian Federal University, Krasnoyarsk (2017)
2. Ermatov, A.B., Muminov, O.Z.: Classification and verification of coordinate-measuring machines. Universum Tech. Sci. Electron. Sci. J. **2**(71) (2020)

3. Kudryavtsev, A.B.: Modern 3D scanning technologies. In: 13th International Research to Practice Conference, Penza, pp. 136–139 (2019)
4. Zvagins, E., Vitols, G.: Comparison of 3D scanning technologies. In: 14th International Research Conference Students on Their Way to Science, Jelgava, p. 39 (2019)
5. Barnes, A.: Digital Photogrammetry. The Encyclopedia of Archaeological Sciences, pp. 1–4. Wiley, Hoboken (2018)
6. Zobov, P.G., Dektyarev, A.V., Morozov, V.N.: Modern 3D scanning methods for dimensional inspection of ship models in case of additive manufacturing. Izvestiya KGTU **53**, 151–161 (2019)
7. YaS, K., Novikov, V.: Method of comparative analysis and dimensional inspection in machine building of cast body blanks, using a 3D scanner. Min. Inf. Anal. Bull. **3**, 92–96 (2017)
8. Rehman, Y., Uddin, H.M.A., Siddique, T.H.M., Jafri, S.R.U.N., Ahmed, A.: Comparison of camera and laser scanner based 3D point cloud. In: 2019 4th International Conference on Emerging Trends in Engineering, Sciences and Technology (ICEEST), pp. 1–5 (2019)
9. Fernandes, J., Deus, A.M., Reis, L., Vaz, M.F., Leite, M.: Study of the influence of 3D printing parameters on the mechanical properties of PLA. In: Proceedings of the 3rd International Conference on Progress in Additive Manufacturing (Pro-AM 2018), pp. 547–552 (2018)
10. Liu, Y., Liang, X., Saeed, A., Lan, W., Qin, W.: Properties of 3D printed dough and optimization of printing parameters. Innov. Food Sci. Emerg. Technol. **54**, 9–18 (2019)
11. Samykano, M., Selvamani, S.K., Kadirgama, K., Ngui, W.K., Kanagaraj, G., Sudhakar, K.: Mechanical property of FDM printed ABS: influence of printing parameters. Int. J. Adv. Manuf. Technol. **102**(9), 2779–2796 (2019)
12. Tignibidin, A.V., Takayuk, S.V.: The use of additive technologies in prototyping. Control of geometric characteristics of parts made of ABS plastic for determining the initial dimensions for printing. Dyn. Syst. Mech. Mach. **3**, 57–65 (2018)
13. Abramenko, A.A.: Calibration of the relative position of a stereo camera and a three-dimensional scanning laser rangefinder. Comput. Opt. **2**, 220–230 (2019)
14. Alexandrov, D.V.: Obtaining solid models using PolyWorks Inspector and Geomagic Design X programs. In: Machinery and Technologies of Mechanical Engineering, pp. 228–235 (2019)
15. Wang, T., Kong, F., Xia, B., Bai, Y., Wang, C.: The method for determining blade inlet angle of special impeller using in turbine mode of centrifugal pump as turbine. Renew. Energy **109**, 518–528 (2017)
16. Eckert, M.: Comparing different types of materials for 3d printing from dimensional respect. Univ. Rev. **15**(1), 63–69 (2021)
17. Alieva, N., Zhurbenko, P., Senchenkova, L.: Building models and creating drawings of parts in the Autodesk Inventor system. Litres (2018)
18. Bintara, R.D., Lubis, D.Z., Pradana, Y.R.A.: The effect of layer height on the surface roughness in 3D Printed Polylactic Acid (PLA) using FDM 3D printing. IOP Conf. Ser. Mater. Sci. Eng. **1034**(1), 012096 (2021)
19. Dozhdelev, A.M., Lavrentiev, A.Yu.: Review of the substrates and coatings of the hot table for 3d printing by the method of layer-foaming. Int. J. Humanit. Nat. Sci. **8-2** (2019)
20. Žarko, J., et al.: Influence of printing speed on production of embossing tools using FDM 3D printing technology. J. Graph. Eng. Des. **1**, 19 (2017)

Multi-agent Approach to Efficient Management of Virtual Power Plants with Distributed Generation

E. Sosnina, A. Shalukho, and N. Erdili(✉)

Nizhny Novgorod State Technical University n.a. R.E. Alekseev,
24, St. Minina, Nizhny Novgorod 603155, N. Novgorod, Russia

Abstract. The article is devoted to the development of a control system for a Virtual Power Plant (VPP) with distributed generation (DG). The general principles of VPP functioning are described. A multi-agent approach to the VPP management that combines DG units, storage units and electricity consumers using a Solid-State Transformer (SST) is proposed. Many agents have been introduced for VPP facilities. The agent interaction principles for the implementation of the VPP intelligent management system have been developed. On the basis of simulation modeling in the RastrWin and Jade software complexes, a study of the effectiveness of using a multi-agent control system for the considered VPP has been conducted. As a result of the first iteration, the use of a multi-agent control system enabled to reduce power losses in the electrical network and increase the utilization rate of the installed capacity of power plants that are part a part/parts of the VPP.

Keywords: Distributed generation · Virtual power plant · Multi-agent control system · Energy exchange

1 Introduction

Distributed generation (DG) plays an important role in Russia's energy strategy. Developments in the field of control and protection devices for electrical networks, the development of digital technologies enable to implement a qualitatively new approach to increasing the efficiency of DG resources based on the "Internet of energy" principles [1].

The "Internet of energy" concept [2, 3] involves the creation of intelligent electrical systems with the possibility of multilateral energy exchange, both at low and at medium voltage between distributed objects. The virtual power plants (VPP) are systems operating according to these principles [4, 5].

VPP is an intelligent electrical system that unites DG installations, storage devices and electricity consumers on the basis of multilateral energy and information exchange for participation to be a success on in the electricity market, provision of provide system

A. A. Radionov and V. R. Gasiyarov (Eds.): RusAutoCon 2021, LNEE 857, pp. 68–79, 2022.
https://doi.org/10.1007/978-3-030-94202-1_7

services and redundancy. The VPP assumes a connection with the power system. However, it can also be used to improve the efficiency of autonomous power supply systems with DG.

The more distributed objects are combined into a VPP, the greater DG efficiency can be achieved. At the same time, the tasks of technical integration and management of the joint operation of DG units and storage devices at a voltage of 0,4 kV, as well as energy exchange between distributed objects at a voltage of 10 (20) kV, become relevant.

A promising solution to the problem of integration of facilities in a VPP is due to the development of solid-state transformer (SST) technologies. The SST is a controlled semiconductor converter [6] which allows a real-time implementation of power flows redistribution functions, the regulation of electrical energy parameters, combining dissimilar DG sources, storage devices and consumers [7].

The complexity of solving the problem of energy exchange control in a VPP lies in various algorithms in functioning of distributed DG sources, combined into a system, and the need to ensure multilateral energy exchange between them. The control methods used in existing energy systems with DG do not take these features into account.

The analysis of algorithms is carried out and it can be applied to control the operating modes of microgrids, in which there is an exchange of power between objects. Thus, the algorithm "Optimization of a swarm of particles by the stochastic method of evolutionary calculations" by J. Kennedy; R. Eberhart, USA, simulates a multi-agent system, where agents-particles move to optimal solutions, while exchanging information with neighbors [8]. "Bee colony" algorithm by D. Karaboga, Turkey, is used to solve discrete (combinatorial) and continuous problems of global optimization [9]. The local energy exchange algorithm based on the theory of coalition games for network microgrids J. May, C. Chen, J. Wang, J. L. Kirtley, USA, was created in 3 stages: - a method based on the theory of auction; - Shapley cost method; - the method of combining and separating [10]. The load balancing algorithm Y. Mensina, W. Setthapuna and W. Rakwichiana, Thailand, is focused on the exchange of power between objects [11]. These algorithms are provided to control energy exchange in microgrids, the analysis results of which will form the basis of the algorithm for controlling energy exchange in a VPP with diverse DG sources.

The research object is a VPP which unites distributed objects 0.4–10 kV with diverse energy sources. The work purpose is to develop a method to manage energy exchange in a VPP based on a multi-agent approach and to study the effectiveness of its application.

2 Materials and Methods

Figure 1 shows the investigated VPP block diagram. The VPP combines 10 micro-networks (energy cells) into a single system at a 20 kV voltage. The topology of the 20 kV electrical network is closed. It corresponds to a part of the MV electrical distribution network test circuit used by CIGRE (Conseil International des Grands Réseaux Electriques) to conduct experimental research and refine control algorithms [12]. Each energy cell is a certain combination of a DG source and a load, combined at 0,4 kV by SST. The energy cells composition, the type and the parameters of the DG sources and the load are given in Table 1.

Table 1. Vpp composition.

Microgrid	Microgrid composition	Parameter value
M1	T2	20/0,4 kV
	L2	200 kW
M2	T3	20/0,4 kV
	L3	900 kW
M3	SPI	400 kW
	L4	200 kW
M4	SPI	200 kW
	L5	400 kW
M5	GPI	1500 kW
	L6	700 kW
M6	SOFC PI	300 kW
	L7	500 kW
M7	DG	600 kW
	L8	200 kW
M8	mini-CHP	500 kW
	L9	100 kW
M9	WPI	300 kW
	L10	300 kW
M10	WPI	1000 kW
	L11	200 kW

We assume that voltage is 20 kV of bus no 1, 2, … , 11, and the SST's voltage levels are 20/0,4 kV. M1, M2, … , M10 – microgrid (energy cell); L1, L2, … , L11 – load; ES – power system; T1, T2, T3 – step-down transformer; SST – solid state transformer; SPI – solar power unit, WPI – wind power unit, GPI – gas piston unit; SOFC PI – solid oxide fuel cell power unit; DG – diesel generator.

The following set of agents is introduced to implement the MACS:

$$Ag_i = < Ag_{ES}, Ag_{GPI}, Ag_{RES}, Ag_{SOFS}, Ag_{DG}, Ag_{\text{mini-CHP}}, Ag_L >, \tag{1}$$

where Ag_{ES} – centralized network agent; Ag_{GPI} – agent of maneuverable gas-fired power plants (GPI); Ag_{RES} – solar or wind power - renewable energy sources (RES) agent; Ag_{SOFC} – agent of energy cells with SOFC; Ag_{DG} – diesel generator agent; $Ag_{\text{mini-CHP}}$ – coal fired power plant agent; AgL – load agent.

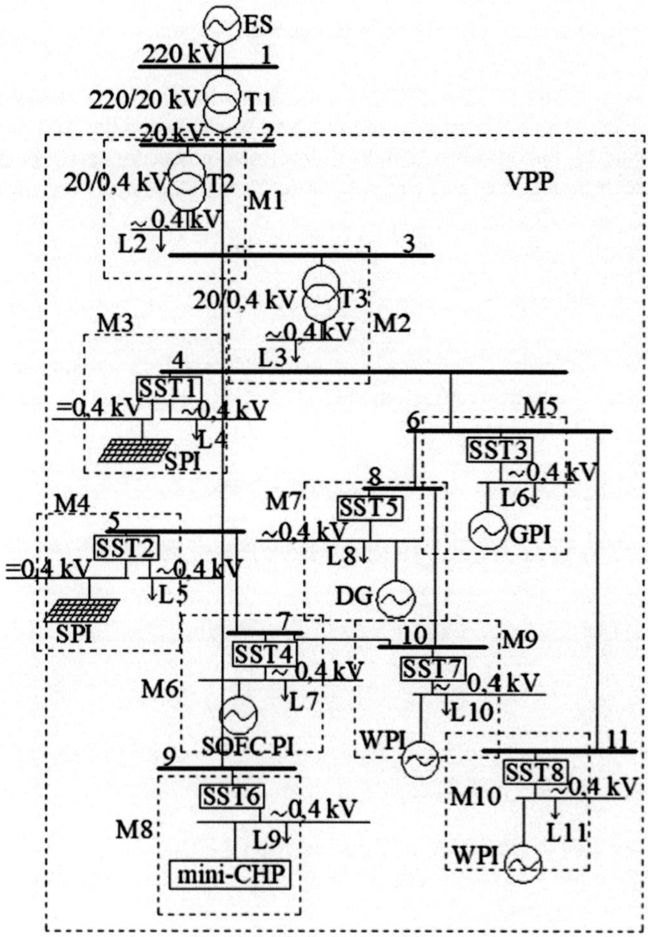

Fig. 1. Structural diagram of the research object.

The MACS work is focused on efficiency for each microgrid (energy cell) achieving while maintaining the power balance in the VPP as a whole. In this case, the generator agent notifies the MACS participants about the amount of excess power and the load agent, based on its criteria, connects with the generation agents suitable for it. The MACS work is based on the following criteria:

Criterion 1 - environmental rating of generation agents (R). Each generation agent is assigned an environmental rating from 1 to 5 (5 - power cells with RES, 4 - power cells with GPI, 3 - power cells with DG, 2 - power cells with mini-CHP, 1 - centralized electric grid). This criterion provides the advantage of electricity consumption from DG sources based on RES:

$$Agn \rightarrow Aggj(R \rightarrow max), \tag{2}$$

where Agn is the load agent; $Aggj$– the j-th generation agent with the corresponding R rating;

Criterion 2 – the length of the power transmission line (l). When both types of agents are declared in the MACS, the lengths of the power transmission line and the connection point are specified. This criterion enables to minimize power losses in the network and ensures that the production of electricity is closer to the place of its consumption:

$$Agn \rightarrow Aggj(l \rightarrow min), \quad (3)$$

where $Aggj$ is the j-th generation agent with the corresponding power transmission line length l.

Criterion 3 - the cost of electrical energy (c). Generation agents set the cost for generated electricity when advertised in the MACS, and load agents choose the cheapest one when criteria '2 and 3 are equal:

$$Agn \rightarrow Aggj(c \rightarrow min), \quad (4)$$

Thus, the load agent when choosing a generation agent analyzes the system of conditions:

$$\begin{cases} Ag_n \rightarrow Ag_{gj}(R \rightarrow \max) \\ Ag_n \rightarrow Ag_{gj}(l \rightarrow \min) \\ Ag_n \rightarrow Ag_{gj}(c \rightarrow \min) \end{cases} \quad (5)$$

It is proposed to use 3 indicators to assess the effectiveness of the MACS use for energy exchange management in VPP:

1) active power loss in the 20 kV VPP power grid ($\Delta P\Sigma$);
2) the coefficient of using the installed power of the DG sources as VPP part (C_{UIP}):

$$C_{UIP} = P_{FACT}/P_{INST}, \quad (6)$$

   where P_{FACT} is the actual generated power of the DG source (for a given steady-state mode), P_{INST} is the installed power of the DG source;
3) Share of RES (eco-friendly) use:

$$C_{RES} = (\Sigma P_{RES} + \Sigma P_{SOFC})/\Sigma Pi \quad (7)$$

The smaller the value $\Delta P\Sigma$ and the higher the values of C_{UIP} and C_{RES} are, the more efficient the operation of VPP is.

3 Agent Functioning Algorithm

The generation and load agents work algorithms have been developed according to the agent interactions (Fig. 2, 3).

There is a condition that generation from agents with RES and SOFCs is not regulated and its value is taken in accordance with the specified generation schedules.

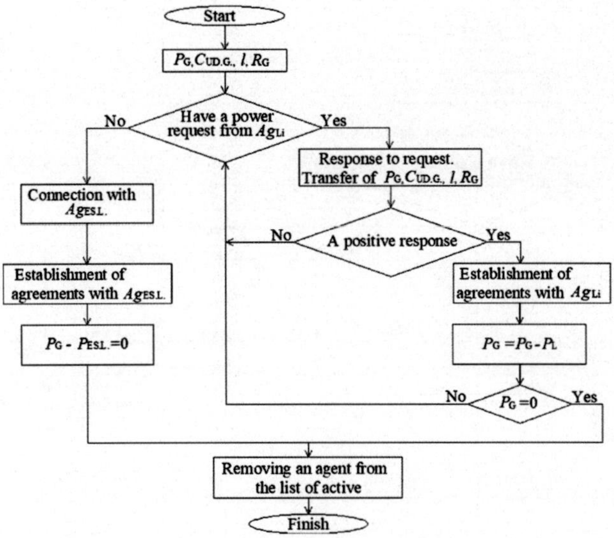

Fig. 2. Generation agents work with RES algorithm.

The algorithm for the work of generation agents with RES begins with checking the condition on the request for power from Ag_{Li}. If the request is confirmed, Ag_{Li} receives a response to the request and the generated power P_{G} is transferred, data at a reasonable cost of generation $C_{\mathrm{UD.G.}}$, l, environmental rating R_{G}. In case of a positive answer, agreements with Ag_{Li} are established. If the generation power is equal to 0, the agent is removed from the active list. In case of a negative answer, or if there is generated power, the condition is checked again at the request of power from Ag_{Li}. If there is no request for capacity from Ag_{Li}, communication with Ag_{ES} takes place and agreements are established with this agent. When the generated capacity is received from the centralized network, the agent is removed from the active list and the algorithm ends.

The algorithm of work of agents with controlled generation begins with checking the condition on the request for capacity from Ag_{Li}. If there is no request for capacity from Ag_{Li}, the agent is removed from the active list and the algorithm ends. When the condition of the request for power from Ag_{Li} is fulfilled, the maximum generated power $P_{\mathrm{G.MAX}}$, data at a reasonable cost of generation $C_{\mathrm{UD.G..}}$, l, environmental rating R_{G}. In case of a positive answer, agreements with Ag_{Li} are established. If there is no $P_{\mathrm{G.MAX}}$ the agent is removed from the list of active ones. In tcase of a negative answer or if there is of $P_{\mathrm{G.MAX}}$, the condition is checked again at the power request from AgLi.

AgGPI, AgDG, $Ag_{\mathrm{mini\text{-}CHP}}$, Ag_{ES} - eliminate active power imbalances arising in the system. Therefore, the power they give out can be adjusted when the agents interact. Ag_{ES} - can regulate the power received or given off to the VPP. Accordingly, this agent can act as both a generation agent and a load agent. Load agent parameters are set using electrical load graphs.

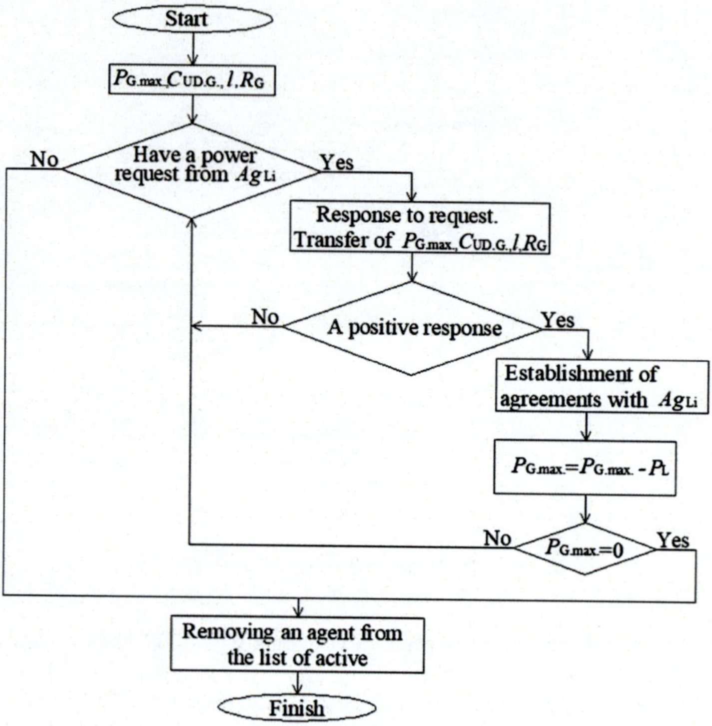

Fig. 3. Algorithm of agents work with controlled generation.

4 Macs Research

To conduct research on the effectiveness of a multi-agent approach to energy exchange control in a VPP with DG, the following models have been developed:

- a simulation model of the electric network of a VPP with DG is in the "RastrWin" software package [13];
- a MACS simulation model is in "Jade" [14].

The simulation model of the VPP with the DG electrical network in "RastrWin" is designed to study the parameters of the steady state and calculate the efficiency indicators (Fig. 4).

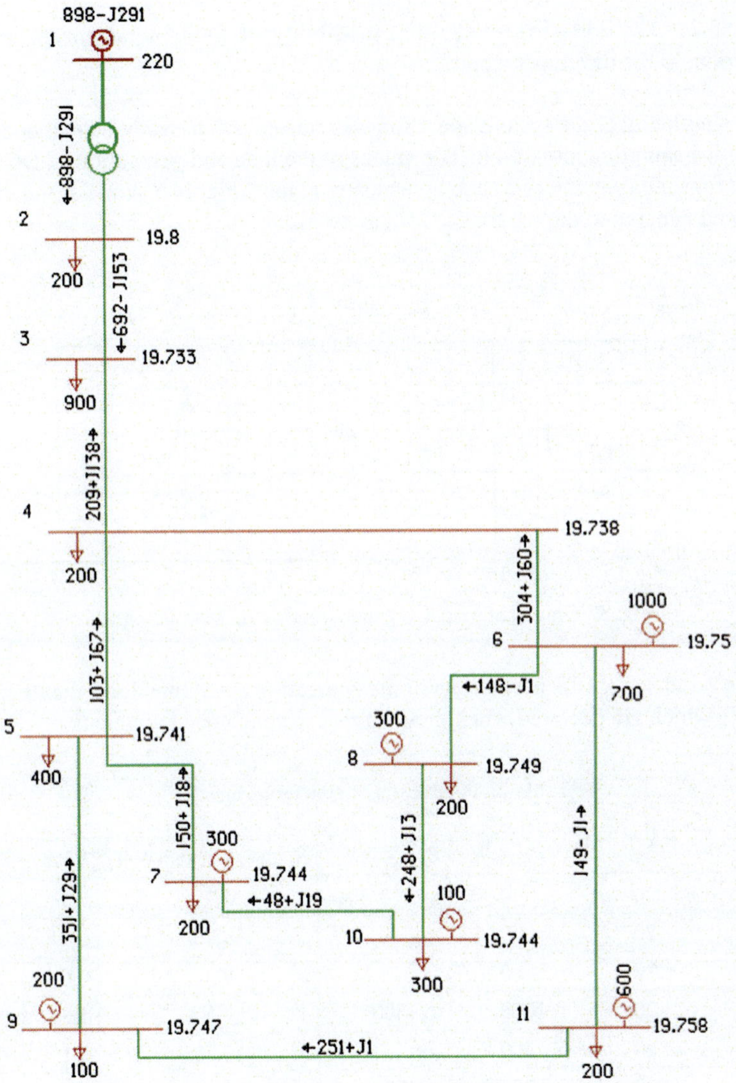

Fig. 4. Simulation model of the VPP with DG electrical network.

The MACS simulation model in "Jade" allows to determine the established connection between load agents and agents of generation of agreements and, on the basis of this, to determine the values of the electric power flows in the VPP. The study of the MACS efficiency is based on a comparison of the system operation, there are two variants:

- Variant 1 - there is no possibility of free multilateral energy exchange between distributed objects (the active power imbalance is covered by the centralized electrical network);

- Variant 2 – the MACS provides the possibility of multilateral energy exchange according to the developed approach.

The simulation of the system operation was carried out in steady-state modes during a day with a sampling rate of 4 h. The values of the load and generation of DG sources were set for each time interval (one hour) using graphs. Figure 5 shows some examples of load and generation curves for the M4 microgrid.

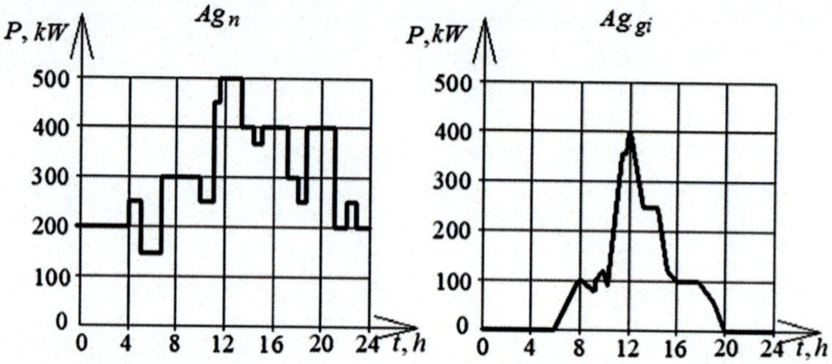

Fig. 5. Graphs of load and generation for the M4 microgrid.

Table 2 and 3 show the parameters of the generation agents and load agents specified in "Jade", which are declared when they interact to conclude agreements.

Table 2. Parameters of generation agents.

Agent	U_{nom}, kV	P_G, kW	$P_{G.adj}$, kW	R, r.u	C, rub/kW·h	l, km	Tire no
$Ag_{ES.G}$	220	0	to 25 MVA	1	7,2	7,5	1
Ag_{RES4}	20	0	–	5	9,75	0	4
Ag_{RES5}	20	0	–	5	9	0,6	5
Ag_{GPI6}	20	1000	to 1500	4	7,5	1,3	6
Ag_{SOFC7}	20	300	–	5	9,75	1,1	7
Ag_{DG8}	20	300	to 600	3	9	1,6	8
$Ag_{mini\text{-}CHP9}$	20	200	to 500	2	7,5	1,2	9
Ag_{RES10}	20	100	–	5	10	1,4	10
Ag_{RES11}	20	600	–	5	6,75	2,7	11

The theoretical values are given as the cost of electricity. The difference in the cost of electricity from different sources enables to consider a more diverse number of VPP participants with their own capabilities and priorities.

The studies of two variants of the system operation have been carried out in a steady state from 0 to 4 h with the help of simulation modeling. The obtained results of performance indicators are presented in Table 4 and 5.

Table 3. Parameters of load agents.

Agent	P_n, kW	$P_{n.adj}$, kW	R_{min}, r.u	C_{max}, rub/kW·h	B_{max}, rub	Tire no
$Ag_{ES.L}$	0	to 25 MVA	1	15	1500	1
Ag_{L2}	200	–	1	11,25	450	2
Ag_{L3}	900	–	1	12	400	3
Ag_{L4}	200	–	1	13,5	500	4
Ag_{L5}	400	–	4	12,75	450	5
Ag_{L6}	700	–	1	16,5	250	6
Ag_{L7}	200	–	3	13,5	200	7
Ag_{L8}	200	–	2	12,75	600	8
Ag_{L9}	100	–	1	12	350	9
Ag_{L10}	300	–	1	11,25	300	10
Ag_{L11}	200		1	10,5	175	11

Table 4. Loss of active power in the 20 kV network.

*N*_st	*N*_end	ΔP without MACS (Variant 1), kW	ΔP with MACS (Variant 2), kW
1	2	0,004	0,001
2	3	4,743	0,817
3	4	0,041	0,787
4	5	0,012	0,213
4	6	0,158	1,006
5	7	0,015	0,097
7	10	0,001	0,032
6	8	0,008	0,003
8	10	0,064	0,242
5	9	0,095	0,214
6	11	0,049	0,01
9	11	0,122	0,209
Total		5,312	3,631

Table 4 shows that the use of MACS allows to reduce the losses of active power in the 20 kV electrical network due to the use of the VPP's own power instead the centralized network's power. The greatest power reduction was achieved between nodes 2–3. Table 5 shows that the use of MACS enabled to ensure the balance of active power in the VPP, while realizing all the active power from sources with RES. At the same time, there are generation agents that are potentially ready to help when the load increases.

Table 5. Efficiency indicators of Mac application

Tire no	Generation source	P_{INST}, kW	$P_{FACT\ 1}$, kW	$P_{FACT\ 2}$, kW	$C_{UIP\ 1}$, r.u	$C_{UIP\ 2}$, r.u
4	SPI (RES)	400	0	0	0	0
5	SPI (RES)	200	0	0	0	0
6	GPI	1500	1000	1500	0,667	1,000
7	SOFC PI	300	300	300	1	1
8	DG	600	300	600	0,500	1,000
9	mini-CHP	500	200	300	0,400	0,600
10	WPI (RES)	300	100	100	0,333	0,333
11	WPI (RES)	1000	600	600	0,600	0,600

when $C_{RES1} = 0{,}29$r.u.; $C_{RES2} = 0{,}29$r.u.

The share environmentally friendly energy sources remained at the same level as without MACS due to the low generation of renewable energy sources and SOFCs at the considered moment of time.

5 Conclusion

The object of the study was the VPP which combines ten micro-networks with a load and various types of power sources by a 20 kV network.

The algorithm for the agent interaction was developed for the studied VPP. A simulation model of a multi-agent control system in "Jade" software complex was developed and studied. The research results showed that the multi-agent control system allows to ensure the balance of active power in the electric network while realizing all the active power from sources with renewable energy sources. As a result of the first iteration, the multi-agent control system enabled to reduce power losses in the electric network and increase the installed capacity utilization rate of power plants that are parts of the VPP.

Acknowledgment. The study was carried out with a grant from the Russian Science Foundation (project No. 20-19-00541).

References

1. Loskutov, A.B., Kulikov, A.L., Ilyushin, P.V.: Ot plana GOELRO k cifrovizacii electroenergeticheskogo kompleksa strani. Electrichestvo **12**, 14–30 (2020)
2. Action plan ("road map") "EnergyNet". National technology initiative. http://www.rvc.ru/nti/roadmaps/dk_energynet_new.pdf
3. The internet of energy architecture. https://medium.com/internet-of-energy/87224da0b72b
4. Sosnina, E., Chivenkov, A., Shalukho, A., Shumskii, N.: Power flow control in virtual power plant LV network. Int. J. Renew. Energy Res. **8**, 328–335 (2018)

5. Sosnina, E.N., Shalukho, A.V., Kechkin, A.Yu.: Optimizaciya electrotehnicheskogo kompleksa virtualnoy electrostancii s istochnikami raspredelennoy generacii. Fedorovskiye chteniya (Fedorov readings), pp. 312–320. MEI, Moscow (2017)
6. Kolar, J.W., Huber, J.E.: Solid-state transformers: key design challenges, applicability, and future concepts. In: APEC 2016 (2016). www.pes-publications.ee.ethz.ch
7. Shamshuddin, M.A., Rojas, F.: Solid state transformers: concepts, classification, and control. Energies **13**(9), 2319 (2020)
8. Kennedy, J., Eberhart, R.: Particle swarm optimization. In: Proceedings of ICNN 1995, pp. 1942–1948 (1995)
9. Karaboga, D.: An idea based on honey bee swarm for numerical optimization Technical report tr06, Erciyes University, Engineering, pp. 1–10 (2005)
10. Mei, J., Kirtley, J.L., Wang, J., Chen, C.: Coalitional game theory based loca power exchange algorithm for networked microgrids February. Appl. Energy **239**, 133–141 (2019)
11. Mensina, Y., Setthapuna, W., Rakwichiana, W.: Simulation for the management of power exchange and payment between renewable energy and electric utility network. In: 11th Eco-Energy and Materials Science and Engineering (11th EMSES), pp. 394–405 (2014)
12. Test Circuits. http://energy.komisc.ru/dev/test_cases#Link_3
13. Software package "RastrWin3". https://www.rastrwin.ru/download/Files/HELP_RastrWin3_29_08_12.pdf
14. Using JADE (Java Agent Development Environment). https://science.donntu.edu.ua/ius/kirgaev/library

Mathematical Modelling of Mechanical Structures and Assembly Processes of Complex Technical Systems

A. Bozhko(✉)

Bauman Moscow State University, 5, building 1, 2nd Baumanskaya Street, Moscow 105005, Russia
bozhkoan@bmstu.ru

Abstract. The paper discusses a hypergraph model of the mechanical structure of a technical system. The model describes the coordination of parts during assembly of a product, obtained by basing on design bases. This model correctly describes the operations and assembly processes of products that have the properties of sequentiality and coherence. The sequential and coherent assembly operations are prevalent in the assembly of modern technical systems: machines and mechanical devices. In terms of this model, the assembly operation is represented as a normal contraction of an edge. The sequence of contractions that transforms the hypergraph into a point is a mathematical description of the assembly process. A theorem on the necessary conditions for contractibility of hypergraphs is presented. It is shown that the necessary conditions are not sufficient. An important theorem on sufficient conditions of contractibility is proved. The concept of an ns-hypergraph is introduced. Ns-hypergraph is a mathematical model of the mechanical structure that cannot be assembled due to structural defects. Computational experiment was carried out to enumerate the ns-hypergraphs of various orders. The proposed apparatus can be used in computer-aided design systems for structural analysis of complex projects and computer-aided planning of assembly processes.

Keywords: Assembly sequence · Coherent assembly operation · Mechanical structure · Connection graph · Hypergraph · Computer-aided design · Computer-aided assembly planning

1 Introduction

Computer aided assembly planning (CAAP) is an important and urgent problem of modern engineering practice and design theory. Various aspects of this problem are actively discussed in modern publications [1–27].

The assembly sequence of a product, the decomposition into assembly units, the organization of assembly production largely depend on the design properties of a product: the mechanical structure, the geometry of parts, materials, etc. The mechanical structure is a set of couplings and matings between the parts that make the product connected and coordinated. Further, couplings and matings will be called mechanical connections.

A. A. Radionov and V. R. Gasiyarov (Eds.): RusAutoCon 2021, LNEE 857, pp. 80–91, 2022.
https://doi.org/10.1007/978-3-030-94202-1_8

In each assembly operation, mechanical connections are realized that connect a part to an assembled fragment of the product. This important property of assembly operations is called coherence [2]. Coherence is the main condition for the existence of design decisions. Its violation automatically excludes the design alternative from consideration.

In CAAP papers, the main mathematical model of mechanical structures of products is a connection graph. This is an undirected graph $G = (X, R)$ in which the set of vertices $X = \{x_i\}_{i=1}^{n}$ correspond to the parts of the product and the edge $r = \{x, y\} \in R$ connects the vertices $x, y \in X$ if and only if there is a mechanical connection between the parts.

This model was first proposed in [3]. This classic work describes a method of computer-aided design of assembly process, based on an expert ordering of the mechanical connections of the product. An expert answers questions about the rational ordering of the mechanical connections. Based on the information received, an assembly sequence of the product is formed. Various methods were developed to minimize the number of questions to the expert [4, 5].

The connection graph under various names is widely used to model assembly processes and operations: Liaison diagram [3], Part mating graph [6], Connective relation graph [7], Component mating graph [8, 9], Connectivity graph [10, 11], Part connectivity graph [12], Mating graph [13], Part liaison graph [14], Product liaison graph [15, 16], Adjacency graph [17], Weighted undirected connected graph [18, 19] and others.

In terms of the connection graph, the coherence condition can be described in two ways: by cutting the graph into connected components and contraction adjacent vertices.

The first method is suggested in [20, 21]. All possible cuts of the graph $G = (X, R)$ into two connected components are searched for, which in their turn are divided into simpler parts. The cutting procedure is presented in the form of AND/OR-graph, in which the root vertex is the graph $G = (X, R)$, the bundles of AND/OR-graph describe the cuts, the inner vertices represent the connectivity components obtained by the cuts, and the leaves are the single-vertex subgraphs [21].

Cutting is a weak model for disassembling the product. So that the set of cuts become a description of a disassembly sequence, several additional conditions and constrains must be checked. The main condition is absence of geometric obstacles (geometric solvability). Methods for analyzing geometric solvability are not discussed in [20, 21]. The authors believe that this can be done using universal collision detection methods [22].

A design system based on cuts of the connection graph is described in [23]. It is shown that the AND/OR graph of cuts, together with additional information about the product, can be used to synthesize assembly sequences and partition the product into assembly units.

The coherence condition for assembly operations is described as contraction of graph vertices in [24, 25]. A sequence of contractions that transforms the source graph into a single-vertex graph is a mathematical description of assembly plan.

Contraction of an edge is a simplified model of an assembly operation. Indeed, all mechanical connections must be realized in an assembled product. This means that in graph terms such a product is represented by a single-vertex graph (a point). It is easy to see that only trees are transformed into a point using the sequence of contractions of edges. If the graph describing the mechanical structure of a product has at least one cycle, then the sequence of contractions transforms it into a single-vertex graph with loops. This contradicts the original interpretation.

2 Hypergraph Models of Product and Assembly Sequence

Coordination of the parts during the assembly of a product is achieved through basing. In general, basing is performed using several mechanical connections (two, three or more). Therefore, basing is a relation of variable arity on the set of the product parts. This relation cannot be correctly described using binary mathematical objects: graphs, matrices, networks, etc.

A hypergraph model of the mechanical structure of a technical system was proposed in [26]. It gives the correct formalization of the basing relationship and the coherence property of assembly operations and plans. Here are the basic definitions and results [27].

Let $X = \{x_i\}_{i=1}^{n}$ the set of the parts of a product. We associate with the product X the hypergraph $H = (X, R)$, in which the vertices X describes the parts, and the hyperedges $r \in R$ – the minimum geometrically defined groups of parts obtained by basing.

Definition 1. Contraction is an operation of combining all the vertices incident to an edge of a hypergraph and removing this edge.

Definition 2. A normal contraction is the contraction of an edge of the second degree.

Definition 3. An assembly operation is called *n*-handed if it requires independent and simultaneous movements of *n* parts that are performed by various working bodies (hands) [28].

Definition 4. A 2-handed assembly operation is called sequential [28].

Figure 1 shows a simple product that can be assembled using simultaneous and independent movement of three parts A, B. and C, implemented by three different working bodies. This is an example of a 3-manual assembly operation. Such operations are called nonsequential.

You can give examples of products that are assembled using n, $n = 3, 4, 5, \ldots$ working bodies [28]. All these examples are geometric configurations that have no practical application.

Apparently, in modern engineering practice there are no design solutions that require nonsequential assembly operations, or their share is vanishingly small. Therefore, further we will discuss only sequential and coherent assembly operations.

Assertion 1. The assembly process of a product $X = \{x\}_{i=1}^{n}$ will be described as a sequence $P(H) = (H_1 \ldots N_N)$ of contractions of the edges of the hypergraph $H = (X, R)$ for which the following conditions are satisfied:

1. $H_1 = H = (X, R)$;
2. H_N – single-vertex graph without loops;

3. Each hypergraph $H_j \in P(H), j = \overline{2, N}$ is obtained from H_{j-1} using the normal contraction;
4. $\forall H_j, H_{j-1} \in P(H), j = \overline{2, N} \Rightarrow |R_j| - 1 = |R_{j-1}|$.

Condition 1 sets the initial state of the product when no mechanical connection is implemented. Condition 2 describes the assembled product in which all mechanical connections have been realized. Condition 3 formalizes the property of sequentiality and coherence of assembly operations. Finally, Condition 4 describes the assembly process without overbasing [29].

Definition 5. A hypergraph H, for which there is the sequence $P(H) = (H_1 ... N_N)$ that meets the conditions 1–4, will be called an *s*-hypergraph.

Figure 2a shows the drawing of intermediate shaft, the connection graph of this product (Fig. 2b) and its *s*-hypergraph H_{sh} (Fig. 2c).

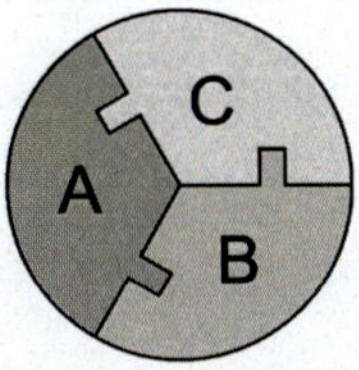

Fig. 1. Example of a 3-handed assembly operation.

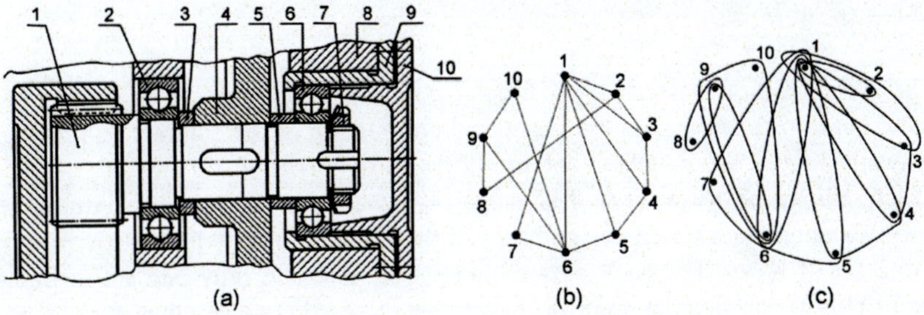

Fig. 2. The intermediate shaft of the gearbox (a), connection graph of the gearbox (b), hypergraph *Hsh* of the gearbox (c).

Figure 3 shows a subset of normal contractions $P(H_{sh})$, which transforms the *s*-hypergraph H_{sh} to the point.

The necessary and sufficient conditions for contractibility of hypergraphs describing mechanical structures of technical systems are of great theoretical and practical interest. These conditions are necessary for evaluation of product assemblability and analysis of design errors at the earliest stages of technical preparation of production.

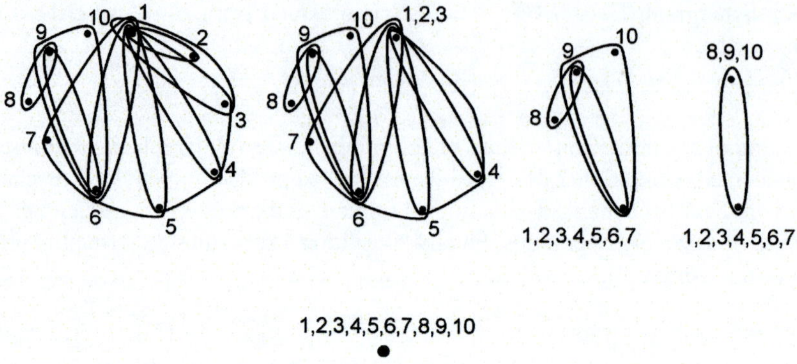

Fig. 3. Sequence of normal contractions of the hypergraph H_{sh}.

3 Contractibility of Hypergraphs

The main theorem on the necessary conditions for the contractibility of hypergraphs is proved in [26].

Theorem 1. If a hypergraph $H = (X, R)$ can be contracted to the point (H is an s-hypergraph), then:

1. among the edges of R there is at least one edge of the second degree;
2. hypergraph H is connected;
3. $|X| = |R| + 1$.

The first condition is necessary to begin the contraction process. The validity of the second condition follows from the successful completion of the contraction procedure. We give substantive arguments in defense of the third condition.

During the assembly process, the first part is installed in an assembly fixture, therefore, it does not need internal design bases of the product. For each next part you need only one set of bases. For each next mounted part one and only one set of bases is required. This means that the number of such sets is exactly one less than the number of parts. These sets correspond to the hyperedges of the hypergraph. From here immediately follows the validity of condition 3 ($|X| = |R| + 1$).

If $|X| > |R| + 1$, then the mechanical structure of the product is undetermined. If $|X| < |R| + 1$, then the structure of the product is overridden, that is, it has extra connections. At the design stage, excessive coordination entails the appearance of unresolvable dimensional chains. At the production stage, it generates overbasing. Both situations are severe design errors requiring amendments to the design.

Definition 6. A hypergraph for which conditions 1–3 of Theorem 1 are satisfied will be called a *ps*-hypergraph.

Unfortunately, the necessary conditions of Theorem 1 are not sufficient. That is, not every *ps*-hypergraph is an *s*-hypergraph. Figure 4 shows samples of hypergraphs that satisfy all the conditions of Theorem 1, but are not contractible. It is easy to see that any sequence of normal contractions transforms these examples not into a point, but into a double triangle.

Noncontractible *ps*-hypergraphs serve as theoretical proof of assembly impossibility for structural rather than geometric reasons. In modern CAD- and CAAP-systems there are no software tools to verify mechanical structures of complex technical systems and detect structural errors. Therefore, the study of noncontractible *ps*-hypergraphs is not only of theoretical interest. It has great practical value.

We introduce a necessary definition.

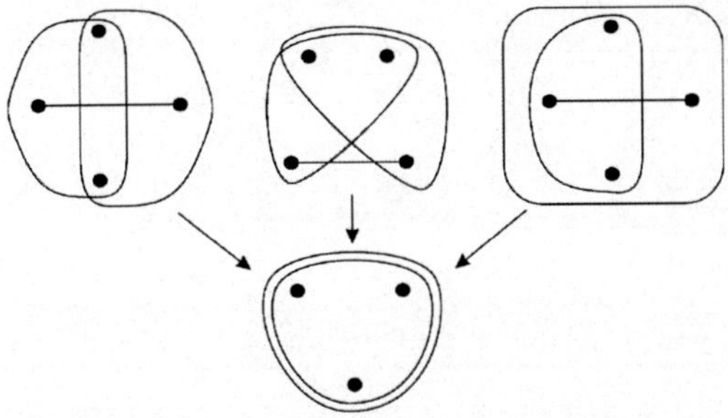

Fig. 4. Noncontractible *ps*-hypergraphs of fourth order.

Definition 7. A hypergraph $H = (X, R)$ is called an *ns*-hypergraph if it:

1. does not contain edges of the second degree;
2. is connected;
3. satisfies equality $|X| = |R| + 1$.

Obviously, such a hypergraph cannot be contracted to the point in the sense of Assertion 1.

Theorem 2. A *ps*-hypergraph is an *s*-hypergraph if and only if it does not contain subgraphs isomorphic to *ns*-hypergraphs.

Proof. Any operation of normal contraction preserves the connectivity of the hypergraph. In addition, the normal contraction reduces both the number of vertices and the number of edges per unit.

At the beginning the *ps*-hypergraph satisfies the equality $|X| = |R| + 1$. Therefore, at any j-th step of the sequence of normal contractions, equality $\left|X_j\right| = \left|R_j\right| + 1$ holds. We perform all possible normal contractions in the hypergraph H. Only two outcomes of this operation are possible: the hypergraph will turn into the point or it will be transformed into a subgraph in which there are no edges of the second degree. This subgraph is connected and satisfies equality $|X_k| = |R_k| + 1$, that is, it is an *ns*-hypergraph. The theorem is proved.

Figure 5 shows *ns*-hypergraphs of the third (a) and fourth orders (b). It can be verified directly that, up to isomorphism, these are the only examples of *ns*-hypergraphs consisting of three and four vertices.

The number of nonisomorphic ns-hypergraphs grows rapidly with increasing number of vertices. Therefore, it is not possible to list these objects manually for high-order hypergraphs. For this, a special program HYPERGEN was developed. It generates all nonisomorphic *ns*-hypergraphs of order $n = 3, 4, 5, \ldots$ Hereinafter nonisomorphic *ns*-hypergraphs will be called forbidden figures (figures).

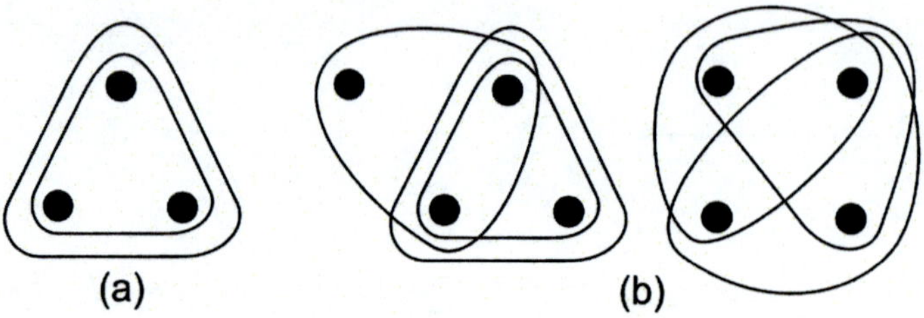

Fig. 5. *Ns*-hypergraphs of the third order (a) and fourth order (b).

4 Computational Experiment

The hypergraphs whose edges have degree three or less are considered. This limitation is associated with computational complexity and practical applicability of the problem. At designing of assembly processes the parts of a product are considered as absolutely solid bodies. This hypothesis about the parts during assembly of products is accepted in all modern CAAP-systems and most of CAD-systems [1, 2]. To determine the position of a solid body in 3D space it is required to deprive it of six degrees of freedom. Therefore, edges of the *ps*-hypergraph describing the mechanical structure of a product cannot have degree above six. In basing schemes, which are used in engineering practice, connection, as a rule, deprive a part of several degrees of freedom at once. The maximum number of different parts that are used for basing is three [29]. This means that in hypergraphs describing the real structures, the degrees of edges do not exceed 3.

HYPERGEN is a console application written in C. It runs on the Linux operating system. The program implements two mechanisms for generating forbidden figures: using combinations and by splitting integers. This duplication is necessary to verify the effectiveness of methods for generating forbidden figures and to control the correctness of the results.

We will denote the vertices of the generated hypergraphs by integers from 1 to V. The algorithm based on combinations creates a basic set of edges, which represents various combinations of E vertices. In graph terms, E represents the degree of an edge. In the general case, this number lies within $3 \leq E \leq (V-1)$.

The cardinality of the set of edges $H(V, E)$ is calculated using the formula of descending factorials $H(V, E) = \frac{V(V-1)\ldots(V-E+1)}{E(E-1)\ldots 1}$. The edges of the generated hypergraphs are presented as an array of vertex numbers incident to the given edge. The address of an edge is its number; this number ranges from 1 to H. The required hypergraphs are formed by combining them with repetitions from H to $(V-1)$ edges.

The described computational procedure can generate isomorphic hypergraphs. The isomorphic hypergraphs are set by comparing vectors of vertex degrees.

Each generated hypergraph is assigned a vector of vertex degrees. In this vector, the i-th coordinate is the degree of the i-th vertex of the hypergraph.

It is easy to see that isomorphic hypergraphs have coincident vectors of degrees. Besides, this technique allows finding and excluding hypergraphs with isolated vertices.

The hypergraph is a forbidden figure in the sense of Theorem 2, if it is connected. The connectivity test was performed using the depth first search algorithm. This algorithm tests the reachability of vertices. If all V vertices of the hypergraph are reachable, then the hypergraph is connected and is the forbidden figure.

The computational experiment showed that the described generation method effectively lists the forbidden figures whose order does not exceed 8. If $V > 8$, then computational time increases rapidly. For example, for $V = 9$, generating the forbidden figures on a computer with a Core i5 central processor and a clock speed of 3.4 GHz took just over 25 h.

To increase the efficiency and verify the results obtained, another computational procedure for generating forbidden figures has been developed. The procedure is based on splitting integers. It is known that in any hypergraph (graph) the sums of degrees of vertices and edges are equal. Any hypergraph can be represented as a bipartite graph, the partites of which indicate the vertices and hyperedges. The edges of bipartite graph represent incidence between these objects. Enumeration of degrees is just recount of edges in the given representation of the hypergraph.

By Definition 7, any forbidden figure consisting of V vertices has exactly $V-1$ edges. In addition, each edge is incident to E vertices. Then the sum of the degrees of all the vertices will be equal $P = \sum_i \deg V_i = (V-1) \times E$. Partitioning of the number $P = \sum_i \deg V_i$ into V terms can be written in the form of a vector $(D_1, D_2, ..., D_V)$ in which D_j is the degree of the vertex V_j. This means that equality $P = \sum_{j=1}^{V} D_j$ holds.

The Hindenburg algorithm [30] was used to generate numerical partitions. This algorithm creates all partitions with the same number of terms and arranges them in lexicographic order. Partition elements represent the degree series of a hypergraph. The final operation is to verify that the hypergraph is connected. This operation is performed, as before, using the depth search.

The computational experiment showed that this procedure for generating forbidden figures has a much higher efficiency than the first procedure.

The main results of the computational experiment are given in Table 1 and Table 2.

The number of forbidden figures grows as a polynomial of a low degree depending on the number of vertices (order) of the hypergraphs (Table 1).

Figure 6 shows all the forbidden figures of the fifth order and gives their vectors of vertex degrees.

Table 2 shows all forbidden figures of the sixth order. The vertices of hypergraphs are indicated in capital latin letters, and the edges are indicated by a list of incident vertices.

Table 1. Enumeration of the forbidden figures consisting of edges of the third degree.

Number of vertices	Number of *ns*-hypergraphs	Number of forbidden figures
3	1	1
4	2	2
5	6	6
6	14	12
7	30	24
8	50	42
9	112	69
10	200	100
11	348	142
12	586	187
13	963	260
14	1547	330
15	2442	441
16	3785	579
17	5785	762
18	8720	970
19	12987	275
20	19125	1611

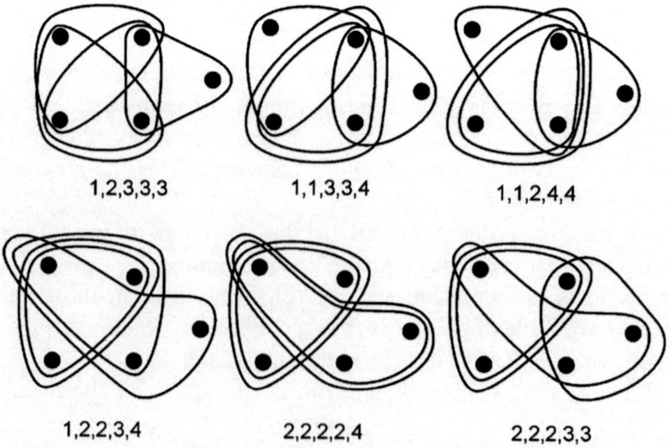

Fig. 6. The forbidden figures of the fifth order and their degree series.

Table 2. The forbidden figures of the sixth order.

Number	Vector of vertex degrees	Edges of *ns*-hypergraph				
1	111255	ABC	ABC	ABD	ABE	ABF
2	111345	ABC	ABC	ABC	ABD	AEF
3	112245	ABC	ABC	ABD	ABD	AEF
4	112335	ABC	ABC	ABC	ADE	ADF
5	112344	ABC	ABC	ABC	ABD	DEF
6	113334	ABC	ABC	ABD	ACD	DEF
7	122235	ABC	ABC	ABD	ADE	AEF
8	122244	ABC	ABC	ABD	ABE	DEF
9	122334	ABC	ABC	ABC	ADE	DEF
10	123333	ABC	ABC	ABD	CDE	DEF
11	222225	ABC	ABC	ADE	ADF	AEF
12	222234	ABC	ABC	ABD	AEF	DEF

5 Conclusion

Assembly sequence and organization of assembly operations depend on the mechanical structure of a product. For modeling mechanical structures in modern CAAP-systems, the connection graph and its various modifications are used. They are binary mathematical models, that, in the general case, are not able to correctly describe properties of assembly operations. The hypergraph model of the mechanical structure of a product is proposed. This model formalizes the basing of the parts during the assembly of the product as a relation of variable arity. In terms of this model, the sequential and coherent assembly operation is the normal contraction of an edge of the hypergraph. The assembly sequence is the set of contractions that transforms the hypergraph to the point. A theorem on the necessary conditions for contractibility of hypergraphs is presented. The theorem has great practical value as it allows to reveal design errors (impossibility of assembly, overbasing, etc.) at the earliest design stages. It is shown that the necessary conditions are not sufficient. The concepts of *s*, *ps*, and *ns*-hypergaphs are introduced. A theorem on sufficient conditions for contractibility is proved. A *ns*-hypergaph is a mathematical description of the structure of a technical system that cannot be assembled due to structural defects. A computational experiment was performed to search for the nonisomorphic *ns*-hypergraphs of various orders. All nonisomorphic *ns*-hypergraphs of the fifth, sixth, and seventh orders are given.

References

1. Ghandi, S., Masehian, E.: Review and taxonomies of assembly and disassembly path planning problems and approaches. Comput. Aided Des. **67**, 58–86 (2015)

2. Delchambre, A.: Computer-Aided Assembly Planning. Springer, Dordrecht (1992). https://doi.org/10.1007/978-94-011-2322-8
3. Bourjault, A.: Methodology of assembly automation: a new approach. In: Radharamanan, R. (ed.) Robotics and Factories of the Future 1887, pp. 37–45. Springer, Heidelberg (1988). https://doi.org/10.1007/978-3-642-73890-6_6
4. De Fazio, T., Whitney, D.: Simplified generation of all mechanical assembly sequences. IEEE J. Robot. Autom. **3**(6), 640–658 (1987)
5. Wilson, R.: Minimizing user queries in interactive assembly planning. IEEE Trans. Robot. Autom. **11**, 308–312 (1995)
6. Sambhoos, K., Koc, B., Nagi, R.: Extracting assembly mating graphs for assembly variant design. J. Comput. Inf. Sci. Eng. **9**(3), 034501 (2009)
7. Cho, D., Cho, H.S.: Inference on robotic assembly precedence constraints using a part contact level graph. Robotica **11**, 173–183 (1993)
8. Irfan, M.A., Bohez, E.: Assembly features: definition, classification and instantiation. In: International Conference on Emerging Technologies. IEEE (2006)
9. Gu, P., Yan, X.: CAD-directed automatic assembly sequence planning. Int. J. Prod. Res. **33**, 3069–3100 (1995)
10. Floriani, L., Nagy, G.: A graph model for face-to-face assembly. In: 1989 Proceedings of the International Conference on Robotics and Automation. IEEE (1989)
11. Shpitalni, M., Elber, G., Lenz, E.: Automatic assembly of three-dimensional structures via connectivity graphs. CIRP Ann. **38**, 25–28 (1989)
12. Lin, A.C., Chang, T.C.: An integrated approach to automated assembly planning for three-dimensional mechanical products. Int. J. Prod. Res. **31**(5), 1201–1227 (1993)
13. Heedong, K., Lee, K.: Automatic assembling procedure generation from mating conditions. Comput. Aided Des. **19**(1), 3–10 (1987)
14. Park, H.-S., Park, J.-W., Park, M.-W., Kim, J.-K.: Development of automatic assembly sequence generating system based on the new type of parts liaison graph. In: Bernard, A., Rivest, L., Dutta, D. (eds.) PLM 2013. IAICT, vol. 409, pp. 540–549. Springer, Heidelberg (2013). https://doi.org/10.1007/978-3-642-41501-2_54
15. Lee, D.H., Kang, J.G., Xirouchakis, P.: Disassembly planning and scheduling: review and further research. Proc. Inst. Mech. Eng. Part B J. Eng. Manuf. **215**(5), 695–709 (2001)
16. Mascle, C., Xing, K.: A liaison model for disassembly-reassembly product eco-design. Int. J. Des. Eng. (IJDE) **2**(3), 346–368 (2009)
17. Roy, U., Banerjee, P., Liu, C.R.: Design of an automated assembly environment. Comput. Aided Des. **21**(9), 561–569 (1989)
18. Cao, Y., Kou, X., Cao, S.: A sub-assembly identification algorithm for assembly sequence planning. In: International Industrial Informatics and Computer Engineering Conference (2015)
19. Vigano, R., Gomez, G.: Assembly planning with automated retrieval of assembly sequences from CAD model information. Assembly Autom. **32**(4), 347–360 (2012)
20. Luiz, S., de Mello, H., Sanderson, A.C.: A basic algorithm for the generation of mechanical assembly sequences. In: Luiz, S., de Mello, H., Lee, S. (eds.) Computer-Aided Mechanical Assembly Planning, vol. 148, pp. 163–190. Springer, Boston (1991). https://doi.org/10.1007/978-1-4615-4038-0_7
21. Homem de Mello, L., Sanderson, A.: A correct and complete algorithm for the generation of mechanical assembly sequences. IEEE Trans. Robot. Autom. **7**(2), 228–240 (1991)
22. Seth, A., Vance, J.M., Oliver, J.H.: Virtual reality for assembly methods prototyping: a review. Virtual Reality **15**(1), 5–20 (2011). https://doi.org/10.1007/s10055-009-0153-y
23. Holland, W., Bronsvoort, W.: Assembly features in modeling and planning. Robot. Comput.-Integr. Manuf. **16**, 277–294 (2000)

24. Naphade, K., Storer, R., Wu, S.D.: Graph-theoretic generation of assembly plans, part I: correct generation of precedence graphs (1999). https://www.researchgate.net/profile/Robert_Storer/publication/. Accessed 25 Nov 2020
25. Naphade, K., Storer, R., Wu, S.D.: Graph-theoretic generation of assembly plans, part II: problem decomposition and optimization algorithms (1999). https://www.researchgate.net/publication/2447296. Accessed 25 Nov 2020
26. Bozhko, A: Math modeling of sequential coherent and linear assembly plans in CAD systems. In: 2018 Global Smart Industry Conference (GloSIC), pp. 1–5 (2018)
27. Bozhko, A.N.: Hypergraph model for assembly sequence problem. In: IOP Conference Series: Materials Science and Engineering, vol. 560, no. 1, p. 012010. IOP Publishing (2019)
28. Natarajan, B.K.: On planning assemblies. In: Proceedings of the Fourth Annual Symposium on Computational Geometry, pp. 299–308 (1988)
29. Whitney, D.E.: Mechanical Assemblies: Their Design, Manufacture, and Role in Product Development. Oxford University Press, New York (2004)
30. Karpenko, A.P., Leshchev, I.A.: Advanced cat swarm optimization algorithm in group robotics problem. Procedia Comput. Sci. **150**, 95–101 (2018). 13th International Symposium "Intelligent Systems - 2018" (INTELS 2018)

Development of the Industrial Room Automation System on the Basis of a Single Computer

R. Nezmetdinov(✉), P. Melikov, and R. Utarbaev

Moscow State University of Technology “STANKIN”, 1, Vadkovsky per., Moscow 127055, Russia

Abstract. The paper proposes an approach to creating an automation system for controlling and monitoring the climatic conditions of a production facility, which is characterized by the use of available microprocessors and mass-produced components. The developed system, including all its software and hardware components, should be flexible in terms of construction, use and readjustment, should be close to analogs existing on the market, such as climate control elements for building automation systems, but have greater functionality and lower cost. The development of such a solution will increase their availability and flexibility by expanding the technology stack applicable to solving such problems.

Keywords: Industry 4.0 · Raspberry PI · IoT sensors · Web development · Climatic control

1 Introduction

The task of controlling and maintaining the microclimate of industrial premises is undoubtedly relevant for modern machine-building production, since it directly affects such indicators as the conditions and nature of work, productivity and quality of work of employees, reliability of work and service life of equipment, as well as its accuracy characteristics [1, 2].

The proposed solution has a modular structure that allows flexible configuration [2, 3] of a set of controlled and adjustable climatic parameters based on the purpose of the production facility and the nature of the equipment installed in it. An important advantage is the presence of a Web interface [3–5] for remote configuration and dispatching.

2 Comparison of Existing Approaches

To form a list of characteristics of the presented solution, which will provide a competitive market advantage and determine the list of requirements for building such a solution, a comparative analysis of the parameters and functionality of existing similar solutions [6] and our own developed solution was carried out, which is presented in the Table 1.

When considering the existing climate control systems on the market, one regularity can be distinguished: the automation core in each of them is a low-power processor,

A. A. Radionov and V. R. Gasiyarov (Eds.): RusAutoCon 2021, LNEE 857, pp. 92–101, 2022.
https://doi.org/10.1007/978-3-030-94202-1_9

which is used as the main computing module in the case of a centralized LanDrive system or as one of the control elements in decentralized control hierarchies (as is the case with LCN and LON systems). This fact significantly limits the flexibility of these systems, as well as the ability to add support for new communication interfaces and control algorithms.

Table 1. Comparison of existing solutions with the developed.

Characteristic	LanDrive	LON	LCN	Own solution
Software	RTOS	RTOS	RTOS	OS Linux
System core	32-bit CPU with tact. frequency 48 MHz	Proprietary processor (8-bit)	Proprietary processor with tact. frequency 10 MHz	32 or 64 bit CPU with tact. Frequency from 700 MHz
Hierarchy type	Centralized	Decentralized	Decentralized	Mixed
FLASH-memory	64 KB	42 KB	32 KB	От 512 MB
Interfaces	RS-232, RS-485, 1Wire, GSM, ZigBee, USB	PCI, USB, Ethernet	BACnet, Modbus, OPC, BMABosch, IP-Symcon, Tobit	UART, GPIO, I2C, 1Wire, Ethernet, etc.
Supply voltage	6–36 V	90–240 or 24 V	90–240 or 24 V	90–240 or 5 V
Operating temperature range	−40 °C…85 °C	−40 °C…85 °C	−40 °C…85 °C	−40 °C…85 °C

Based on the analysis of similar solutions, the requirements for the developed system were formulated:

- use of a full-fledged Linux operating system with a real-time extension, which makes it possible to implement complex control algorithms and control over the network;
- hardware platform with significantly greater power and flexibility for executing control programs and providing additional services;
- support for a wide range of communication interfaces, performance characteristics that are not inferior to those of existing systems.

3 Comparison of Existing Approaches

The structural diagram of both the software and hardware parts of the industrial premises automation system is based on the principle of using common software and hardware modules, among which the following main software components can be distinguished: the main automation program written in the C programming language, which directly controls and controls all systems production premises through the existing communication interfaces, and interacting with the Nginx web server [7–9] and the php-script processor, as well as the mjpg-streamer utility, which allows you to take pictures from the camera and transfer them to the Internet (Fig. 1).

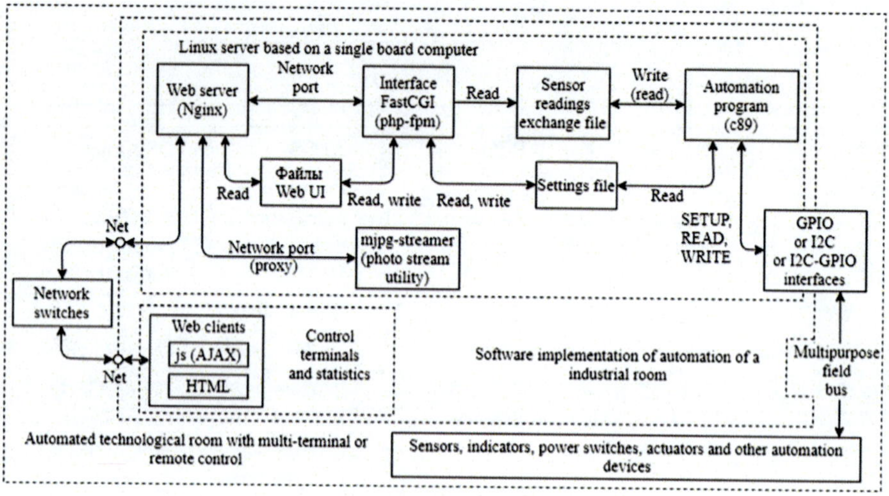

Fig. 1. Developed block diagram of the software part.

Due to the large number of interfaces, it is possible to connect a large variety of devices. If there is a need to use other interfaces, such as RS-485, Modbus, ZigBee, then they can be implemented using appropriate adapters of physical layer interfaces, for example, by connecting such adapters to the USB port and using widely available drivers and API libraries provided for these modules [10].

Sensors are built in the form of modules that have analog and threshold digital outputs. The analog signal is proportional to the concentration of gases to which the sensor is sensitive. The sensitivity of the module (the level of the digital output triggering threshold) is adjusted by a pull-down resistor. The digital output of the sensor data is implemented by a simple boundary comparator and is of no interest for the designed system [11].

The main database of the modules is a tin oxide sensor, which is sensitive to the content of specific gases in the air.

The main control algorithm of the system (Fig. 2) is based on the principle of sequential cyclic control, support for multi-terminal control is provided. For each of the monitored parameters, a check is made for compliance with the specified range of values, and in case of going beyond the interval, the actuators are turned on/off in accordance with the rule base and the operating mode. There are three operating modes: automatic control; control from a local console and control from a remote terminal in the WEB space [12, 13].

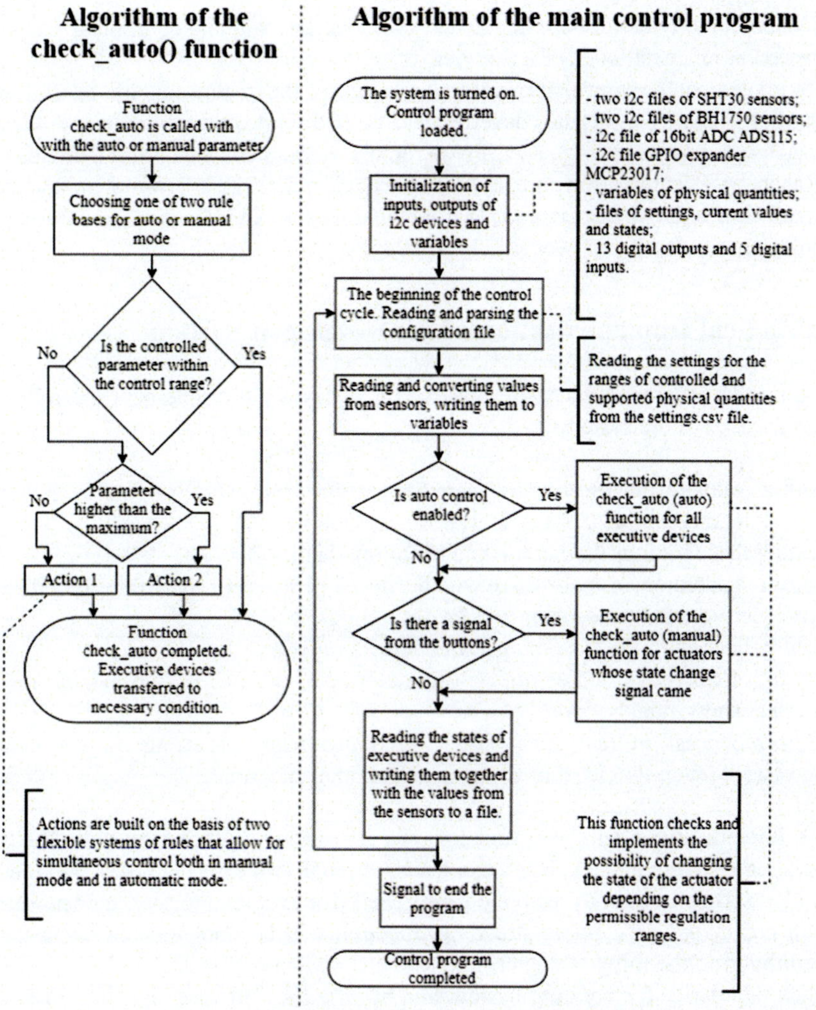

Fig. 2. Basic control and monitoring algorithm.

4 Selection of Basic Hardware Modules for Climate Control and Management

To meet the needs for automation of a typical production facility, the choice of the most important parameters for the control and management of climate was determined: illumination, humidity, temperature and the content of pollutants in the air. To realize the possibility of not only monitoring, but also direct control of these parameters, it is necessary to have executive modules in the system [14].

The following sensors were selected as basic sensor modules: a BH1750 illumination sensor, an SHT30 temperature and humidity sensor, MQ series air pollutants (emitted during the operation of various technological equipment in the room). These monitored parameters directly affect both health and safety of a person and equipment, including the period or characteristics of its service.

To implement the control of monitored parameters, the system contains the following modules: I^2C analog-to-digital converter ADS1115, discrete power output module based on relays, I^2C bus buffers P82B715PN to remove sensors and execution modules from the Raspberry Pi automation core, I^2C GPIO expander MCP23017 (optional - to expand the number of inputs-outputs), and other modules such as an indication module, camera module, buttons, buzzer, power supply and others.

5 Practical Implementation of the Developed System

Based on the developed hardware model, it is necessary to implement the following main functions of the system:

- reading values from sensors connected via various hardware interfaces, as well as converting these values, if necessary;
- reading these settings from a file and operating them;
- control of actuators in accordance with the developed control algorithms based on the data received from the sensors and the set settings;
- local indication of the system operation and its functions;
- control of executive bodies not only in automatic, but also in two manual modes: physical and virtual buttons;
- photo-broadcast in real time, sound accompaniment of events in the system, convenient reconfiguration of the system and other functions.

To implement the described functions, an approach was chosen to write the main automation program in the C language, which, through two exchange files, will interact with PHP scripts or directly with the Linux shell (for example, to send the "poweroff" system command). The photo-streaming function will be implemented by installing and configuring the mjpg-streamer utility. The C language was chosen because of the greatest popularity for creating system and application software, and like PHP, it is the most popular when implementing the server side of an application. However, as an alternative, development in Python is possible, where it will be even easier to implement such functionality due to the wide availability of ready-made libraries [15, 16].

Lightweight files in CSV format (Comma-Separated Values - values separated by commas) are the main unified means of data exchange between the server part and the main control program.

To unify the architecture of the main control program, the functionality implemented by it is divided into separate functions that ensure the operation of one or another mechanism - for example, the read/write function of settings, the function of initialization of inputs/outputs, the function of reading and converting values from the light sensor, and others.

The server-side API, implemented in PHP, is also divided by functional purpose, for example, the buttons.php call is responsible for processing the buttons for switching executive devices [17].

The terminal client part is written using HTML markup language, CSS styling language and client-side scripts in JavaScript. This approach is not optimal, since it is time consuming in development and changeover. It is more preferable to use specialized frameworks like Graphana, Node-Red (or web application development frameworks like Angular or React). Some frameworks even have built-in communication mechanisms such as MQTT, sockets, and others [18].

6 Creation of a Prototype and Testing

Figure 3 shows the developed WEB-interface for management and monitoring. The main screens here are the "Terminal" page, which allows you to monitor the main indicators of the system and manually control the executive devices, and the "Settings" page, where you can configure the system and enable/disable automatic control algorithms.

During testing of the developed panel "Terminal", built on the designed algorithms of the system's operation, no deviations of the actual behavior of the system from the expected were revealed.

As laid down in the algorithm of the control system, with any change in the state of any of the executive devices, the corresponding light comes on or goes out, and at the same time the buzzer of the local control system gives a sound signal. Control through virtual buttons, as designed in the algorithm, works both with the automatic mode turned off, and with it.

The display of the current microclimatic indicators with the indication of going out of the permissible ranges works properly, as well as the display of the current state of the executive devices [19]. The photo broadcast works stably and with good quality even through remote network nodes, while it is broadcast in HD quality with a bit rate of 2.3 Mb/s at a frame rate of 15 per second.

Since the remote terminal performs only the function of interacting with the user, and the local control system implements the automatic control logic, the automatic control and maintenance of parameters within the specified limits work properly [20].

The "Statistics" screen (Fig. 4) presents a graphical representation of the monitored parameters with the possibility of increasing the zones.

The graph display is working properly. Display ranges are automatically selected; the possibility of increasing the zones (approach/stretch) is present.

The tests carried out showed that all hardware and software elements of the system function in a given mode and without failures during long-term operation in various operating conditions, such as: instability of the supply voltage, work under intense sunlight, work in an environment with a high content of pollutants.

The software part also works correctly: the readings from the sensors are displayed correctly, the settings affect the operation of visualization and automatic control algorithms in a given mode, all methods of system control work and do not interfere with mutual operation: local control of physical buttons, WEB control in manual, automatic and mixed modes.

The laboratory bench is shown in Fig. 5.

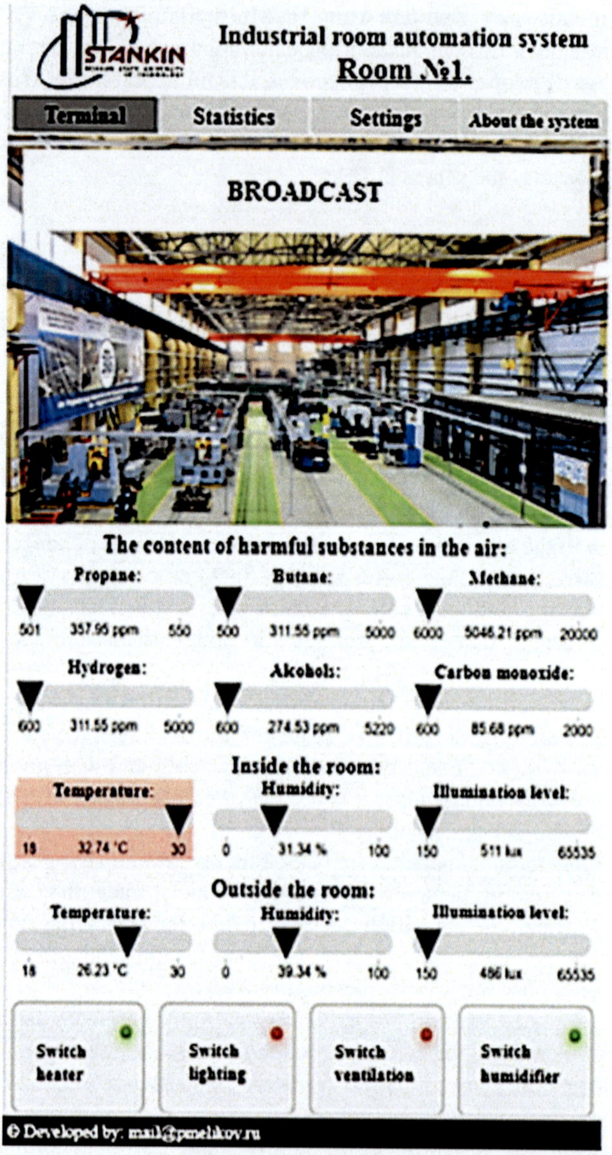

Fig. 3. Terminal page.

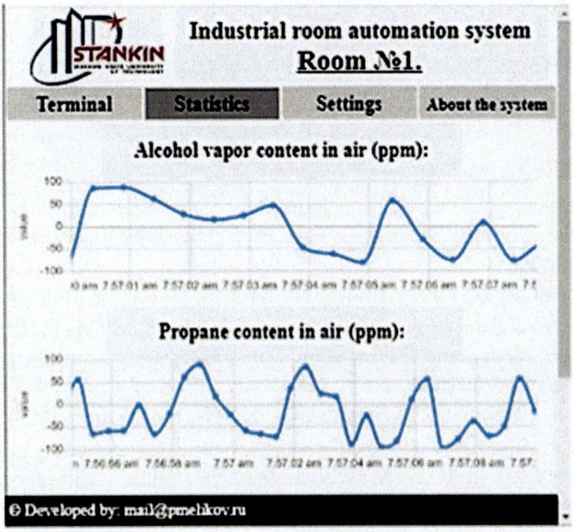

Fig. 4. Statistics page.

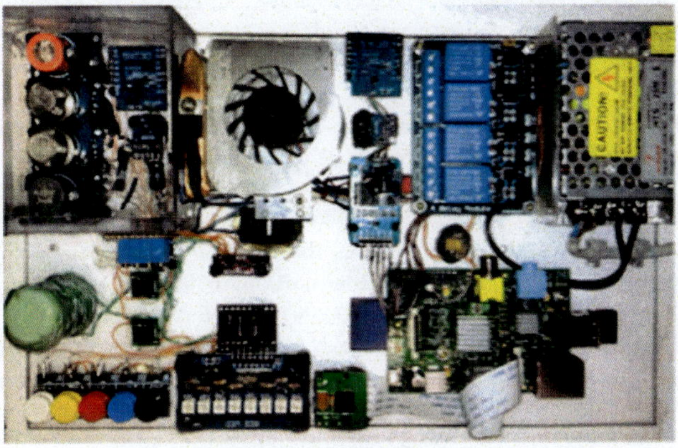

Fig. 5. Statistics page.

7 Conclusion

The proposed system has a number of advantages over existing counterparts: a variety of available hardware interfaces and ways to expand it; great flexibility of implementation and speed of changeover due to the many available software tools and programming languages, achieved using a full-fledged Linux operating system, the integration of IoT technology [21] allows you to provide the system with modules for remote monitoring, control and dispatching, as well as integrate it with other existing solutions.

At the moment, the developed solution is at the stage of testing in real operating conditions, and the components of the created MVP (minimum viable product, minimum viable product) have been partially transferred to production.

Acknowledgments. This research was supported by Fund for Assistance to Small Innovative Enterprises in Science and Technology (agreement 16266ГУ/2021 from 18 May 2021).

References

1. Martinov, G.M., Kozak, N.V., Nikishechkin, P.A.: Approach to solving the task of backup on machine tools with CNC. Russ. Eng. Res. **40**, 1024–1029 (2020). https://doi.org/10.3103/S1068798X20120096
2. Kovalev, I., Kvashnin, D., Chervonnova, N., Nikich, A.: Application of agile methodology at industrial manufacturing as part of the Industry 4.0. In: ICMTMTE 2020. IOP Conference Series: Materials Science and Engineering, vol. 971, p. 052034 (2020). https://doi.org/10.1088/1757-899X/971/5/052034
3. Nezhmetdinov, R., Kovalev, I., Chervonnova, N., Nezhmetdinova, R., Khoury, A.A.l.: An approach to the development of logical control systems for technological equipment in the concept of Industry 4.0. In: ICMTMTE 2020, MATEC Web of Conference, vol. 329 p. 03044 (2020). https://doi.org/10.1051/matecconf/202032903044
4. Martinova, L.I., Kozak, N.V., Kovalev, I.A., Ljubimov, A.B.: Creation of CNC system's components for monitoring machine tool health. Int. J. Adv. Manuf. Technol. **117**(7–8), 2341–2348 (2021). https://doi.org/10.1007/s00170-021-07107-1
5. Kovalev, I., Issa, A., Nikishechkin, P., Chervonnova, N., Petrov, A.: Development of a data collection system for a CNC system using cloud FRED technology and OPC UA specification. In: ICMTMTE 2020, MATEC Web of Conference, vol. 329 p. 03043 (2020). https://doi.org/10.1051/matecconf/202032903043
6. Tikhonov, A.F., Demidov, S.L., Smelyakov, A.L.: Automation of engineering systems to ensure optimal parameters of the microclimate of a manufacturing enterprise. Mech. Constr. **12**, 56–58 (2013)
7. Nikishechkin, P.A., Chervonnova, N.Yu., Nikich, A.N.: Approach to development of specialized terminals for equipment control on the basis of shared memory mechanism. In: Radionov, A.A., Karandaev, A.S. (eds.) RusAutoCon 2019. LNEE, vol. 641, pp. 181–188. Springer, Cham (2020). https://doi.org/10.1007/978-3-030-39225-3_20
8. Martinov, G.M., Kovalev, I.A., Chervonnova, N.Y.: Development of a platform for collecting information on the operation of technological equipment with the use of Industrial Internet of Things. In: IOP Conference Series: Materials Science and Engineering, vol. 709, no. 4 (2020). https://doi.org/10.1088/1757-899X/709/4/044063
9. Martinova, L.I., Pushkov, R.L., Fokin, N.N.: Development of standardized tools for shopfloor programming of turning and turn-milling machines. In: IOP Conference Series: Materials Science and Engineering, vol. 709, no. 4 (2020). https://doi.org/10.1088/1757-899X/709/4/044064
10. Nikishechkin, P.A., Chervonnova, N.Y., Nikich, A.N.: An approach of developing solution for monitoring the status and parameters of technological equipment for the implementation of Industry 4.0. In: IOP Conference Series: Materials Science and Engineering, vol. 709, no 4 (2020). https://doi.org/10.1088/1757-899X/709/4/044065
11. Martinov, G., Nikishechkin, P., Al Khoury, A., Issa, A.: Control and remote monitoring of the vertical machining center by using the OPC UA protocol. In: IOP Conference Series: Materials Science and Engineering, vol. 919, p. 032030, pp. 1–8 (2020). https://doi.org/10.1088/1757-899X/919/3/032030
12. Martinova, L., Obukhov, A., Sokolov, S.: Practical aspects of ensuring accuracy of machining on CNC machine tools within framework of "smart manufacturing". In: 2020 International Russian Automation Conference, pp. 898–902 (2020). https://doi.org/10.1109/RusAutoCon49822.2020.9208079

13. Martinov, G., Pushkov, R., Martinova, L., Kozak, N., Evstafieva, S.: Approach to development of HMI screens for CNC with dynamic kinematics. In: ICMTMTE 2020, MATEC Web of Conference, vol. 329, p. 03026, pp. 1–6 (2020). https://doi.org/10.1051/matecconf/202032903026
14. Kovalev, I., Nezhmetdinov, R., Nikishechkin, P.: Approach to assessing the possibility of functioning of CNC and PAC systems on various software and hardware platforms. In: International Multi-Conference on Industrial Engineering and Modern Technologies, pp. 1–5 (2019). https://doi.org/10.1109/FarEastCon.2019.8933999
15. Kovalev, I.A., Babin, M., Nikishechkin, P.: Development of a method for the determination and registration of unauthorized data transmission channels at industrial manufactories. In: MATEC Web of Conference, vol. 298 (2019). https://doi.org/10.1051/matecconf/201929800110
16. Chekryzhov, V., Kovalev, I.A., Grigoriev, A.S.: An approach to technological equipment performance information visualization system construction using augmented reality technology. In: International Conference on Modern Trends in Manufacturing Technologies and Equipment, pp. 1–7 (2018). https://doi.org/10.1051/matecconf/201822402093
17. Grigoriev, S., Martinov, G.: An approach to creation of terminal clients in CNC system. In: 3rd Russian-Pacific Conference on Computer Technology and Applications, pp. 1–4 (2018). https://doi.org/10.1109/RPC.2018.8482153
18. Martinova, L.I., Fokin, N.N.: An approach to creation of a unified system of programming CNC machines in the dialog mode. In: International Conference on Modern Trends in Manufacturing Technologies and Equipment, vol. 224, pp. 1–5 (2018). https://doi.org/10.1051/matecconf/201822401101
19. Kovalev, I., Nezhmetdinov, R., Kvashnin, D.: Big data analytics of the technological equipment based on Data Lake architecture. In: MATEC Web of Conference, vol. 298 (2019). https://doi.org/10.1051/matecconf/201929800079
20. Martinov, G.M., Nikishechkin, P.A., Grigoriev, A.S., Chervonnova, N.: Organizing interaction of basic components in the CNC system AxiOMA control for integrating new technologies and solutions. Autom. Remote Control **80**(3), 584–591 (2019). https://doi.org/10.1134/S0005117919030159
21. Martinov, G.M., Al Khoury, A., Issa, A.: An approach of developing low cost ARM based CNC systems by controlling CAN drives. In: International Conference on Modern Trends in Manufacturing Technologies and Equipment, vol. 224, pp. 1–6 (2018). https://doi.org/10.1051/matecconf/201822401020

Control System with a Predictive PID-Controller with a First-Order Filter: Estimation of the Efficiency for Thermal Proceses

E. Merzlikina[1](✉), G. Sviridov[1], and Hoang Van Va[2]

[1] NRU Moscow Power Engineering Institute, 14, Krasnokazarmennaya, Moscow 111250, Russia
MerzlikinaYI@mpei.ru

[2] EVN Information Technology, Hanoi, Vietnam

Abstract. This paper considers the problem of the control system transient process quality improvement, which is very important in thermal engineering even now. The way of improvement under consideration is the use of a predictive control algorithm. The control algorithm described is a PID control law with a first order filter and a linear prediction module; the control system is also equipped with an auto-tuning module, that allows to retune the system if necessary. The auto-tuning module uses a fast auto-tuning algorithm calculating the process model parameters using the process reaction of a rectangular impulse. The PID algorithm coefficients are calculated using indirect optimality indexes. The system works with a typical thermal process with a second order transfer function with a time delay. Calculations and simulation of the system are carried out in Mathcad and Matlab/Simulink. Some examples are considered, the predictive control system performance is estimated and recommendations on the prediction time interval determination are given. These recommendations may be further used while developing this system using modern controllers and controller programming software.

Keywords: Predictive control · Prediction · PID-controller · Auto-tuning · Filter · Time delay

1 Introduction

Nowadays in many industries the demands to the control system performance are rather strict. For example, the temperature of the superheated steam at thermal power units can change in a very narrow interval. The things mentioned mean that the performance of the control system should be high and it is still important to improve the performance. There are some features of control plants and processes that can influence the control system performance badly, and one of them is the time delay. It is typical for some processes, for example, for thermal processes at thermal power plants, heating systems, industrial thermal technologies, etc.

Generally, two ways can be used to improve the control system performance. First, one can use a more advanced control algorithm (for example, one can use PID instead

A. A. Radionov and V. R. Gasiyarov (Eds.): RusAutoCon 2021, LNEE 857, pp. 102–112, 2022.
https://doi.org/10.1007/978-3-030-94202-1_10

of PI and so on), second, one can use a more complex control system structure, for example, a cascade system [1, 2]. The cascade systems are standard engineering solution in many cases nowadays, for example, at thermal power plants [3], but they have certain drawbacks in comparison with single-loop systems. Single-loop systems are easy to use and tune, there are a lot of tuning methods and auto-tuning algorithms [2, 4] that may be easily implemented by means of modern programmable controllers. That is why, from the one side, the abilities of the standard linear control algorithms are under-used, and from the other side, it is possible to improve the performance of the single-loop systems using auto-tuning algorithms [4] and predictive algorithms [1, 5]. This paper is going to consider a single-loop control system with a predictive PID-controller, auto-tuning module and a thermal process.

There are a lot of adaptation and auto-tuning methods [2, 4, 6, 7], all of them have their advantages and disadvantages. In order to get the process model and tune the control system the method may use the process transient response [2] or frequency response [2, 4] or more complicated things such as artificial neural networks [8]. In this paper a method described in [4] will be used, this method is called AT-1 and it has two important advantages, it allows to get coefficients of the considerably complex model (the second order with the time delay and the four coefficients may be determined independently) and it works faster than other methods because it performs only one iteration. This method was used earlier in some systems where the process model was used, for example, in a single-loop control system with a PID-algorithm, a thermal process and a Smith predictor [9].

Predictive control systems are widely described in the technical literature, for example, in [1, 5, 6]. There are a lot of articles considering using the predictive systems with the thermal processes [5, 7, 8, 10, 11], and all of them show that the predictive systems of various kinds are effective for thermal processes. But generally they use either rather complex algorithms, such as MPC [11] or neural networks [8]. Some articles consider standard linear control algorithms, such as [10, 12, 13], but they do not consider the systems with the PID control law with filters, although these systems are typical for industry [2, 4]. In this situation it seems logical to consider a control system with the predictive PID control law with a filter. Such systems with the second order filters were briefly considered in [14], but the PID law with the first order filter is even more widely used.

This article considers a single-loop control system with a PID control law with the first order filter. The system is equipped with an auto-tuning module using the fast auto-tuning algorithm AT-1 and works with a typical thermal process.

2 System Description and Problem Statement

As has already been mentioned, the paper considers a single-loop control system with a PID control law and an auto-tuning module. The PID is equipped with a first order filter. The prediction module is in the feedback. The structure of the system is given in Fig. 1.

The PID control law with the first order filter has the following transfer function:

$$W_{PID}(s) = K_p \cdot (1 + \frac{1}{T_i s} + \frac{T_d s}{T_f s + 1}) \tag{1}$$

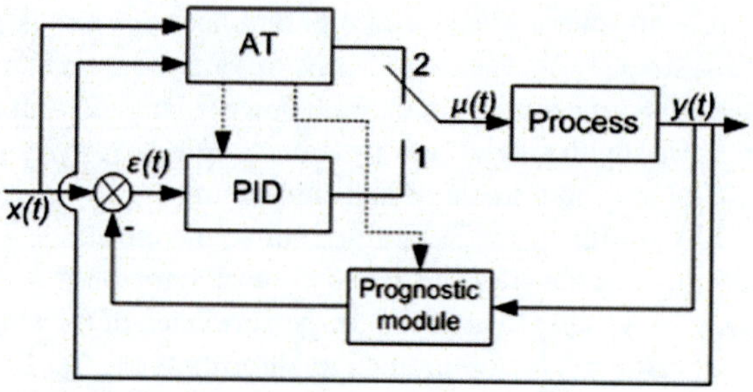

Fig. 1. The structure of the control system.

where: K_p – the gain of the PID control law, T_i – the integrating time constant, T_d – the derivative time constant, T_f – the filter time constant. Tuning methods for this control law are described in many books and papers, for example, in [2, 4, 15]. In this paper we will use the method based on the ideas shown in [4, 16], the method will be described below.

The process model has the second order transfer function with the time delay:

$$W_{mod}(s) = \frac{K_{mod} \cdot e^{-\tau_{mod} s}}{(T_{1mod} s + 1)(T_{2mod} s + 1)} \tag{2}$$

where: K_{mod} – the gain of the process model; T_{1mod}, T_{2mod} – the time constants of the process model; τ_{mod} – the delay time of the process model.

The system shown in Fig. 1 operates as a common single-loop system with a PID control algorithm in the normal working mode. If it is necessary to retune the system one switches the switch shown in Fig. 1 from position 1 to position 2 and the auto-tuning module began to work. The general structure of the auto-tuning module is shown in Fig. 2.

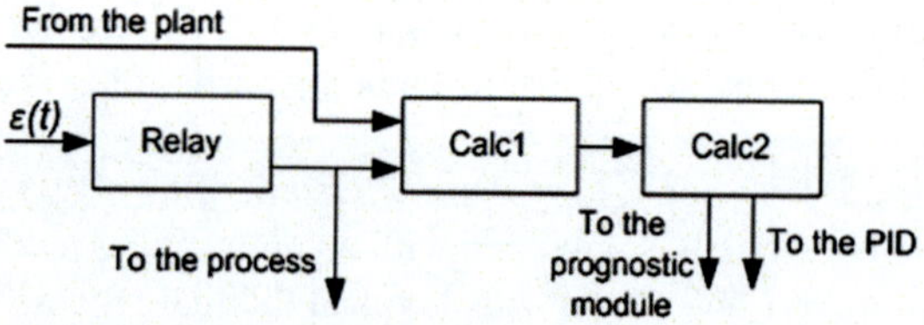

Fig. 2. The structructure of the auto-tuning module.

The general working algorithm of the autotuning module is simple. The relay forms the rectangular impulse on the input of the process. The impulse response goes to the Calc1 module and the module calculates the coefficients of the process model with the transfer function (2). Then the coefficients are sent to the Calc2 module that calculates the PID-law parameters, the transfer function of the PID-law is (1). The PID parameters are calculated using indirect frequency optimality indicators. The calculation methods mentioned are described in detail in [4, 16].

The prediction module operates according to the linear prediction algorithm, described by the following formula [10, 12, 13, 16]:

$$y_{pred}(t) = y(t) + y'(t) \cdot \tau_{pred} \tag{3}$$

where: $y(t)$ – the process output signal (see Fig. 1), $y'(t)$ – the derivative of the process output signal, $y_{pred}(t)$ – the output of the prediction module, τ_{pred} – the prediction time. The linear prediction algorithm was chosen because it is simple, easy to implement by means of modern controller programming systems and generally provides good results.

There is an undetermined parameter τ_{pred} in the linear prediction algorithm. In the papers, for example, [10, 12, 13] there are recommendations on how to choose this parameter. Although these recommendations are given for the P, PI and the so-called ideal PID-algorithm (or the PID-algorithm with the derivative term without the filter at all). This kind of PID-algorithm is not typical for the modern controllers, where the PID-algorithms with the first order filter or the second order filter are normally used. The PID-law with the prediction module and the second order filter is considered in [14], but it is impossible to say that the recommendations for this case will work properly for the PID-algorithm with the second order filter.

So, the problem statement can be represented as follows. One should consider the control system shown in Fig. 1 and described above and study how the prediction module influences the system transient process quality. If the influence is positive it is necessary to give recommendations (at least approximate) on how to choose the prediction time interval τ_{pred}. The process under consideration is a typical thermal process with the transfer function (2).

3 The Process Characteristics

The process under consideration has transfer function (2), this kind of transfer function is typical for non-integration thermal processes, for example, in the temperature control sections. Several processes with different time constants and delay times were considered during the preliminary calculations, Table 1 shows the parameters of some of them. The table shows the parameters of the process (line "Process") and parameters of the model (line "Model") obtained by the auto-tuning module using the method described in [2, 4].

Table 1. Control process parameters.

	Item	K	T_1	T_2	τ
1	Process	1	16	80	6
	Model	0.999	17.7	77.6	6.3
2	Process	1	12	120	4
	Model	1.12	12.1	134.5	3.9

An example of the step response of process 1 and the model of the process are given in Fig. 3. As one can see from the figure the curves almost coincide with each other so the auto-tuning system works efficiently. The step responses of this type are rather wide-spread in thermal power engineering [3, 17].

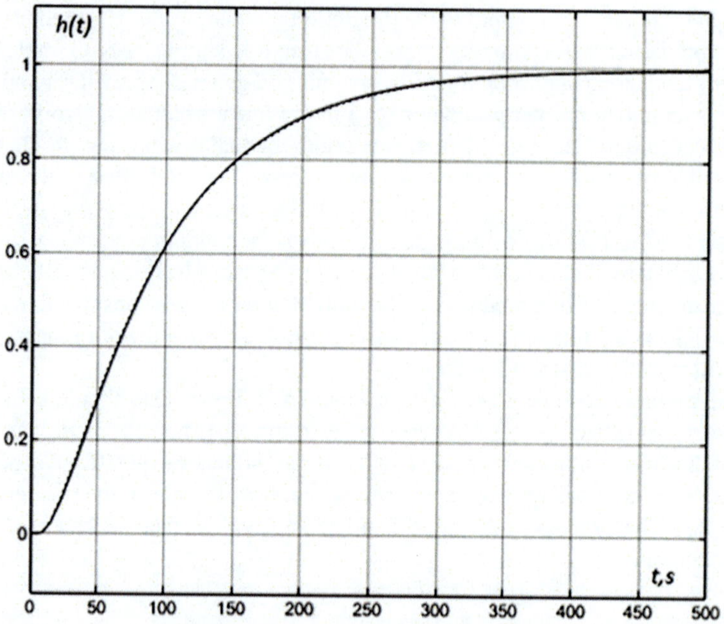

Fig. 3. The step response of process 1.

4 PID Algorithm Tuning and Transient Processes

First let us consider an ordinary single-loop system with a PID control law with a first order filter, the transfer function (1) is given above. The calculation of the control law parameters are carried out by the auto-tuning module on the basis of the method based on [2] and described in [4, 18]. The method is based on the fact that the vector of the frequency response of the closed optimally tuned system is always in the more or less same position if the frequency is the resonance one (or close to it). So, one can give the magnitude and the phase of the vector and consider them as the indirect frequency optimality criteria, and the auto-tuning module calculates the PID-law coefficients that can provide the stated frequency response vector position on the resonance frequency. The parameters calculated are given in Table 2. The parameters provide a damping ratio value of 0.9.

The transient processes in the closed loop system with the PID control law are given in Fig. 4 and 5. Figure 4 shows the system response when the set point changes.

Figure 5 shows the system response to the step signal on the input of the control process.

Table 2. Control algorithm parameters.

	K_p	T_i	T_d	T_f
1	14.796	16.814	10.395	1.25
2	32.71	12.18	7.796	0.863

From the given processes we can see that the system provides the above mentioned damping ratio.

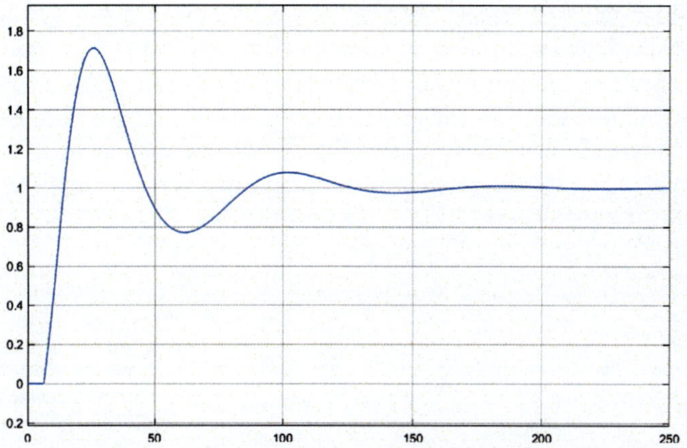

Fig. 4. The transient process when the set point changes.

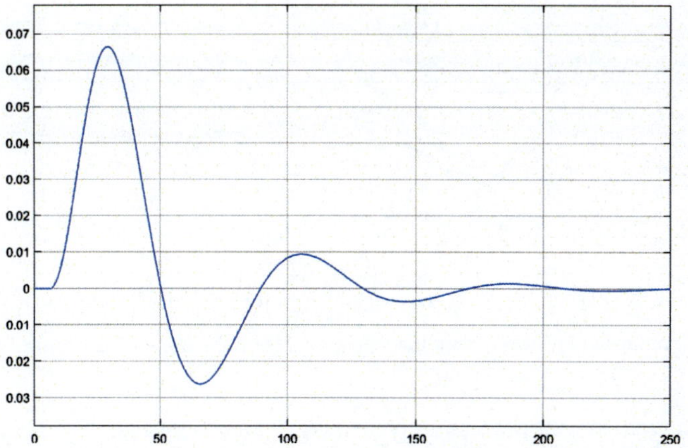

Fig. 5. The transient process when the step signal is on the input of the control process.

5 Control System with the Predicitive PID Algorithm

When one considers the control system with the PID control law with a prediction module one faces a very important question of how to choose the prediction time interval. The use of the prediction module helps to provide higher transient process quality but the improvement of the quality depends on the value of the time interval. It is logical to assume that the value of the time interval may depend on the process model. There are some recommendations in the literature, for example, [10, 12, 13, 19] that generally state that the prediction time interval τ_{pred} should be where $\tau_{pred} = 0.2(T_{1mod} + T_{2mod})$. The problem is that this recommendation is for the systems with the predictive PID-law without the filter. As it has already been checked, the formula does not work for the PID control algorithm with the second order filter, the time interval obtained by means of this formula is too high and the control system becomes unstable [14]. That is why it is necessary to check the recommendations for the control system under consideration.

In order to estimate the transient process quality in the predictive system let us obtain the transient processes for various values of the prediction time interval. The transient processes are obtained for the case when the single step is on the input of the control process. The transient processes for the system with process 1 (see Table 1) are given in Fig. 6.

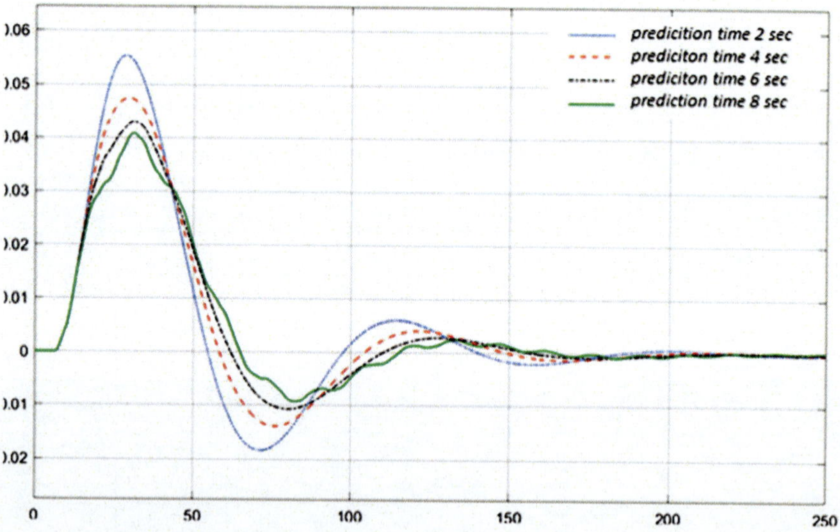

Fig. 6. The transient process when the step signal is on the input of the control process

As one can see from the processes given in Fig. 6 even a considerably small prediction time (such as 2 s) may provide quality improvement and the longer the prediction interval the better the transient process quality is. The problem observed is the same as the problem with the system with the PID with the second order filter: when the prediction time interval is too high, one can see high frequency oscillations in the transient process and with a higher prediction time interval the process may become unstable. Even if the process is stable high frequency oscillations are highly undesirable, because they influence the reliability of the equipment badly.

In order to explain the high frequency oscillations let us obtain the Nyquist curves for the open systems with and without the prediction. The curves are obtained for the system with process 1 and given in Fig. 7.

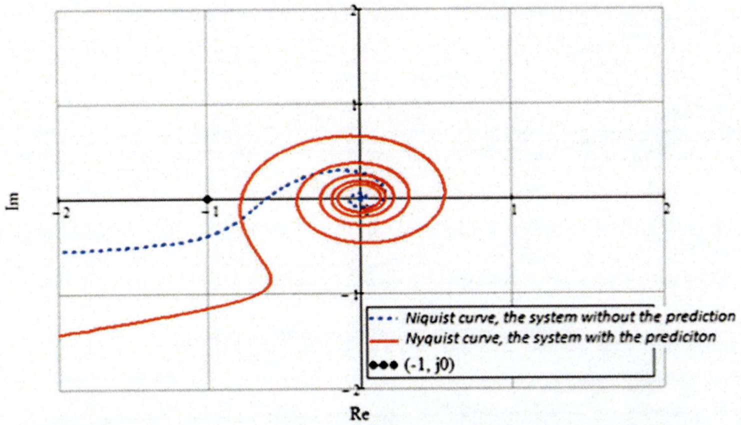

Fig. 7. Nyquist curves for the systems with and without prediction.

From Fig. 7 one can see that because of the introduction of the prediction module the Nyquist curve on low frequencies becomes farther from the point with coordinates $(-1, j0)$ and it becomes closer to the point on high frequencies.

If one calculates the prediction time interval according to the recommendations given in [10, 12, 13, 19], for example, for the system with process 1, one obtains the following:

$$\tau_{pred} = 0.2 \cdot (T_{1\,\mathrm{mod}} + T_{2\,\mathrm{mod}}) = 0.2 \cdot (17.7 + 77.6) = 19.06 \text{ s}$$

It is possible to see from Fig. 6 that the high frequency oscillations start to appear when $\tau_{pred} = 8$ s and it means that the prediction time interval obtained (19.06 s) is too high and the system is unstable if this value is used. For the system with process 1 $\tau_{pred} = 6$ s is optimal. Let us compare the transient processes in the systems with and without the prediction. The processes are shown in Fig. 8. The quality indexes are shown in Table 3.

In Table 3 y_{din} is the dynamic deviation (the maximal deviation from the set point), t_{tran} is the duration of the transient process (in seconds) and Ψ is the damping ratio. One can see that the use of the prediction module improves the quality considerably. Calculations for other processes gives the same results. Now it is necessary to formulate at least approximate recommendations on the calculation of the prediction time interval so that the auto-tuning module could calculate the prediction time as well.

Using the calculations obtained one can conclude that the optimal prediction time interval for the case under consideration is equal to the time delay (or a little bit lower to be on the safe side, but not higher). The preliminary formula may be as follows:

$$\tau_{pred} = 0.95 \cdot \tau_{mod} \tag{4}$$

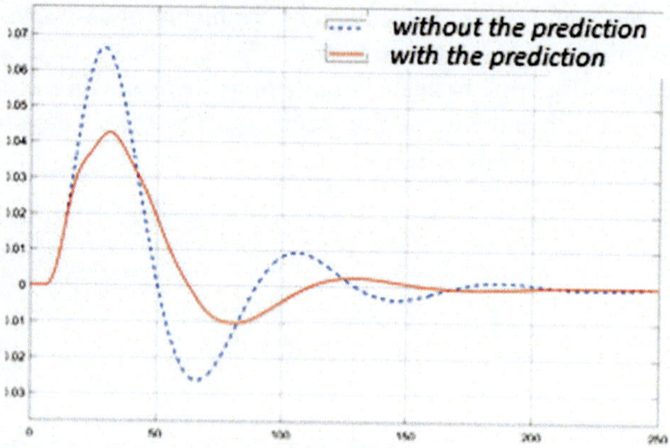

Fig. 8. Comparison of the transient processes in the systems with and without the predicition module.

Table 3. Transient process qualityi indexes

System	y_{din}	t_{tran}	Ψ
Without the prediction	0.066	170	0.86
With the prediction	0.043	140	0.93

This formula may be easily used in programming controllers using modern software, for example, CODESYS [20]. The authors plan to test the system described using real hardware and software in the future.

6 Conclusion

In this paper the authors considered the control system with the predictive PID control law with the first order filter and estimated its efficiency for the system for the thermal processes with time delay. The system described is effective and provides the considerable transient process quality improvement. The authors also formulated the recommendation on the calculation of the prediction time interval τ_{pred} for this system, the formula obtained may be easily implemented using modern controller programming software. The advantage of the developed system is that it does not require to install any additional equipment, so the only work that has to be carried out is improving the controller program.

References

1. Hu, Y., Jia, X., Lei, X., Hou, P., Bai, J.: Multiple model switching DMC-PID cascade predictive control for SCR denitration systems. In: 2018 Chinese Automation Congress, pp. 2573–2577 (2018). https://doi.org/10.1109/CAC.2018.8623613

2. Rotach, V.: Automatic Control Theory. MPEI Publishing House, Moscow (2004)
3. Bilenko, V.A.: Multi-loop automatic control systems with several control inputs and their application for maintaining steam temperature in once-through boilers. Therm. Eng. **58**(10), 850–858 (2011). https://doi.org/10.1134/S004060151110003X
4. Kuzishchin, V.F., Tsarev, V.S.: Algorithms for accelerated automatic tuning of controllers with estimating the plant model from the plant response to an impulse disturbance and under self-oscillation conditions. Therm. Eng. **61**(4), 281–290 (2014). https://doi.org/10.1134/S0040601514040041
5. Ping, M., Qian, Z., Nannan, L.: Study of a dynamic predictive PID control algorithm. In: 2015 Fifth International Conference on Instrumentation and Measurement, Computer, Communication and Control, pp. 1418–1423 (2015). https://doi.org/10.1109/IMCCC.2015.302
6. Zhou, K., Yu, H.: Application of fuzzy predictive-PID control in temperature control system of Freeze-dryer for medicine material. In: 2011 Second International Conference on Mechanic Automation and Control Engineering, pp. 7200–7203 (2011). https://doi.org/10.1109/MACE.2011.5988712
7. Wang, Y., Zou, H., Tao, J., Zhang, R.: Predictive fuzzy PID control for temperature model of a heating furnace. In: 2017 36th Chinese Control Conference, pp. 4523–4527 (2017). https://doi.org/10.23919/ChiCC.2017.8028070
8. Du, C., Ling, H.: Generalized predictive control algorithm applied to thermal power units based on PID neural network. In: 2010 2nd International Conference on Advanced Computer Control, pp. 38–42 (2010). https://doi.org/10.1109/ICACC.2010.5486737
9. Kuzishchin, V.F., Merzlikina, E.I., Van Va, H.: Study of the efficiency of the control system with smith predictor using a simulator based on controller owen PLC. In: 2nd International Conference on Industrial Engineering, Applications and Manufacturing, pp. 1–4 (2016). https://doi.org/10.1109/ICIEAM.2016.7910912
10. Pikina, G.A., Kuznetsov, M.S.: Tuning methods for typical predictive control algorithms. Therm. Eng. **59**(2), 154–158 (2012). https://doi.org/10.1134/S0040601512020139
11. Marzaki, M.H., Jalil, M.H.A., Shariff, H.M., Adnan, R., Rahiman, M.H.F.: Comparative study of Model Predictive Controller (MPC) and PID Controller on regulation temperature for SSISD plant. In: 2014 IEEE 5th Control and System Graduate Research Colloquium, pp. 136–140 (2014). https://doi.org/10.1109/ICSGRC.2014.6908710
12. Pikina, G.A.: Implementation of the predictive control principle in automatic control systems. In: XII Russian Conference on Control Problems, pp. 200–211 (2014)
13. Pikina, G.A., Pashchenko, F.F.: The predictive principle in control systems with standard lows. Procedia Comput. Sci. **150**, 403–409 (2014)
14. Merzlikina, E., Van Va, H., Farafonov, G.: Automatic control system with an autotuning module and a predictive PID-algorithm for thermal processes. In: 2021 International Conference on Industrial Engineering, Applications and Manufacturing, pp. 525–529 (2021). https://doi.org/10.1109/ICIEAM51226.2021.9446467
15. Zhang, W., Yang, M.: Comparison of auto-tuning methods of PID controllers based on models and closed-loop data. In: Proceedings of the 33rd Chinese Control Conference, pp. 3661–3667 (2014). https://doi.org/10.1109/ChiCC.2014.6895548
16. Kuzishchin, V.F., Merzlikina, E.I., Van Va, H.: Indirect frequency optimality indicators for automatic tuning of controllers: consideration of the dependence on the process model parameters. Bull. MPEI **6**, 106–114 (2019)
17. Ismatkhodzhaev, S.K., Kuzishchin, V.F.: Enhancement of the efficiency of the automatic control system to control the thermal load of steam boilers fired with fuels of several types. Therm. Eng. **64**(5), 387–398 (2017). https://doi.org/10.1134/S0040601517050032

18. Kuzishchin, V.F., Petrov, S.V.: A procedure for tuning automatic controllers with determining a second-order plant model with time delay from two points of a complex frequency response. Therm. Eng. **59**(10), 779–786 (2012). https://doi.org/10.1134/S0040601512100084
19. Pikina, G.A., Kuznetsov, M.S.: Typical predictive control algorithms. Therm. Eng. **58**(4), 336–341 (2011). https://doi.org/10.1134/S0040601511040100
20. Programming software CODESYS. https://www.codesys.com/

Intelligent Support for Medical Decision Making

E. I. Kiseleva[1(✉)] and I. F. Astachova[2]

[1] Voronezh State Pedagogical University, 86, Lenin st, Voronezh 394043, Russia
[2] Voronezh State University, 1, Univercity sq, Voronezh 394018, Russia

Abstract. This paper presents the development and study of a model for formalizing the process of making a diagnosis using artificial intelligence methods. Currently, various artificial neural networks and expert systems have been created and are used for diagnosis. Analysis of these works has shown that these methods show good results, but have a number of drawbacks, the most significant of which is the complexity of organization and the significant time required to train a neural network. Thus, the problem is to develop new algorithms that have a probability of making an accurate diagnosis, comparable with artificial neural networks and expert systems, while having a shorter training time. One of the ways to solve this problem is to develop a model for diabetes diagnosis based on an artificial immune system. The purpose of this work is to develop and study of a model for formalizing the process of diagnosis using methods of artificial intelligence. The paper reviews a model of the diagnosis process: pre-diabetes (impaired glucose tolerance, impaired fasting glycemia), type I diabetes, type II diabetes. The problem of diagnosing the disease can be regarded as a classification problem. In this paper, the process of diagnosis was examined as a division of test data and patient history into four classes corresponding to one of the diagnoses: pre-diabetes (impaired glucose tolerance, impaired fasting glycemia), type I diabetes, type II diabetes. An artificial immune system and Kohonen artificial neural network were used to solve this problem.

Keywords: Diabetes mellitus · Artificial neural network · Artificial immune system

1 Introduction

At presentation a diagnosis, the medical practitioner has to process a large amount of information. This increases the physician's information load, which leads to physical and psychological fatigue, errors in selecting and administering treatment, or delaying the process of making an accurate diagnosis. Therefore, it is clear that there is currently a trend toward increasing the number of diagnostic medical information systems (MIS) being developed. In addition, Government Decree No. 555 of May 05, 2018 "On the Unified State Health Information System" promotes the growth of health information system implementations.

MIS helps medical workers, facilitates their work, and improves the quality of medical services [1–7].

A. A. Radionov and V. R. Gasiyarov (Eds.): RusAutoCon 2021, LNEE 857, pp. 113–120, 2022.
https://doi.org/10.1007/978-3-030-94202-1_11

This paper discusses algorithms that allow for differential diagnosis of type I and type II diabetes mellitus, impaired glucose tolerance, and fasting glycemia disorders.

The optimal treatment regimens for patients with diabetes mellitus (DM) have become an increasingly urgent task over the years. In spite of the efforts of health care organizations in many countries, the number of people with diabetes is steadily increasing. The disease is the fourth leading cause of premature death in the world. Today it affects about 422 million people, which is 6.028% of the world's population. In Russia about 8 million people suffer from diabetes. Over the last decades, physicians have recorded a steady increase in the incidence of diabetes in all age groups. In the past the disease was more widespread among the people older than 40 years old, today even children and teenagers suffer from it. Researches show that each age group is specific in its own way about the course of the disease. The causes of the disease are not fully investigated. However, scientists believe that the main source of this trend is a sedentary lifestyle and a negative environmental situation.

There are a number of domestic and foreign studies that tried to diagnose type I or type II diabetes mellitus, and they used artificial intelligence technologies as their basis. Among them are the works of E.A. Pustozerov, T.A. Obelets, Kiran Tangod, O.P. Shesternikova, Dilip Kumar Chubi, O.M. Alade, J. Vijayashri, J. Jayashri [8–15].

The analysis of these works has shown that these methods show good results, but have a number of drawbacks, the most significant of which is the complexity of organization and the long time required to train a neural network. Therefore, the problem is to develop new algorithms that have a probability of making an accurate diagnosis, comparable with artificial neural networks and expert systems, while having a shorter training time. One of the ways to solve this problem is to develop a model of diabetes diagnosis based on an artificial immune system.

1.1 Objective of the Work

The objective of this work is to develop and investigate a model for formalizing the process of diagnosis using artificial intelligence methods.

1.2 Materials and Methods

Let's consider a model of the diagnosis process: pre-diabetes (impaired glucose tolerance, impaired fasting glycemia), type I diabetes, type II diabetes.

The problem of diagnosing the disease can be regarded as a classification problem. In this work, the process of diagnosis was considered as a division of test data and patient history into four classes, corresponding to one of the diagnoses: pre-diabetes (impaired glucose tolerance, impaired fasting glycemia), type I diabetes, type II diabetes.

An artificial immune system and Kohonen artificial neural network were used to solve this problem.

Artificial immune system represents an idealized version of its natural counterpart and reproduces the key components of the natural process: selection of the best antibodies of the population depending on their affinity to the antigen, cloning of antibodies, mutation of antibodies [16–23, 24].

In the proposed artificial immune system, a vector g, is considered as an antigen, which components are real and Boolean values reflecting the data obtained during the collection of anamnesis and clinical examinations of the patient whose diagnosis needs to be determined. $g = (g_1, g_2 \ldots g_{15})$, where g_1 is the patient's gender, g_2 is a Boolean variable reflecting the fact that the patient had a child over 4 kg, g_3 – is age, g_4 – is weight, g_5 – is height, g_6 – is body mass index, g_7–g_9 are Boolean variables reflecting whether the patient had relatives with diabetes, polydipsia and polyuria, g_{10} – is fasting plasma glucose level, g_{11} – is plasma glucose level 2 h after use of oral glucose tolerance test, g_{12} – is glucose level at random determination, g_{13} – HbA1c, g_{14} – is insulin, g_{15} – C - peptide.

The antibody is a vector, $l = (l_1, l_2 \ldots l_{16})$, where l_1–l_{15} is a sequence of real and Boolean values similar to the antigen, l_{16} – is the patient's diagnosis that corresponds to such indicators. The components of antibody and antigen vectors we will call genes. Antibodies belong to one of four classes, according to the diagnosis: type 1 or type 2 diabetes, impaired glucose tolerance, impaired fasting glycemia. The task of the immune system is to determine which class the antigen belongs to.

The learning algorithm of the artificial immune system can be represented as follows.

1. The user enters the name of the diagnosis to be taught to the system in the text field and initiates training.
2. A group of antibodies $l_i, i = 1 \ldots n$, is created, each of which receives the name of the diagnosis entered by the user. The values of the vector components of each antibody are set randomly.
3. An antigen $g = (g_1, g_2 \ldots g_{15})$ is randomly selected from the training dataset, with values corresponding to the diagnosis entered by the user.
4. The antibody-antigen affinity function is calculated according to the following rule:
5. $A = k/15$, where k – is the number of antibody genes x_i, satisfying the condition
6. $|l_i - g_i| \leq \alpha, i = 1, \ldots, 15, \alpha = 0, 05$.
7. For t antibodies with affinity to the antigen exceeding the set threshold p, a cloning procedure is applied, in which m copies of each antibody are created.
8. A mutation operator is applied to antibody clones, consisting of randomly selecting genes and making random changes in their values.
9. The affinity of antibody clones is calculated.
10. Destroying l $(l > mt)$ antibodies with the lowest affinity.
11. Population number is restored by generating randomly $n - l$ new antibodies.
12. Steps 4–9 are repeated until the population stabilizes over a number of cycles.
13. Steps 3–9 are repeated until all antigens from the training sample have been used.
14. All antibodies from the population are added to a separate group to be used for further diagnosis.

The described process is repeated if the immune system needs to be trained for other diagnoses.

The diagnosis is obtained on the basis of the following sequence of steps:

1. A vector containing the data of the patient to be diagnosed is presented to the system as an antigen.

2. A number of antibodies belonging to different classes of possible diagnoses are randomly selected from the population of antibodies.
3. The number of antibodies in the selected population belonging to class d_i^1 is counted, where $i = 1,...,4$.
4. Steps 4–10 of the previous algorithm are repeated for the selected population until its stabilization for a certain number of cycles is achieved.
5. The number of antibodies in the selected population belonging to the class number id_i^2 where $i = 1,...,4$ is selected.
6. For each class we calculate the rati $v_i = \frac{d_i^2}{d_i^1}, i = 1, \ldots 4$.

If only one of these relations is greater than 1, the class number corresponds to the diagnosis; otherwise, steps 2–6 are repeated, but only antibodies of classes for which $v_i > 1$ are included in the population.

The diagnosis problem can be considered as a classification problem. The classification problem was solved using the Kohonen neural network.

In the Kohonen network the number of inputs of each neuron is equal to the dimensionality of parameters of the classified object. In our case, we will classify the results of analysis and patient data. As defined earlier, patient data contains 15 parameters, i.e. each neuron has fifteen inputs. The number of neurons is equal to the number of diagnoses that can be given to patients. In this paper, the patients will be divided into four groups:

1. patients with type I diabetes mellitus;
2. patients with type II diabetes mellitus;
3. patients with impaired fasting glycemia;
4. patients with impaired glucose tolerance.
5. The training algorithm for the Kohonen network consists of the following steps.
6. Network Initialization.
7. Assigning small random values to the weights of the network $W_{ij}, i = \overline{1, n}, j = \overline{1, m}$. The following values are set: α_0 – the initial learning rate and D_0 – the maximum distance between the weight vectors (columns of the matrix W).
8. Presenting the network with a new input signal X from the training sample.
9. Calculating the distance from input X to all neurons of the network:
10. $d_j = \sum_{i=1}^{n} \left(X_i - W_{ij}^N\right)^2, j = \overline{1, m}$
11. Selection of the neuron k, $1 \leq k \leq m$ with the shortest distance d_k from the input to the network neurons.
12. Adjustment of weights of the k-th neuron and all neurons that are at a distance not exceeding $D_N : W_{ij}^{N+1} = W_{ij}^N + \alpha_N \left(X_i - W_{ij}^N\right)$.
13. Decreasing values of α_N, D_N.

Steps 2–7 are repeated until the weights stop changing (or until the total change of all weights is less than the value set by the user).

After the network is trained, the diagnosis is made by feeding the test vector to the network input and calculating the distance from it to each neuron, followed by selecting the neuron with the shortest distance as the class indicator. The number of the selected class corresponds to the diagnosis of the patient.

1.3 Results

The accuracy of diagnosis using an artificial neural network and an artificial immune system was compared.

Data from 186 patients with known final diagnoses were used to organize the software package. The training sample included 100 records, the control sample included 86 records.

Patients' diagnoses that were obtained during the work of the program complex were compared with the known final diagnoses, and then the percentage of coincidence with those obtained during the work of the program was calculated (Table 1).

Table 1. Comparison of diagnosis results

Method	Percentage of correct diagnoses
Linear classifier (neural network with 1 layer)	88%
Kohonen neural network	92%
Artificial immune system with initial population n = 80	95%
Artificial immune system with initial antibody population n = 100	96%
Artificial immune system with initial antibody population n = 200	96%

As can be seen from the results of the experiments, the best results were shown by the artificial immune system with an initial population of more than 100 antibodies.

A software package that implements the developed algorithms and models **was developed**. The development environment C++ Builder and Paradox DBMS were chosen to implement the tasks. The functional diagram of the software product is shown in Fig. 1.

The software package is divided into 2 parts, the user (patient) part and the administrator (doctor) part. The first part of the system is for the work of the administrator (doctor), it is designed to accumulate and store information, set up training and making a diagnosis using the artificial neural network and artificial immune system. The second part provides the user's (patient's) work and is intended for making a diagnosis. The structural diagram of interaction between the main modules is shown in Fig. 2.

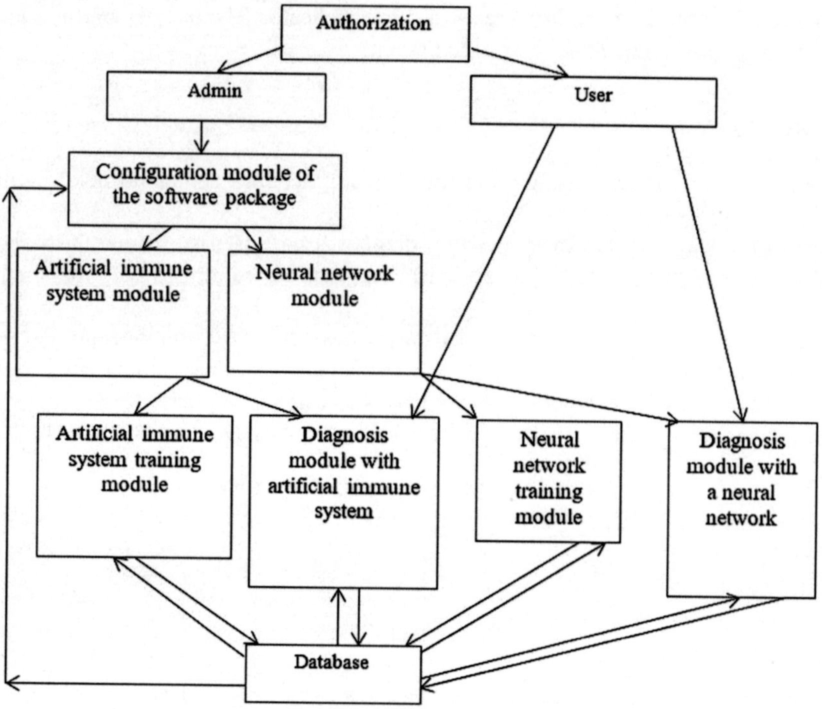

Fig. 1. Functional diagram of the software package.

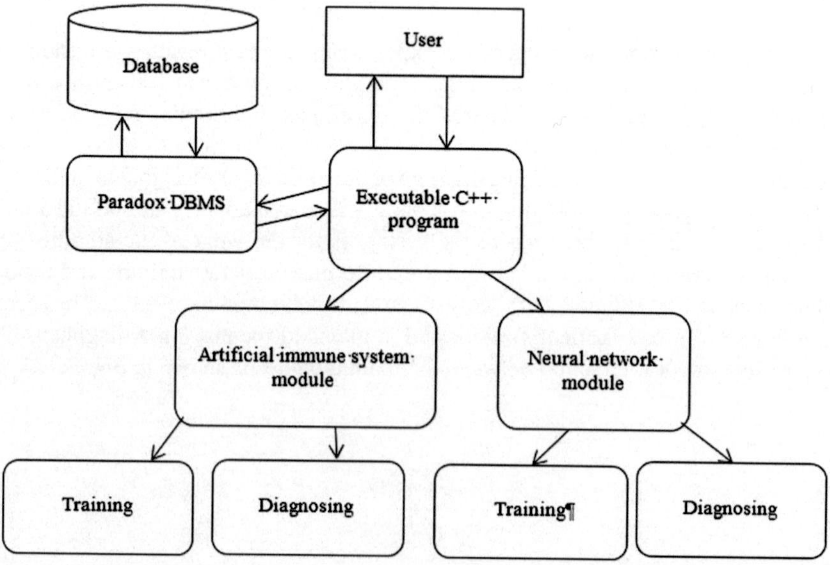

Fig. 2. Block diagram of the interaction between the modules of the software package.

1.4 Summary

The following main results are obtained in this paper.

- A formal model of the diagnosis process was developed.
- Diagnosis algorithm was developed: prediabetes (impaired glucose tolerance, impaired fasting glycemia), type I diabetes, type II diabetes, distinguished by the use of artificial immune system.
- The algorithm and model of Kohonen neural network, allowing to solve the problem of prediabetes state (disorder of glucose tolerance, disorder of fasting glycemia), type I diabetes, type II diabetes were developed.
- Comparison of the effectiveness of diagnosis using neural networks and artificial immune system was made.
- A software package implementing the described algorithms was created.

References

1. Haglin, J.M., Jimenez, G., Eltorai, A.E.M.: Artificial neural networks in medicine. Health Technol. **9**(1), 1–6 (2018). https://doi.org/10.1007/s12553-018-0244-4
2. Soumya, C.V., Ahmed, M.: Artificial neural network based identification and classification of images of Bharatanatya gestures. In: Innovative Mechanisms for Industry Applications, pp. 162–166 (2017)
3. Nowikiewicz, T., Wnuk, P., Małkowski, B., Kurylcio, A., Kowalewski, J., Zegarski, W.: Application of artificial neural networks for predicting presence of non-sentinel lymph node metastases in breast cancer patients with positive sentinel lymph node biopsies. Arch. Med. Sci. **13**(6), 1399–1407 (2017)
4. Sheikhtaheri, A., Sadoughi, F., Hashemi Dehaghi, Z.: Developing and using expert systems and neural networks in medicine: a review on benefits and challenges. J. Med. Syst. **38** (2014). Article number: 110. https://doi.org/10.1007/s10916-014-0110-5
5. Ultsch, A., Korus, D., Kleine, T.O.: Integration of neural networks and knowledge-based systems in medicine. In: Barahona, P., Stefanelli, M., Wyatt, J. (eds.) AIME 1995. LNCS, vol. 934, pp. 425–426. Springer, Heidelberg (1995). https://doi.org/10.1007/3-540-60025-6_170
6. Summers, R.M.: Deep learning and computer-aided diagnosis for medical image processing: a personal perspective. In: Lu, L., Zheng, Y., Carneiro, G., Yang, L. (eds.) Deep Learning and Convolutional Neural Networks for Medical Image Computing. ACVPR, pp. 3–10. Springer, Cham (2017). https://doi.org/10.1007/978-3-319-42999-1_1
7. Wang, J., Shi, M., Zheng, P., Xue, S., Peng, R.: Quantitative analysis of Ca, Mg, and K in the roots of Angelica pubescens f. biserrata by Laser-induced breakdown spectroscopy combined with artificial neural networks. J. Appl. Spectrosc. **85**(1), 190–196 (2018). https://doi.org/10.1007/s10812-018-0631-7
8. Sejdinović, D., et al.: Classification of prediabetes and type 2 diabetes using artificial neural network. In: Badnjevic, A. (ed.) CMBEBIH 2017, vol. 62, pp. 685–689. Springer, Singapore (2017). https://doi.org/10.1007/978-981-10-4166-2_103
9. Alade, O.M., Sowunmi, O.Y., Misra, S., Maskeliūnas, R., Damaševičius, R.: A neural network based expert system for the diagnosis of diabetes mellitus. In: Antipova, T., Rocha, Á. (eds.) MOSITS 2017. AISC, vol. 724, pp. 14–22. Springer, Cham (2018). https://doi.org/10.1007/978-3-319-74980-8_2

10. Srivastava, S., Sharma, L., Sharma, V., Kumar, A., Darbari, H.: Prediction of diabetes using artificial neural network approach. In: Ray, K., Sharan, S.N., Rawat, S., Jain, S.K., Srivastava, S., Bandyopadhyay, A. (eds.) Engineering Vibration, Communication and Information Processing. LNEE, vol. 478, pp. 679–687. Springer, Singapore (2019). https://doi.org/10.1007/978-981-13-1642-5_59
11. Li, X.: Artificial intelligence neural network based on intelligent diagnosis. J. Ambient. Intell. Humaniz. Comput. **12**(1), 923–931 (2020). https://doi.org/10.1007/s12652-020-02108-6
12. Asad, M., Qamar, U.: A review of continuous blood glucose monitoring and prediction of blood glucose level for diabetes type 1 patient in different prediction horizons (PH) using artificial neural network (ANN). In: Bi, Y., Bhatia, R., Kapoor, S. (eds.) IntelliSys 2019. AISC, vol. 1038, pp. 684–695. Springer, Cham (2020). https://doi.org/10.1007/978-3-030-29513-4_51
13. Raihan, M., Alvi, N., Tanvir Islam, M., Farzana, F., Mahadi Hassan, M.: Diabetes mellitus risk prediction using artificial neural network. In: Uddin, M.S., Bansal, J.C. (eds.) IJCCI 2019. AIS, pp. 85–97. Springer, Singapore (2020). https://doi.org/10.1007/978-981-15-3607-6_7
14. Arul Kumar, D., Jayanthy, T.: Application of back propagation artificial neural network in detection and analysis of diabetes mellitus. J. Ambient. Intell. Humaniz. Comput. **12**(7), 7063–7070 (2020). https://doi.org/10.1007/s12652-020-02371-7
15. Jayashree, J., Kumar, S.A.: Linear discriminant analysis based genetic algorithm with generalized regression neural network – a hybrid expert system for diagnosis of diabetes. Program. Comput. Softw. **44**, 417–427 (2018). https://doi.org/10.1134/S0361768818060063
16. Şahan, S., Polat, K., Kodaz, H., Güneş, S.: The medical applications of attribute weighted artificial immune system (AWAIS): diagnosis of heart and diabetes diseases. In: Jacob, C., Pilat, M.L., Bentley, P.J., Timmis, J.I. (eds.) ICARIS 2005. LNCS, vol. 3627, pp. 456–468. Springer, Heidelberg (2005). https://doi.org/10.1007/11536444_35
17. Lin, H., Su, C., Wang, P.: An application of artificial immune recognition system for prediction of diabetes following gestational diabetes. J. Med. Syst. **35**, 283–289 (2011). https://doi.org/10.1007/s10916-009-9364-8p283-289
18. Chikh, M.A., Saidi, M., Settouti, N.: Diagnosis of diabetes diseases using an Artificial Immune Recognition System2 (AIRS2) with fuzzy K-nearest neighbor. J. Med. Syst **36**, 2721–2729 (2012). https://doi.org/10.1007/s10916-011-9748-4
19. Wu, J.-Y.: Hybrid artificial immune algorithm and CMAC neural network classifier for supporting business and medical decision making. In: Tang, J., King, I., Chen, L., Wang, J. (eds.) ADMA 2011. LNCS (LNAI), vol. 7121, pp. 41–54. Springer, Heidelberg (2011). https://doi.org/10.1007/978-3-642-25856-5_4
20. Xiaoyang, F., Zhang, S.: An improved artificial immune recognition system based on the average scatter matrix trace criterion. In: Tan, Y., Shi, Y., Ji, Z. (eds.) Advances in Swarm Intelligence, pp. 284–290. Springer Berlin Heidelberg, Berlin, Heidelberg (2012). https://doi.org/10.1007/978-3-642-30976-2_34
21. Astachova, I.F.: The application of artificial immune system for parallel process of calculation and their comparison with existing methods. In: Journal of Physics: Conference Series, vol. 1202, p. 012003 (2019). https://doi.org/10.1088/1742-6596/1202/1/012003
22. Kashirina, I.L.: Neural network technologies. VGU, Voronezh (2008)
23. Astahova, I.F., Kiseleva, E.I.: Algorithm for using an artificial immune system to optimize the target component of the educational information system. Voronezh State University Bulletin. System Analysis and Information Technology **2**, 61–65 (2017)

Modeling and Simulation

Mathematical Model for Evaluation of the Parameters Influence on the Productivity of the Flexible Production Site

V. V. Dyadichev[1](✉), S. S. Stoyanchenko[2], A. V. Dyadichev[1], and S. Ye. Chornobay[1]

[1] V.I. Vernadsky Crimean Federal University, 2, Pavlenko Street, Simferopol 295007, Russia
[2] Lugansk National University named after Vladimir Dal, 20-A, Molodezhnyi Sq., Lugansk 91000, Ukraine

Abstract. The paper researches the regularities of functioning processes of the multiposition conveyors which are widely used in industry. The process of manufacturing systems' functioning is characterized by a significant degree of uncertainty and is influenced by many accidental factors. The working efficiency of the manufacturing systems under study in many respects is determined by arranging the process of multiposition conveyor's functioning. The main factors, determining the conveyor's efficiency, are brought to light. The objective of the current research is the identification of the main factors, deter-mining the efficiency of the multiposition conveyor, and the analysis of the character of the conveyor's efficiency dependency on the exposed factors. To evaluate the influence of reliability parameters of a multiposition conveyor on the efficiency of a flexible manufacturing section a mathematical model was developed.

Keywords: Reliability parameters · Maintainbility parameters · Multiposition conveyor · Flexible manufacturing section

1 Introduction

The analysis of manufacturing processes of the machine building and instrument making enterprises shows that the product processing time is not more that 10% of the entire time of manufacturing a product. The rest time is spent on the processes of transporting and intermediate storage. In modern manufacturing systems multiposition conveyors are widely used. Such conveyors are used for products' in-process float at the enterprises of light and food industry. The process of manufacturing systems' functioning is characterized by a significant degree of uncertainty and is influenced by many accidental factors. The working efficiency of the manufacturing systems under study in many respects is determined by arranging the process of multiposition conveyor's functioning.

1.1 Analysis of Scientific Publications

Many works [1–16] are dedicated to the study of arranging an effective interaction of conveying and manufacturing equipment. The work [3, 4] scrutinizes the issues of

A. A. Radionov and V. R. Gasiyarov (Eds.): RusAutoCon 2021, LNEE 857, pp. 123–129, 2022.
https://doi.org/10.1007/978-3-030-94202-1_12

manufacturing systems' structures synthesis. These works suggest analytical models of the conveyor's work. Such models are built on the basis of different assumptions of the processes going on in the system. Developing analytical models is distinguished by a significant labour-intensiveness. Besides, such models often present the real processes inadequately. For the effective management of the multiposition conveyor's work it is necessary to find out the main factors determining its efficiency.

1.2 The Objective of the Work

The objective of the current research is the identification of the main factors, determining the efficiency of the multiposition conveyor, and the analysis of the character of the conveyor's efficiency dependency on the exposed factors.

2 Research

In the construction of the multiposition conveyor we may distinguish two main parts (Fig. 1): single-channel and multi-channel. At the failure of the single-channel part the downtime of the whole conveyor happens. The failure of one position of the multi-channel part leads to the decrease of the conveyor's efficiency. The strategy of the conveyor's control has to foresee its stop after the failure of a definite amount of the positions of the multi-channel part. The number of the failure positions of the multi-channel part, when the conveyor's stopping occurs for the failures' elimination, will be further called a critical number.

The choice of the critical number of the multi-channel part's positions determines the strategy of controlling the multiposition conveyor. The optimum control strategy has to provide a maximum efficiency of the conveyor. A great number of factors influences the choice of the control strategy. They include the number of positions of the multi-channel part, the duration of the working cycle, the reliability parameters of the conveyor. The first parameter characterizes the level of the equipment reliability; the second parameter characterizes the level of its maintainability.

The following formula can be used to evaluate the efficiency of the conveyor:

$$P = \frac{\sum_{i=0}^{k} t_i(N-i)}{TN} 100, \qquad (1)$$

where – P is the evaluation of the conveyor's efficiency (%); k is a critical number of the failure working positions; N is a total number of the working positions of the conveyor; T is the time of the conveyor's work (min); t_i is the time of the conveyor's work with i failure working positions (min).

The presented formula shows that for evaluating the efficiency a value, equal to the ratio of the real efficiency to the conveyor's efficiency with absolutely reliable working positions, is used. This value can be measured by percent. It characterizes the efficiency losses, conditioned by the failure of the conveyor's working positions. The efficiency losses proper Z (%) are calculated by the formula:

$$Z = 100 - P. \qquad (2)$$

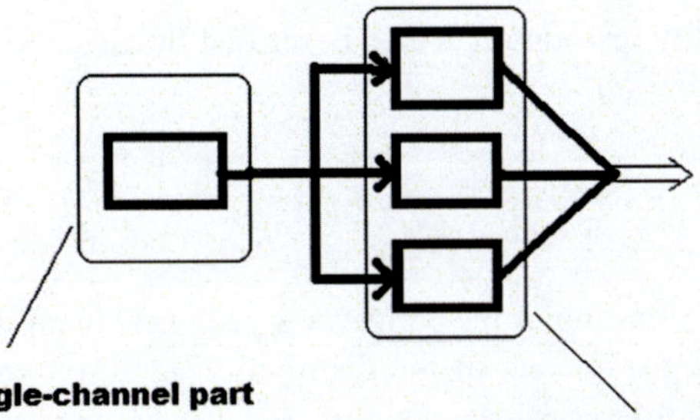

Fig. 1. The structural scheme of the multi-position conveyor.

The analysis of the functioning process of the multiposition conveyor shows that the efficiency losses, connected with the failure of working positions, can be divided into two types. The losses of the first type (Z1) are caused by the downtime of the conveyor's working positions during its work with faultless working positions. The evaluation of this type efficiency losses can be done by the formula (3). The variables from the formula (1) are used in this formula, too:

$$Z1 = \frac{\sum_{i=0}^{k} t_i \cdot i}{TN} 100. \tag{3}$$

The losses of the second type are connected with the downtime of all the conveyor's positions during its stop for maintaining the failure working positions. We will designate the losses of the second type by Z2. These losses are represented by the formula:

$$Z2 = \frac{\sum_{i=0}^{l} t_{ni} \cdot k}{TN} 100, \tag{4}$$

where – l is the amount of the conveyor's stops for the restoration of the operability; k is a critical number of the conveyor's failure working positions; t_{ni} is the duration of the operability restoration of k positions of the conveyor at i stop; N is the total number of the conveyor's positions: T is the time of the conveyor's work (min).

The issue of the losses Z1 and Z2 influence on the conveyor's efficiency is of great interest. To find out this dependency an analytical model on the basis of Markovian chain theory was developed. To evaluate the reliability of the technological equipment such parameters are often used: an average intensity of rejections λ (min-1) and an average duration of the operability restoration μ (min) [2]. It is known that the time of the faultless work of technological equipment with sufficient for practical aims approximation can be described by an accidental value, distributed according to the exponential law:

$$P(t) = e^{-\int_{0}^{t} \lambda(t)dt}, \tag{5}$$

where t is the time of faultless work of the conveyor's position (min); λ(t) is the intensity of the rejections' flow as the time function (min-1).

The time of operability restoration is also described by an accidental value with an exponential law of distribution.

$$G(t) = e^{-\int_0^t \frac{1}{\mu(t)} dt}, \tag{6}$$

where μ(t) is an average duration of the conveyor's position operability restoration.

We will consider the parameters of the distribution law λ(t) and μ(t) permanent, i.e. not depending on the time. Besides, we will consider the parameters of rejections' flows of all the conveyor's positions identical and equal to λ (min-1). An average time of the operability restoration of each position will be considered equal to μ (min). The indicated assumptions are widely used during calculations of the technological equipment reliability [3, 4, 17–20] and bring negligibly small distortions into the obtained results.

At the indicated assumptions for describing the functioning process of the multi-position conveyor an analytical model, with the use of discreet permanent Markovian chains theory, can be developed. To develop such a model it is necessary to construct a great number of states of the modeled system and describe the matrix of transitions probabilities between the system's states.

The researched system, presenting a multiposition conveyor, may be in k different states S_i (i = 1..k). K is a critical number of the positions. Each discreet state is characterized by the number of working position failures. The numbers of the states are 0, 1, 2,…k. The transition from i state (i = 0..k − 1) into j – e (j = i + 1) occurs with the permanent intensity (N − I) • λ. Here N is the number of the conveyor's positions. Each transition of such a type corresponds to the failure of one conveyor's position. The transition from the state S_k into the state S0 occurs with the probability μ^{-1} • (0,1 • k). Such a transition corresponds to the restoration of operability K of the failure conveyor's positions. Other types of transitions in the studied system are not acceptable. The matrix of probabilities of the acceptable transitions in Markovian chain of the studied system can be described by the following ratio:

$$P_{i,j} = \begin{cases} (N-i)\cdot\lambda, & i = \overline{0..k-1}, j = i+1, \\ \mu^{-1}(0, 1\cdot k), & i = k, j = 0. \end{cases} \tag{7}$$

The graph of transitions, presented in Fig. 2, corresponds to such a matrix.

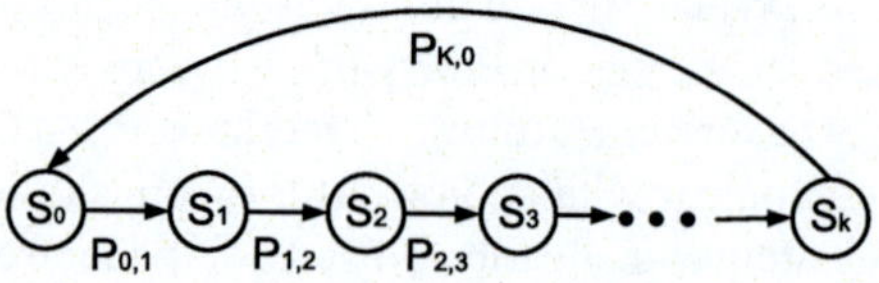

Fig. 2. The graph of Markovian chain transitions.

The system of Chapman – Kolmogorov equations can be written for Markovian chain:

$$\sum_{i=0}^{k} T_i \cdot p_{i,j} = T_j, \quad j = \overline{0..k}, \tag{8}$$

The solution to this equation system allows finding the permanent probabilities (T) of the system's being in each discreet state. The probability of the system's being in one of the states can be viewed as a relative part of time of the conveyor being in that state. In this case each state of the conveyor can be assigned a corresponding efficiency of the conveyor. The conveyor's efficiency, corresponding to the state So, equals 1. The efficiency, corresponding to the state Si (at I > 0 and I < k), will equal:

$$P_i = \frac{N - i}{N}. \tag{9}$$

At last the conveyor's efficiency at the sate Sk will be equal to 0, as this state corresponds to the downtime of the conveyor due to the maintenance of the failure positions.

Thus it is possible to determine the conveyor's efficiency (P) by the formula:

$$P = \sum_{i=0}^{k} T_i \cdot P_i, \tag{10}$$

Using the presented ratio, it is possible to determine the expected efficiency of the conveyor depending on the adopted control strategy (critical number K) and the reliability parameters of the conveyor.

3 The Results of the Research

Figures 3, 4 and 5 show the obtained dependencies of the conveyor's efficiency losses on the adopted control strategy and the liability level of its positions. To evaluate the reliability level the value $(\lambda \bullet \mu) - 1$ was used.

The presented dependencies show that at low values k the losses of the second type prevail, and at high values k the losses of the first type prevail. The obtained results allow determining an optimum strategy for controlling the conveyor (the amount of failed positions k), and evaluate the influence of reliability parameters of the conveyor on its efficiency.

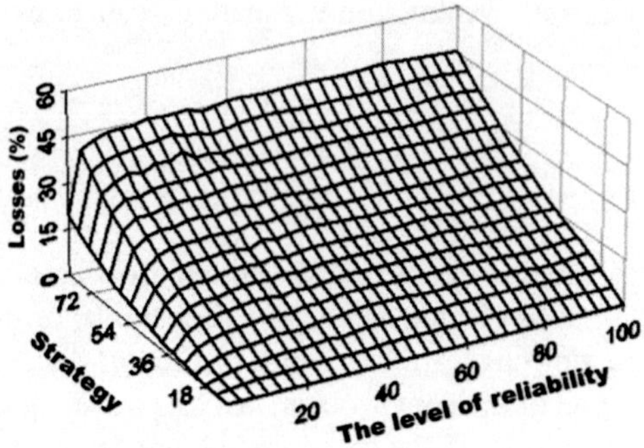

Fig. 3. The dependence of the first type losses on the adopted control strategy and reliability parameters of the conveyor position.

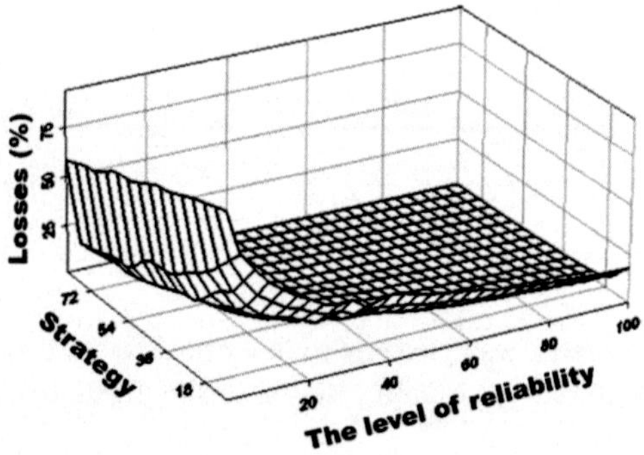

Fig. 4. The dependence of the second type losses on the adopted control strategy and reliability parameters of the conveyor position.

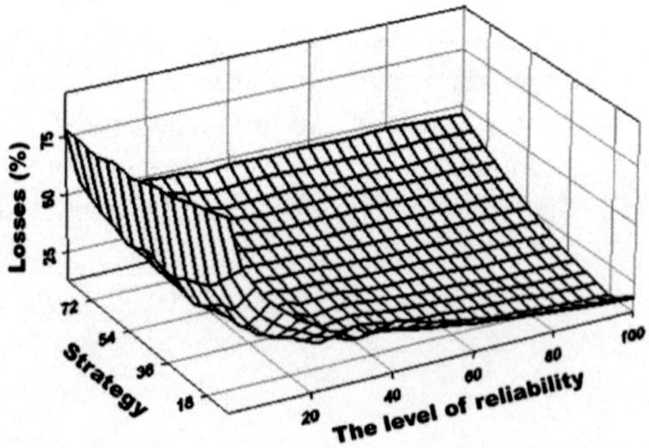

Fig. 5. The dependence of efficiency losses on the adopted control strategy and reliability parameters of the conveyor position.

Acknowledgement. The study was carried out with the financial support of the Ministry of Education and Science of the Russian Federation within the framework of the scientific project FZEG-2020-0030.

References

1. Golinkevich, T.: Applied Theory of Reliability. Vysshaya Shkola, Moscow (1989). (in Russian)
2. Gavrish, A.P.: Robotized Machine Building Construction Complexes of Mechanical Processing. Technika, Kiev (1984). (in Russian)
3. Chaar, J.K., Volz, R.A.: A specification language for planning and fault recovery in manufacturing systems. Int. J. Flex. Manuf. Syst. **5**(3), 209–253 (1993). https://doi.org/10.1007/BF01328742
4. Miriyala, K., Viswanadham, N.: Reliability analysis of flexible manufacturing systems. Int. J. Flex. Manuf. Syst. **2**, 145–162 (1989). https://doi.org/10.1007/BF00222708
5. Frankovič, B., Labátová, S.: Approach to determination of FMS reliability by computer simulation. In: Proceedings of the 6th International Conference on Artificial Intelligence and Information-Control Systems of Robots, pp. 269–276. Smolenice Castle, Slovakia (1995)
6. Kemeny, J.G., Snell, J.L.: Finite Markov Chains. Springer, New York (1976)
7. Stewart, W.J.: Introduction to the Numerical Solution of Markov Chains. Princeton University Press, Princeton (1995)
8. Kallenberg, O.: Foundations of Modern Probability. Springer, New York (1997)
9. Gamerman, D.: Markov Chain Monte Carlo. Stochastic Simulation for Bayesian Inference. CRC Press, Boca Raton (1997)
10. Gilks, W.R., Richardson, S., Spiegelhalter, D.J.: Markov Chain Monte Carlo in Practice. Chapman & Hall, Boca Raton (1996)
11. Bolch, G., Greiner, S.: Queueing Networks and Markov Chains, 2nd edn. Wiley, Hoboken (2006)
12. Doob, J.L.: Stochastic Processes. Wiley, New York (1953)
13. Hinkelmann, K., Kempthorne, O.: Design and Analysis of Experiments. I and II, 2nd edn. Wiley, Hoboken (2008)
14. Kempthorne, O.: The Design and Analysis of Experiments (1979). Corrected Reprint of (1952) Wiley ed. by Robert E. Krieger
15. Lentner, M., Thomas, B.: Experimental Design and Analysis, 2nd edn. Valley Book Company, Blacksburg (1993)
16. Lindman, H.: Analysis of Variance in Complex Experimental Designs. W. H. Freeman & Co. Hillsdale/Erlbaum, San Francisco (1974)
17. Lu, R., Gross, L.: Simulation modeling of a pull and push assemble-to-order system. In: The European Operational Research Conference, The Netherlands, Rotterdam (2001)
18. Qiao, G., Lu, R., Riddick, F.: Flexible modeling and simulation for mass customization manufacturing. In: Proceedings of the 2003 IIE Annual Conference, OR, Portland (2003)
19. Qiao, G., McLean, C., Riddick, F.: Simulation system modeling for mass customization manufacturing. In: Proceedings of the 2002 Winter Simulation Conference, San Diego, California (2002)
20. Talavage, J., Hannam, R.G.: Flexible Manufacturing Systems in Practice: Application, Design and Simulation, p. 358. Marcel Dekker Inc., New York (1999)

The Dynamic Defect Models for Rotor Mechanical Assemblies of Rolling Stock

V. Tetter(✉), A. Tetter, and I. Denisova

Omsk State Transport University (OSTU), 35, Marksa pr., Omsk 644046, Russia

Abstract. The dynamic and static defect models for rotor mechanical assemblies have been compared in the article. The article gives reasons for the significance of dynamic modeling to define remaining assembly life. The article points out that it is feasible to use empiric data as a foundation for creating models. The article gives the example of a small gear defect dynamic modeling for a complete wheelset. The approximate amount of work on creating dynamic models has been defined in the article. The article proposes the methods for implementing dynamic defect models.

Keywords: Model · Defect · Rolling stock · Diagnostics · Spectrum · Wavelet

1 Introduction

Using the vibration diagnostic equipment (VDE) to determine the technical condition of the rotor mechanical assemblies (RMA) of rolling stock (roller and friction bearings, toothed-wheel gearing) is regulated by the standardized documents of Joint Stock Company "Russian Railways" (JSCo "RZD") [1, 2]. Such equipment is used for the incoming and outgoing inspection.

2 Static Defect Models in the Time and Frequency Domains

2.1 Time Parameters and Characteristic Frequencies

For trouble shooting and identifying the degree of defect growth all VDE manufacturers use static models as a set of vibration signal parameters in the time and frequency domains.

Such parameters in the time domain are:

Root-mean-square value:

$$X_{rmsv} = \sqrt{\frac{1}{T}\int_{t_0}^{t_0}[X(t)]^2 dt} \tag{1}$$

A. A. Radionov and V. R. Gasiyarov (Eds.): RusAutoCon 2021, LNEE 857, pp. 130–139, 2022.
https://doi.org/10.1007/978-3-030-94202-1_13

Peak factor: it is defined as the ratio of the maximum (peak) signal value to RMA value of the vibration level:

$$K_{peak} = \frac{X_{\max}}{X_{rmsv}}, \tag{2}$$

Kurtosis factor (KF) [3]: It can be calculated using the following formula:

$$F_K = \frac{\int_{t_0}^{t_0+T} (X - X)^2 P(x)dx}{\sigma^4}, \tag{3}$$

where $X(t)$ – signal amplitude from the vibration sensor (time signal), $P(x)$ – probability function of the random value (time signal), T – monitoring period, t_0 – time of the monitoring starting, Xcp – root-mean-square deviation of the time signal.

The prime tool for identifying RMA defects is the spectral analysis for vibration signals which are received from the rotating assemblies during diagnostics.

2.2 The RMA Model

In general, RMA model of the frequency domain can be described as two groups of components [4]: the first group involves specific periodic components (including subharmonics and superharmonics) generated by the individual bearing or toothed-wheel gearing elements during their operation; the second one involves all other components, including "background", noise, impulse noise and the components generated by other elements during their operation.

$$P(f) = P_1(f) + P_2(f) + P_3(f) + P_4(f) + Pz \tag{4}$$

where, P_1 – harmonic components which are multiple to the characteristic outer race frequency; P_2 – harmonic components which are multiple to the characteristic inner race frequency; P_3 – harmonic components which are multiple to the characteristic cage frequency; P_4 – harmonic components which are multiple to the characteristic roller frequency; P_z – noise terms.

Formulas for calculating P_1; P_2; P_3 and P_4 (respectively: f_H; f_B; f_C; f_{TK}) are given, for example, in the literature [5].

Harmonics with frequencies which are multiple to the roller pass frequency outer race kf_H, where

$$kf_H = \frac{1}{2} kf_{BP}\left(1 - \frac{d_{TK}}{d_c}\cos(\alpha)\right)z, \tag{5}$$

where d_{TK} – roller diameter; d_C – cage diameter; α – contact angle of bearing rollers and bearing races; z – number of rollers in one row of the bearing; k – harmonic number (whole number).

Harmonics with frequencies which are multiple to the roller pass frequency inner race kf_B, where

$$kf_B = \frac{1}{2} kf_{BP}\left(1 + \frac{d_{TK}}{d_c}\cos(\alpha)\right)z, \tag{6}$$

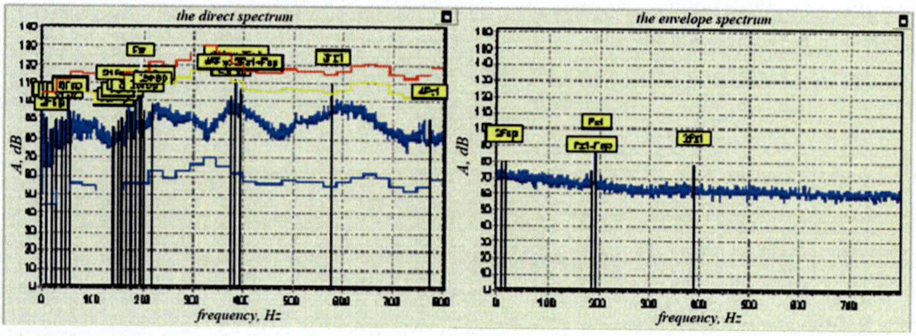

Fig. 1. The example of the assembly defect static model in the frequency domain (intensely developed *small gear defect in a complete wheelset*) - the direct spectrum and the envelope spectrum.

Harmonics with frequencies which are multiple to the cage frequency kfc, where

$$kf_C = \frac{1}{2} kf_{BP} \left(1 - \frac{d_{TK}}{d_c} \cos(\alpha)\right), \tag{7}$$

Harmonics with frequencies which are multiple to the roller frequency kf_{TK}, where

$$kf_{TK} = \frac{1}{2} kf_{BP} \frac{d_c}{d_{TK}} \left(1 - \frac{d_{TK}^2}{d_c^2} cos^2(\alpha)\right) \tag{8}$$

The number of additional groups of harmonics to be considered could be quite large - more than 10.

To identify individual RMA defects that the models obtained after the vibration signals processing in the frequency domain are compared with the reference models. According to the comparison of the results, a conclusion is made about the RMA technical condition.

2.3 The Static (Single-Dimensional) in Time Model

In this case, the technical condition is evaluated according to the static (single-dimensional) in time RMA model in the frequency domain (this is how all the VDE operate at Russian Railways). Static models do not provide essential information for predicting the remaining life of the investigated assembly. In order to perform that, it is necessary to know how each of the typical defects develops through time. With regard to rolling stock assemblies, it is reasonable to determine how the the defect growth depends on the distance run. Therefore, it is necessary to develop dynamic models.

3 Dynamic Defect Models in the Time and Frequency Domains

3.1 The Concept of Dynamic Model

In [6], it is suggested to introduce a multi-parameter vibration analysis including the function of time. Such an approach allows one to estimate the dynamics of individual spectral components related to diagnosing the evidence of defects. The trend of the envelope spectrum component of high-frequency vibration of the bearing assembly is given as an example. The envelope spectrum component is responsible for such defect as an uneven roller and cage wear. This trend was obtained for a specific bearing in a three-year time span.

In order to introduce digitalization and predictive analytics at Russian Railways it is proposed to develop and put into diagnosing practice the defect growth dynamic models in the significant RMA of the rolling stock.

3.2 The Options for Creating Models and Experiments

The dynamic model in this case should be understood as the dependence of the degree growth of every identified defect on time or, more preferably, on the distance run.

There are two ways to obtain dynamic models - analytical methods and using empirical data basis. According to the authors the first way is not acceptable due to a number of factors with unpredictable influence on the device under test during its operation (for example, impact loads that occur when there is a flat spot on the running surface of the wheel, expansion joint gaps, the violation of loading rules for freight cars, etc.) and the need to introduce a large number of assumptions.

The analysis of open information sources showed the absence of publications on the dynamic modelling for the developing rolling stock RMA. The authors in cooperation with their colleagues attempted to refine a technique for creating dynamic defect models based on the results of diagnostic and future disassembly analysis of the complete wheelsets of the motor-cars in the ED4M electric multiple unit.

During the experiments with the help of the vibration diagnostic equipment "Prognoz" the bearings and gear defects of the complete wheelsets were detected. The capabilities of the VDE allow to detect up to 12 defect types and the degree of each defect growth without disassembling. The degree of each detected defect growth is determined by the modulation depth of the largest characteristic harmonic component according to formulas 5–8 [7].

The target of the study was the wheel-motor drive unit of the ED4M electric multiple unit with an average daily run of 1012 km (operation on the West-Siberian railway).

Regular monitoring of the detected defect growth degree allowed us to develop a dynamic defect growth model depending on time and run.

Figures 2 and 3 demonstrate the defect dynamics in the small gear of the gear unit respectively depending on time and run.

The defect from the "light" scale turns into the "unacceptable" scale after 25 thousand kilometers of the distance run. Or, like in this particular case, in 25 days. From the graph shown in Fig. 1, it can be seen that the defect had been detected long before the moment when its growth degree required to roll out the complete wheelset and replace the gear. The gear with the incipient defect ran 60 thousand km and only after that the rapid defect growth began. The run of the electric multiple unit from the last large-scale regular maintenance to putting a complete wheelset out of operation due to the intensive toothed-wheel gear defect growth was 160336 km.

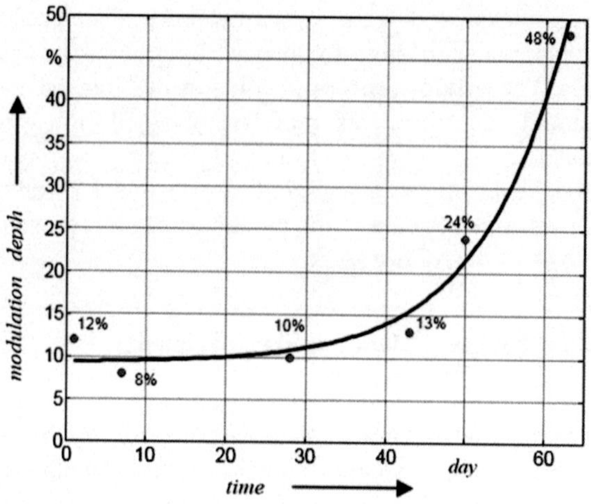

Fig. 2. The value of the defect growth depending on time.

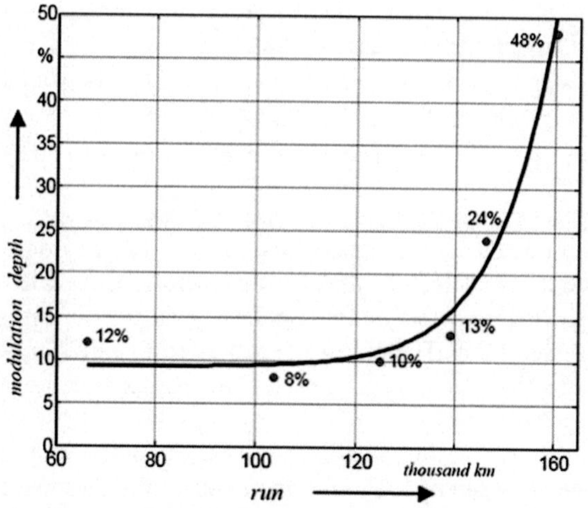

Fig. 3. The value of the defect growth depending on run.

3.3 Complexity of Experiments

In order to establish the typical dependences of the defects growth on time or distance run, it is necessary to have data on a large number of the results of diagnosing assemblies with detected defects and the results of disassembling these assemblies. Each breakdown requires at least 10 selection. For 12 typical defects, the number of the selections will be 120. This is only for one of the complete wheelset elements. The complete wheelset contains five various types of rotating elements (small gear, large gear, axle bearings, motor-anchor bearings, small gear bearing). So, for the empirical defects growth modelling of one type of the complete wheelset it is necessary to conduct about 3000 diagnostic sessions and 600 disassemblies taking into account that each model will be built according to five reference points in time or distance run.

For the car wheel pairs, the amount of experiments will be far less.

As a result of experimental work the generalized dynamic of defect models can be obtained. It should be considered that even for the same type elements operating in different conditions, dynamic models possess a different nature. The operating conditions that affect the nature of the model include:

- grading and condition of the track in field;
- climatic conditions of the region of operation;
- the nature of the goods transported and loading technology.

In this regard, the generalized dynamic models can be adjusted (in terms of the defect growth rate), taking into account local conditions.

4 Dynamic Models Based on the Wavelet Transform

4.1 Fourier Transform and Wavelet Transform

Another trend in RMA defect dynamic modeling is using the wavelet transform of the vibration signals. At the beginning of the article it was already mentioned the static in the time RMA model in the frequency domain which represents the amplitude-frequency spectrum - the result of Fourier transforming a single time sample of the vibration signal.

Fourier transform can be presented as a sum of harmonic components, each of them has its own amplitude, frequency and phase shift.

$$f(\omega t) = A_0 + A_1 \sin[\omega t + \varphi_1] + A_2 \sin[2\omega t + \varphi_2] + A_3 \sin[3\omega t + \varphi_3] + \ldots \quad (9)$$

where A_0 – constant component amplitude; A_1 – sin(ωt + φ1) – fundamental harmonic; $A_2, A_3, A_4, \ldots$ – the corresponding harmonic amplitudes.

One-dimensional signal wavelet transform *f(x)* is a two-dimentional function:

$$W_\psi(a, b)f(x) = \frac{1}{\sqrt{C_\psi}} \int_{-\infty}^{+\infty} \frac{1}{\sqrt{|a|}} \psi\left(\frac{b-x}{a}\right) f(x)dx \quad (10)$$

where the kernel Ψ is called a wavelet, b – a shift, a – a scale or a bar.

The normalization factor is equal to:

$$C_\psi = 2\pi \int_{-\infty}^{+\infty} \frac{|\psi'(\omega)|^2}{\sqrt{|a|}} d\omega < \infty \quad (11)$$

where $\Psi(\omega)$ – Wavelet transform Ψ made with the help of Fourier transform.

When converting a time vibration signal using the fast Fourier transform (FFT), the obtained amplitude-frequency spectrum does not contain information on the harmonic component phases (Fig. 1). This is the static (point) RMA model in the frequency domain for a particular point in time (specific time sample).

Using the wavelet transform allows to trace the change of the spectral model throughout the entire time sample. In other words, it is possible to trace changes in the spectral components of the amplitude-frequency spectrum and the analysis of the RMA condition could be carried out in the three-dimensional space (the three-dimensional spectrogram, Fig. 4).

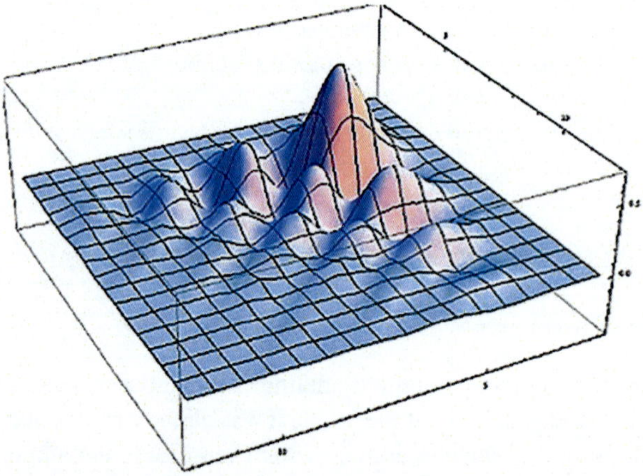

Fig. 4. Example of a three-dimensional spectrogram as a result of wavelet transform.

Wavelet transform (WT) of a regular signal is a generalized Fourier series according to the system of basic functions. The continuous (integral) wavelet transform is the *s(t)* signal scalar product of the two-parameter wavelet function $\Psi_{a,\,b}(t)$ of the selected type.

The integral function transformation of the function *s(t)* takes the form:

$$S_{\psi}(a, b) = \int_{-\infty}^{+\infty} S(t)\psi_{a,b}(t)dt \tag{12}$$

where, *a* – time scale parameter, inversely proportional to the frequency and responsible for the wavelet width; *b* – shift parameter determining the wavelet position on the time axis.

Wavelet function $\Psi a,b\ (t)$ of the affected set is obtained from one maternal function Ψ by stretch or compression and subsequent shift

$$\psi_{a,b}(t) = \frac{1}{\sqrt{a}}\psi\frac{(t-b)}{a} \tag{13}$$

The multiplier $1/\sqrt{a}$ determines that the integral energy of each wavelet $\Psi a,\ b\ (t)$ does not depend on *a*.

The function with two parameters $S_{\Psi}(a,b)$ gives the information on the change in the relative contribution of components of different scales in time and is called the spectrum of wavelet transform coefficients. The scale is similar in meaning to the concept of frequency in Fourier transform [8].

4.2 Dynamic Model at a Particular Point in Time

Continuous wavelet transform is conducted in a particular time interval. For a vibration signal the result of the transformation will be a three-dimensional spectrogram, which will represent a three-dimensional model in the amplitude - frequency - time period. Such model is a dynamic model at a particular point in time, it represents the technical condition of the investigated assembly in a very short time period and it will be the model "at the point" about the defect growth time interval.

5 Time Samples Joining

5.1 Three-Dimensional Defect Growth Model in Time Period

It is proposed to create a three-dimensional defect growth model in time period with the help of wavelet transform, i.e. the investigated time interval should cover the time from the beginning of the defect growth to its transition to an unacceptable defect. It is almost impossible to record, save and process a continuous time signal within an interval equal to the time of full defect growth. It takes an extended period of time. To create a wavelet defect growth model in time period it seems advisable to use four or five time samples of a vibration signal reflecting the condition of a defect-free assembly, an assembly with an incipient, moderate, severe and unacceptable defect. Next, it is proposed to join these time samples into a unified time function. The necessary condition for joining is to ensure the smoothness and continuity of the function at the place of joining. Such condition is the continuity of the derivative at the position of the joint - on the interval between two local maxima of the opposite sign (the interval *ab* at the position of the joint of the two functions):

5.2 The Smoothness of the Function

The function f is smooth on (a, b) if it is continuous on the segment (a, b) and has such continuous derivative $f(x)$ that there are limits $f(a+0) = A, f(b-0) = B$ [9].

It makes sense to join at the points where f (Δt1; Δt2; Δt3; Δt4; Δt5) = 0.

According to the first Bolzano – Cauchy theorem such a point must exist on the segment ab of the smooth f (t) function:

If the function f (t) is continuous on the segment ab, the function values b at the ends of the segment have different signs f(a) > 0, f (b) < 0 or f(a) < 0, f(b) > 0, then there is a point ξ ∈ (a, b), in which the value of the function is zero f (ξ) = 0 [10].

5.3 Joining Functions

Figures 5a and 5b show the examples of the incorrect joining. Figure 5c shows the example of the desired joining.

Incorrect joining of time samples will cause appearing a powerful stray noise in the high-frequency domain after the Fourier transform in the amplitude-frequency spectrum,. This fact will crucially distort the dynamic defect growth model in time period. The authors do not issue the challenge to describe the algorithm for implementing the desired joining - this is a particular mathematical problem which probably has more than one alternative solution.

6 Joining of Three-Dimensional Spectrograms

Theoretically, there is another approach to implement a dynamic defect growth model in time period using wavelet transform. As it was already suggested above, four or five time samples of the vibration signal are used and the wavelet transform is conducted separately for each time sample. As a result, three-dimensional spectrograms are obtained, one of the versions is shown in Fig. 4 (some scientific works on the wavelet transform describe other methods of graphical interpretation). Then it is assumed that the three-dimensional platforms are joined separately. The joining conditions require a solid mathematical study and their determination is not the issue of the current discussion. According to the

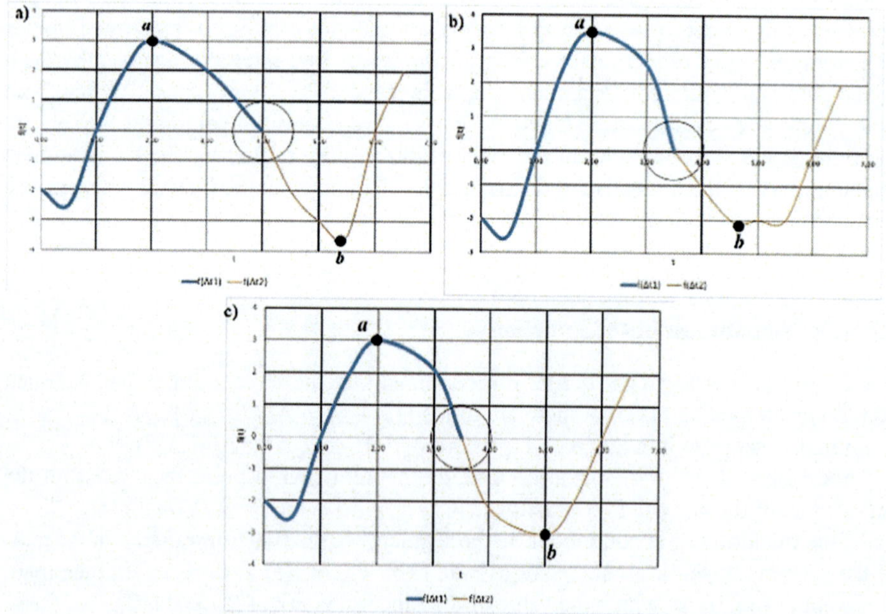

Fig. 5. a) Incorrect joining - *f(t)* function is not smooth on the *ab* segment; b) Incorrect joining - *f(t)* function is not smooth on the *ab* interval; c) desired joining - *f(t)* function is smooth on the *ab* segment.

preliminary estimates the joining of three-dimensional spectrograms will be connected with additional restrictions that can lead to data corruption or data loss and it will require much more complex implementation algorithms.

7 Conclusions

1. The expediency of using vibration diagnostics for the RMA dynamic defect growth models is established.
2. The practicability of using empirical data as a basis for modeling is shown.
3. The example and the method of small gear defect dynamic modeling for the complete wheelset are given.
4. The estimated amount of work on dynamic modeling is determined.
5. Dynamic defect models will make it possible to forecast of the remaining life of the complete wheelset if only a nascent defect is detected.

References

1. PKB CT.060050: Vibration diagnostics of locomotive components. Russian Railways, Moscow (2012)
2. The guide to vibration diagnostics of axle box bearings of wagon wheelsets. JSC "Russian Railways", Moscow (2010)
3. Tetter, V.Yu.: Kurtosis factor as a diagnostic obearing defect symptom – Nika. Control Diagnostics **3**, 28–34 (2010)
4. Tetter, V.Y.: "Standard" defects for diagnosing rotor mechanical assemblies. The Measurement World **10**, 14–19 (2007)
5. Barkov, A.V., Barkova, N.A., Azovtsev, A.Y.: Monitoring and diagnostics of rotor machines by vibration, S-Petersburg, p. 158 (2012)
6. Barkov, A.V., Barkova, N.A.: Vibration diagnostics of machines and equipment. Vibration analysis, p. 152. North-West training center, S-Petersburg (2013)
7. Barkov, A.V., Barkov, N.Ah.: Intelligent monitoring and vibration diagnostics systems **9**, 115–156 (1999)
8. Schoberg, A.G.: Modern methods of image processing modified by the wavelet transform, p. 125. Publishing House of the Pacific State University, Khabarovsk (2014)
9. Nikolsky, S.N.: Course of mathematical analysis, p. 592. FIZMAT-LIT, Moscow (2001)
10. Shilov, G.E.: Mathematical analysis (one variable functions), p. 528. Science, Moscow (1969)

Modeling Spiral Dispenser Operation Based on Structural Transformations

M. A. Novoseltseva, S. G. Gutova, and E. S. Kagan(✉)

Kemerovo State University, 6, Krasnaya Street, Kemerovo 650000, Russia

Abstract. Preparing homogeneous compositions from various materials and mixing them is a widely used process in food industry. The quality of the finished product often depends on the process. Increasing requirements for the quality of mixtures demand the need for studying the performance characteristics of various mixing equipment types. One of these approaches is mathematical modeling. In this paper, we consider the simulation of a spiral dispenser operation, providing a preliminary initial stage of high-quality components mix. Fluctuations in the dispenser lead to stochastic changes in its output signal, which in turn complicates the modeling process. To solve this, we use sequential structural transformations of the signal to filter fluctuations, and the continuous fractions apparatus to construct a mathematical model of the dispenser output signal in the form of a harmonic component with a certain amplitude and a circular frequency. The highly accurate results achieved allow us to simulate the processes at the input of continuous-type mixers, evaluate the efficiency of mixing processes and assess the impact of technological parameters on the quality of mixtures.

Keywords: Spiral dispenser · Multisinusoidal signal · Structural transformation · Continued fraction

1 Introduction

Modern food industry is in constant development, since health and proper nutrition are largely correlated. The quality of products is at the forefront, since it ensures normal human activity, increasing their efficiency, resistance to various diseases and adaptation to environment. In particular, many Russian regions experience a significant nutritional shortage of certain vitamins and micronutrients. Therefore, combined foods, eliminating the lack of various substances are introduced into the diet. Such substances are usually contained in products in small quantities (less than 1%), so it is necessary to carry out their uniform distribution over the volume of the product (with different physical and chemical properties) at the output. To obtain such uniform distribution, various (centrifugal, helical, vibrational) continuous mixers (CM) allowing to obtain good quality mixtures are often used [1]. In addition, due to the development of food nanotechnologies, there are increased requirements for the quality of mixtures, while the content of nanomaterials in the total volume of the mixture can be very insignificant.

A. A. Radionov and V. R. Gasiyarov (Eds.): RusAutoCon 2021, LNEE 857, pp. 140–150, 2022.
https://doi.org/10.1007/978-3-030-94202-1_14

2 Setting the Task: Modeling the Dispenser Operation Process

Improving the quality of the mixture obtained under working conditions determines the interest of researchers in mathematical modeling of the process. Using a cybernetic approach, the CM is represented as a dynamic system subject to some input influences (Fig. 1).

A mandatory part of this system is the presence of a number of dispensers that form the CM input signal summing their output components. The quality of compositions prepared in such CM depends not only on the internal mixing processes, but also on the dosage. No existing dispensers can provide a continuous flow of bulk material in strictly specified quantities at any given time. Consequently, when the components enter the CM, certain in their ratio deviations will be observed. To obtain a given ratio of components in the finished mixture, the CM, in addition to the qualitative mixing of the components, must provide smoothing of the fluctuations of the dispensers. Obviously, the characteristics of the dispenser (size and form of nominal values performance deviations) affect the kinetics of the mixing process and, ultimately, the quality of the ready mix [1–4].

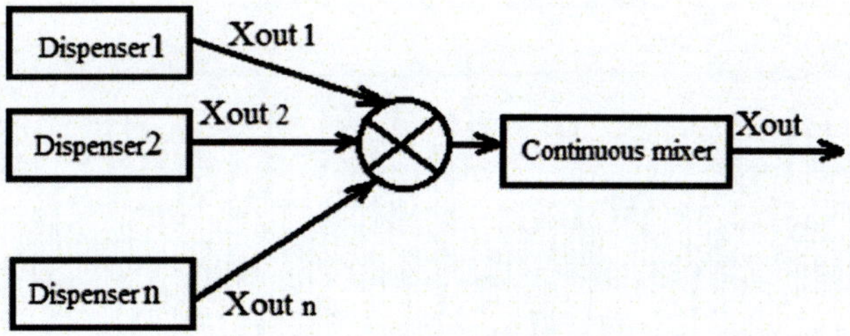

Fig. 1. Mixing unit structural and functional diagram.

As noted earlier [2], special conditions can be added to the regular fluctuations in the dispenser output signal. These may be: material hang-up, focal consolidation, humidity, etc., which lead to stochastic changes in the dispenser output signal. The presence of stochasticity leads to the complication of mathematical modeling processes and the need to use new and more advanced approaches.

Studying production processes of mixing bulk materials in various devices is of great interest. Various approaches to modeling these processes have been developed. For example, in [1, 3], the signals in the simulation are considered as deterministic and do not take into account the fluctuations described above. In addition, random Markov chain apparatus was previously used to describe the production process [4–10], but the determination of probabilities in the transition matrix is somewhat difficult. A number of experimental studies were carried out in [11–16], graphical data about the screw dispenser output signal were also obtained, but only vague practical recommendations regarding the design of the system were given.

In [17], physical and chemical studies of dispensers that can be used at the stage of their design were carried out. In [18], it is emphasized that the processes in the dispenser cannot be described deterministically. But at the same time, the available formulas for calculating process indicators – either theoretical [19, 20] or empirical [21] – are not considered accurate and have not found application in industry. The authors advise to use the only method for studying and predicting processes in dispensers, which is laboratory testing.

Given the incompleteness of research and the lack of effective study methods, we consider the task of modeling the processes of dispenser operation in mixing units to be urgent.

3 Applying Structural Transformations for Multi-sinusoidal Signal Identification

Consider the construction of an output signal mathematical model, exemplified by a spiral dispenser. Figure 1 displays a spiral dispenser output signal when potassium iodide is fed at the input. As seen from the Fig. 2, it can be assumed that the output signal pulsations obey the multisinusoidal law [4], which has an offset equal to the average mass consumption of the material.

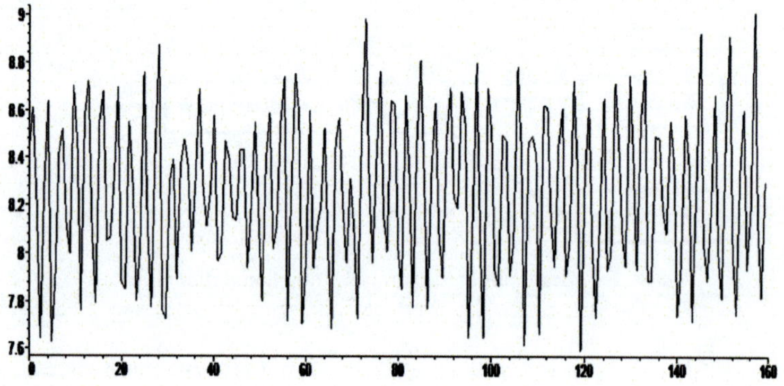

Fig. 2. The spiral dispenser output signal with potassium iodide fed at the input.

We will use the random process structural analysis [22, 23] to construct a model of a noise-contaminated multisinusoidal signal. Let the model of a multisinusoidal signal have the following form:

$$x(t) = \mu + \sum_{i=1}^{n} C_i \sin(\omega_i t) + a(t) \tag{1}$$

that is, it is the sum of *n* harmonics with ω_i unknown frequencies, C_i amplitudes, μ constant displacement, and the a(t) noise component. The digitized signal values with the Δt for $k = 0, 1, 2, \ldots$ discretization step are determined by the ratio

$$x(k\Delta t) = \mu + \sum_{i=1}^{n} C_i \sin(\omega_i k\Delta t) + a(k\Delta t) \tag{2}$$

The structural function [22, 23] of some random x(t) signal is defined as follows:

$$C_x(t,\ t+\tau) = M\{x(t) - x(t+\tau)\}^2 \tag{3}$$

Structural functions reflect the presence of oscillating components of the analyzed random signal. This can be useful for determining the multisinusoidal components of the process parameters.

With digitized measurements of a random signal available, the structural function is constructed using the formula

$$C_x(k) = \frac{1}{N_1 - k}\sum_{i=1}^{N_1-k}(x(i) - x(i+k))^2 \tag{4}$$

where N_1 is the number of measurements of the x(t) signal. Further on, the operation of finding the structural function according to the (4) formula will be called the first structural transformation and denoted as $c_x^1(k) = c_x(k)$.

In (4), the initial data for finding the structural function is the x(t) signal itself. Then, by analogy, the second structural transformation can be introduced:

$$C_x^2(k) = \frac{1}{N_2 - k}\sum_{i=1}^{N_2-k}\left(C_x^1(i) - C_x^1(i+k)\right)^2 \tag{5}$$

where the initial data is the structural function of the x(t) random signal, that is, the first structural transformation.

The third structural transformation has the form

$$C_x^3(k) = \frac{1}{N_3 - k}\sum_{i=1}^{N_3-k}\left(C_x^2(i) - C_x^2(i+k)\right)^2 \tag{6}$$

Hence, the m-th structural transformation is represented as

$$C_x^m(k) = \frac{1}{N_m - k}\sum_{i=1}^{N_m-k}\left(C_x^{m-1}(i) - C_x^{m-1}(i+k)\right)^2 \tag{7}$$

where $N_m < N_{m-1}$. Empirical research shows that while modeling a noise-contaminated multisinusoidal signal, it is advisable to subject the original process to three consecutive structural transformations. This is due to the fine selectivity of the structural transformation with respect to the highest amplitude harmonic [22]. We will construct a model of the third structural signal transformation based on signal values, using the theory of continuous fractions and, in particular, V. Viskovatov's modified algorithm [24, 25].

There is a theory [26], according to which any reaction is the result of an elementary impact on an object at some point in space. The unit step function is applied to the input of the object, which can mean loading the mixed component into the dispenser. It is noteworthy that the amplitude of such an elementary effect does not have to be equal to 1, which does not affect the construction of the model.

The identifier matrix is calculated according to the values of the third structural transformation (6):

$$\begin{matrix} (-1)-string \\ 0-string \\ 1-string \\ 2-string \\ 3-string \\ 4-string \end{matrix} \begin{pmatrix} 1 & 1 & 1 & \ldots & 1 & \ldots \\ C_x^3(\Delta t) & C_x^3(2\Delta t) & C_x^3(3\Delta t) & \ldots & C_x^3(n\Delta t) & \ldots \\ \alpha_1(0) & \alpha_1(\Delta t) & \alpha_1(2\Delta t) & \ldots & \alpha_1(n\Delta t) & \ldots \\ \alpha_2(0) & \alpha_2(\Delta t) & \alpha_2(2\Delta t) & \ldots & \alpha_2(n\Delta t) & \ldots \\ \alpha_3(0) & \alpha_3(\Delta t) & \alpha_3(2\Delta t) & \ldots & \alpha_3(n\Delta t) & \ldots \\ \alpha_4(0) & \alpha_4(\Delta t) & \alpha_4(2\Delta t) & \ldots & \alpha_4(n\Delta t) & \ldots \end{pmatrix} \tag{8}$$

where 0-string represents the values of the third structural transformation $C_x^3(n\Delta t)$, obtained on the basis of (6), while the $\alpha_m(n\Delta t)$ elements are sequentially determined using the ratio:

$$\alpha_m(n) = \frac{\alpha_{m-2}(n+1)}{\alpha_{m-2}(0)} - \frac{\alpha_{m-1}(n+1)}{\alpha_{m-1}(0)} \tag{9}$$

where $\alpha_0(n) = C_x^3(n\Delta t)$, $\alpha_{-1}(n) = 1$, m = 1, 2, 3,..., n = 0, 1, 2,...

If the modeled process has the (2) form and is sequentially transformed structurally according to (6), then in the (8) matrix the 4th string will become null.

Further signal modeling is determined by the elements of the (8) identifier matrix first column, which generate the partial numerators of the proper C-fraction [25]. This allows us to obtain a model of the third structural transformation the signal in the form of the object discrete transfer function (DTF) [22]:

$$G(z) = \frac{C_x^3(\Delta t)z^{-1}}{1 + \frac{\alpha_1(0)\,z^{-1}}{1+\frac{\alpha_2(0)\,z^{-1}}{1+\alpha_3(0)\,z^{-1}}}} \tag{10}$$

The signal is periodic if the DTF contains two complex-conjugate poles. Otherwise, the signal does not belong to the type mentioned and has no period.

In the case of complex-conjugate poles, the resulting DTF model (10) allows us to determine the harmonic parameters with the maximum amplitude. Initially, the ω_1 circular frequency of this harmonic is found

$$\omega_1 = \frac{1}{\Delta t}\arg(z_1) \tag{11}$$

where $z_1 = u \pm iv$ are DTF poles (10). Then the harmonic amplitude is found by the formula

$$C_1 = \sqrt[8]{\max\left(C_x^3(k\Delta t)\right)/2}, \tag{12}$$

where k = 1,..., N_3.

The amplitude of this harmonic is maximum. The harmonic has the form

$$s_1(k\Delta t) = C_1 \sin(\omega_1 k\Delta t) \tag{13}$$

Since the number of harmonics of the original signal is unknown, the following procedure must be performed. Using the resulting harmonic model (13), we subtract it from the original x(kΔt) signal, thus obtaining the x1 (kΔt) signal

$$x_1(k\Delta t) = x(k\Delta t) - s_1(k\Delta t) \tag{14}$$

After that, the entire procedure should be repeated with the (14) signal to identify its remaining harmonics. The next harmonic will have the largest amplitude value among the remaining ones. The procedure is repeated until the elements of the 4th row of the ID-matrix (8) cease to be equal to 0. Thus, the criterion for stopping the procedure of identifying harmonic components in the original signal is the appearance of non-zero elements in the 4th string of the identifier matrix.

4 Modeling Spiral Dispenser Operation

Consider the procedure for modeling the spiral dispenser output signal with potassium iodide fed at the input. 200 signal values were taken with a discretization step of Δt = 0.8 s. The signal graph is shown in Fig. 2. The graph of this signal structural function is shown in Fig. 3.

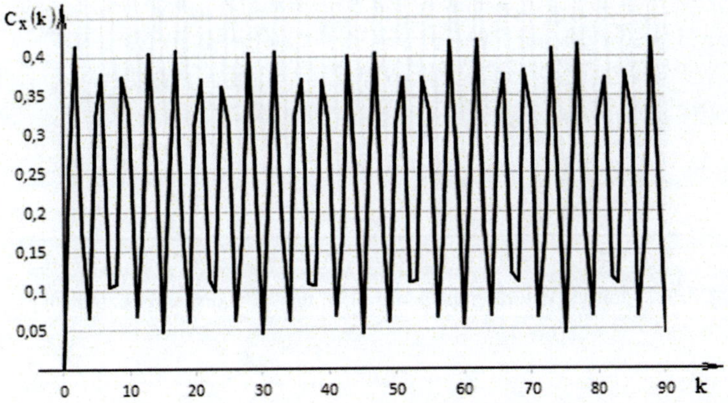

Fig. 3. The structural function of spiral dispenser output signal with potassium iodide fed at the input.

Figures 4 and 5 show the second (5) and third (6) structural transformations of the signal.

The identifier matrix will be calculated on the basis of the third structural transformation values in Table 1.

Further, we use only the elements of the first column of the matrix:

$$G(z) = \frac{0.00143z^{-1}}{1 + \frac{-0.78712\,z^{-1}}{1+\frac{2.26489z^{-1}}{1-1.27147\,z^{-1}}}}. \tag{15}$$

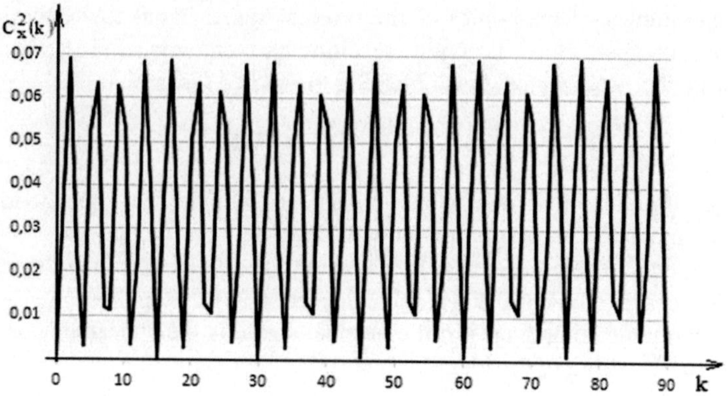

Fig. 4. The second structural transformation of the spiral dispenser output signal.

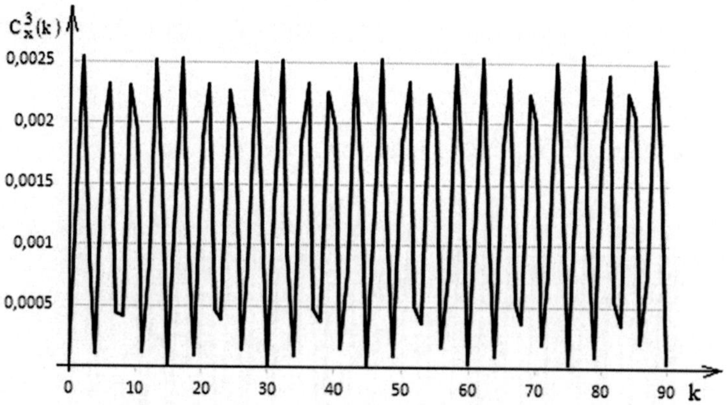

Fig. 5. The third structural transformation of the spiral dispenser output signal.

Table 1. The third structural transformation values.

1	1	1	1	1	1
0.00143	0.00255	0.00089	0.00011	0.00192	0.00233
−0.78712	0.37606	0.92385	−0.349104	−0.63097	0.69595
2.26489	1.79764	−0.36737	0.54749	2.51514	1.206437
−1.27147	−1.01150	0.20179	−0.30887	−1.41683	−0.68224
0	0	0	0	0	0

We transform the continued fraction and obtain a fractional rational expression:

$$G(z) = \frac{z + 0.99342}{z^2 + 0.20630z + 1.00080} \tag{16}$$

The DTF poles (16) are equal to $z = -0.10315 \pm j0.99507$. Next step is defining the signal parameters. According to (11), the circular frequency of the harmonic is $\omega_1 = 2.09261$, hence its period: $T_1 = 3.0020$.

The harmonic amplitude is found by the (12) formula: $C_1 = 0.43466$. Then the harmonic with the maximum amplitude is determined by the equation

$$x_1(t) = 0.43466 \cdot sin(2.09261t) \tag{17}$$

Using the model (17), we subtract it from the original signal x(t). Then, we repeat the procedure for identifying the next harmonic using the same algorithm.

The first, second and third structural transformations are in Figs. 6, 7 and 8, respectively.

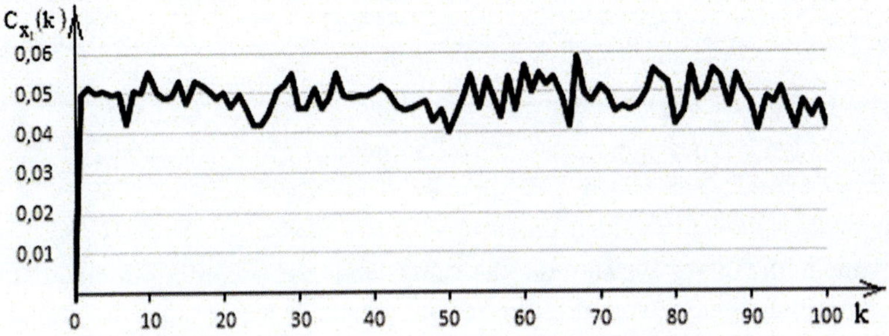

Fig. 6. The first structural transformation of the signal (17).

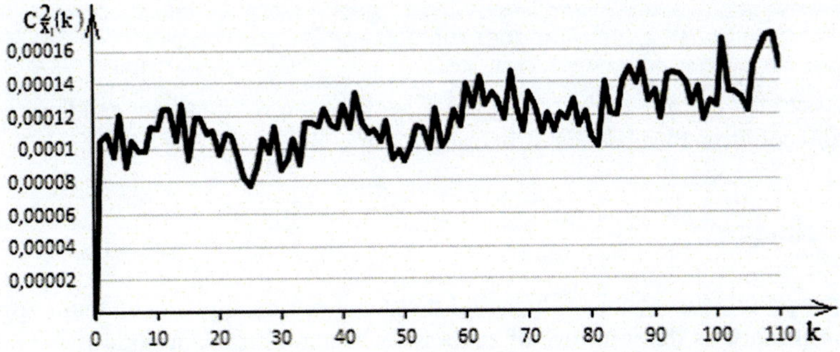

Fig. 7. The second structural transformation of the signal (17).

On the basis of the third structural transformation, we will fill in and further calculate the identifier matrix in Table 2.

Since there are no zero elements in the 4th row of the identifier matrix, there are no more harmonic components in the signal.

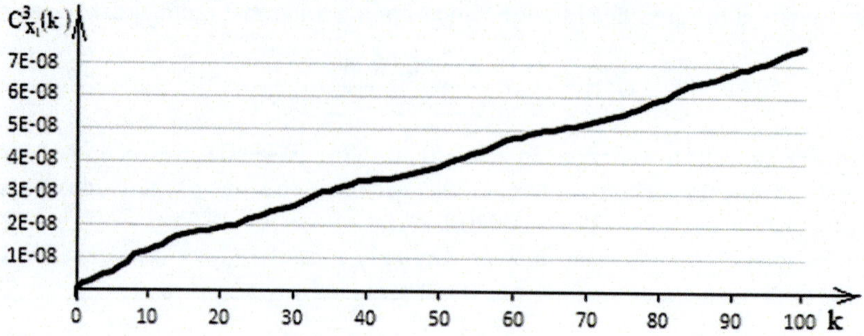

Fig. 8. The third structural transformation of the signal (17).

Table 2. Values for matrix.

1	1	1	1
$0.20204 \cdot 10^{-8}$	$0.27352 \cdot 10^{-8}$	$0.41489 \cdot 10^{-8}$	$0.53290 \cdot 10^{-8}$
−0.35376	−1.05347	−1.637567	−1.74763
−1.62414	−2.57556	−2.30260	−4.86801
1.39211	3.21130	1.94289	4.35310
−0.72000	0.02209	−0.12971	0.56584

To estimate the offset, we subtract the values obtained according to the (17) model from the original signal. Averaging the results obtained, we see that the signal offset corresponding to the average material mass consumption is $\mu = 8.29561$. Thus, the model of the spiral dispenser output signal with potassium iodide fed at the input contains only one harmonic component and has the form

$$X_{\text{mod}}(t) = 8.29561 + 0.43466 \sin(2.09261). \tag{18}$$

Subtract (18) from the original signal. Checking the residuals for randomness and correlation confirms the absence of harmonic components.

5 Conclusion

This paper presents a model of the spiral dispenser output signal. The main approach to its construction is the joint use of successive structural transformations for filtering noise fluctuations, as well as further expansion of the third structural transformation into a continued fraction to estimate the structure and parameters of the signal. The resulting model is the initial stage of modeling the mixing unit operation using the CM and a number of dispensers that, summing their output components, form the CM input signal.

References

1. Borodulin, D.M.: Development and mathematical modeling of non-interacting centrifugal mixing units for bulk materials processing. Generalized theory and analysis (Cybernetic approach). Kemerovo Technological Institute of Food Industry, Kemerovo (2013)
2. Makarov, Y.: Apparatuses for Mixing Bulk Materials. Mashinostroenie, Moscow (1973)
3. Ivanets, V.N., Fedosenkov, B.A.: Processes of dosing bulk materials in continuous mixing units – generalized theory and analysis: monograph. Kemerovo (2002)
4. Selivanov, Yu.T., Pershin, V.F.: Calculation and Design of Circulating Mixers of Bulk Materials without Internal Mixing Devices. Mashinostroenie-1 Publishing House, Moscow (2004)
5. Gusev, Y., Karasev, I.N., Kolman-Ivanov, E.E., MakarovYuI, M.M.P., Rasskazov, N.I.: Design and Calculation of Chemical Production Machines: A University Textbook. Mashinostroenie, Moscow (1985)
6. Pershin, V.F.: Modeling the process of mixing bulk material in the cross-section of a rotating drum. Theor. Found. Chem. Technol. **20**(4), 508–513 (1986)
7. Satomo, I.: Mixing solids: translated from Japanese. Puranto Kogaku **10**(5), 63–69 (1968)
8. Makarov, Y.: Fundamentals of calculation of mixing processes of bulk materials. Research and development of mixing devices: Author's Abstract of dis. Doctor of technical Sciences. Moscow (1975)
9. Koga, D.: Investigation of the mixing process of particles with different densities in a horizontal drum mixer: translation from Japanese. Rikakogu kenkyuse hokoku **56**(5), 95–102 (1980)
10. Sayyad, T.A.l., Pershin, V., Pasko, A., Pask, T.: Virtual modelling of particles two-Step feeding. J. Phys. Conf. Ser. **1084**(1), 012005 (2018)
11. Bates, L.: Entrainment pattern of screw hopper dischargers. Trans. ASME J. Eng. Ind. **91**(2), 295–302 (1969)
12. Ola, D.C.: Screw feeder flow profile of agro-food bulk solids. Laboratory stand review. Bull. Transilvania Univ. Brasov Forestry, Wood Ind. Agric. Food Eng. Ser. II **10**(2), 133–140 (2017)
13. Mehos, G., Morgan, D.: Hopper design principles. Chem. Eng. **123**(1), 58 (2016)
14. Schumacher, W.: Zum förderverhalten von bunkerhalten von bunkerabzugsschnecken nit vollblattwendeln. Diss. Doctoral thesis, Rheinish-Westfälishen Technischen Hochschule Aachen, Germany (1987)
15. Vetter, G.: Handbuch Dosieren. Vulkan Verlag Essen (1994)
16. Vollmann, A.: Untersuchung der Schüttgutförderung in geneigten Schneckenförderern. Dissertation TU München, Germany (2000)
17. Mehos, G., Morgan, D.: Hopper design principles (2016). https://www.chemengonline.com/hopper-design-principles/?printmode=1
18. Bortolamasi, M., Fottner, J.: Design and sizing of screw feeders. Partec (2001)
19. Roberts, A.W.: Aspects of Attrition and Wear in Enclosed Screw A.W. Conveyors. In: Proceedings of the Technical Program Powder and Bulk Solids Conference and Exhibition. – REED EXHIBITION COMPANIES, p. 225 (1993)
20. Roberts, A.W.: Predicting the volumetric and torque characteristics of screw feeders. Bulk Solids Handling **16**(2), 233–244 (1996)
21. Screw Conveyors for Bulk Materials - ANSI/CEMA Standard #350 (2019). https://cemanet.org/wp-content/uploads/2019/06/ANSI-CEMA-350-FinalReview.pdf
22. Novoseltseva, M.A.: Detection of noise-contaminated signal hidden periodicities using a structural function model in a continuous fraction form. Bull. Kemerovo State Univ. J. Theor. Appl. Res. **4**, 79–84 (2010)

23. Romanenko, A.F., Sergeev, G.A.: Problems of random processes applied analysis. Sovetskoe Radio, Moscow (1968)
24. Novoseltseva, M.A., Gutova, S.G., Kazakevich, I.A.: Structural and parametric identification of a multisinusoidal signal model by using continued fractions. In: 2018 International Russian Automation Conference (RusAutoCon), Sochi, pp. 1–5 (2018)
25. Kartashov, V.Ya., Gutova, S.G.: Continued fractions and their application to the problems of technical cybernetics: tutorial. Ministry of Education and Science of the Russian Federation, Kemerovo State University, Kemerovo
26. Logov, A.B., Zamaraev, R.Yu., Logov, A.A.: Principles of monitoring unique objects. Part 1. Criteria and boundaries on the phase plane. Min. Inf. Anal. Bull. (Sci. Tech. J.) **17**(12) (2009)

Study of Gas Giant Satellites System

A. Chernenkii(✉)

Peter the Great St. Petersburg Polytechnic University,
29, Polytechnicheskaya Street, St. Petersburg 195251, Russia
andrey@qmd.spbstu.ru

Abstract. The paper discusses the forthcoming exploration of gas giant' satellites in our Solar system, that will become important for humanity in the near future. In the entire history of cosmonautics, less than 10 spaceships have been directed to gas giants' planets in order to study their satellites. All these flights were unmanned, and there were no sample capsule returns. All information came through a network of communication satellites. This information indicates many promising satellites for exploration and colonization. The reason for a poor study is the huge distance between our planet and gas giants. The development and construction of a spaceship for astronaut's flights and scientific equipment is a difficult and expensive task. To deal with this task we should preliminarily use computer modeling. The software that we propose for this is the Kerbal Space Program 1.7. The current paper was designed based on this program, which is very beneficial for the development of "space" literacy, especially for young people. The tasks of this research are differed and cover the choice of a flight trajectory and types of spacecrafts and equipment needed for satellites exploration. The algorithm for project realization is shown, and steps of its implementation are described by simulation.

Keywords: Gas giants' satellites · Simulation · Kerbal space program

1 Introduction

A gas giant is a type of planets that consists of various gases only. In our solar system there are 2 planets of this type: Jupiter and Saturn.

Each of these planets has plenty of satellites, which, unlike the planets themselves, are solid-state structures. According to preliminary studies [1, 2] the composition of these satellites soil contains a number of valuable (from the standpoint of Earthlings) minerals and natural resources that are becoming scarce on our planet, in particular, fuel-containing ore. This rises the interest, including commercial one, in studying gas giants satellites and, in perspective, their exploration in order to build space bases for the extraction of minerals and fuel [3]. Actuality, such projects are of great demand, since the gas giants have many satellites that are potentially suitable for resource extraction or colonization [4–8].

Nowadays several projects on gas giants' satellites exploration are deployed. There are a lot of various research tasks, since satellites differ in size and composition. The

A. A. Radionov and V. R. Gasiyarov (Eds.): RusAutoCon 2021, LNEE 857, pp. 151–164, 2022.
https://doi.org/10.1007/978-3-030-94202-1_15

study of gas giants' satellites needs tremendous expenses, but results assumed to be terrific [9–12].

Nevertheless, it is not advisable to equip expensive expeditions for the exploration and development of such remote space objects without conducting preliminary surveys and evaluating the possibility of successful projects implementation. Computer simulations can help you do this.

In order to avoid errors, modeling methods are traditionally used, especially in the development and implementation of previously unknown objects. One of the most common tools for space modeling is the Kerbal Space Program, based on which the author developed and teaches the discipline "Space modeling" for pupils at the Highest Engineering school of Peter the Great St. Petersburg Polytechnic University [13].

Kerbal Space Program is an engineering game aimed at familiarizing with the complexities of space exploration. The first version of the program was developed in 2010 by the Mexican company "Squad", the lead developer is Felipe Falanghe. In 2019, the project has been supported by the American developer of the GTA series of games. Now the program goes beyond applying only by interested users. In the modern version of the educational paradigm, it is proposed to include elements of computer games and simulations in the content of training courses. Kerbal Space Program is declared as one of the "Top 10 games for people with high IQ".

The study of gas giants in our Solar system is a continuous process. Up to now, the main emphases of researches are concentrated around Jupiter. The first scientist, who started the systematic observation of this gas giant in 1610, was the Italian astronomer Galileo, and he discovered the largest Jupiter satellites – Io, Europa, Ganymede, Callisto. Much later, in the years of 1972 and 1974 two spaceships "Pioneer" passed by Jupiter. In 1977 spaceship "Voyager" has been lunched; it reached Jupiter in 2 years. In 1989 the spaceship Galileo started its flight to Jupiter, and in 1995 it launched a Zond, which began to collect information about the Jupiter atmosphere. Later scientists started systematic studies of the giant using the Hubble telescope [14, 15].

Here, in this paper, we'll use the advantages of the Kerbal Space Program to demonstrate a step-by-step work on the project of gas giants' satellites system exploration. We are focused on Jupiter satellites, thus, we have to find an analog of Jupiter in the Kerbin solar system. For this, we must analyze the available data on Jupiter and previous flights, possible flight trajectories, and to choose the spacecrafts, that we'll be able to deliver all equipment to the orbit of the analog of Jupiter needed for satellites' study and a satellite base construction.

2 Satellite Base Construction Algorithm

This project is aimed at a step-by-step study of the gas giant's satellites. Strategic objectives of the project: a comprehensive study of each satellite (structure, size, composition, presence of the atmosphere, the composition of the atmosphere (if any), satellite gravity, etc.); search for sources of fuel-containing ore or minerals, especially pure (without impurities); search for potential sites for the foundation of a colony, which will become another base for space exploration and, in perspective, will help solve the problem of overpopulation of our own planet [16]. In frames of this work, we are considering a private project aimed at the construction of the fuel production complex.

Project tasks and the algorithm for their deployment are shown in Fig. 1.

3 Jupiter and the Flight Trajectory Description

3.1 Selection of the Gas Giant Analog Under Study

Jupiter is the largest planet in the Solar system. The main data about the planet, which are important for our study, we took form [17]:

- the average distance from Sun is 7.783×10^8 km;
- average orbital speed is 13.1 km/s;
- the orbital period is 11.86 years;
- diameter is 142.984 km (which is 11.2 times more than the Earth diameter);
- mass is appr. 1.9×10^{27} kg (about 318 Earth masses);
- the average density is 1326 kg/m^3;
- surface gravity is 2.36 g;
- the average temperature at cloud tops is –108 °C (165 K);
- atmospheric composition: H_2 – 86.2%, He – 13.6%; CH_4 – 0.2%, ammonia, water vapor, and other gases;
- Jupiter has 63 satellites.

Based on this data, we selected the planet Jool in the Kerbin solar system, which is the closest analog of Jupiter.

3.2 Flight Trajectory

The choice of the flight path from the Earth to the Jupiter orbit affects the calculation of the required rocket fuel amount and, as a result, the choice of the launch vehicle and its layout with additional fuel tanks. Since our expedition assumes the delivery of a large amount of equipment to the Jupiter orbit, the solution to the problem of economical fuel consumption becomes particularly urgent.

In our case, the optimal trajectory is computed on the criterion of maximizing the delivered spacecraft mass.

From observations of the movement of natural satellites, as well as from the analysis of spacecraft trajectories, it is possible to restore the gravitational field of Jupiter. This is noticeably different from the spherical-symmetric one due to the rapid rotation of the planet. The gravitational potential is usually represented as a Legendre polynomial equation [18]:

$$V_{ext}(r,\theta) = -\frac{GM}{r}\left[1 - \sum_{i=1}^{\infty}\left(\frac{R_{eq}}{r}\right)^i J_i P_i(\cos\theta)\right] \quad (1)$$

where G is the gravitational constant; M is the mass of the planet; R is the distance to the center of the planet; R_{eq} is the equatorial radius of the planet; θ is the polar angle; P_i is the Legendre polynomial of the i-th order; J_{i} are the decomposition coefficients;

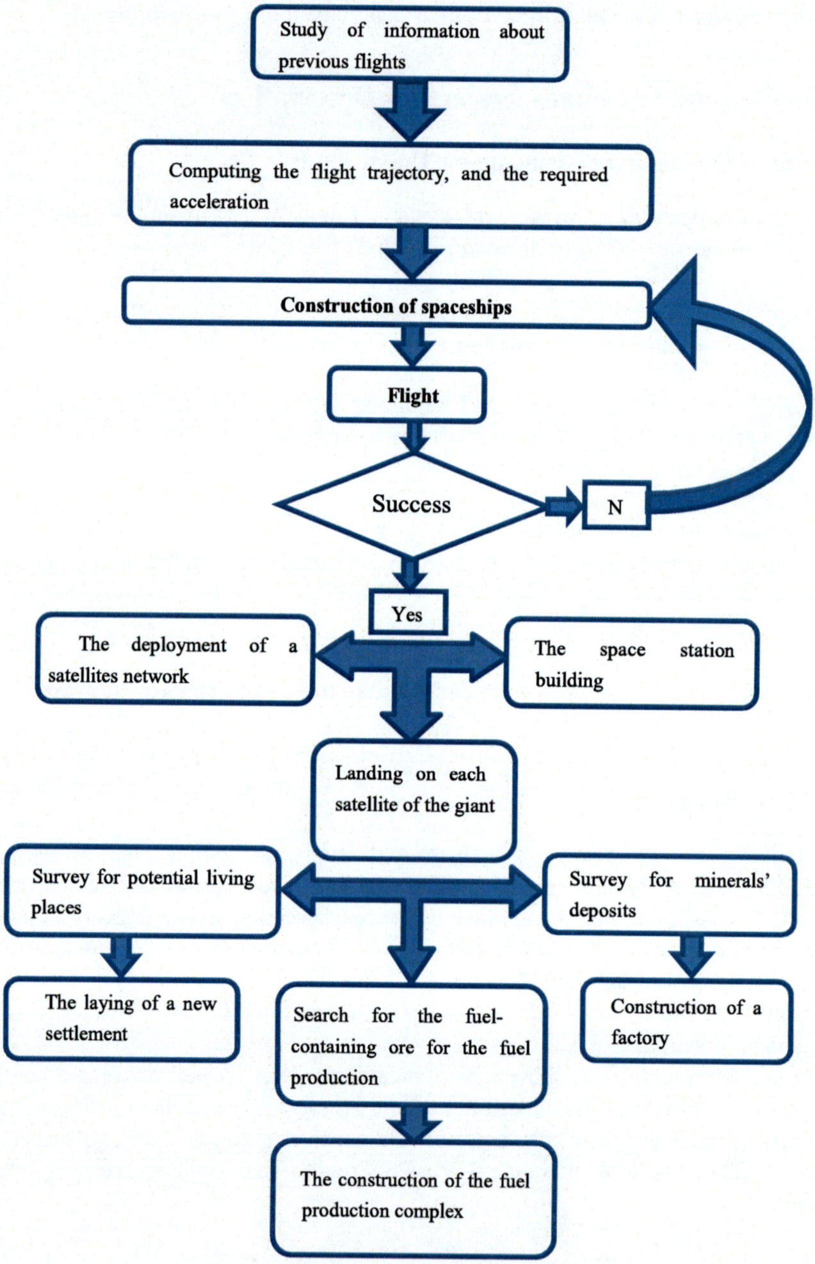

Fig. 1. Algorithm for project deployment.

depending on i, these coefficients take rather different values, in particular: $J_2 = 1.4697 \cdot 10^{-2}$, $J_4 = -5.84 \cdot 10^{-4}$, $J_6 = 0.31 \cdot 10^{-4}$ [19].

The condition for the hyperbolic excess velocity optimality at the start from the Earth includes the value of the spacecraft mass derivative after the separation of over clocking engines by the value of the hyperbolic excess velocity V_∞ at the start from the Earth (2):

$$a_{ob} \times \lambda V(T_0) - \lambda t_m \times w_b \times \frac{dM_0}{dV_\infty} = 0 \tag{2}$$

Besides, to optimize the heliocentric flight of the spacecraft to Jupiter, a gravitational maneuver near the Earth can be used. To do this, we need to define the following parameters:

- start date;
- vector of hyperbolic excess velocity at the start from the Earth;
- a program for turning the rocket engine on and off during a heliocentric Earth–to Earth flight;
- a program for orienting the rocket engine thrust vector in the active sections of the spacecraft flight.

Optimal start date condition formula:

$$\frac{[\lambda r(T_0), \lambda V(T_0)]}{\lambda V(T_0)} + \Pi_1(T_0) = 0, \tag{3}$$

where $\Pi_1(T_0) = \begin{cases} \Pi(T_0), & if\ \Pi(T_0) > 0, \\ 0, & if\ \Pi(T_0) \leq 0, \end{cases}$

$$\Pi = \frac{a_{ob}}{1 - \frac{a_{ob}}{w_b}} \lambda V + \lambda t_m, \tag{4}$$

a_{ob} is the dimensionless initial reactive acceleration, w_b is the dimensionless specific impulse of the rocket propulsion system, t_m is the engine operating time, λt_m is the conjugate time variable.

It should also be borne in mind that in space, in addition to the rocket engines thrust forces, there are other forces that could be used to give the necessary acceleration to the spacecraft. We are talking about accelerations when flying near a large gravitational object, which saves fuel. A good example is the flight path of the Galileo spacecraft to Jupiter (Fig. 2), which was well thought out not only from the position of the spacecraft launch date, but also based on the analysis of the relative position of the space objects in Solar system along the flight path.

As one can see from Fig. 2, the "Galileo" flight trajectory included several flybys:

- Venus flyby,
- Earth flyby,
- Gaspra flyby,
- Earth flyby,

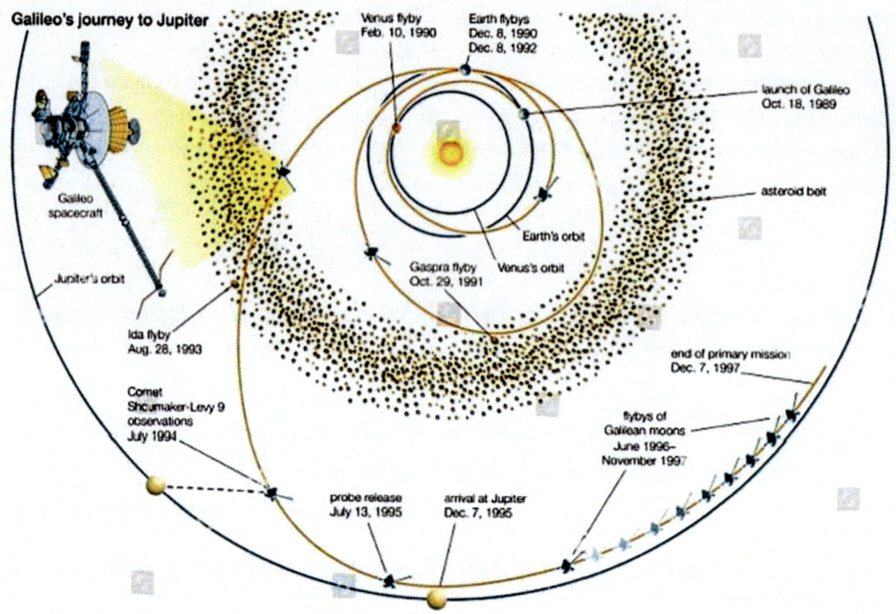

Fig. 2. "Galileo" flight trajectory.

- Ida Flyby,
- Flybys of the Galilean moons [20].

All these flybys gave the additional acceleration to "Galileo", which allowed reaching the Jupiter orbit on time.

The described approach is taken into consideration while simulation of our expedition to the planet Jool in the Kerbin solar system.

4 Specification of Required Spaceships and Equipment

After analyzing publications on the gas giants, we come to the implementation of project stages.

This project is realized in Kerbal Space Program. The base planet in Kerbin, a Gas giant planet, is Jool – an analog of Jupiter. This planet has 5 natural satellites: Oceanic moon Leit, Icy moon Val, Rocky Moon Tilo, Captured Asteroid Bop, Satellite Jool, Pol [21].

4.1 Selection of Spaceships

The project is rather complicated, thus we must undertake the system approach [22]. The "Many Communications" vehicle is used for the transportation of 12 communication satellites that will allow to maintain communication between spaceships (Fig. 3).

"Living module of the station" (Fig. 4) consist of: landing module Mk2, 2 storage area PPD-10; Storage module (2.5 m) includes re-translator RA-15 and 4 nuclear reactors [23, 24].

Fig. 3. "Many Communications" vehicle.

Fig. 4. The living module of the station.

"Folded station" (Fig. 5). It was decided to transport the other station modules together, since they are rather small and do not weigh much. "Folded station" includes:

- Module "Science": MPL-LG-2 mobile laboratory, large improved gyrodine, 2 Z-4K rechargeable batteries, large docking node "Grip-o-Tron", Rocomax branded adapter, multi-point connector Rocomax Muzel, 4 docking nodes "Grip-o-Tron", and 4 searchlights Mk-1;

- "Dome" module: Rocomax proprietary adapter, Mk2 lander, and PPD-12 Dome module;
- Module "Transition": 2 structural fuselages, 4 searchlights Mk-1, and 2 docking nodes "Grip-o-Tron";
- "Accumulator" module: 5 Z-1k batteries, SC-9001 small scientific module, 2 88-88 communotrons, and 2 "Grab-o-Tron" docking nodes";
- Module "Power": 2 long segments of the modular beam, a segment of the modular beam, 2 adapters to the modular beam, 2 docking nodes "Grip-o-Tron", 6 complexes of solar panels "Gigantor XL", and 2 thermal control systems (large);
- "Collector" module: RS-001S remote control unit, built-in improved gyrodyne, Z-1k battery, 2 FL-R25 monofuel tanks, 4 Mk-1 searchlights, and 8 RV-105 DCS engine blocks [25, 26].

"Vertical take-off device" (Fig. 6). A 3-seat vehicle with a docking station and a small scientific module. The main application is the study of small satellites. Composition: Mk-1 crew cabin, built-in lock Grip-o-Tron [27–29], SC-9001 small science module, FL-R25 monofuel tank, 3 Z-1k batteries, 2 RA-2 repeaters, 4 portable reactors, 7 FL-T200 fuel tanks, 3 CR-7 RAPIER engines, 4 LV-909 "Terrier" engines, 8 LT-1 landing supports, and 16 RV-105 DCS engine blocks.

Fig. 5. Folded station.

"Long-haul" (Fig. 7, at the left). This is a mobile base in space. Composition: Mk-1 cabin, Mk-1 crew cabin, 4 PPD-10 hitchhiker cabins, 5 "Grab-o-Tron" docking nodes, and 3 repeaters (one RA-100 – for communication with the MCC, two of RA-2 – for communication with other vehicles).

"Super Truck" (Fig. 7, at the right). This device is designed for interplanetary cargo transportation. It is equipped with 3 cargo compartments with docking nodes, 1 cargo compartment with reactors and batteries, and a passenger compartment for 16 astronauts.

Fig. 6. Vertical take-off device.

"The descent of the trio with science" (Fig. 8). The device is designed for the landing of large satellites. It consists of: a 3-seats cabin; a docking station; 2 batteries; a cargo compartment with scientific equipment, communications, and reactors (Fig. 9); 5 engines for a soft landing on satellites; and 4 landing gears.

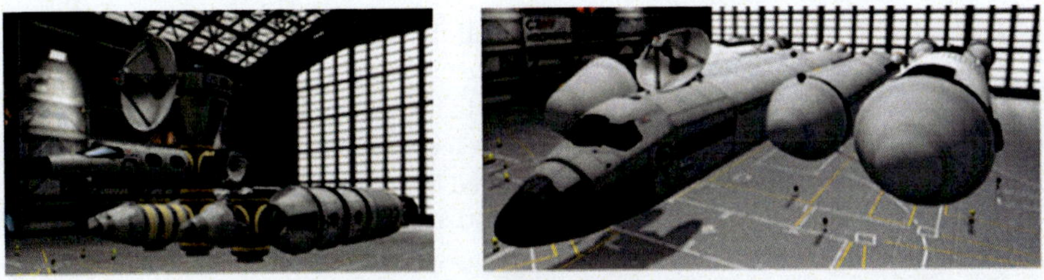

Fig. 7. Long-haul and super truck.

Fig. 8. The descent of the trio with science.

Fig. 9. Cargo compartment with scientific equipment, communications, and reactors.

5 Project Realization

In the previous part all spaceships needed for the project were selected. Here we will follow the process of the project implementation and the project results.

1. The flight was completed successfully by all the spaceships. The "Descent of the Trio with Science" spacecraft has been placed in the central cargo compartment of the "Super-Cargo" ship on the Kerbin orbit. In it's Two other compartments – spare parts for the starships repair and equipment for the base construction start – were placed into this spacecraft, and were delivered to the Jool orbit
2. After arrival at the specified orbit, the "Many Communications" spacecraft reset the repeaters in different-altitude orbits. This made it possible to create a communication network between the piloted spaceships.
3. At the same time, the construction of the orbital station has begun. The first docking of "Living module of the station" and "Folded station" was successful. The attempt to use the non-pilot module "Collector" fault – the connection with the module was lost.
4. New spaceship was build and delivered – piloted Collector.
5. With the help of a manned collector crew, the station was built (Fig. 10). After completion of the assembly work, the large laboratory started with a cycle of experiments. Figure 10 shows the docking of a "Vertical Take-off Vehicle" to the station.
6. Next, it was decided to begin with exploring gas planet satellites. The orbit of the "Long-haul" was changed – a maneuver was introduced into the orbit of the "Long - haul" to fly to the nearest satellite, Jool-Leit. That was a two-stage maneuver using the Tailor gravity field (Fig. 11).
7. When our spaceship reached the selected satellite orbit, it became evident that 98% of the satellite surface is an ocean. To continue the study spaceship "Vertical Take-off Vehicle" was requested.
8. Numerous studies of this satellite have shown its unsuitability for colonization. There are several reasons: the lack of flat areas for a safe spaceships' landing, the weak atmosphere consisting of toxic gases, permanent seismic activity due to the proximity to the gas giant. Therefore, it was decided to send the cargo vehicle to the next satellite – Vall.
9. Cargo vehicle reached the selected satellite orbit. We decided to undertake only one landing in case of finding the appropriate area for the touchdown.

Fig. 10. Completed station and "Vertical Take-off Vehicle".

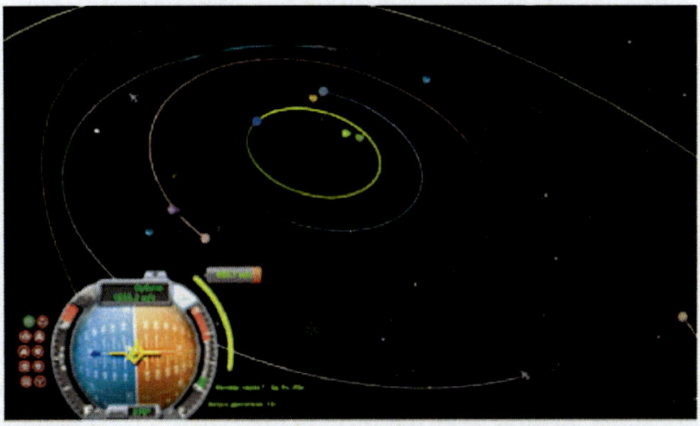

Fig. 11. Two-stage maneuver.

10. After landing, the crew conducted research with instruments. It turned out that this satellite consists of cry volcanoes. Many volcanoes are active, and there is heavy seismic activity, difficult terrain, and large voids under the surface of the planet. It is dangerous to stay on the surface. By this, the study of the second satellite has been completed.
11. After a successful flight to the third satellite, the Lander successfully set in one of the craters. The study with the help of scientific instruments showed: there is a high pressure, weak seismic activity, and the presence of fuel-retaining ore. Visually, the surface has many wide flat areas. The third satellite can become a good platform for fuel production base construction.
12. The approach to the next moon and its inspection showed that this moon is a large asteroid. The surface is very uneven, and has clearly visible traces of collisions with meteorites. The touchdown is quite easy, but there are no prospects for exploration.
13. The last (fifth) satellite. The «Long-haul» successfully flew to its orbit. An inspection of the surface from orbit showed that it is unsuitable for landing – the terrain consists of deep gorges and high sharp rocks.

14. After studying all gas giant satellites, it was decided to build a fuel recovery complex and lay a settlement on Tailo. The construction process of the complex is similar to the process presented in the author's report at the RusAutoCon2019 conference [30].

Summarizing all above, we must make the necessary notes:

- the first impression about a satellite is usually wrong. We need to undertake complex studies for correct decision-making;
- we'll need several spaceships for the implementation of the project, as well as an artificial satellites network;
- since we are planning manned flights, special training for the selected crew members, taking into account the long-range flight and irrevocability, should be provided on using and maintaining of all spaceships types; touchdown on non-atmosphere space objects with different sizes; research procedures on the space objects surface; evacuation.

All these must assure successful project realization.

6 Conclusion

The solution to the problem to which this work is devoted belongs to the distant future. It is necessary to recognize that the modern development of technology and technology does not yet allow implementing such a project in life. Therefore, preliminary modeling is required to assess the feasibility of the project. Here, this problem is solved using the Kerbal Space Program simulator. This program allowed us to get as close as possible to the conditions for real project implementation, to choose the appropriate action program, as well as to determine and select the composition of the necessary equipment and types of spacecraft that should be used at different stages of the gas giant satellites' exploration and utilization. The basic calculations on the choice of flight parameters from the Earth to the Jupiter orbit are given, and the optimal flight trajectory to the satellite of the gas planet is modeled using the gravitational fields forces of space objects along the spacecraft way in order to obtain additional acceleration, which economize the rocket fuel consumption considerably. The specification of spacecrafts and equipment for performing the assigned mission is determined, and their layouts, modeled in the selected program, are demonstrated. The main conditions for the successful implementation of the project are formulated.

References

1. Sharma, A.: Earlier Missions Exploring Jupiter, Technology on Science (2019). https://technologyonscience.blogspot.com/2019/01/missions-jupiter-blogspot.html. Accessed 01 Apr 2021
2. Guillot, T.: A comparison of the interiors of Jupiter and Saturn. Planet. Space Sci. **47**(10–11), 1183–1200 (1999). https://doi.org/10.1016/S0032-0633(99)00043-4

3. Cook, J.-R.C.: Clay-Like Minerals Found on Icy Crust of Europe. Jet Propulsion Laboratory, Pasadena, California, NASA (2013). https://www.jpl.nasa.gov/news/clay-like-minerals-found-on-icy-crust-of-europa. Accessed 01 Apr 2021
4. 50 most interesting facts about the planet Jupiter. http://obshe.net/posts/id2042.html. Accessed 01 Apr 2021
5. Space projects for the study of Jupiter's moons. https://scientifically.info/news/2014-07-01-2820. Accessed 01 Apr 2021
6. The life on Jupiter's moons: is it possible, and when will people go there to live? https://hightech-fm.turbopages.org/hightech.fm/s/2021/01/18/jupiters-moons. Accessed 01 Apr 2021
7. Satellite Io. https://spaceworlds.ru/solnechnaya-sistema/planeta-jupiter/io-sputnik.html. Accessed 03 Apr 2021
8. Jupiter's satellites. https://spacegid.com/sputniki-yupitera.html. Accessed 03 Apr 2021
9. The ocean on Jupiter's moon could support life, NASA planetary scientists say. https://in-space.ru/okean-na-sputnike-yupitera-mozhet-podderzhivat-zhizn-zayavili-planetologi-nasa/. Accessed 03 Apr 2021
10. NASA has selected four possible deep space exploration missions as part of the Discovery program. https://naked-science.ru/article/cosmonautics/nasa-vybralo-chetyre-missii-po-izucheniyu-dalnego-kosmosa-v-ramkah-programmy-discovery. Accessed 15 Apr 2021
11. Korablev, O.: The Jupiter system is a completely inhospitable world. https://ria.ru/20210216/korablev-1597526535.html. Accessed 15 Apr 2021
12. Jupiter Exploration. http://minspace.ru/Education/edu5astr/Jupit/2000s.html. Accessed 03 Apr 2021
13. Chernenkii, A.V.: Space modeling. Kerbal Space Program. Tutorial, St. Petersburg Polytechnic University, p 470 (2017)
14. Jupiter is a formidable giant. http://galspace.spb.ru/index44-jup.html. Accessed 02 Mar 2021
15. Decades of Jupiter exploration. https://rusnasa.ru/main/1859-desyatiletiya-issledovaniya-yupitera.html. Accessed 02 Apr 2021
16. In Russia, the Jupiter satellite is considered as a place to create a habitable base. https://naked-science.ru/article/cosmonautics/v-rossii-rassmatrivayut-sputnik-yupitera-v-kachestve-mesta. Accessed 15 April 2021
17. Jupiter data. https://slideplayer.com/slide/15041456/91/images/3/Jupiter+Data+Extremely+fast+spinning+5+AU%2C+12+years+Largest+Planet.jpg. Accessed 01 Apr 2021
18. Sochi, T.: Introduction to Differential Geometry of Space Curves and Surfaces. Createspace, 1st edn., p. 197 (2017)
19. Konstantinov, M.S., Ngoc, N.D.: Optimization of the trajectory to Jupiter, taking into account the possible temporary shutdown of the engine. Moscow Aviation Institute, MAI 12bf420af73aa42c804a7e115025e9e7.pdf
20. Galileo's journey to Jupiter. https://editorial01.shutterstock.com/wm-preview-1500/2556679a/88b2d537/education-shutterstock-editorial-2556679a.jpg. Accessed 02 Mar 2021
21. Kerbal Space Program (description). https://www.youtube.com/watch?time_continue=7&v=IdIr78zJF3Y. Accessed 18 Apr 2021
22. Chernenkaya, L.V., Desyatirikova, E.N., Chepelevm, S.A., Sergeeva, S.I., Slinkova, N.V.: Optimal planning of distributed control systems with active elements. In: Proceedings 2017 IEEE II International Conference on Control in Technical Systems (CTS), St. Petersburg, Russia, pp. 39–42 (2017). https://doi.org/10.1109/CTSYS.2017.8109482
23. NASA has tested a portable nuclear reactor. https://progress.online/kosmos/3342-nasa-protestirovala-portativnyy-yadernyy-reaktor. Accessed 15 Apr 2021
24. The history of Russian space nuclear installations. https://sdelanounas.ru/blogs/29489/. Accessed 15 Apr 2021
25. DCS engine block. https://pbs.twimg.com/media/D6PJBfiUUAAs6Lz.jpg. Accessed 15 Apr 2021

26. Semikhatov Research and Production Association on Automation. https://www.npoa.ru/about/. Accessed 15 Apr 2021
27. About docking of Soyuz spaceships. https://www.roscosmos.ru/26709/. Accessed 16 Apr 2021
28. Teryoshin, V.N., Sukhanov, D.V.: Device for docking and undocking of space rockets. Patent RU2457985C1 (2010). https://patents.google.com/patent/RU2457985C1/en. Accessed 15 Apr 2021
29. Syromayatnikov, V.: Docking is always an Event. Sci. Life **1**, 44–49 (1988). https://epizodsspace.airbase.ru/bibl/n_i_j/1988/1/1-styk.html. Accessed 15 Apr 2021
30. Desyatirikova, E.N., Chernenkii, A.V.: Design of mobile manufacturing system for extraction of fuel from lunar soil. In: Proceedings 2019 International Russian Automation Conference (RusAutoCon), p. 8, Sochi, Russia, 14 Sept 2019. https://doi.org/10.1109/RUSAUTOCON.2019.8867780

Mathematical Model of Accidental Gas Leakage from Underwater Pipelines

S. Podvalny[1], E. Kutsova[2], and E. Vasiljev[1](✉)

[1] Voronezh State Technical University, 84, 20 letiya Oktyabrya Street, Voronezh 394071, Russia
[2] Subsidiary Open Joint Stock Company Gazproektengineering, Building 3, 34/63, Obrucheva Street, Moscow 117342, Russia

Abstract. The paper deals with the problem of determining the parameters of the transported gas in case of partial destruction of subsea sections of gas pipelines. The solution to this problem is proposed to be obtained on the basis of creating a mathematical model describing the relationship of pressure, velocity and gas temperature at each point of the subsea main pipeline, including at the place of emergency outflow. To build such a model, the equations of the continuity of the gas flow in all sections of the gas pipeline, the equations characterizing the change in the momentum of this flow, as well as the equations of heat content for the molar gas flow rate for the pipeline sections before and after the rupture were used. The process of submerged gas outflow through a partially destroyed pipeline wall is considered as an adiabatic process obeying Saint-Venant's law and occurring at the local speed of sound. For the case of the location of the rupture in the deep-water section of the main pipeline, a mode of subsonic gas outflow is provided as its adiabatic movement through a short pipe without friction. A numerical check of the performance of the proposed model was carried out for a subsea gas pipeline with a variable profile. Quantitative distributions of the characteristics of the gas flow along the main pipeline with fractures with different areas are also obtained. The combination of these distributions makes it possible to automatically identify the parameters of emergency gas outflows in subsea gas pipelines. According to the values of gas pressure, velocity and temperature at the inlet and outlet of the controlled section, the area of local destruction of the main pipeline and the location of the rupture on the route are calculated. These parameters make it possible to form such a control of the compressor stations, which is adequate to the current abnormal state of the pipeline.

Keywords: Mathematical modeling · Subsea gas pipeline · Emergency gas outflow

1 Introduction

The transportation of natural gas by subsea pipelines is currently recognized as a promising method of supplying it to consumers and economically preferable in comparison with the above ground or underground laying of gas pipelines through areas with a high population density and lands for agricultural and environmental purposes [1–5].

A. A. Radionov and V. R. Gasiyarov (Eds.): RusAutoCon 2021, LNEE 857, pp. 165–173, 2022.
https://doi.org/10.1007/978-3-030-94202-1_16

The design and construction of underwater sections of gas pipelines requires a technical and environmental examination of the consequences of emergency gas outflows through pipeline ruptures.

Such an examination presupposes the presence of mathematical models of the corresponding modes of gas transportation [6–12]. This work is devoted to the development of an analytical model of a submerged gas outflow in a subsea gas pipeline, and contains an example of using this model for the numerical study of the specified emergency mode.

2 Problem Statement

Modeling and analysis of subsea gas pipelines in the scientific literature is mainly carried out either for the nominal mode of gas transportation, or for the emergency mode, but limited by the actual rupture zone. In particular, in [13], the model of the main operating mode of a subsea gas pipeline is considered in detail, taking into account the inertial forces during gas movement at high pressures, and the incorrectness of the assumptions about the constancy of the gas density is substantiated.

Articles [14–16] are devoted to the analysis of the gas plume arising from the underwater gas outflow from the rupture and do not affect the process of its transportation along the pipeline. In [17], a limited model is used for calculating the gas leakage rate, built on the basis of experimental data. In [18–21], it is indicated the need for a systematic approach to the analysis of the state of gas in the pipeline as a single system.

The proposed paper differs from the well-known works by the joint consideration of both the process of gas movement along the pipeline and the process of its underwater outflow from the rupture. This consideration of the emergency mode makes it possible to formulate a unified system of thermodynamic equations and obtain quantitative characteristics of the gas state at any point of the pipeline, including in the section of the rupture.

3 Mathematical Model

As a model of gas outflow through the local destruction of the pipeline wall, it is proposed to use a system of equations based on the laws of conservation of mass, momentum and energy of gas at any point of the gas pipeline [22–26], supplemented by relations characterizing the adiabatic outflow of gas without friction through the wall at the point of rupture as through a short pipe:

1. The conditions for the continuity of the gas flow in all sections of the pipeline (the law of conservation of mass):

$$r_1 w_1 S = r_a w_a S; \tag{1}$$

$$r_a w_a S = r_r w_r S_r + r_2 w_2 S; \tag{2}$$

$$r_2 w_2 S = Q. \tag{3}$$

Conditions (1) and (3) correspond to the beginning and end of the considered section of the pipeline, and (2) - to the place of rupture of the pipeline wall.

2. Equations characterizing the change in the amount of gas movement along the pipeline:

$$\rho_1 w_1 S(w_\alpha - w_1) = (P_1 - P_\alpha)S - \lambda \frac{\rho_1 w_1 S}{2D} \int_0^{\alpha \cdot L} w(x)dx - g\rho_1 w_1 S \int_{h_1}^{h_\alpha} \frac{\sin\beta(h)dh}{w(h)}. \tag{4}$$

Equation (4) describes the change in the momentum of the gas mass, numerically equal to its second flow rate $\rho_1 w_1 S$, along the section of the pipeline between its beginning and the place of rupture.

$$\rho_2 w_2 S(w_2 - w_\alpha) = (P_\alpha - P_2)S - \lambda \frac{\rho_2 w_2 S}{2D} \int_{\alpha \cdot L}^{L} w(x)dx - g\rho_2 w_2 S \int_{h_\alpha}^{h_2} \frac{\sin\beta(h)dh}{w(h)}. \tag{5}$$

Equation (5) describes the change in the momentum of the gas along the section of the pipeline between the place of rupture and its end.

Note that Eqs. (4) and (5) do not involve the momentum of a part of the gas flowing through the rupture due to the assumed orthogonality of the gas velocity vectors in the pipeline and in the rupture section.

3. The heat content equations for the molar flow rate (the law of conservation of energy) are also compiled for the pipeline sections before and after the rupture place, but they take into account a part of the total energy (internal, kinetic and potential) taken by the molar flow rate ν_r in the outflow jet (7):

$$\left(C_P T_1 + \frac{\mu w_1^2}{2} + \mu g h_1\right)\nu_1 = \left(C_P T_\alpha + \frac{\mu w_\alpha^2}{2} + \mu g h_\alpha + C_T \frac{\pi D \mu}{\rho_1 w_1 S} \int_0^{\alpha \cdot L} (T(x) - T_{\text{amb}})dx\right)\nu_\alpha, \tag{6}$$

$$(C_P T_\alpha + \frac{\mu w_\alpha^2}{2} + \mu g h_\alpha)\nu_\alpha = (C_P T_r + \frac{\mu w_r^2}{2} + \mu g h_\alpha)\nu_r + (C_P T_2 + \frac{\mu w_2^2}{2} + \mu g h_2 + \\ + C_T \frac{\pi D \mu}{\rho_2 w_2 S} \int_{\alpha \cdot L}^{L} (T(x) - T_{\text{amb}})dx\Bigg)\nu_2. \tag{7}$$

4. The underwater (submerged) gas outflow through a partially destroyed pipeline wall will be considered as an adiabatic process obeying the relation:

$$\frac{T_r}{T_\alpha} = \left(\frac{P_r}{P_\alpha}\right)^{\frac{\gamma-1}{\gamma}}, \tag{8}$$

where T_α, T_r – the thermodynamic temperature of the gas at the point α of the pipeline, located at a distance αL from its beginning (L is the length of the pipeline; $\alpha = [0;1]$), and in the outlet section of the rupture, respectively; P_α, P_r – absolute gas pressure in the indicated sections; γ – the adiabatic exponent.

In the event that at the point of rupture the ratio of the pressure in the pipeline to the external pressure P_{ex} is equal to or greater than the critical value:

$$\frac{P_\alpha}{P_{ex}} = \left(\frac{\gamma+1}{\gamma}\right)^{\frac{\gamma}{\gamma-1}}, \tag{9}$$

then there is an outflow that obeys Saint-Venant's law:

$$\frac{P_r}{P_\alpha} = \left(\frac{2}{\gamma+1}\right)^{\frac{\gamma}{\gamma-1}}, \tag{10}$$

and occurring with the local speed of sound w_r:

$$w_r = \sqrt{\frac{\gamma R T_r}{\mu}}, \tag{11}$$

where R – the universal gas constant; μ – the molar mass of the gas.

Since modern pipelines are laid along deep-water sea sections, and in case of emergency outflows, the gas pressure in the pipeline, including at the point of rupture α, significantly decreases compared to the nominal, then the case of failure to meet the critical ratio should be taken into account (9).

For this case, it is proposed to consider the mode of subsonic gas outflow as its adiabatic motion through a short pipe without friction.

Then, in accordance with the law of conservation of gas energy during its outflow through the wall, we obtain the relations (see also (7)):

$$C_P T_\alpha = C_P T_\alpha + \frac{\mu w_r^2}{2}; \tag{12}$$

$$P_r = P_{amb}, \tag{13}$$

A complete list of the designations used in expressions (1)–(13) and their dimensions:

g – acceleration of gravity, m/s^2;

T_2, T_α, T_r, $T(x)$, T_{amb} – thermodynamic gas temperature: at the end of the section; at point α of pipeline rupture; in the outer section of the rupture; current temperature at a distance of x meters from a certain reference point; average ambient temperature along the length of the section, K, respectively;

w_1, w_α, w_r, w_2 – gas velocity in the above mentioned sections of the pipeline and at the end of the section, m/s;

P_1, P_α, P_r, P_2 – absolute gas pressures at the beginning of the pipeline, in the place of its damage, in the outlet section of the rupture and at the end of the pipeline, Pa;

h_1, h_α, h_2 – leveling levels of the beginning of the pipeline, the place of rupture and the end of the pipeline, m;

D – inner diameter of the pipeline, m;

S, S_r – internal cross-sectional area of the pipeline and the destroyed part of the wall, m^2;

L – pipeline length along its geometric axis, m;

L_α – distance from the beginning of the pipeline to the point of rupture, m; $L_\alpha = \alpha L$, $\alpha = [0;1]$;

Q_1, Q_r, Q – gas consumption at the beginning of the pipeline, through the rupture section, and at the end of the pipeline, kg/s; $Q_1 = Q_r + Q$;

C_P – molar heat capacity of gas at constant pressure, J/(mol·K);

C_T – heat transfer coefficient from gas to external environment, J/(s·m^2·K);

ν_1, ν_α, ν_r, ν_2 – molar gas flow rates in the corresponding sections, mol/s, determined from the expressions:

$$\nu_1 = \frac{\rho_1 w_1 S}{\mu}; \quad \nu_\alpha = \frac{\rho_\alpha w_\alpha S}{\mu} \quad \ldots, \quad \nu_2 = \frac{\rho_2 w_2 S}{\mu}, \tag{14}$$

where ρ_1, ρ_α, ρ_r, ρ_2, – gas density in the considered sections, kg/m^3; μ – molar mass of gas, kg/mol;

$$\rho_1 = \frac{\mu}{V\mu_1}; \quad \rho_\alpha = \frac{\mu}{V\mu_\alpha} \quad \ldots, \quad \rho_2 = \frac{\mu}{V\mu_2}, \tag{15}$$

$V_{\mu 1}$, $V_{\mu\alpha}$,..., $V_{\mu 2}$ – molar volumes, m^3/mol;

$$V\mu_1 = V\mu_n \frac{P_n T_1}{P_1 T_n}; \ldots; V\mu_2 = V\mu_n \frac{P_n T_2}{P_2 T_n}, \tag{16}$$

$V_{\mu n}$ – molar volume of gas at temperature $T_n = 273$ K and pressure $P_n = 1.01 \cdot 10^5$ Pa, m^3/mol; γ – adiabatic exponent; R – universal gas constant, J/(mol·K); β – current angle of inclination of the pipeline axis to the horizon, rad; λ – internal viscous friction coefficient in gas flow.

4 Numerical Simulation Results

Let us consider the results of the numerical solution of Eqs. (1)–(13) using the difference scheme [27–29] using the example of a gas pipeline with the following parameters:

$w_1 = 4$ m/s; $w_2 = 6$ m/s; $P_1 = 25 \cdot 10^6$ Pa; $P_2 = 14{,}5 \cdot 10^6$ Pa; $T_1 = 323$ K; $T_2 = 281$ K; $T_{amb} = 278$ K; $C_T = 0.909$ J/(s·m^2·K); $\lambda = 0.010$; $\beta = -1.2/200$ rad $= (-1.08/\pi)$ deg at x $\leq 200 \cdot 103$ m; $\beta = 1.2/200$ rad $= (1.08/\pi)$ deg at $x > 200 \cdot 10^3$ m; $L = 400 \cdot 10^3$ m; $S = 0.2$ m^2; $D = 0.505$ m; $C_P = 35.6$ J/(mol·K); $C_V = 27.3$ J/(mol·K); $R = 8.31$ J/(mol·K); $\mu = 16.04 \cdot 10^{-3}$ kg/mol; $V_{\mu n} = 22.4 \cdot 10^{-3}$ m^3/mol; $Q = \rho_1 w_1 S = 119.85$ kg/s; $h_1 = h_2 = 0$; $\alpha = 0.4$; $h_\alpha = -960$ m; $P_{ex} = 97.5 \cdot 10^5$ Pa.

The profile of the subsea section of the gas pipeline is shown in Fig. 1.

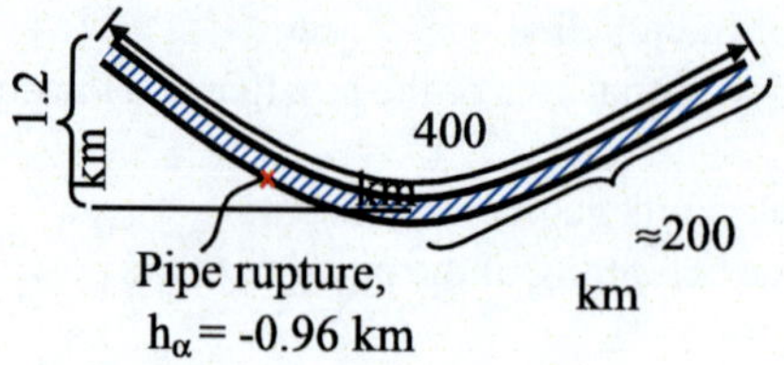

Fig. 1. Profile of the subsea section of the gas pipeline.

The distributions of the main thermodynamic characteristics of the gas: velocity w, temperature T and pressure P along the pipeline for various values of the rupture area S_r are shown in Fig. 2, Fig. 3, Fig. 4, where the indicated values without the index correspond to the normal mode of gas transportation, and with the *emer* index - to the emergency mode of operation with the destruction of the S_r area at the point α.

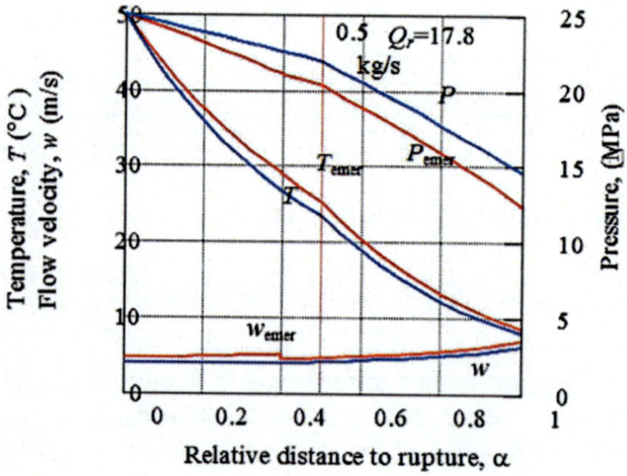

Fig. 2. Distribution of gas characteristics during sound outflow through rupture with an area Sr = 5 cm^2.

When the rupture area $S_r \approx 16$ cm^2 is reached, a transition to the subsonic outflow mode occurs (Table 1 and Fig. 4).

Note that, in all flow regimes, the following conditions were met: $T_1 = \text{const}$, $P_1 =$ const and $Q = \text{const}$

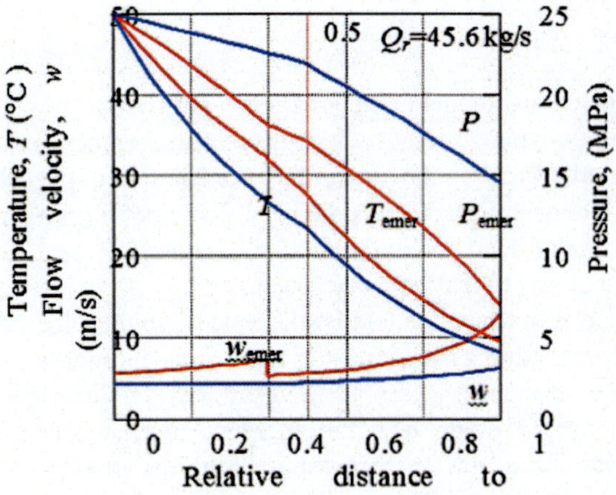

Fig. 3. Distribution of gas characteristics during sound outflow through rupture with an area $S_r = 15\ cm^2$.

Table 1. Gas characteristics in the rupture section.

Gas characteristics			
Rupture area, S_r (cm^2)	*Gas consumption, Q_r* (kg/s)	*Gas velocity in the outer section of the rupture, w_r* (m/s)	*Pressure, P_r* (MPa)
5	17.8	420.6	11.5
10	33.0	421.9	10.7
15	45.6	422.7	9.85
17	49.7	413.8	9.75
20	55.5	397.8	9.75

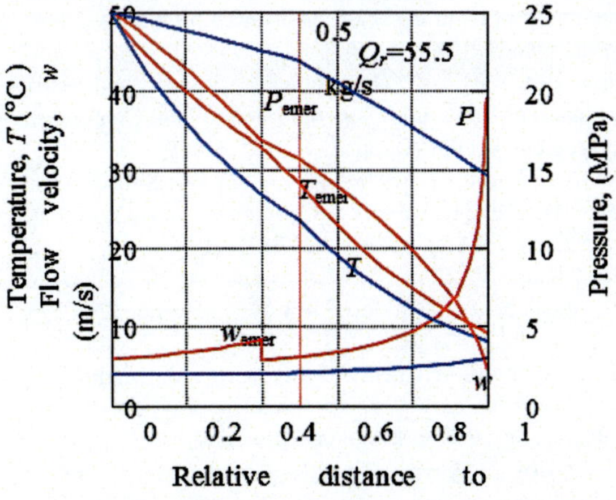

Fig. 4. Distribution of gas characteristics during subsonic outflow through rupture with an area $S_r = 20\ cm^2$.

5 Conclusions

The proposed mathematical model of an emergency gas outflow in subsea gas pipelines is based on the thermodynamic relationship of pressure, temperature and gas velocity along the entire length of the pipeline, including the location of the submerged rupture. Establishing this relationship makes it possible to conduct a joint analysis of the gas movement in the pipeline itself and in the section of its rupture, and, as a result, ensures the adequacy of the model to real physical processes.

For cases of deep-water pipeline laying, the model provides for a description of the subsonic gas outflow regime as its adiabatic movement through a short pipe without friction.

The results of numerical modeling of the emergency outflow process made it possible to obtain the distribution of gas characteristics along the pipeline and in the rupture section for sonic and subsonic outflow.

In particular, the existence of a critical value of the discontinuity area separating the regions of subsonic and sonic outflows is shown.

The combination of these distributions makes it possible to automatically identify the parameters of emergency gas outflows in subsea gas pipelines. According to the values of gas pressure, velocity and temperature at the inlet and outlet of the controlled section, the area of local destruction of the main pipeline and the location of the rupture on the route are calculated. These parameters make it possible to form such a control of the compressor stations, which is adequate to the current abnormal state of the pipeline.

References

1. Shilin, M., et al.: Environmental safety of the Nord Stream 2 marine gas pipeline (Rus. section). In: 2018 IEEE/OES Baltic International Symposium (BALTIC), pp. 1–8. IEEE (2019)
2. El-Reedy, M.A.: Subsea pipeline design and installation. In: Offshore Structures (Second Edition). Elsevier, Gulf Professional Publishing, pp. 609–647 (2020)
3. Relva, S.G., Oliveira da Silva, V., Peyerl, D., Gimenes, A.L.V., Udaeta, M.E.M.: Regulating the electro-energetic use of natural gas by gas-to-wire offshore technology: case study from Brazil. Utilities Policy **66**, 101085 (2020)
4. Hai, C., Xiaojian, J., Bin, Y., Bin, Z.: The application of safety and security system in the long distance landing subsea pipeline. In: 2019 16th International Conference on Service Systems and Service Management (ICSSSM), pp. 1–4. IEEE (2019)
5. Mai, C., Pedersen, S., Hansen, L., Jepsen, K.L., Yang, Z.: Subsea infrastructure inspection: a review study. In: 2016 IEEE International Conference on Underwater System Technology: Theory and Applications (USYS), pp. 71–76. IEEE (2017)
6. Tian, Y., Liu, M., Zhang, S., Zhou, T.: Risk assessment of submarine pipelines based on multi-source data model. In: 2019 Chinese Control Conference (CCC), pp. 7192–7196. IEEE (2019)
7. Kenny, S.: Offshore pipelines—elements of managing risk. Methods Chem. Process Saf. **2**, 289–325 (2018)
8. Poberezhnyi, L., Maruschak, P., Prentkovskis, O., Danyliuk, I., Pyrig, T., Brezinová, J.: Fatigue and failure of steel of offshore gas pipeline after the laying operation. Arch. Civ. Mech. Eng. **16**(3), 524–536 (2016). https://doi.org/10.1016/j.acme.2016.03.003
9. Li, X., Yang, M., Chen, G.: An integrated framework for subsea pipelines safety analysis considering causation dependencies. Ocean Eng. **183**, 175–186 (2019)

10. Li, X., Chen, G., Chang, Y., Xu, C.: Risk-based operation safety analysis during maintenance activities of subsea pipelines. Process Saf. Environ. Prot. **122**, 247–262 (2019)
11. Behari, N., Sheriff, M.Z., Rahman, M.A., et al.: Chronic leak detection for single and multiphase flow: A critical review on onshore and offshore subsea and arctic conditions. J. Nat. Gas Sci. Eng. **81**, 103460 (2020)
12. Ismail, S., Najjar, S., Sadek, S.: Reliability of offshore pipelines subject to upheaval buckling. In: 2015 International Mediterranean Gas and Oil Conference (MedGO), pp. 1–4. IEEE (2015)
13. Kurbatova, G.L., Ermolaeva, N.N.: The mathematical models of gas transmission at hyperpressure. Appl. Math. Sci. **8**(121–124), 6191–6203 (2014)
14. Xinhong, L., Guoming, C., Renren, Z., Hongwei, Z., Jianmin, F.: Simulation and assessment of underwater gas release and dispersion from subsea gas pipelines leak. Process Saf. Environ. Prot. **119**, 46–57 (2018)
15. Wu, K., Cunningham, S., Sivandran, S., Green, J.: Modelling subsea gas releases and resulting gas plumes using computational fluid dynamics. J. Loss Prev. Process Ind. **49B**, 411–417 (2017)
16. Yousef, Y.A., Imtiaz, S., Khan, F.: Subsea pipelines leak-modeling using computational fluid dynamics approach. J. Pip. Syst. Eng. Pract. **12**(1), 04020056 (2021)
17. Liu, C., Liao, Y., Wang, S., Li, Y.: Quantifying leakage and dispersion behaviors for sub-sea natural gas pipelines. Ocean Eng. **216**, 108107 (2020)
18. Syed, M.M., Lemma, T.A., Seshu, S.K., Ofei, T.N.: Recent developments in model-based fault detection and diagnostics of gas pipelines under transient conditions. J. Nat. Gas Sci. Eng. **83**, 103550 (2020)
19. Podvalny, S.L., Vasiljev, E.M.: Multi-alternative control of large systems. In: 13th International Scientific-Technical Conference on Electromechanics and Robotics "Zavalishin's Readings", MATEC Web of Conferences, vol. 161, p. 02003 (2018)
20. Podvalny, S.L., Vasiljev, E.M.: A multi-alternative approach to control in open systems: origins, current state, and future prospects. Autom. Remote Control **76**(8), 1471–1499 (2015)
21. Podvalny, S.L., Vasiljev, E.M., Barabanov, V.F.: Models of multi-alternative control and decision-making in complex systems. Autom. Remote Control **75**(10), 1886–1891 (2014). https://doi.org/10.1134/S0005117914100166
22. Cleaver, R.P., Halford, A.R.: A model for the initial stages following the rupture of a natural gas transmission pipeline. Process Saf. Environ. Prot. **95**, 202–214 (2015)
23. Agarwal, A., Molwane, O.B., Pitso, I.: Analytical investigation of the influence of natural gas leakage & safety zone in a pipeline flow. Mater. Today: Proc. **31–1**, 547–552 (2021)
24. Wang, S., Wu, G., Wang, Y., Zhou, Y., Qi, H., Li, D.: Transient modeling of natural gas flow in double-skin pipeline leakage. In: Better Pipeline Infrastructure for a Better Life, International Conference on Pipelines and Trenchless Technology, Wuhan, China, Proceedings Article published, pp. 854–860 (2012)
25. Alamian, R., Behbahani-Nejad, M., Ghanbarzadeh, A.: A state space model for transient flow simulation in natural gas pipelines. J. Nat. Gas Sci. Eng. **9**, 51–59 (2012)
26. Bhardwaj, U., Teixeira, A.P., Soares, C.G.: Uncertainty quantification of burst pressure models of corroded pipelines. Int. J. Press. Vessels Pip. **188**, 104208 (2020)
27. Seleznev, V.E.: Numerical recovery of gas flows in pipeline systems. J. Appl. Math. **2012**, 180645 (2012)
28. Pryalov, S.N., Seleznev, V.E.: Special numerical analysis methods for mathematical models of gas flow in trunkline network. Appl. Math. Sci. **8**(33–36), 1763–1780 (2014)
29. Seleznev, V.E.: Numerical gas flow recovery at complex pipeline network. Appl. Math. Sci. **7**(17–20), 945–958 (2013)

Check for updates

Models for Determining the Cost of Services for the Reduction of Power Losses in a Network Organization with Reactive Power Compensation in a Consumer Network

A. Kuznetsov(✉) and D. Rebrovskaya

Ulyanovsk State Technical University, 32, Northern Venets Street, Ulyanovsk 432027, Ulyanovsk Region, Russia
kav2@ulstu.ru

Abstract. The article shows that accurate models for determining the cost of services for the reduction of power losses in a network organization with reactive power compensation (RPC) in a consumer network did not exist for a long time. The development of such models will help to increase the quality of the state management in terms of compensating the consumer for the costs of installing and operating compensating devices (CD). Currently, in order to carry out a feasibility study for the installation of a CD of a certain capacity, the consumer can use, depending on the capabilities, a mathematical or a programme model, or a polynomial model proposed in this article. The creation and the improvement of the model are associated with the need to activate consumers in terms of the PRC.

Keywords: Reactive power compensation · Reduction of power losses · Network organization · Electricity consumer · Mathematical model · Power coefficient

1 Introduction

When compensating devices (CD) are installed in the network of a grid company, the reduction of energy losses is carried out only in the company network [1]. In the consumer's network the levels of energy losses remain the same. Therefore the power of CDs is not used effectively [2, 3]. The distribution of the same CD power in the consumer network increases the effect of reducing losses. Unfortunately, organizational and legal problems associated with the departmental affiliation of networks to various owners hinder an effective solution for the reactive power compensation (RPC). There were several attempts to create a control mechanism for RPC in the Russian Federation. However, the implementation of this mechanism failed [4].

The reason for this is the legal barriers associated with the application of the stimulating tariff in the form of discounts or surcharges to the basic tariff for the consumption of reactive power and electricity [4]. An analysis of literary sources shows that one of the reasons for the legal inconsistency of the mechanism is the lack of an accurate, simple

A. A. Radionov and V. R. Gasiyarov (Eds.): RusAutoCon 2021, LNEE 857, pp. 174–182, 2022.
https://doi.org/10.1007/978-3-030-94202-1_17

and accessible tool for consumers to determine the cost of services for the reduction of losses in a network organization.

The installation of a CD is associated with the implementation of an investment project. If compensating devices are used in the consumer network, then the consumer acts as the employer of the investment project. He pays all expenses associated with the implementation of the project [5, 6].

The lack of an accurate model for determining annual savings leads investors to doubt the return on the funds spent on the installation of a CD in the consumer network. As a consequence, the installation of compensating devices is not carried out and the network losses fail to be reduced [7].

The tool for determining the annual savings and the cost of services to reduce power losses can be based on the model of an adjacent network organization.

2 Methods

The implementation of the installation project is evaluated according to a number of technical and economic indicators. The most important of them are the payback period PB, the net present value NPV, the index of profitability PI, the internal rate of return IRR, the discounted payback period DPB. The investor makes a decision on financing the project based on an analysis of these indicators. Should the investor be dissatisfied with the indicators, such a decision is not made. The installation of compensating devices is not carried out and the network losses are not reduced. The calculation of project performance indicators includes the amount of annual savings, which determines their values. The higher the values of annual savings, the higher the values of indicators, hence, the higher the probability of a positive decision on financing the project.

Annual savings, which consist of the value of losses reduction resulting from the installation of a CD, are a profitable part of the investment project regarding the installation of a CD. In this case, annual savings are represented by the sum of the losses reduction value in the consumer's network and of losses reduction value in the network organization. The losses reduction value in a network organization is the cost of the losses reduction service in the network organization. Calculating the annual savings in the consumer network presents no difficulties. To do this, the consumer can obtain information about the network configuration, loads, operating modes, etc. The calculation of the annual savings in a network organization for a consumer is associated with the need to obtain a large amount of information from an adjacent network organization and to carry out calculations.

These calculations are quite complex and require a large amount of the initial data on the network configuration, branch resistances, loads in the nodes of the substitution circuit, as well as time. As a rule, such data is not available to the consumer. In some cases, it may constitute a trade secret. Consumers do not have the opportunity to make calculations that make it possible to evaluate annual savings as a result of reducing electricity losses in an adjacent network organization.

In such cases, it becomes necessary to model networks that are outside the scope of the calculation [8]. The use of models has an advantage over classical calculations. They are built and operated in the conditions of limited information, which is easier to obtain from an adjacent network organization, and they make it possible not only to make specific calculations, but also to conduct research, analyze and identify regular relationships between the parameters of the organization, and create simplified engineering methods.

Such a model for assessing the reduction of power losses in a grid of a network organization when installing a CD in consumer networks is described in [8]. In this model, the electric network is reduced to an equivalent circuit consisting of one equivalent consumer, with a power equal to the sum of the powers of all consumers connected to the considered electric network, of one equivalent resistance of the supply branch connected to the power source. This well-known model was created and used during the planned economy. There was no need to think about a specific contribution of each individual consumer to the reduction of losses. The average values of the parameters of the RPC for consumers quite satisfied both the consumers and the network organization. With the transition to market relations, taking into account the legal aspects of the consumer, it became necessary for each consumer to know their specific contribution to the reduction of losses in the network organization. There was a need to develop a new mathematical model. For the first time such a model was proposed in [9, 10].

The mathematical model [9] is an equivalent circuit of a network organization in the form of three resistances connected by a "star" scheme. One of the resistances is the resistance of branch 1, feeding the considered consumer of electricity with power P1; the second resistance is the equivalent resistance of branch 2, feeding an equivalent consumer, with power P2 equal to the sum of the powers of all consumers in the network organization, except for the power of the first one. The third resistance is the equivalent resistance of the supply branch 3 and is connected to a power supply.

The expression for determining the reduction of power losses $\delta \Delta P^*{}_{CD}$ in accordance with [9] has the following form:

$$\delta \Delta P^*_{CD} = 1 - \left(\frac{(1 + tg^2\phi_{1new})}{(1 + tg^2\phi_1)} \cdot \frac{\Delta P_1}{\sum_1^3 \Delta P_i} + \frac{\Delta P_2}{\sum_1^3 \Delta P_i} + \frac{(1 + tg^2\phi_{3new})}{(1 + tg^2\phi_3)} \cdot \frac{\Delta P_3}{\sum_1^3 \Delta P_i} \right) \tag{1}$$

where ΔP_1, ΔP_2, ΔP_3 are losses in the branches of the equivalent circuit before the RPC; ΔP_{1new}, ΔP_{2new}, ΔP_{3new} are losses in the branches of the equivalent circuit after RPC (new); $tg\varphi_1$, $tg\varphi_3$ are power factors in the branches of the circuit before the RPC; $tg\varphi_{1new}$, $tg\varphi_{3new}$ are power factors after RPC.

It was proposed to adopt the following notations in [9]:

$$P_1^* = \frac{P_1}{P_1 + P_2}, \quad Q_{CD}^* = \frac{Q_{CD}}{Q_1}, \tag{2}$$

where P_{1*} is a relative value (in the form of a fraction) of the consumer power in the first branch; Q_{CD*} is the degree of reactive compensation power at the consumer in the same branch due to the installation of CD.

According to the accepted notations, the quantities in the expression (1) will be calculated as follows:

$$tg\phi_{1new} = (1 - Q_{CD}^*) tg\phi_1, \tag{3}$$

$$tg\phi_3 = P_1^* \cdot tg\phi_1 + (1 - P_1^*)tg\phi_2, \tag{4}$$

$$tg\phi_{3new} = P_1^*(1 - Q_{CD}^*)tg\phi_1 + (1 - P_1^*)tg\phi_2. \tag{5}$$

Through transformations the model variables are represented as follows:

$$P_1^*;\ \frac{\Delta P_1}{P_1};\ \frac{\Delta P_2}{P_2};\ \frac{\Delta P_3}{P_1 + P_2};\ tg\phi_1;\ tg\phi_2;\ Q_{CD}^*. \tag{6}$$

As shown in [11], adjacent consumers can change the consumption of reactive power by installing new CDs or turning them off as a result of failure or dismantling. Each change requires a new calculation, the results of which will differ from the previous one. The model does not take into account the dynamics of the changes of the RPC status in a network organization. In [12] it is proposed to transform the parameters of the equivalent circuit taking into account the possible dynamics of changes in the parameters of the RPC for adjacent consumers.

The transformation is carried out to the form in which all possible changes in the reactive power flows from the RPC in the networks of adjacent consumers will have already occurred. This is possible for the moment in time when all consumers will have fulfilled the conditions of the PRC, prescribed by regulatory documents. These conditions are the normalized values of the degree of the RPC for each consumer.

The creation and improvement of the model is associated with the need to activate consumers in terms of the RPC. First of all, it is necessary to inform wider scientific communities, electricity consumers and investors about the existence of the model. The purpose of this activity is to reduce electricity losses in the electricity system. The achievement of the goal is facilitated by the improvement of the model in the direction of its simplification and the creation of simplified engineering techniques.

In many cases, the consumer is unable to use either the mathematical or the programme model. Such consumers need a simple and affordable tool that does not require high qualifications, significant time for adoption, or material resources for the training and for purchasing the program. Such a tool can be a linear polynomial model created on the basis of the programme model.

The simplification of the model is possible through transforming the mathematical model into a polynomial one by applying the mathematical theory of experimental design [13]. To do this, a series of computational experiments are carried out using a mathematical model with a certain combination of upper and lower limits for the variation of factors. Based on the results of the computational experiments, the coefficients of the polynomial model are calculated and checked for the adequacy of the programme model.

3 Result

The analysis of the mathematical model made it possible to reduce the number of variables included in its composition from seven to four, to change the set of variables, to introduce restrictions on the intervals of variation of factors, and to determine constants.

As a result of the analysis, the mathematical model is transformed to a form:

$$\delta \Delta P_{CD}^{*} = f\left(P_1^*;\ \frac{\Delta P_1}{P_1};\ \frac{\Delta P_2}{P_2};\ \frac{\Delta P_3}{P_1 + P_2};\ tg\varphi_{1,NAT}\right) \tag{7}$$

The following values are accepted as the model constants:

$$tg\phi_2 = tg\phi_{2,\,\mathrm{lim}} = 0,4; \tag{8}$$

$$tg\phi_{1,\,\mathrm{HOB}} = tg\phi_{1,\,\mathrm{lim}}(0,35;\ 0,4). \tag{9}$$

ΔP_1, ΔP_2, ΔP_3 in (7) are losses in the branches of the equivalent circuit before RPC; $P_1{}^*$ is a relative value (in the form a fraction) of consumer power in the first branch $P_1^* = \frac{P_1}{P_1+P_2}$.

The variable $tg\phi_{1,\,\mathrm{new}} = tg\phi_{1,\,\mathrm{lim}}$, the value of which is taken as a constant, is the value of the power factor that the consumer aims at when installing a CD. If the consumer receives power at a voltage below 1 kV, its value should be taken as a constant equal to 0.35. At a voltage of 1–35 kV this value becomes equal to 0.4 [14].

In expression (7) $tg\phi_{1,NAT}$ is the natural power factor of the first consumer.

As a result of the analysis of the domain of the factors, we obtain a complete set of variables with ranges of change for which cases of using incorrect parameter values that go beyond the limits of real values will be excluded in the polynomial model. The set of variables with their ranges of change will be as follows:

$$\begin{array}{l} 0.05 \le P_1^* \le 1; \quad 0.01 \le \frac{\Delta P_1}{P_1} \le 0.05; \quad 0.01 \le \frac{\Delta P_2}{P_2} \le 0.05; \\ 0.01 \le \frac{\Delta P_3}{P_1+P_2} \le 0.05; \quad 0.6 \le tg\varphi_{1,\,NAT} \le 1. \end{array} \tag{10}$$

The algorithm of the programme model (7) is presented in Fig. 1.

According to the mathematical model algorithm, a series of computational experiments were carried out. The values of each of the factors in the experiment were set in accordance with the planning matrix of a full-factor experiment for five factors.

A full-factor computational experiment was conducted for each constant value $tg\phi_{1,\,\mathrm{new}} = tg\phi_{1,\,\mathrm{lim}}$, and a regression equation was compiled. Based on the results of a computational experiment using a programme model, two polynomial regression equations are obtained with constant values $tg\phi_{1,\,\mathrm{new}} = tg\phi_{1,\,\mathrm{lim}} = 0.35$; 0.4.

The form of each of the equations is presented as follows:

$$\delta \Delta P_{CD}^{*} = a_0 + \sum_{i=1}^{n} a_i \cdot x_i + \sum_{i=1}^{n} \sum_{j=i+1}^{n} a_{ij} \cdot x_i \cdot x_j \tag{11}$$

In these equations: a_0, a_i are the free term and the coefficients of the regression equation in named units a_0, a_1, a_2, a_3, a_4, a_5; a_{ij} is the coefficient of the pair interaction a_{12}, a_{13}, a_{14}, a_{15}, a_{23}, a_{24}, a_{25}, a_{34}, a_{35}, a_{45}; n is the number of coefficients in the regression equation, where $n = 5$; x_i, x_j, x_l are values of the factors in coded units $i = 1, 2, 3, 4, 5$, $i + 1 \le j \le 5$, $j + 1 \le l \le 5$.

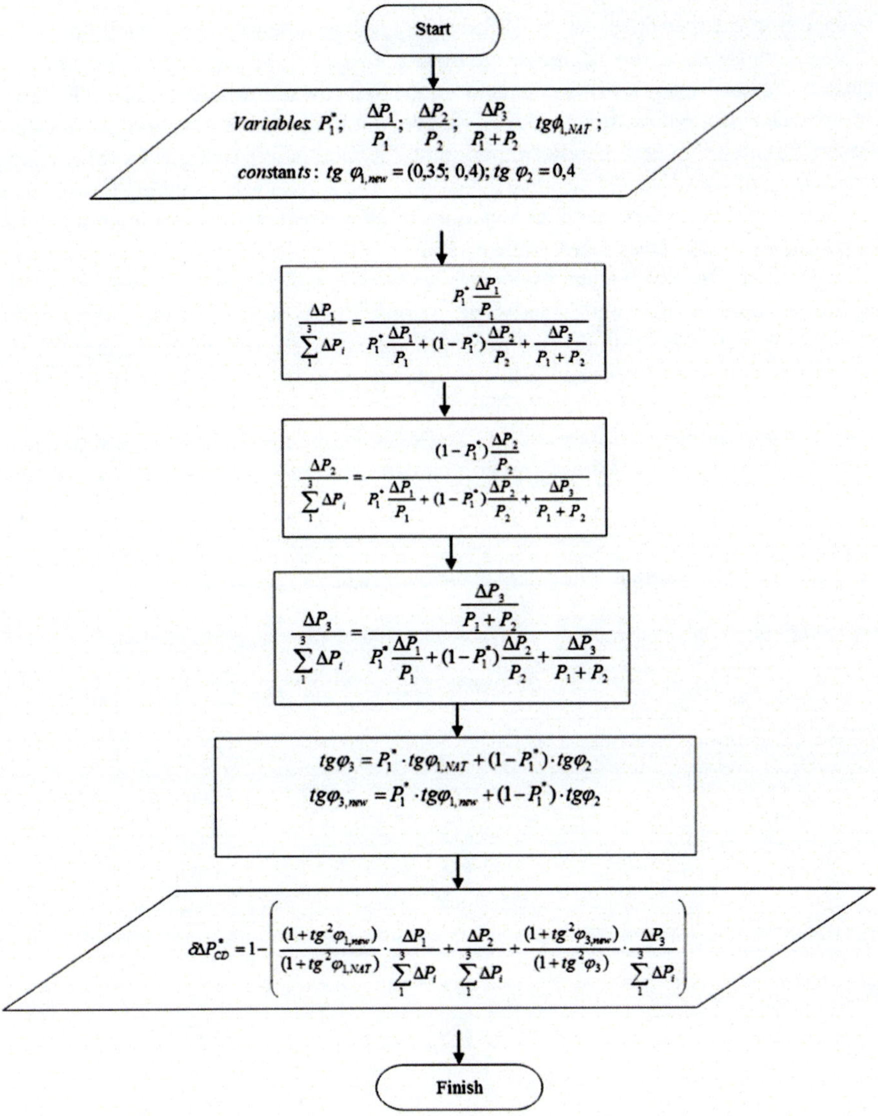

Fig. 1. Block diagram of the calculation algorithm $\delta \Delta P^*_{KY}$ with a reduced number of variables.

Equations in the form (9) can be used for the calculation of $\delta \Delta P^*_{CD}$ with the value of the constant $tg\phi_{2,} = tg\phi_{2,\lim} = 0,4$. The coefficients of the regression equations obtained by the computational experiments are presented in Table 1.

For the final statement about the adequacy of these equations, it is necessary to assess the statistical significance of the coefficients according to the Student's test and to check for adequacy according to the Fisher test [15]. Suppose that the variance is determined by the spread of the parameters of the object in a given tolerance field. To determine the variance of the experiment, the so-called parallel experiments were carried out for an arbitrarily chosen first row of the design matrix. In such cases, the main level of factor values is the level specified for each factor in the first row of the design matri. The lower and upper levels are determined taking into account the possible variation in the values of each parameter. In each parallel experiment, the factors were assigned a value taking into account the positive and negative deviations. The spread was set taking into account the possible field of tolerances and deviations of all independent factors. For all factors, the possible deviation was taken equal to 3%.

In each parallel experiment, the factors were assigned a value at a given level, taking into account positive and negative deviations. Based on the results of the parallel experiments, the variance in the first row of the planning matrix and the variance of the experiment were calculated, and also the Student criterion, the variance of adequacy, and the Fisher criterion were determined.

As a result, all the coefficients of the regression equations appear to be significant. Another conclusion is that the regression equations are adequate to the mathematical model.

Table 1. The values of the coefficients of the model.

The coefficients of the model	The values of the coefficients of the model $\delta\Delta P^{*}_{CD} = a_0 + \sum_{i=1}^{n} a_i \cdot x_i + \sum_{i=1}^{n}\sum_{j=i+1}^{n} a_{ij} \cdot x_i \cdot x_j$	
	$tg\phi_{1,new} = 0,35$	$tg\phi_{1,new} = 0,4$
a_0	16,234	14,995
a_1	14,435	13,358
a_2	0,324	0,300
a_3	−0,327	−0,298
a_4	−0,078	−0,077
a_5	6,997	7,225
a_{12}	−0,324	−0,300
a_{13}	0,327	0,298
a_{14}	0,078	0,077
a_{15}	6,209	6,422
a_{23}	−0,124	−0,115
a_{24}	−0,132	−0,122
a_{25}	0,139	0,144
a_{34}	0,132	0,122
a_{35}	−0,143	−0,146
a_{45}	−0,031	−0,034

4 Conclusion

Any of the models reviewed can be used as the basis for developing an accurate, simple and affordable tool for the consumer to determine the cost of services for the reduction of power losses in a network organization. The models of an adjacent network organization are convenient because they deal with a limited amount of information. For a long time, such models did not exist. Currently, in order to carry out a feasibility study for the installation of a CD of a certain capacity, the consumer can use, depending on the capabilities, a mathematical or a programme model, or a polynomial model [16].

The creation of a mathematical model and its improvement make it possible to increase the quality of the government management and avoid legal issues associated with the implementation of the mechanism of compensating consumers for the cost of installing and operating a CD; to increase the accuracy of determining the annual savings of the investment project regarding the installation of a CD, and provide the investor with confidence in the return of the funds spent on the project.

Acknowledgement. This work is supported by the Russian Federal Property Fund (contract № 18-48-730025/18 of June 11, 2018).

References

1. Bakshaeva, N.S., Suvorova, I.A., Cherepanov, V.V.: Voltage quality improving in power distribution networks with abruptly variable load by application of reactive power series compensation devices. In: 2017 International Conference on Industrial Engineering, Applications and Manufacturing (ICIEAM), St. Petersburg, pp. 1–5 (2017). https://doi.org/10.1109/ICIEAM.2017.8076281, https://ieeexplore.ieee.org/document/8076281
2. Vorotnitskiy, V.E.: Energy efficiency and reactive power compensation in electrical networks. Probl. Solutions Energosovet **1**(47), 44–54 (2017)
3. Vorotnitskiy, V.E., Shakaryan, Y., Sokur, P.V.: On the development and coordination of services for reactive power compensation. Energoekspert **5**(40), 32–37 (2013)
4. Kuznetsov, A.V.: The use of incentive tariffs to improve the quality of electric power and reduce the flow of reactive power in the electric power system. Electrics **12**, 12–17 (2012)
5. Fei, D., et al.: Research on the configuration strategy for reactive power compensations in offshore wind farm clusters. In: 4th International Conference on Power and Renewable Energy (ICPRE), Chengdu, China, pp. 127–131 (2019). https://doi.org/10.1109/ICPRE48497.2019.9034811, https://ieeexplore.ieee.org/document/9034811
6. Mestrineretal, D.: ITER reactive power compensation systems: analysis on reactive power sharing strategies. In: IEEE International Conference on Environment and Electrical Engineering and 2019 IEEE Industrial and Commercial Power Systems Europe (EEEIC/I&CPS Europe), Genova, Italy, pp. 1–6 (2019). https://doi.org/10.1109/EEEIC.2019.8783219, https://ieeexplore.ieee.org/document/8783219
7. Tian, Y., Li, Z.: Research status analysis of reactive power compensation technology for power grid. In: Condition Monitoring and Diagnosis (CMD), Perth, WA, pp. 1–7 (2018). https://doi.org/10.1109/CMD.2018.8535641, https://ieeexplore.ieee.org/document/8535641
8. Zhelezko, Y.S.: Reactive Power Compensation and Power Quality Improvement, p. 224. Energoatomizdat, Moscow (1985)

9. Kuznetsov, A.V., Argentova, I.V.: The mathematical model for evaluating the reduction of power losses in a network organization with reactive power compensation in a consumer network. Electr. Eng. **10**, 68–73 (2016)
10. Kuznetsov, A.V., Argentova, I.V., Rebrovskaya, D.A.: The programme model for evaluating the reduction of power losses in a network organization while reactive power compensation is in a consumer network. Ind. Energy **6**, 48–54 (2016)
11. Kuznetsov, A.V., Rebrovskaya, D.A.: Process control of reactive power compensation in electrical power system. In: Matematicheskie metody v tekhnike i tekhnologiyakh – MMTT-31 XXXI Mezhdunarodnaya nauchnaya konf. St. Petersburg, pp. 71–74 (2018)
12. Kuznetsov, A.V., Rebrovskaya, D.A.: Refinement of the model for evaluating the reduction of power losses in a network organization while reactive power compensation is in a consumer network. Ind. Energy **10**, 31–36 (2018)
13. Adler, Y.P., Granovsky, Y.V., Markova, E.V.: Theory of experiment: past, present, future. Knowledge, Moscow, p. 64 (1982)
14. On the procedure for calculating the ratio of consumption of active and reactive power for individual power receivers (groups of power receivers) of electric power consumers. Order of the Ministry of Energy of the Russian Federation dated June 23, 2015 N 380. SPS Consultant Plus. http://www.pravo.gov.ru, http://publication.pravo.gov.ru/Document/View/0001201507270034
15. Kuznetsov, A.V., Rebrovskaya, D.A., Yurenkov, Y.: Transformation of a Model for estimation of the power losses enhancement in an allied grid company, with power factor compensation in the consumer's network. Russ. Electromechanics **4**, 82–89 (2019)
16. Brusilowicz, B., Szafran, J.: Comparison of reactive power compensation methods. In: 2016 Electric Power Networks (EPNet), Szklarska Poreba, pp. 1–6 (2016). https://doi.org/10.1109/EPNET.2016.7999369, https://ieeexplore.ieee.org/document/7999369

Comparison of a Vapor Compression Unit with an Absorption Chiller

I. Maslov(✉), G. Maslova, and M. Novoselova

Kazan State Power Engineering University, 26, 2-ya Yugo-Zapadnaya Street, Kazan 420034, Russia

Abstract. Reliability of electricity supply is the main and most important task of the energy industry in Russia. With the introduction of new equipment, such as gas turbine and combined cycle plants, the problem of effective energy saving becomes an important component. At the moment, it can be solved by increasing the efficiency of generating equipment. This article discusses the possibility of using a vapor compression unit, an absorption refrigeration machine in the cycle of a thermal power plant. The main energy characteristics of the equipment, advantages and disadvantages are considered. A comparison of absorption chiller and vapor compression unit is considered.

Keywords: Absorption chiller · Energy saving · Energy efficiency · Power increase · Refrigeration units

1 Introduction

Replacement of morally and physically obsolete steam power equipment is carried out under the DPM and DPM-stroke programs [1–36]. According to these programs, gas turbine and combined cycle power plants with a capacity of 16 to 350 MW are installed at thermal power plants. Currently, more than 60% of electricity is generated at thermal power plants Fig. 1. Changes in electricity and capacity consumption by UPS of Russia from 2010 to 2020 are shown in Fig. 2.

The positive aspects of installing gas turbine units include high maneuverability, low unit costs, quick start-up. But when operating gas turbines, there is one significant drawback - it is a decrease in power when the ambient temperature rises [6–9]. This is due to an increase in the cost of compressing air in the compressor of a gas turbine (GTU) plant. Figure 3 shows the dependence of the generated power generated by the GTE-160 gas turbine at different ambient temperatures. When the air temperature reaches 30 °C, the power generated by the gas turbine plant decreases by 31 MW.

Figure 4 shows the monthly temperature change in Kazan in 2020.

Figure 4, it can be concluded that a decrease in power will be observed in the summer months.

Practice shows that the efficiency of investment in efficiency and energy saving is much higher than the construction of new generating facilities. A generally accepted

A. A. Radionov and V. R. Gasiyarov (Eds.): RusAutoCon 2021, LNEE 857, pp. 183–192, 2022.
https://doi.org/10.1007/978-3-030-94202-1_18

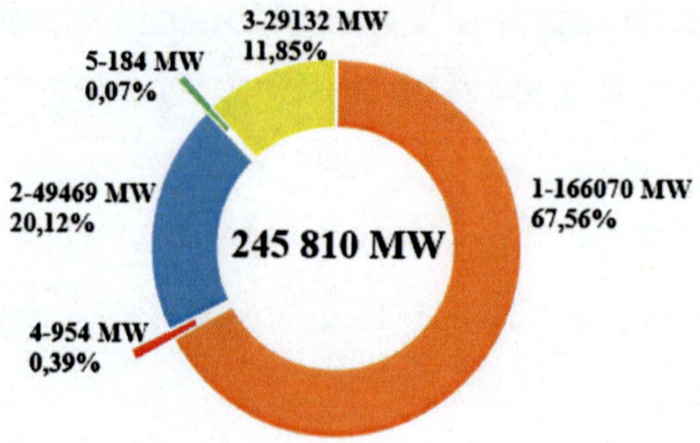

Fig. 1. The structure of the installed capacity of power plants in Russia. 1 - thermal power plants; 2 - hydroelectric power plants; 3 - nuclear power plants; 4 - solar power plants; 5 - wind farms.

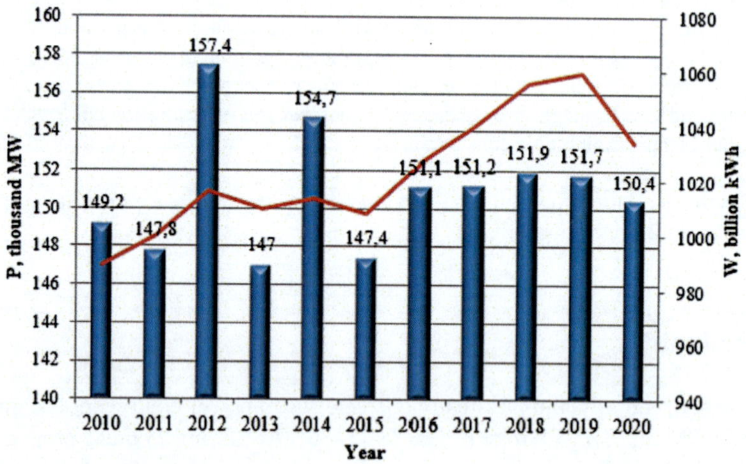

Fig. 2. Changes in electricity and capacity consumption by UPS of Russia from 2010 to 2020.

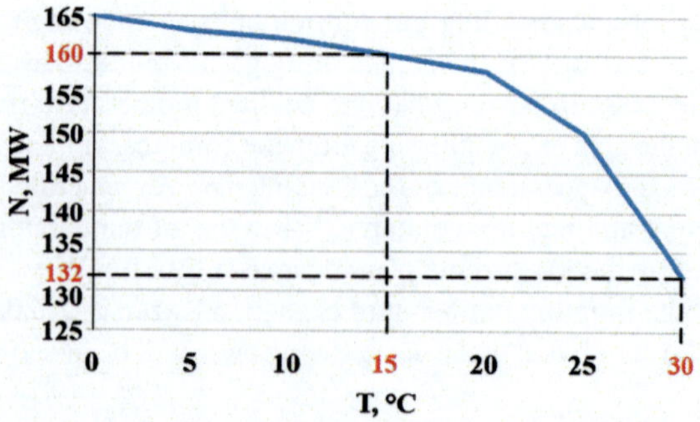

Fig. 3. Dependence of the generated power generated by the GTE-160 gas turbine at different ambient temperatures.

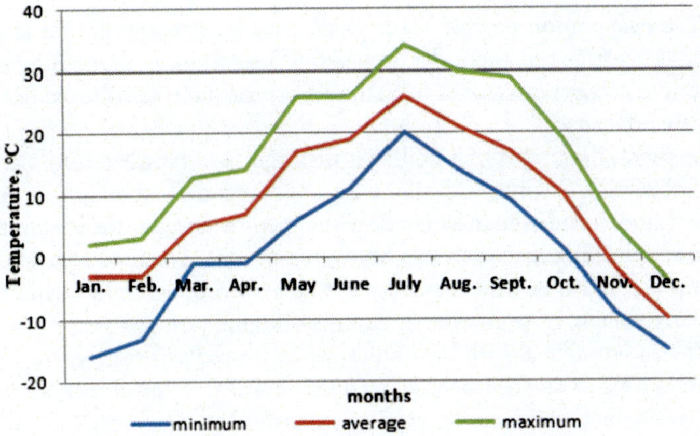

Fig. 4. Monthly change in temperature in Kazan in 2020.

approach to increasing the efficiency of electricity production is the method of combined production of thermal, eclectic energy, as well as industrial cold [10–17].

At the moment, two technologies can be distinguished for generating cold: 1. Absorption refrigeration machines; 2. Vapor compression refrigeration units.

To determine the most efficient and less costly technology for operation as part of a gas turbine cycle, a comparison of the Absorption chiller and a vapor compression refrigeration machine is presented below. Chillers with screw compressors are the most advanced and have reached a unit capacity comparable to the Absorption chiller, therefore, the expediency of the Absorption chiller is best compared with these machines. This paper considers a water-cooled chiller. We are considering a cooling tower for water cooling [10, 18–22]. With such a scheme, the heat received from the refrigerant is transferred to the water circulating in the cooling circuit. The water and refrigerant circuits are closed, with the possibility of make-up. The main elements of the chiller are the compressor, condenser and evaporator. The applied refrigerant in the gaseous state is compressed by the compressor to the pressure required for the operating mode, while the temperature of the refrigerant rises. Next, it is necessary to cool the refrigerant, while condensation occurs. Then the refrigerant enters the evaporator and, when sufficient for condensation, the phase state changes, that is, the refrigerant boils, while boiling, it cools with water. Reducing the consumption of electrical energy is the main advantage of using an Absorption chiller. In this heat engine, cooling is achieved at the expense of not electrical (as in a chiller), but thermal energy. Thermal energy can be obtained from the extraction of steam turbines, extraction of steam boilers, utilization of auxiliary steam at a thermal power plant. Absorption chiller can be used both as part of a refrigeration system and in heat supply systems [1, 2, 23–27].

2 Absorption Chiller Classification

The absorption machine works on the principle of a steam condensing refrigeration unit. In this installation, the refrigerant is evaporated by absorption (absorption effect) by the absorbent. The evaporation process takes place with the absorption of a large amount of heat. Then the refrigerant vapor due to heating (heating is carried out by an external source of thermal energy) is released from the absorbent and enters the condenser, where it subsequently condenses.

At the moment, the absorption chiller is distinguished between direct and indirect heating. Direct heating absorption chiller uses natural gas and diesel fuel as heat. Such a fuel supply scheme can be used in isolated and hard-to-reach areas for container-type gas turbines. Indirect heating machines use hot water, steam. If there is a surplus of steam at the power plant, it can be sent for disposal to the absorption chiller. Also absorption chiller is distinguished: 1. Ammonia; 2. Bromistolithium [29, 36–39].

In the lithium bromide absorption chiller, water is used as a refrigerant, and lithium bromide LiBr is used as an absorbent. In an ammonia absorption chiller, ammonia NH_3 is used as a refrigerant, and water is used as an absorbent. The most widespread at the moment are the lithium bromide absorption chiller.

If we consider the number of absorption chiller stages, then one-, two-, three-stage absorption chiller are distinguished. The difference lies in the efficiency of work, the more stages the absorption chiller has, the more efficient its work, but the cost of the installation increases with the increase in stages.

In a single-stage absorption chiller ("single effect", in the literature sometimes the term "single-circuit" is used), the refrigerant moves sequentially through the four main components of the machine - the evaporator, absorber, desorber and condenser. The refrigeration cycle of a single-stage absorption chiller is shown in Fig. 5. It is very similar to the refrigeration cycle of a vapor compression refrigeration machine. The diagram of a single-stage absorption chiller is shown in Fig. 6. The refrigerant evaporates when the pressure in the evaporator drops. 1. This process involves the absorption of heat. Unlike a vapor compression refrigeration machine, the process of lowering the pressure in the evaporator occurs not due to the operation of the compressor, but due to the volumetric absorption (absorption) of the refrigerant by the liquid absorbent in absorber 2. Then the absorbent with the absorbed refrigerant (binary solution) enters the desorber 3. In the stripper, the binary solution is heated by combustion of gas, steam, etc., as a result of which the refrigerant is released from the absorbent. The lean absorbent from the stripper is returned to the absorber. The refrigerant enters condenser 4 under high pressure, where it passes into the liquid phase with the release of heat, and then through the expansion valve 5 enters the evaporator, after which a new cycle begins.

A change in the concentration of the refrigerant in the absorber and stripper is accompanied by a change in the saturation temperature. To reduce energy losses during the circulation of the absorbent, a recuperative heat exchanger is installed between the absorber and the stripper [28, 30–35].

An ideal single-stage absorption chiller could provide a cooling effect equal to the amount of thermal energy supplied to the generator, however, due to thermodynamic losses in real installations, the cooling effect will always be lower than the consumption of thermal energy.

If we consider the number of absorption chiller stages, then one-, two-, three-stage absorption chiller are distinguished. The difference lies in the efficiency of work, the

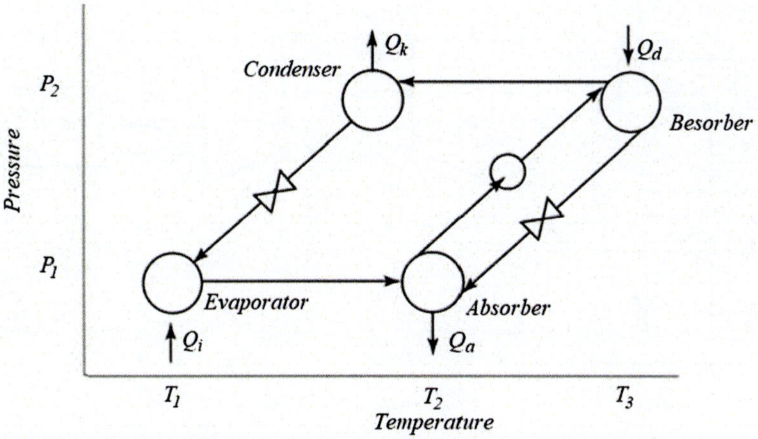

Fig. 5. Refrigeration cycle of a single stage absorption chiller.

more stages the absorption chiller has, the more efficient its work, but the cost of the installation increases with the increase in stages.

The two-stage steam-powered absorption chiller is the most widespread, due to its efficiency and quick payback period. The operation of a single-stage absorption chiller is characterized by a high rate of corrosion destruction. In a two-stage scheme, this is excluded due to the technology of distribution and separation of streams. This avoids dangerous boundary conditions in which the lithium bromide solution reaches its maximum concentration and maximum temperature simultaneously.

3 The Principle of Operation of a Two-Stage Absorption Chiller

In the generator of the first stage (the high-temperature generator 0 evaporates from the lithium bromide solution, then the steam is directed to the second-stage generator (low-temperature generator). couple.

A diagram of the Absorption chiller connection with the GTU cycle is shown in Fig. 6. This paper does not consider a specific gas turbine, the connection principle is the same for all turbines, for stationary and conversion gas turbines.

Designations in the diagram, Fig. 6: CGT-compressor of a gas turbine unit; CC - combustion chamber; GT gas turbine; WHB - waste heat boiler; EG - exhaust gases.

In this paper, a diagram of the absorption chiller connection (Fig. 6) is considered, with the outlet of cold air to the gas turbine compressor. Cold air mixes with hot air, thereby increasing the power of the gas turbine.

A two-stage absorption chiller with a steam flow rate of 1200 kg/h, a steam pressure of 3 kgf/cm^2 from the station auxiliaries collector is considered.

For comparison, a Thermax absorption chiller and a J&E Hall chiller were selected. The main characteristics are presented in Table 1.

To compare the absorption chiller and the chiller, it is necessary to compare the operating and capital costs, Table 2.

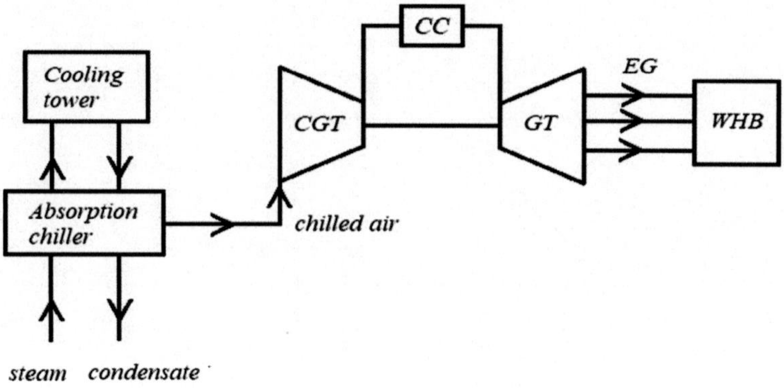

Fig. 6. Wiring diagram of absorption chiller with GTU cycle.

Table 1. Main characteristics of Thermax absorption chiller and J&E Hall chiller.

Item name	Absorption chiller	Chiller
Cooling capacity, kW	1044	1012,6
Chilled water consumption, m^3/h	180	175
Cooling water consumption, m^3/h	300	-
Steam consumption, kg/h	1200	-
Total consumption Ne, kW	-	221
Specific refrigerating capacity, kW/kg/h	0,87	0,88

To estimate operating costs, it is necessary to determine the cost of electricity during the year, the cost of steam production is not taken into account (waste steam). Cooling costs in the condenser are assumed to be 40 kW.

$$Ez = Ndv \cdot t \cdot k, \tag{1}$$

where: Ez - operating costs, rubles; Ndv - electric motor power, kW; t - unit operation time, h; k - the cost of 1 kW of consumed electricity, rubles.

$$Ez_{absorption\ chiller} = 1.2563 \text{ million rubles} \tag{2}$$

$$Ez_{chiller} = 7.120 \text{ million rubles} \tag{3}$$

Obviously, the cost of electricity consumed by absorption chiller is much lower than that of a chiller. Given this, we can conclude. that the installation of absorption chiller is economically more profitable. However, the use of a chiller is also effective when it is installed in a single gas turbine system, purely for generating electricity. Another advantage of the chiller operation is fast regulation, since during absorption chiller operation it is necessary to select the steam temperature for operation, the steam supply

Table 2. Operating and capital costs of the ABCM and chiller.

Item name	Absorption chiller	Chiller
Installation cost, commissioning, million rubles	1,6	1,1
Transportation costs, mln rubles	0,8	0,8
Installation cost, million rubles	17,4	11,3
Total costs, million rubles	19,8	13,2

control range is very small. It is necessary to monitor the temperature and pressure of steam in order to exclude condensation of steam in the steam supply pipelines to the absorption chiller.

4 Conclusion

In conclusion, I would like to note the following:

1. With the joint work of absorption chiller and GTU, the energy characteristics of the energy-saving system are improved.
2. The use of absorption chiller in comparison with the chiller is more efficient, since the waste steam is utilized, the cost of operating the electric motors is minimal.
3. For absorption chiller, low operating costs are typical

References

1. Marin, G.E., Osipov, B.M., Mendeleev, D.I.: Research on the influence of fuel gas on energy characteristics of a gas turbine. E3S Web Conf. **124**, 05063 (2019). https://doi.org/10.1051/e3sconf/201912405063
2. Mendeleev, D.I., Marin, G.E., Akhmetshin, A.R.: The implementation and use of gas turbines with absorption refrigerating machine in the technological schemes of thermal power plants. In: 2019 International Multi-Conference on Industrial Engineering and Modern Technologies, FarEastCon, NSPEC Accession Number: 19229407 (2019). https://doi.org/10.1109/FarEastCon.2019.8934431
3. Marin, G.E., Mendeleev, D.I., Akhmetshin, A.R.: Analysis of changes in the thermophysical parameters of the gas turbine unit working fluid depending on the fuel gas composition. INSPEC Accession Number: 19229280 (2019). https://doi.org/10.1109/FarEastCon.2019.8934021
4. Bahrami, S., Ghaffari, A., Genrup, M., Thern, M.: Performance comparison between steam-injected gas turbine and combined cycle during frequency drops. Energies **8**, 7582–7592 (2015)
5. Hou, G., Gong, L., Dai, X., Wang, M., Huang, C.: A novel fuzzy model predictive control of a gas turbine in the combined cycle unit. Complexity **2018**, 6468517 (2018)
6. Benrajesh, P., John Rajan, A.: Design and analysis of a two-stage adsorption air chiller. IOP Conf. Ser. Mater. Sci. Eng. **197**, 012030 (2017). https://doi.org/10.1088/1757-899X/197/1/012030

7. Zheng, L.K., Cronly, J., Ubogu, E., Ahmed, I., Zhang, Y., Khandelwal, B.: Experimental investigation on alternative fuel combustion performance using a gas turbine combustor. Appl. Energy **238**, 1530–1542 (2019)
8. Lokini, P., Roshan, D.K., Kushari, A.: Influence of swirl and primary zone airflow rate on the emissions and performance of a liquid-fueled gas turbine combustor. J. Energy Resour. Technol. Trans. ASME **141**(6), 9 (2019)
9. Khudair, O.A., Abass, K.A., Abed, N.S., Ali, K.H., AbdulAziz, S., Shaboot, A.C.: Theoretical investigation for the effect of fuel quality on gas turbine power plants. In: Al-Haitham, I. (ed.) 1st International Scientific Conference on Biology Chemistry Computer Science Mathematics and Physics (IHSCICONF) Minist S.T. 1003: Journal of Physics Conference Series (2017)
10. Junquera, I., et al.: Home energy management systems and electric vehicles: challenges and opportunities. Renew. Energy Power Qual. J. **1**(15), 334–339 (2017)
11. Soluyanov, Y.I., Fedotov, A.I., Ahmetshin, A.R.: Calculation of electrical loads of residential and public buildings based on actual data. IOP Conf. Ser. Mater. Sci. Eng. **643**(1), 012051 (2019). https://doi.org/10.1088/1757-899X/643/1/012051
12. Vedernikov, A.S., Goldshtein, V.G.: Increase of power efficiency of electric networks with two-chain transmission line. In: 6th International Scientific Symposium on Electrical Power Engineering, ELEKTROENERGETIKA, pp. 231–233 (2011)
13. Gracheva, E., Alimova, A.: Calculating probabilityof faultless work of shop nets with the help of coefficients of ratio. In: 2019 International Russian Automation Conference, RusAutoCon, INSPEC, 19090885 (2019)
14. Alajmi, A.E.S.E.T., Adam, N.M., Hairuddin, A.A., Abdullah, L.C.: Fuel atomization in gas turbines: a review of novel technology. Int. J. Energy Res. **43**(8), 3166–3318 (2019)
15. Taufan, A., Djubaedah, E., Manga, A., Nasruddin, A.: Experimental performance of adsorption chiller with fin and tube heat exchanger. In: AIP Conference Proceedings, vol. 2001, p. 020012 (2018)
16. Mendeleev, D.I., Galitskii, Y.Y., Marin, G.E., Akhmetshin, A.R.: Study of the work and efficiency improvement of combined-cycle gas turbine plants. E3S Web Conf. **124**, 05061 (2019). https://doi.org/10.1051/e3sconf/201912405061
17. Akhmetshin, A., Marin, G., Mendeleev, D.: Modeling of asynchronous motor operation modes for the correct selection of voltage regulation devices. E3S Web Conf. **178**, 01015 (2020). https://doi.org/10.1051/e3sconf/202017801015
18. Nurin, F.N., Amano, R.S.: Review of gas turbine internal cooling improvement technology. J. Energy Res. Technol. **143**(8), 080801 (2021). https://doi.org/10.1115/1.4048865
19. Gofman, A.V., Vedernikov, A.S., Vedernikova, E.S.: Increasing the accuracy of the short-term and operational prediction of the load of a power system using an artificial neural network. Power Technol. Eng. **46**(5), 410–415 (2013). https://doi.org/10.1007/s10749-013-0370-0
20. Addabbo, T., et al.: Ion sensor measurement systems: application for combustion monitoring in gas turbines. IEEE Trans. Instrum. Meas. **69**(4), 1474–1483 (2020). https://doi.org/10.1109/TIM.2019.2961483
21. Akhmetshin, A., Mendeleev, D., Marin, G.: Improvement of electricity quality indicators in electric networks with voltage of 0.4–10 kV. In: 2020 International Russian Automation Conference, RusAutoCon 2020, pp. 454–458, INSPEC Accession Number: 20007605 (2020). https://doi.org/10.1109/RusAutoCon49822.2020.9208158
22. Rigo-Mariani, R., Zhang, C., Romagnoli, A., Kraft, M., Ling, K.V., MacIejowski, J.: A combined cycle gas turbine model for heat and power distribution considering grid constraints. IEEE Trans. Sustain. Energy 11(1), 448–456 (2020). https://doi.org/10.1109/TSTE.2019.2894793

23. Nazarychev, S.A., Akhmetshin, A.R., Gaponenko, S.O.: Full compensation of reactive power in electric networks 0.4–10kV. In: 11th Scientific Technical Conference on Low Temperature Plasma during the Deposition of Functional Coatings, Journal of Physics: Conference Series, vol. 1588, no. 1, p. 012036 (2020). https://doi.org/10.1088/1742-6596/1588/1/012036
24. Pilarsky, S., Stanishevsky, M., Villeneuve, F., Varro, D.: On artificial intelligence for modeling and design. Space exploration in the design of gas turbines. 22nd ACM/IEEE 2019 International Conference on Model Driven Design Languages and System Companions, MODELS -C 2019, Germany, 8904682, pp. 170–174 (2019). https://doi.org/10.1109/MODELS-C.2019.00029
25. Soluyanov, Y.I., Fedotov, A.I., Soluyanov, D.Y., Akhmetshin, A.R.: Experimental research of electrical loads in residential and public buildings. IOP Conf. Ser. Mater. Sci. Eng. **860**(1), 012026 (2020). https://doi.org/10.1088/1757-899X/860/1/012026
26. Liu, Y., Banerjee, A., Hanachi, H., Kumar, A.: Data-driven model selection study for long-term gas turbine degradation of gas turbines. In: 2019 IEEE International Conference on Health Prediction and Management, ICPHM 2019, Art. no 88194332019, United States (2019). https://doi.org/10.1109/ICPHM.2019.8819433
27. Gaponenko, S.O., Kondratiev, A.E., Shakurova, R.Z.: Improving the efficiency of energy complexes and heat supply systems using mathematical modeling methods at the operational stage. In: International Scientific and Technical Conference Smart Energy Systems, E3S Web of Conferences, vol. 124, p. 05029 (2019). https://doi.org/10.1051/e3sconf/201912405029
28. Gaponenko, S.O., Kondratiev, A.E., Tazeev, I.R.: Assessment of natural oscillation frequencies of rotor for development of hard-bearing balancing machine. In: Radionov, A.A., Kravchenko, O.A., Guzeev, V.I., Rozhdestvenskiy, Y.V. (eds.) ICIE 2019. LNME, pp. 249–257. Springer, Cham (2020). https://doi.org/10.1007/978-3-030-22041-9_29
29. Diaz-Herrera, P.R., Alcaraz-Calderon, A.M., Gonzalez-Diaz, M.O., Gonzalez-Diaz, A.: Capture level design for a natural gas combined cycle with post-combustion CO_2 capture using novel configurations. Energy **193**, 116769 (2020). https://doi.org/10.1016/j.energy.2019.116769
30. Diaz-Herrera, P.R., Ascanio, G., Romero-Martinez, A., Alcaraz-Calderon, A.M., Gonzalez-Diaz, A.: Theoretical comparison between post-combustion carbon capture technology and the use of blue and green H_2 in existing natural gas combined cycles as CO_2 mitigation strategies: a study under the context of Mexican clean energy regulation. Int. J. Hydrogen Energy **46**(2), 2729–2754 (2021). https://doi.org/10.1016/j.ijhydene.2020.10.076
31. Bonfillo, A., Cachacarne, S., Invernizzi, M., Lanzarotto, D., Palmieri, A., Procopio, R.: Sliding mode control approach for gas turbine power generators. IEEE Trans. Energy Convers. **34**(2), 921–932 (2019). https://doi.org/10.1109/TEC.2018.2879688
32. Ravikumar, C.G., Bosley, B., Clarke, T., Garcia, J.: Generation control system: Using isochronous load sharing principles with gas and steam turbine generators. IEEE Ind. Appl. Mag. **25**(2), 36–44 (2019). https://doi.org/10.1109/MIAS.2018.2875127
33. Islamova, S.I., Timofeeva, S.S., Ermolaev, D.V., Khamatgalimov, A.R.: Kinetic analysis of the thermal decomposition of lowland and high-moor peats. Solid Fuel Chem. **54**(4), 154–162 (2020). https://doi.org/10.3103/S0361521920030040
34. Goldstein, V., Shishkov, E., Vedernikov, A., Kolcun, M.: Calculation of steady state of multi-chain overhead power transmission line in phase. In: 7th International Scientific Symposium on Electrical Power Engineering, ELEKTROENERGETIKA, pp. 145–148 (2013)
35. Karaeva, J.V., Timofeeva, S.S., Bashkirov, V.N., Bulygina, K.S.: Thermochemical processing of digestate from biogas plant for recycling dairy manure and biomass. Biomass Convers. Biorefinery 1–11 (2020). https://doi.org/10.1007/s13399-020-01138-6

36. Goldshteyn, V.G., Vedernikov, A.S., Tanaev, N.V.: Mathematical models of short-circuit and break modes in electric networks with multi-circuit overhead lines. In: 2020 International Ural Conference on Electrical Power Engineering, UralCon, Art. no. 9216271, pp. 304–312 (2020). https://doi.org/10.1109/UralCon49858.2020.9216271
37. Suslov, K., et al.: Development of the methodological basis of the simulation modelling of the multi-energy systems. E3S Web Conf. **124**, 01049 (2019). https://doi.org/10.1051/e3sconf/201912401049
38. Erickson, D.K., Anand, G., Makar, E.: Absorption refrigeration cycle, turbine inlet air conditioning. Int. J. Air Conditioning Refrig. **23**(1), 1550003 (2015). https://doi.org/10.1142/S2010132515500030
39. Subbarao, R., Sarat, K.S.: Gas turbine unit analysis for distributed power generation alongside heating, cooling and air conditioning. J. Distrib. Gener. Altern. Energy **32**(2), 56–72 (2017). https://doi.org/10.1080/21563306.2017.11869109

Influence of Perovskite Layer Parameters and Back Contact Material on Characteristics of Solar Cells

A. Sayenko[1](✉), S. Malyukov[1], and A. Palii[2]

[1] Southern Federal University, 105/42, Bolshaya Sadovaya Street, Rostov-on-Don 344006, Russia
avsaenko@sfedu.ru

[2] Polytechnical Institute (Branch) of Don State Technical University, 109 a, Petrovskaya Street, Taganrog 347904, Russia

Abstract. The article is devoted to the numerical simulation of a perovskite solar cell with the FTO/TiO_2/$CH_3NH_3PbI_{3-x}Cl_x$/Cu_2O/Me structure in the SCAPS-1D program. The effect of the thickness and defect concentration in the perovskite layer $CH_3NH_3PbI_{3-x}Cl_x$, as well as the work function of the back contact material (Me), on the photoelectric characteristics of a solar cell has been studied. It was found that the optimal thickness of the $CH_3NH_3PbI_{3-x}CI_x$ layer is 600–700 nm, and the defect concentration should be less than 10^{14} cm^{-3}. It is shown that the work function of the back contact material must be greater than or equal to 5 eV to create solar cells with high efficiency. A maximum efficiency of 21.55% was obtained (short circuit current 24.87 mA/cm^2, open circuit voltage 1.1 V, fill factor 78.82%) for the structure of a perovskite solar cell with a carbon (C) back contact. The results can be used in the design and manufacture of efficient and inexpensive perovskite solar cells.

Keywords: Solar cell · Numerical simulation · Perovskite · Defect concentration · Layer thickness · Back contact · Work function · Efficiency

1 Introduction

Thin film solar cells containing organometallic compounds with a perovskite structure ($CH_3NH_3PbI_{3-x}Cl_x$) as a photoactive material show an energy conversion efficiency of more than 20%. The high efficiency of these solar cells is a consequence of such properties of perovskites as high absorption coefficient, high mobility and long diffusion length of charge carriers. The technology of manufacturing perovskite solar cells does not require energy-intensive and complex technological processes, which makes it possible to create light, inexpensive and flexible film devices [1–3]. Despite the high efficiency of perovskite solar cells, it is still far from the theoretical maximum (31%). One of the reasons is the recombination of charge carriers, which reduces the fill factor and the open circuit voltage in the solar cell. Another reason is optical losses, imperfect conducting layers of the n- and p-type, as well as ineffective collection of charge carriers by contacts

A. A. Radionov and V. R. Gasiyarov (Eds.): RusAutoCon 2021, LNEE 857, pp. 193–202, 2022.
https://doi.org/10.1007/978-3-030-94202-1_19

[4–10]. Nonradiative recombination is the main recombination mechanism in perovskite solar cells, which limits their energy conversion efficiency. Nonradiative recombination occurs when an electron (or hole) captured by a defect (energy level in the band gap of a perovskite) recombines with a hole (or electron) in the valence (or conduction band) of the perovskite. In thin films of polycrystalline perovskite, defects are located mainly at grain boundaries and on the film surface [4].

The investigated planar n-i-p structure of a perovskite solar cell on a glass substrate includes an electron conducting layer (TiO_2), a photoactive layer ($CH_3NH_3PbI_{3-x}Cl_x$), and a conductive hole layer (Cu_2O). The p-type semiconductor Cu_2O is a promising alternative for replacing the expensive organic compound Spiro-OMeTAD, since it has a suitable arrangement of energy bands (band gap 2.17 eV), high mobility of charge carriers (up to 110 $cm^2/V{\bullet}s$), as well as non-toxicity and low cost [8]. FTO (SnO_2:F) and gold (Au) are widely used as front and rear contacts in the structure of a solar cell. Replacing the expensive Au back contact (work function 5.1 eV) is an important task. Among the materials of the back contact, nickel (Ni), carbon (C), and copper (Cu) are promising with the work function of 5.15 eV, 5 eV, and 4.65 eV, respectively [6, 11, 12].

In this work a model of a perovskite solar cell with the FTO/TiO_2/$CH_3NH_3PbI_{3-x}Cl_x$/Cu_2O/Me structure was created in the SCAPS-1D numerical simulation program. The effect of the thickness and concentration of defects in the $CH_3NH_3PbI_{3-x}Cl_x$ perovskite layer and the work function of the rear contact material on the photoelectric characteristics of solar cells has been studied.

2 Numerical Model

Numerical simulation is a necessary stage in the development of semiconductor devices, including solar cells, which can reduce the cost of experimental research and optimize the characteristics of devices. There are several programs for the development and research of solar cells, including AMPS-1D, SCAPS-1D, PC1D, AFORS-HET and others [13–15].

SCAPS-1D is a one-dimensional numerical simulation of solar cells. SCAPS-1D is based on an unsteady diffusion-drift system of semiconductor equations, which includes the continuity equations and the Poisson equation [14–18]:

$$\frac{\partial}{\partial x}\left[\mu_n\left(-n\frac{\partial\varphi}{\partial x}+\varphi_t\frac{\partial n}{\partial x}\right)\right]+G-R=\frac{\partial n}{\partial t}, \tag{1}$$

$$\frac{\partial}{\partial x}\left[\mu_p\left(p\frac{\partial\varphi}{\partial x}+\varphi_t\frac{\partial p}{\partial x}\right)\right]+G-R=\frac{\partial p}{\partial t}, \tag{2}$$

$$\frac{\partial}{\partial x}\left(\varepsilon\frac{\partial\varphi}{\partial x}\right)=\frac{q}{\varepsilon_0}(n-p-N_D+N_A-p_t+n_t), \tag{3}$$

where n, p is electrons and holes concentration; μ_n, μ_p is electrons and holes mobility; φ is electric potential; φ_t is temperature potential; q is elementary charge; ε is relative dielectric constant; ε_0 is dielectric constant; G is optical generation rate of electron-hole pairs; R is recombination rate of electron-hole pairs; N_D, N_A is donor and acceptor dopant concentration; n_t, p_t is the density traps for electrons and holes.

In the simulation, a perovskite solar cell with a nip structure (Fig. 1) was considered, consisting of three layers and two contacts: front contact (FTO), electron conducting layer (TiO_2), photoactive layer ($CH_3NH_3PbI_{3-x}Cl_x$), hole conducting layer (Cu_2O) and back contact (Au). The main physical parameters of materials used in modeling a solar cell are given in Table 1 and Table 2 [17–22]. For all layers, the effective cross section for the capture of electrons and holes by a defect was taken to be $2{\cdot}10^{-14}$ cm^2, and the thermal velocity of charge carriers was 10^7 cm/s. The absorption coefficient for each layer was 10^5 cm^{-1} with the standard spectrum of the AM1.5G photon flux density. The type of crystal lattice defects (recombination centers) was set to be neutral, and their local energy level was set in the center of the band gap; therefore, the recombination mechanism was described according to the Shockley-Reed-Hall theory [20]. The value of the series resistance was 1 Ω·cm, and the shunting resistance was 10^6 Ω·cm. Contact voltage ranged from 0 V to 1.2 V.

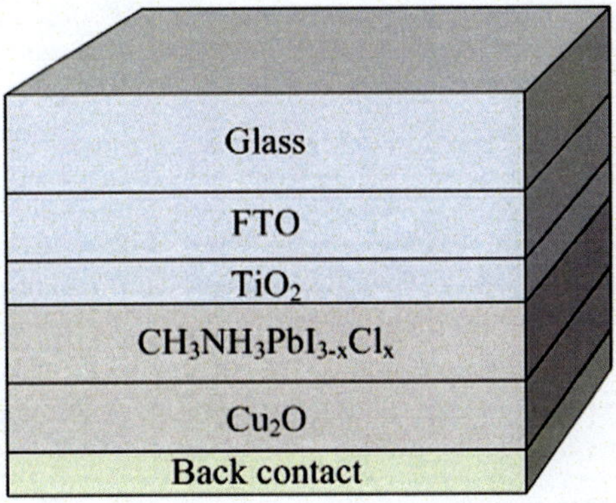

Fig. 1. Schematic representation of the simulated structure of a perovskite solar cell.

Table 1. Physical parameters of TiO_2 and $CH_3CN_3PbI_{3-x}Cl_x$ layers.

Parameters	TiO_2	$CH_3CN_3PbI_{3-x}Cl_x$
Thickness (nm)	50	400
N_A (cm^{-3})	–	–
N_D (cm^{-3})	10^{17}	–
E_g (eV)	3,2	1,55
χ (eV)	4,0	3,9
ε	9	6,5
N_C/N_V (cm^{-3})	$2{,}2{\cdot}10^{18}/1{,}8{\cdot}10^{19}$	$2{,}2{\cdot}10^{18}/1{,}8{\cdot}10^{19}$
μ_n/μ_p (cm^2/V·s)	20/10	2/2
σ_n/σ_p	$2{\cdot}10^{-14}/2{\cdot}10^{-14}$	$2{\cdot}10^{-14}/2{\cdot}10^{-14}$
N_t (cm^{-3})	10^{15}	$2{,}5{\cdot}10^{13}$

Table 2. Physical parameters of the Spiro-OMeTAD and Cu_2O layers.

Parameters	Spiro-OMeTAD	Cu_2O
Thickness (nm)	250	250
N_A (cm^{-3})	10^{19}	10^{18}
N_D (cm^{-3})	–	–
E_g (eV)	2,9	2,17
χ (eV)	2,2	3,2
ε	3	7,11
N_C/N_V (cm^{-3})	$2{,}2{\cdot}10^{18}/1{,}8{\cdot}10^{19}$	$2{,}2{\cdot}10^{18}/1{,}8{\cdot}10^{19}$
μ_n/μ_p ($cm^2/V{\cdot}s$)	$10^{-4}/10^{-4}$	80/80
σ_n/σ_p	$2{\cdot}10^{-14}/2{\cdot}10^{-14}$	$2{\cdot}10^{-14}/2{\cdot}10^{-14}$
N_t (cm^{-3})	10^{15}	10^{15}

3 Simulation Results

To confirm the adequacy and accuracy of the simulation results, a solar cell model was created in the SCAPS-1D program with the FTO/TiO_2 (50 nm)/$CH_3NH_3PbI_{3\text{-}x}Cl_x$ (400 nm)/Spiro-OMeTAD (250 nm)/Au structure and the simulation results were compared characteristics with the experimental data presented in [23, 24]. The simulation results show close agreement with the experimental data (Table 3).

Table 3. Theoretical and experimental parameters of solar cells.

	J_{sc}, mA/cm^3	V_{oc}, V	FF, %	η, %
FTO/TiO_2/ $CH_3NH_3PbI_{3\text{-}x}Cl_x$/ Spiro-OMeTAD/Au [24]	22,75	1,13	75,01	19,30
FTO/TiO_2/ $CH_3NH_3PbI_{3\text{-}x}Cl_x$/ Spiro-OMeTAD/Au	22,39	1,12	76,21	19,19
FTO/TiO_2/ $CH_3NH_3PbI_{3\text{-}x}Cl_x$/ Cu_2O/Au	22,53	1,12	80,79	20,47

In Fig. 2 shows the current-voltage characteristics obtained as a result of modeling the structures of solar cells with a conductive hole layer Spiro-OMeTAD and Cu_2O. It is shown that a solar cell with a hole conducting Cu_2O layer has better parameters compared to the Spiro-OMeTAD layer and has an efficiency (η) equal to 20.47%.

The main factor affecting the characteristics of perovskite solar cells is the thickness of the perovskite layer ($CH_3NH_3PbI_{3\text{-}x}Cl_x$), since it absorbs solar radiation and generates

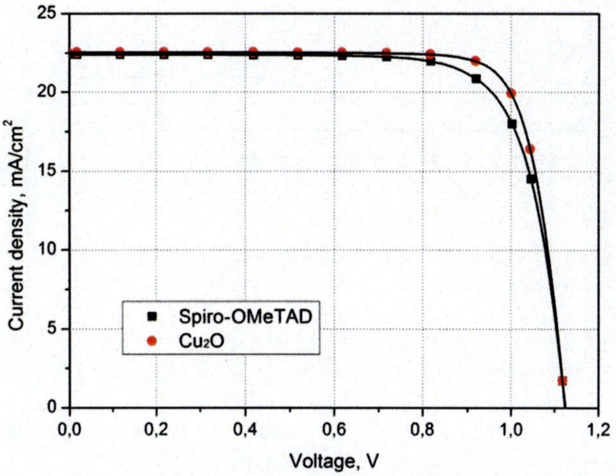

Fig. 2. Current-voltage characteristics of solar cells with hole conducting layers Spiro-OMeTAD and Cu_2O

electron-hole pairs. To study the effect of the thickness of the perovskite layer on the characteristics of a solar cell, modeling was carried out with a change in thickness in the range from 100 nm to 1000 nm and other parameters given in Table 1.

Figure 3 it can be seen that the efficiency of the solar cell increases sharply (from 12.94% to 21.5%) with an increase in the thickness of the perovskite layer to 600 nm, then slightly increases to 700 nm and decreases after 700 nm. With an increase in the thickness of the perovskite layer, a greater number of photons are absorbed, which leads to the generation of a larger number of excess charge carriers and, accordingly, an increase in the short-circuit current density from 12.92 mA/cm^2 to 25.61 mA/cm^2 (Fig. 3). An increase in the thickness of the perovskite layer also leads to a slight decrease in the open circuit voltage by 0.1 V (from 1.18 V to 1.08 V), which is associated with an increase in the density of the dark saturation current (J_0) due to an increase in the probability of carrier recombination. This can be explained by the dependence of the open circuit voltage (V_{OC}) on the saturation dark current density and photogenerated short circuit current density (J_{SC}) [20–22]:

$$V_{OC} = \frac{AkT}{q} \ln\left(\frac{J_{SC}}{J_0} + 1\right), \tag{4}$$

where A is the diode ideality coefficient, kT/q is the temperature potential. The open circuit voltage is limited by the value of the saturation dark current density, which increases with increasing thickness of the perovskite layer. Thus, the thickness of the perovskite layer ($CH_3NH_3PbI_{3-x}Cl_x$) equal to 600–700 nm is optimal for obtaining highly efficient solar cells.

In addition to the thickness of the perovskite layer, the characteristics of solar cells are also significantly affected by the concentration of defects in the perovskite crystal lattice, since a high concentration of defects leads to a higher recombination rate due to the formation of point vacancies, rapid degradation of the perovskite layer, and deterioration

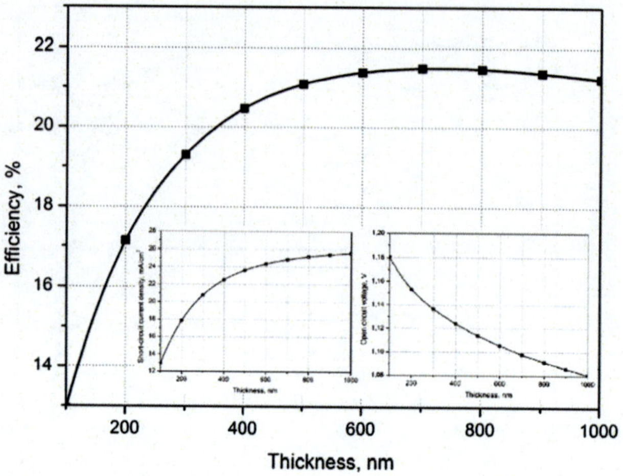

Fig. 3. Dependences of efficiency, short circuit current density and open circuit voltage of solar cell on the perovskite layer thickness.

of the characteristics of a solar cell [4]. Shockley-Reed-Hall recombination through local levels created by lattice defects is the main mechanism in perovskite solar cells, since the perovskite layer usually has a high concentration of defects. The recombination rate (R) of Shockley-Reed-Hall is determined by the formulas [17–20]:

$$R = \frac{n \cdot p - n_i^2}{\tau_n \cdot \left(n + N_c e^{\frac{-E_c + E_t}{kT}}\right) + \tau_p \cdot \left(p + N_v e^{\frac{-E_t + E_v}{kT}}\right)} \tag{5}$$

$$\tau_{n,p} = \frac{1}{\sigma_{n,p} \cdot \upsilon_{n,p} \cdot N_t}, \tag{6}$$

where τ_n, τ_p is electrons and holes lifetime; $\sigma_{n,p}$ is the effective cross section for the capture of electrons and holes by a defect; E_c, E_v is energy levels of the bottom of the conduction band and the top of the valence band; E_t is the local energy level created by defects; $\upsilon_{n,p}$ is the thermal velocity of electrons and holes; N_t is the concentration of defects.

The diffusion length ($L_{n,p}$) of electrons and holes in a perovskite is determined using the equation [19–25]:

$$L_{n,p} = \sqrt{D_{n,p} \cdot \tau_{n,p}}, \tag{7}$$

$$D_{n,p} = \frac{kT}{q} \mu_{n,p}, \tag{8}$$

where $D_{n,p}$ is diffusion coefficient of electrons and holes.

To study the effect of the defect concentration in perovskite on the characteristics of a solar cell, modeling was carried out with a change in the concentration of defects in the range fromот 10^{13} cm^{-3} to 10^{17} cm^{-3}, a thickness of 700 nm, and other parameters given in Table 1.

In Table 4 shows the values of the diffusion length and lifetime of electrons and holes obtained by formulas (7) and (8) from the defect concentration in perovskite and used in modeling.

Table 4. Values of the diffusion length and lifetime of charge carriers on the defect concentration in the perovskite.

N_t, cm^{-3}	10^{13}	10^{14}	10^{15}	10^{16}	10^{17}
$L_{n,p}$, nm	1600	510	160	51	16
$\tau_{n,p}$, ns	500	50	5	0,5	0,05

Figure 4 that when the defect concentration in the perovskite changes from 10^{13} cm^{-3} to 10^{17} cm^{-3} open circuit voltage decreases from 1.14 V to 0.68 V, short circuit current density with 24.89 mA/cm^2 to 3.67 mA/cm^2, and efficiency from 22.66% to 0.66%. Thus, the characteristics of the solar cell are significantly reduced with an increase in the concentration of defects in the perovskite. To obtain high efficiency, the defect concentration should be less 10^{14} cm^{-3} by improving the crystal structure, i.e. optimization of the perovskite layer formation technology. The simulation obtained the best efficiency of the order of 21.5% at a practically attainable defect concentration in perovskite 2, $5 \cdot 10^{13}$ cm^{-3}.

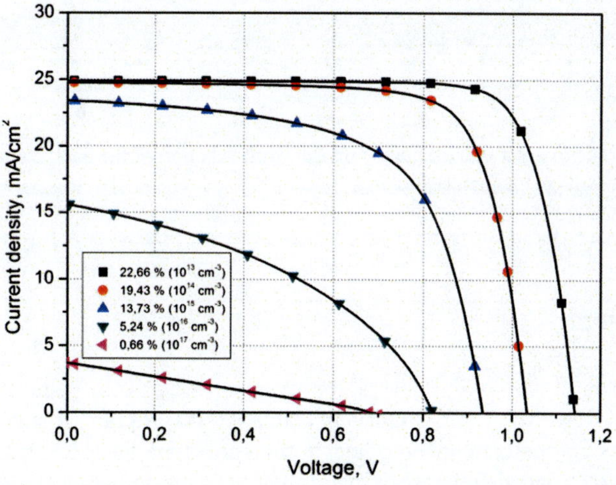

Fig. 4. Current-voltage characteristics of solar cells at different defect concentration in the perovskite layer.

Replacing the expensive Au back contact (5.1 eV) in a solar cell is also an important task. As this contact, it is necessary to use a material with a certain work function to obtain an ohmic contact with the Cu_2O layer. In Fig. 5 shows the dependence of the solar cell efficiency on the work function of the back contact material and their current-voltage characteristics. As a back contact material (Me) when simulating a solar cell with the structure FTO/TiO_2/ $CH_3NH_3PbI_{3-x}Cl_x$ (700 nm)/Cu_2O (200 nm)/Me used Ni (5,15 eV), Au (5,1 eV), C (5 eV), Ag (4,74 eV), Cu (4,65 eV) and Al (4,3 eV). It was found that an increase in the work function leads to a significant increase in the open circuit voltage from 0.76 V to 1.1 V (Fig. 5, b). The short circuit current remains practically unchanged. (24.87 mA/cm^2). The efficiency of a solar cell increases approximately up to the work function of 5 eV (Fig. 5, a) due to a decrease in the potential Schottky barrier at the interface Cu_2O/Me, which contributes to a more efficient transfer of holes in Cu_2O to the back contact.

Thus, the work function of the back contact must be greater than or equal to 5 eV, which is necessary to obtain high characteristics of the solar cell (efficiency 21.55%). Most suitable as back contact for structure FTO/TiO_2/$CH_3NH_3PbI_{3-x}Cl_x$ (700 nm)/Cu_2O (200 nm)/Me is carbon (C) with a work function of 5 eV.

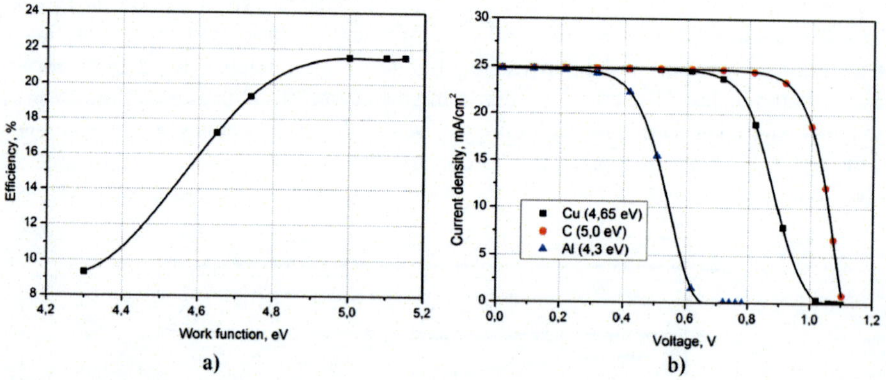

Fig. 5. Dependence of solar cell efficiency on the work function of the back contact material (a) and their current-voltage characteristics (b).

4 Conclusion

The work created a model of a perovskite solar cell with the structure FTO/TiO_2/$CH_3NH_3PbI_{3-x}Cl_x$/Cu_2O/Me in a SCAPS-1D program. A study of the effect of the thickness and defect concentration in the perovskite layer has been carried out $CH_3NH_3PbI_{3-x}Cl_x$, and also the work function of the back contact material (Me) on the photovoltaic characteristics of the solar cell. It was found that the optimal layer thickness $CH_3NH_3PbI_{3-x}Cl_x$ is 600–700 nm, and the concentration of defects should be less 10^{14} cm^{-3}.

It is shown that the work function of the rear contact material must be greater than or equal to 5 eV to create solar cells with high efficiency. Maximum efficiency obtained is 21.55% (short circuit current 24.87 mA/cm^2, open circuit voltage 1.1 V, fill factor 78.82%) for the structure of a perovskite solar cell with a carbon (C) back contact. The results can be used in the design and manufacture of efficient and inexpensive perovskite solar cells.

Acknowledgment. The reported study was funded by RFBR according to the research project № 19-29-03041_mk.

References

1. Kim, H., Lim, K.-G., Lee, T.-W.: Planar heterojunction organometal halide perovskite solar cells: roles of interfacial layers. Energy Environ. Sci. **9**, 12–30 (2016)
2. Jung, H.S., Park, N.-G.: Perovskite solar cells: from materials to devices. Small **11**, 10–25 (2015)
3. Sun, W., Choy, K.-L., Wang, M.: The role of thickness control and interface modification in assembling efficient planar perovskite solar cells. Molecules **24**, 3466–3480 (2019)
4. Sherkar, T.S., et al.: Recombination in perovskite solar cells: significance of grain boundaries, interface traps, and defect ions. ACS Energy Lett. **2**, 1214–1222 (2017)
5. Liu, M., Johnston, M.B., Snaith, H.J.: Efficient planar heterojunction perovskite solar cells by vapour deposition. Nature **501**, 393–398 (2013)
6. Sirotinskaya, S., Schmechel, R., Benson, N.: Influence of the cathode microstructure on the stability of inverted planar perovskite solar cells. RSC Adv. **10**, 23653–23661 (2020)
7. De Wolf, S., et al.: Organometallic halide perovskites: sharp optical absorption edge and its relation to photovoltaic performance. J. Phys. Chem. Lett **5**, 1035–1039 (2014)
8. Tominov, R.V., et al.: The effect of growth parameters on electrophysical and memristive properties of vanadium oxide thin films. Molecules **26**, 118 (2021)
9. Tominov, R.V., et al.: Synthesis and memristor effect of a forming-free ZnO nanocrystalline films. Nanomaterials **10**, 1007 (2020)
10. Vakulov, Z., et al.: Piezoelectric Energy Harvester Based on LiNbO3 Thin Films. Materials **13**, 3984 (2020)
11. Kudryashov, D.A., et al.: Nanoscale Cu2O films: radio-frequency magnetron sputtering and structural and optical studies. Semiconductors **51**(1), 111–115 (2017)
12. Behrouznejad, F., Shahbazi, S., Taghavinia, N., Wud, H.-P., Diau, E.-G.: A study on utilizing different metals as the back contact of CH3NH3PbI3 perovskite solar cells. J. Mater. Chem. A **4**, 13488–13498 (2016)
13. Burgelman, M., Nollet, P., Degrave, S.: Modelling polycrystalline semiconductor solar cells. Thin Solid Films **361–362**, 527–532 (2000)
14. Minemoto, T., Murata, M.: Device modeling of perovskite solar cells based on structural similarity with thin film inorganic semiconductor solar cells. J. Appl. Phys. **116**, 054505 (2014)
15. Raoui, Y., Ez-Zahraouy, H., Tahiri, N., El Bounagui, O., Ahmad, S., Kazim, S.: Performance analysis of MAPbI3 based perovskite solar cells employing diverse charge selective contacts: simulation study. Sol. Energy **193**, 948–955 (2019)
16. Malyukov, S.P., Sayenko, A.V., Ivanova, A.V.: Numerical modeling of perovskite solar cells with a planar structure. IOP Conf. Ser. Mater. Sci. Eng. **151**, 012033 (2016)
17. Rozhko, A.A., Petrov, V.V., Sayenko, A.V.: Study of the effect of the thickness of the photosensitive layer of perovskite on its efficiency using SCAPS-1D software. IOP Conf. Ser. Mater. Sci. Eng. **1035**, 012032 (2021)
18. Sayenko, A.V., Malyukov, S.P., Palii, A.V., Goncharov, E.V.: Influence of a Cu2O hole-transport layer on perovskite solar cells characteristics. Appl. Phys. **2**, 45–51 (2021)

19. Du, H.-J., Wang, W.-C., Zhu, J.-Z.: Device simulation of lead-free CH3NH3SnI3 perovskite solar cells with high efficiency. Chin. Phys. B **25**(10), 108803 (2016)
20. Takashi Minemoto, Y., Kawano, T.N., Chantana, J.: Numerical reproduction of a perovskite solar cell by device simulation considering band gap grading. Opt. Mater. **92**, 60–66 (2019)
21. Singh, N., Agarwal, A., Agarwal, M.: Numerical simulation of highly efficient lead-free all-perovskite tandem solar cell. Sol. Energy **208**, 399–410 (2020)
22. Abdelaziz, S., Zekry, A., Shaker, A., Abouelatta, M.: Investigating the performance of formamidinium tin-based perovskite solar cell by SCAPS device simulation. Opt. Mater. **101**, 109738 (2020)
23. Kawano, Y., Jakapan, C., Takahito, N., Takashi, M.: Influence of halogen content in mixed halide perovskite solar cells on cell performances through device simulation. Solar Energy Mater. Solar Cells **205**, 110252 (2020)
24. Zhou, H., et al.: Photovoltaics. Interface engineering of highly efficient perovskite solar cells. Science **345**(6196), 542–546 (2014)
25. Hima, A., Lakhdar, N., Benhaoua, B., Saadoune, A., Kemerchou, I., Rogti, F.: An optimized perovskite solar cell designs for high conversion efficiency. Superlattices Microstruct. **129**, 240–246 (2019)

Integrated Approach to Combinatorial and Logic Graph Problems

V. Kureichik(✉), D. Zaruba, and V. Kureichik Jr.

Southern Federal University, 105/42, Bolshaya Sadovaya Street, Rostov-on-Don 344006, Russia
vkur@sfedu.ru

Abstract. The article considers a new approach to combinatorial and logic graph problems, which belong to the class of NP-complex optimization problems. To address them effectively, the authors have suggested a new strategy based on the integrated approach. A distinctive feature of this approach is a division of search process in several levels. At the first level, areas with a high value of the objective function are allocated by an artificial bee colony optimization method. At the second level, the obtained results are improved by evolutionary and genetic algorithms. To implement this approach, it is developed an integrated search architecture, which can obtain sets of quasi-optimal solutions in polynomial time. As an example of the combinatorial and logic graph problem, a partitioning of elements can be given. The authors have suggested an evaluation function as a ratio of the total number of internal edges to the total number of connecting edges. A software application has been developed on the C#. A computational experiment has been carried out on the basis of IBM benchmarks. Conducted experimental investigation have shown that the quality of solutions obtained by the integrated approach on average exceeds the quality of solutions obtained by hMetis and PGAComplex by 5–7%. The time complexity of the developed integrated approach is $O(n^2)$ in the best case and $O(n^3)$ in the worst case.

Keywords: Combinatorial and logic problems · Graph · Bee colony optimization · Evolutionary algorithm · Genetic algorithm · Graph partitioning

1 Introduction

The present day, meeting the information needs of the users is one of the key problems in terms of solving the intelligent assistant systems that provide safety and effectiveness of search and cognitive activity in the Internet space [1–3]. The problem of discovering direct knowledge in the distributed educational Internet resources when modeling the scenarios of user interaction is very important. This problem can be considered as semantical and knowledge-based as it is related to the recognition of information objects (queries based on the information needs), connected with the task the user needs to solve [4–7]. In terms of obtaining the direct knowledge in the semantic search task, challenges can occur due to the dynamic character of the users' information needs and uncertainty of the level of their information awareness while forming the search queries.

A. A. Radionov and V. R. Gasiyarov (Eds.): RusAutoCon 2021, LNEE 857, pp. 203–213, 2022.
https://doi.org/10.1007/978-3-030-94202-1_20

The paper proposes a method for building the semantic net based on the ontological description of the structure of knowledge in the information space, where we try to carry out the search procedures. The ontological approach allows us to provide an integrated representation of the models of information needs, search queries domain, information models of the search images of Internet resources for their future comparison and intelligent processing [4, 8–10]. To increase the accuracy and reduce the structural and semantic conflicts in the process of information search, we propose a method for cluster analysis of the structure of high-dynamical Internet objects. Clustering can provide the division of the vector space of Internet objects features into semantic clusters with constraints on the feature of selecting the hidden patterns in the user queries, which indicates the possibility of content risks [11–13]. The suggested approach can improve the effectiveness of the contextual access to the resources in the open information and education environment and provide more accurate user preferences.

2 Problem Description

A graphs combinatorial and logical problem means a problem in which it is necessary to find the optimal solution. This decision is made optimal on the basis of a criterion (measure of evaluation of the investigated phenomenon) or an objective function (OF). Currently, there are a large number of graphs combinatorial and logical problems, and they have a different nature. However, their setting is similar. First, the initial set of alternative options or solutions M is specified. In this solution space M, constraints D are set, which must be satisfied by optimal solutions. Finally, the optimality criterion F is specified. The optimality criterion is a function defined on the set of feasible decisions and taking real non-negative numbers. In the general case, the statement of the graph combinatorial and logic problem can be represented as a tuple of length three: M, D, F. Consequently, the mathematical model of the problem consists of three components: the objective function, constraints and boundary conditions.

3 The Integrated Approach for Combinatorial and Logical Problems

All graph combinatorial and logic problems are NP-complete and NP-hard, i.e. there are no methods for their optimal solution in polynomial time. In this regard, the problem of the practical solvability of these problems arises, that is, it becomes necessary to find practical and effective methods for their solution. Currently, there are two approaches to solve these problems. The first approach is to simplify the algorithms, i.e. to reduce their computational complexity. The second approach is to simplify the problems by reducing the dimension or their decomposition [11].

Using the ideas of these two approaches, we propose the following strategy to find effective solutions based on multilevel optimization. The strategy is shown on Fig. 1 [12, 13]. Multilevel approaches are easily adaptable to the external environment, since they can embed various heuristics to improve the results.

The advantages of such approaches are quite obvious and this kind of research is being carried out by G. Karypis [14], V. Kumar [15], B. Hendricson [16], R. Leland [17], S. Barnard [18] etc. In this regard, the author proposes an integrated architecture based on methods of evolutionary modeling and bioinspired search for the effective solution of graph combinatorial and logic problems.

This strategy can implement different methods at different search levels. This allows the developers to reduce the dimension of the problem at the first level, and to perform effective optimization procedure based on a fast evolutionary or genetic algorithm at the second level.

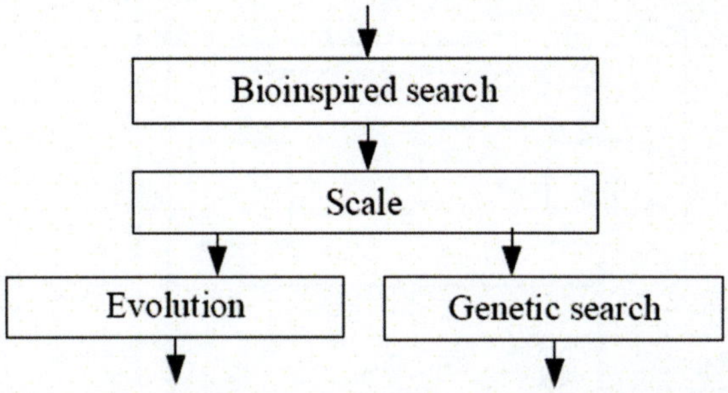

Fig. 1. The "bioinspired search-evolution-genetic search" strategy.

To implement this strategy, the authors propose the integrated search architecture based on the methods of evolutionary modelling and bioinspired search which is shown in Fig. 2 [7, 8, 19].

Let us describe each block of the architecture in more detail. The input data are the initial data of the combinatorial and logical problem. At the first stage, a modified bees optimization algorithm is applied.

This method consists of four key procedures which is shown in Fig. 3 [20, 21].

There are generation of the initial population on the basis of well-known principles [22], getting sites with a high objective function value, screening elite sites and its neighborhoods to increase search efficiency, and random search of new sites to avoid pre-convergence of the algorithm.

Let us describe the modified bees algorithm in detail. It is shown in Fig. 4.

Initially, initial information about the problem and main parameters are input in the algorithm. Each alternative solution is represented as a bee which save coordinates of the site. Next, the initial population of bees is generated. Bees are distributed among the sites depending on the objective function value. Scout bees starts search to find new sites in the neighborhood of the elite ones. After that researcher bees go there, and bees with the high objective function are selected. To avoid getting the algorithm into local optima, foragers are dispatched to perform a random search. Then their objective function values are assessed, and a new population of bees is formed. Finding the site with the highest amount of nectar corresponds to finding the global optimum. Note that the bee optimization method is not prone to looping in local optima, since it has a random search block in its structure. Moreover, obtaining effective results is based on

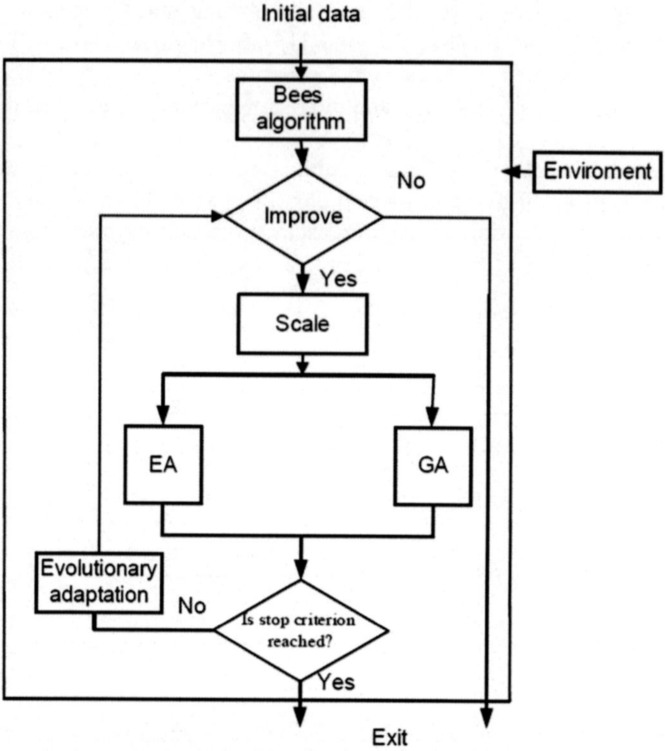

Fig. 2. Integrated search architecture.

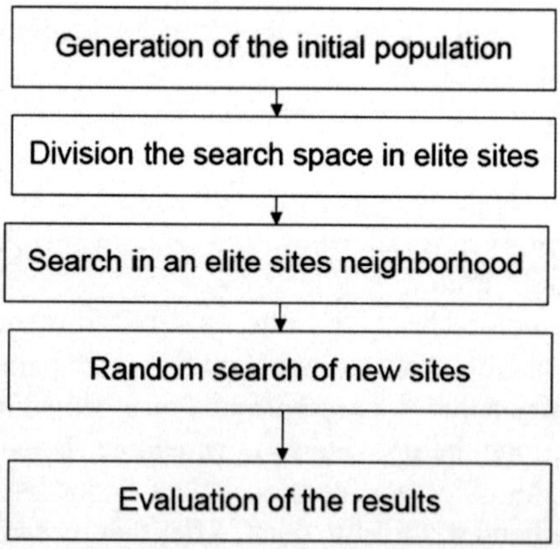

Fig. 3. Key procedures of the modified bee algorithm.

the decisions of all agents of the bee colony. Note that the proposed method and the developed algorithm for bee optimization allows you to dynamically divide the search space into subdomains with a high CF value, which significantly reduces the time of its operation.

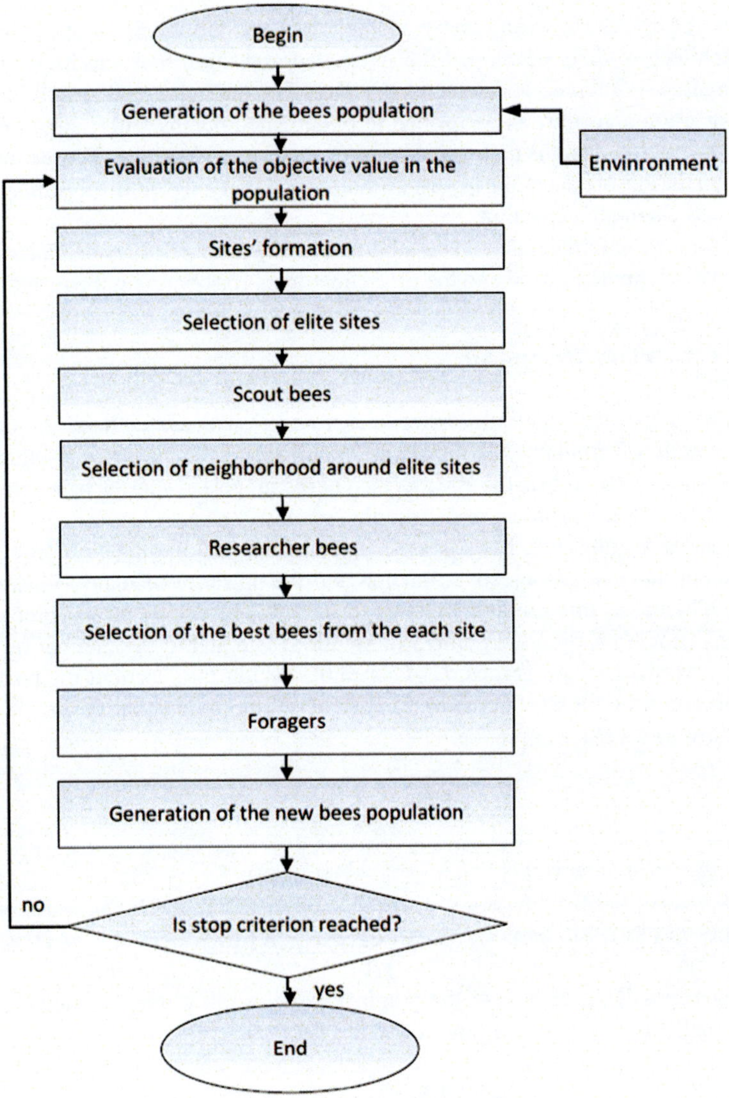

Fig. 4. Modified bees algorithm.

Next, it is checked whether an improvement is necessary. if not, then the output, if yes, then the results are transferred to the scale block, where the decision maker chooses which algorithm will be applied for further search [4, 23]. Note that, based on this architecture, two types of search can be implemented: combined and integrated. For the combined search, after the bee optimization algorithm, only an evolutionary algorithm is performed. This algorithm is based on only one operator - mutation and its various modifications. This method can quickly obtain sets of quasi-optimal solutions in reasonable time. In case that unsatisfactory results are obtained, the genetic algorithm is connected through the evolutionary adaptation block. On the basis of the presented search architecture, it is developed the integrated algorithm that can implement two types of search: combined and integrated search. For the combined search, firstly the bee optimization algorithm is performed to obtain subdomains with a high OF value, and then the improvement is performed by an evolutionary algorithm. For the integrated search, after the bee optimization algorithm and evolutionary algorithm, the improvement is made by the genetic algorithm.

The developed architecture can be modified to any number of levels, depending on the computer technology used and the time constraints for obtaining the result.

4 The Objective Function

The quality of the developed algorithm is assessed in accordance with the objective function, which is formulated taking into account a specific combinatorial-logical optimization problem. To confirm the effectiveness of the suggested approach, let us consider a specific optimization problem of partitioning graphs into parts.

Partitioning a graph into parts is a discrete conditional optimization problem due to the discontinuity of its objective function and the presence of many constraints on variables. Therefore, this problem belongs to the class of combinatorial problems. In partitioning problems, the total number of solutions is equal to the number of permutations of n vertices of the graphs, i.e. C_n = n!. Taking into account the restrictions on the formation of subsets, the total number of solutions is equal to the number of combinations of n vertices in m, i.e.

$$C_n^m = \frac{n!}{m!(n-m)!} \tag{1}$$

m is a number of subgraphs [1].

Let us present the formulation of a combinatorial-logical problem. Partitioning of the graph $G = (X, U)$ in parts $G_i = (X_i, U_i), X_i \subseteq X, U_i \subseteq U, i \in I = \{1, 2, \ldots, l\}$ (l is a number of parts) is to find such a set of parts G_i and G_j so that the number of connecting edges $|U_{i,j}| = K_{i,j}$ of the graph G satisfies the given optimality criterion [9, 23]. Usually, the partitioning criterion K is calculated as follows:

$$K = \frac{1}{2}\sum_{i=1}^{n}\sum_{j=1}^{n} K_{i,j}, i \neq j \tag{2}$$

The goal of optimization is $K \to min$. To estimate the efficiency of partitioning, we introduce the partitioning coefficient $\mu(G)$ that is calculated as the total number of interior ribs to the total number of connecting ribs.

$$\mu(G) = \frac{|U| - K}{K} \tag{3}$$

When carrying out a computational experiment, this coefficient is used as an estimate for comparing various algorithms to divide a graph and any of its parts. Note that the larger the value of μ (G), the more optimal the graph partition.

5 Experiments

To demonstrate the efficiency, as well as the computational characteristics of the integrated approach, a software application has been developed.

Experimental studies are planned as follows:

- Conducting a series of experiments with static values of control parameters to determine the time and space costs depending on the amount of input data.
- Search for a solution on a set of tests using the developed multilevel algorithm to determine the quality of the solution.

The experiments have been conducted for IBM test circuits [18, 19]. The description of the IBM test circuits package is shown in Table 1. Partition has been implemented in 4 parts by hMetis [24], PGAComplex and developed combined (CA) and integrated (IA) algorithms.

Table 1. IBM test citcuits.

Test circuit	Number of elements	Number of nets
Ibm01	12506	14111
Ibm02	19342	19584
Ibm03	22853	27401
Ibm04	27220	31970
Ibm05	28146	28446
Ibm06	32332	34862
Ibm07	45639	48117
Ibm08	51023	50513
Ibm09	53110	60902
Ibm010	68685	75196
Ibm11	70152	71076
Ibm12	70439	77241

Several series of experiments were carried out for the purpose of a clear demonstration of the effectiveness, as well as to determine the optimal values of the characteristics of the developed algorithms.

The hMetis algorithm belongs to the class of multilevel hypergraph partitioning algorithms. Unlike traditional partitioning methods, where the original graph is partitioned directly, the hMetis algorithm is executed in three stages. At the first stage, the dimension of the original hypergraph is reduced by combining the vertices or edges in such a way that the quality of the partition of the resulting hypergraph is not significantly lower than the quality of the partition of the original hypergraph. Then the given Fiduccia-Mattheyses (FM) hypergraph is partitioned by the algorithm. At the final stage, the resulting hypergraph is expanded to its original dimension and then the partition improvement algorithm is applied to reduce the number of interblock connections (FM algorithm).

The PGA Complex algorithm also belongs to the class of multilevel partitioning algorithms. Its key feature is the ability to control the speed of combining vertices and edges while reducing the dimension of the hypergraph and, consequently, the number of hierarchical levels.

A series of experimental studies of the partitioning algorithm was carried out with fixed values of the parameters of the multilevel search and the methods included in its composition.

The dependences of the quality and CPU time of these algorithms on the number of elements in circuits are shown in Table 2 and Fig. 5 and 6.

Table 2. Comparison of the partitioning results by the hMetis, PGAComplex, CA and IA.

c	hMetis		PGAComplex		CA		IA	
	μ(G)	t, (s)	μ(G)	t, (s)	μ(G)	t, (s)	μ(G)	t, (s)
ibm01	21.33	65	21.27	78	21.21	62	21.39	75
ibm02	22.74	213	22.76	226	22.67	210	22.86	224
ibm03	19.43	192	19.48	214	19.35	190	19.52	198
ibm04	17.32	214	17.27	226	17.27	207	17.41	224
ibm05	22.47	972	22.53	986	22.41	967	22.54	991
ibm06	14.64	1115	14.61	1121	14.57	1108	14.71	1123
ibm07	15.87	1509	15.83	1517	15.78	1501	15.96	1518
ibm08	20.33	1453	20.32	1468	20.27	1442	20.43	1469
ibm09	18.43	1724	18.47	1739	18.38	1713	18.57	1741
ibm10	21.13	1678	21.09	1691	21.02	1667	21.26	1687
ibm11	16.74	2156	16.68	2168	16.64	2148	16.88	2173
ibm12	15.84	2247	15.79	2259	15.73	2241	15.93	2263

here $\mu(G)$ is an objective function showing the ratio of the total number of internal ribs to the total number of connecting ribs, calculated by (3), t-time in seconds.

Having analysed the table and histograms, it can be concluded that the fastest optimization algorithm is the combined algorithm based on the bees and evolutionary algorithms. At the same time its quality is 10% worse than that of other presented algorithms.

Having analysed the table and histograms, it can be concluded that the fastest optimization algorithm is the combined algorithm based on the bees and evolutionary algorithms. At the same time its quality is 10% worse than that of other presented algorithms. The integrated algorithm, based on the bee optimization method, evolutionary

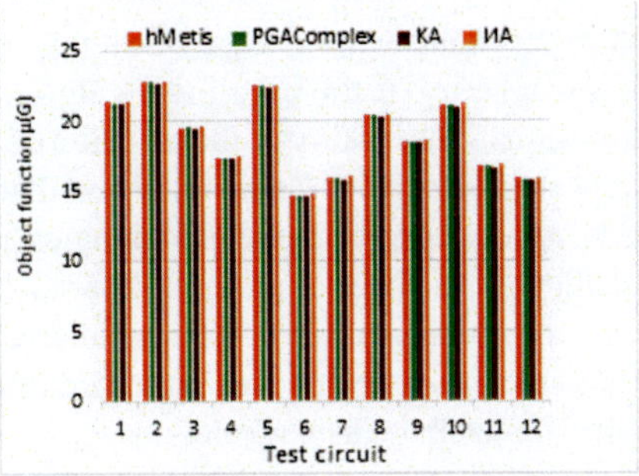

Fig. 5. Histogram of comparison of objective function values from IBM test circuits.

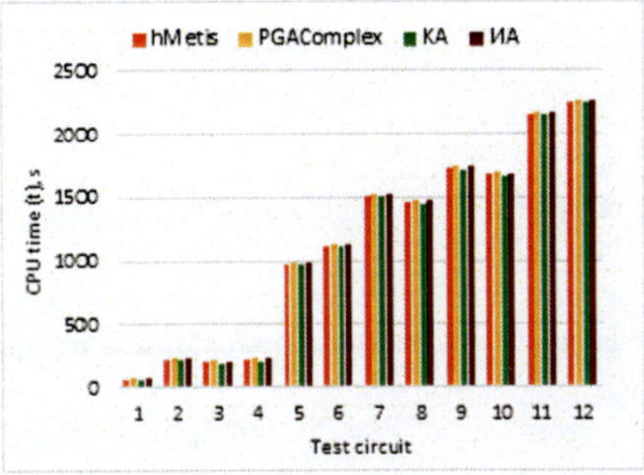

Fig. 6. Histogram of comparison of CPU time from IBM test circuits.

and genetic search algorithms, requires more CPU time, but in terms of quality, on average, 5 to 7% surpasses the partitioning results obtained using the well-known hMetis, PGAComplex algorithms.

6 Conclusion

The paper considers the approach to solve graphs combinatorial-logical problems. A new strategy based on the integrated approach is proposed for their effective solution. A distinctive feature of the approach is the division of the search process into several levels. At the first level, subareas with a high value of the objective function are distinguished based on the bee optimization method. On the second level, the obtained solutions are improved on the basis of evolutionary and genetic algorithms. To implement this approach, the integrated search architecture has been developed and described. It allows one to obtain sets of quasi-optimal solutions in polynomial time and avoid looping in local areas. To confirm the suggested approach, it has been considered the partitioning problem. On the basis of the software application the computational experiment has been carried out on test examples (benchmarks) from IBM. Experimental studies have shown the advantage of the developed integrated approach in comparison with known methods for graphs combinatorial-logical problems.

The series of tests and experiments carried out made it possible to refine the theoretical estimates of the time complexity of the partitioning algorithms and their behavior for circuits of various dimensions. The quality of the partition obtained by the developed integrated algorithm is, on average, from 5 to 7 percent higher than the partitioning results obtained by the well-known algorithms hMetis, PGAComplex with a comparable CPU time. This indicates the effectiveness of the proposed approach. The time complexity of the developed integrated algorithm is $O(n^2)$ at the best case, and $O(n^3)$ at the worst one. This testifies not only to the efficiency of the created multilevel algorithm, but also to the efficiency of the software package as a whole.

Acknowledgment. The reported study was funded by RFBR according to the research project № 19-01-00059.

References

1. Sherwani, N.: Algorithms for VLSI Physical Design Automation, 3rd edn. Kluwer Academic Publisher, Philip Drive Norwell (2013)
2. De Jong, K.: Evolutionary computation: recent development and open issues. In: Proceedings 1st International Conference, Evolutionary Computation and Its Application, Moscow, pp. 7–18 (1996)
3. Karpenko, A.: Modern algorithms of search optimization. Algorithms inspired by nature. Moscow, Russia, p. 446 (2014)
4. Abraham, A., Ramos, V., Grosan, G.: Swarm Intelligence in Data Mining. Springer, Heidelberg (2007)
5. Hassanien, E.: Emary, E.: Swarm Intelligence. Principles Advances, and Applications. CRC Press, Boca Raton (2015)
6. Kureichik, V., Bova, V., Kureichik, Jr. V.: Hybryd approach for computer-aided design problems. In: 2019 International Seminar on Electron Devices Design and Production, SED 2019 – Proceedings, 8798406 (2019)
7. Kureichik, V., Zaruba, D.: Integrated approach for partitioning of electronic computer equipment blocks. In: Proceedings - 2019 International Russian Automation Conference, RusAutoCon (2019)
8. Kureichik, V., Kuliev, E., Kureichik, V.: Integrated algorithm for elements placement on the printed circuit board. IOP Conf. Ser. Mater. Sci. Eng. **734**(1), 012146 (2020)
9. Kureichik, V., Zaruba, D., Kureichik, V., Jr.: Hybrid approach for graph partitioning. Adv. Intell. Syst. Comput. **573**, 64–73 (2017)
10. Kureichik, V.M., Kureichik, V.V.: A genetic algorithm for graph partitioning. J. Comput. Syst. Sci. Int. **38**(4), 580–588
11. Karypis, G., Kumar, V.: Multilevel k–way partitioning scheme for irregular graphs. J. Parallel. Distrib. Comput. **48**, 96–129 (1998)
12. Hendrickson, B., Leland, R.: A multilevel algorithm for partitioning graphs. In: Proceedings of the 1995 ACM/IEEE Conference on Super Computing, pp. 626–657 (1995)
13. Kureichik, V., Kureichik, V., Bova, V.: Placement of VLSI fragments based on a multilayered approach. Adv. Intell. Syst. Comput. **464**, 181–190 (2016)
14. Zheng, D., Wang, M., Gan, Q., Song, X., Zhang, Z., Karypis, G.: Scalable graph neural networks with deep graph library. In: 14th ACM International Conference on Web Search and Data Mining, WSDM 2021 20121:1141–1142 (2021)

15. Stephan, J., Fink, J., Kumar, V., Ribeiro, A.: Hybrid architecture for communication-aware multi-robot systems. In: 2016 IEEE International Conference on Robotics and Automation, ICRA 2016, pp. 5269–5276 (2016)
16. Mattson, T., et al.: Standards for graph algorithm primitives. In: IEEE High Performance Extreme Computing Conference, HPEC 2013, Waltham, MA, 10 September 2013–12 September 2013 (2013)
17. Berthold, B., Bihl, T.J., Cox, C., Jenkins, T.A., Leland, L.: Probabilistic reasoning for real-time UAV decision and control. In: Signal Processing, Sensor/Information Fusion, and Target Recognition XXVIII 2019, 15 April 2019 –17 April 2019 (2019)
18. Barnard, S., et al.: Large-scale distributed computational fluid dynamics on the information power grid using Globus. 7th Symposium on the Frontiers of Massively Parallel Computation, Frontiers 1999, 21 February 1999 - 25 February 1999 (1999)
19. Karaboga, D.: An idea based on honey bee swarm for numerical optimization. Erciyes University, Engineering Faculty, Computer Engineering Department, p. 110(2005)
20. Zaruba, D., Zaporozhets, D., Kureichik, V.: Artificial bee colony algorithm—A novel tool for VLSI placement. Adv. Intell. Syst. Comput. **450**, 433–442 (2016)
21. Kureichik, V.M., Kureichik, V.V.: A genetic algorithm for graph partitioning. J. Comput. Syst. Sci. Int. **38**(4), 580–588 (1999)
22. Alpert, C.J.: The ISPD-98 circuit Benchmark suit. In: Proceedings of the ACM/IEEE International Symposium on Physical Design, April 1998, pp. 80–85 (1998)
23. Karypis, G., Kumar, V.: Analysis of Multilevel Graph Partitioning. Department of Computer Science, University of Minnesota, Minnesota (1995)
24. Karypis, G., Kumar, V.: METIS: A Software Package for Partitioning Unstructured Graphs, Partitioning Meshes, and Computing Fill–Reducing Orderings of Sparse Matrices Version 5.1.0. Department of Computer Science and Engineering University of Minnesota Minneapolis (2013)

Ontological Tools for Modeling the Quality of Radiopharmaceuticals Production

S. Larin[1], R. Bildanov[1,2(✉)], and A. Smagin[1]

[1] Ulyanovsk State University, 42, Leo Tolstoy Street, Ulyanovsk 432000, Russia
[2] TRINITI JSC, 12, st. Pushkovs, Moscow 108840, Russia

Abstract. This article considers the implemented means of modeling and recording the parameters of technological processes, documentation, and support of the life cycle of radiopharmaceutical production. Simulation tool (imitator) is a hardware-software complex for recording, storage, and analysis of data on the technological process of radiopharmaceutical production. The purpose of the functioning of this system is also the automation of production processes of forming accompanying documentation, control of production processes. The means of ontological support make it possible to identify possible risks in the production processes of radiopharmaceuticals, based on partially entered indicators and previous checks, as well as to form a database of precedents to ensure compliance of the manufacturing process with regulatory documents, information support and recording of technological parameters during manufacturing, including document flow and process control is necessary. For this purpose, a new information system for recording technological process parameters, documentation, and support of the radiopharmaceutical production life cycle was developed.

Keywords: Technological processes · Life cycle · Radiopharmaceutical production

1 Introduction

The purpose of the work is to ensure the quality of the manufactured radiopharmaceutical product (RFP) using a given technological process (TP). The quality of the product is understood as complete compliance of the characteristics (parameters) of the manufactured product with the requirements for its declared quality, which is achieved by the correct (error-free) performance of all stages and steps of the technological process. To this end, the technological process must be represented at the operational level, i.e. it must consist of a set of operations that can be fully controlled with the required degree of accuracy.

A large number of scientists have contributed to the development of the idea of ontology. On the one hand, the ideas of Enterprise Ontology were proposed [1]. Then METHODOLOGY methodology for creating ontological systems was created [2]. The ideas of introducing ontology systems into engineering were also developed [3]. CYC methodology was developed [4]. The ideas of ontology were also introduced in natural language processing [5] and knowledge base systems [6].

A. A. Radionov and V. R. Gasiyarov (Eds.): RusAutoCon 2021, LNEE 857, pp. 214–223, 2022.
https://doi.org/10.1007/978-3-030-94202-1_21

In Russia, the ideas of ontology have also been constantly developing. Vittich V. suggests an ontological approach to building information models [7]. Borgest N. develops the ideas of solving design problems with the help of ontologies [8]. Burdo G. Introduces the ideas of ontology in CAD [9, 10]. Such scientists as, Ogroskin V., Andrich O., etc. offer ontologies for engineering systems [11, 12]. Smirnov considered the implementation of ontologies in computer modeling [13–16]. P. Sosnin considered ontology in the context of professional experience and question-response environments [17, 18].

2 Theoretical Part

For a description of TP of RFP manufacturing considering its listed features and characteristics the event-process sequence of TP is the most adequate one and it is expressed by the chain event - transition - operation - transition - event.

Under event we will understand the result of operation completion at each step, under transition - the transition to the implementation of the next in order operation i.e., to the next step, under operation - the very technological operation, defined in this step. In a general way, the chain can be represented as links, which are connected to the source of control of these links, which can be the TP map. Each processing state must be characterized by a certain set of parameters. Let Si be the state of the process at the i point in time where i = 1, 2, 3,....n, then Ti = {p1, p2.....pm} is the set of parameters characterizing the i state.

The number of parameters m for each state is fixed, finite, and has a discrete value, which is fixed by sensors or measured by a lab technician. All parameters fixed with a given degree of accuracy are controlled by comparing them with the limits of intervals, which for them are predetermined by the technology of radiopharmaceutical production.

The tool for checking the performance of the stages of RFP production includes the following components, Fig. 1.

- An operator (lab technician) entering new metrics on the stages and operations of RFP production.
- An ontology describing the RFP development process.
- A precedent database containing the data on the entered indices and the results of verification.
- A imitator with a graphical interface that provides input of new indicators by the operator (laboratory technician) and performs verification of the entered indicators.

Interaction of components with the imitator is carried out using data query languages.

- Ontology and imitator - SPARQL query language that works with data represented by RDF model.
- Precedent database and imitator - SQL query language for interacting with relational databases.

The imitator performs the following functions and consists of modules that perform these functions, Fig. 2.

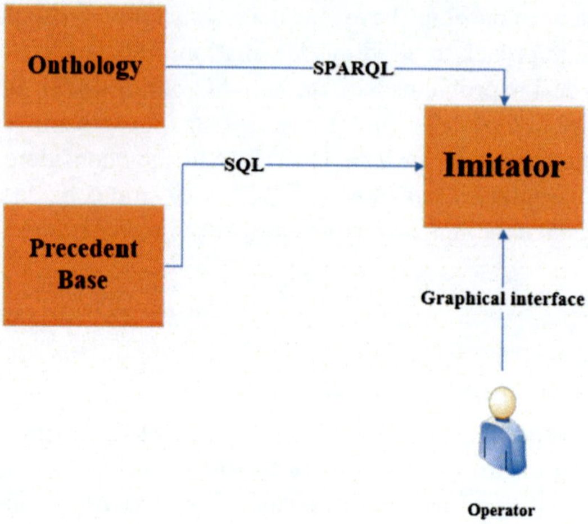

Fig. 1. Flowchart of the RFP production step verification tool.

- Operator input and validation of new indicators.
- Connection to the ontology and data exchange with it using the SPARQL language.
- Connection to the precedent database and data exchange with it using the SQL language.
- Formation of check matrices based on the classes, objects, and their attributes represented in the ontology.
- Checking of the indicators entered by the operator and recording the results in the precedent database.

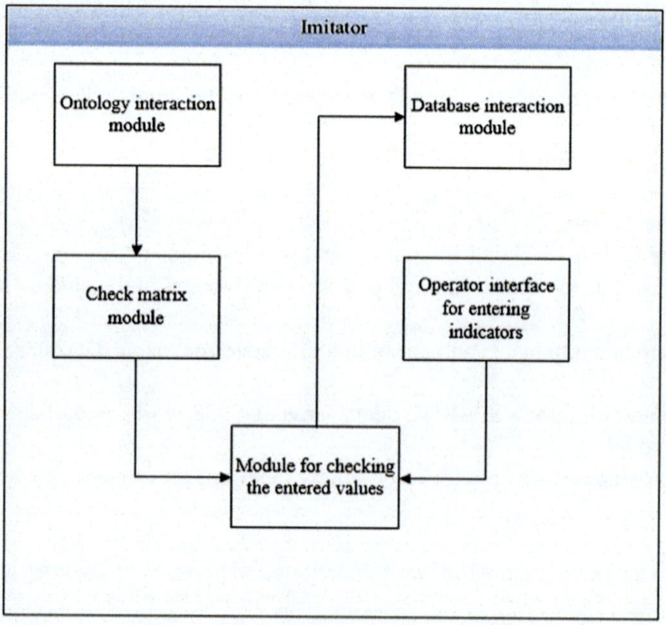

Fig. 2. Schematic diagram of the imitator.

3 Practical Part

Below is an algorithm of the imitator, which includes the following steps, Fig. 3.

- Connecting to the ontology and the precedent database.
- Formation of checking matrices based on the data from the ontology.
- Entering indicators using the operator interface (input indicators is carried out by setting a "tick" in those operations of production, which have a Boolean nature (done/not done) and entering the numerical values of other indicators. If any value has not been entered, the interface "highlights" this indicator in red and offers to enter its value).
- Verification of the production stages by multiplying the vector-string of the entered indicators by the corresponding verification matrix.
- Displaying the result of the check on the screen and recording it in the precedence database.

The precedents database contains 5 tables, Fig. 4.

- Operators (users) with the ability to enter indicators.
- Full list of RFP production operations with linkage by stages.
- RFP production history.
- Previously entered data on the RFP production operations.
- RFP releases with the description of check results, check date, and indicators entered during this check.

The precedent database allows predicting the possibility of correct RFP production based on partially entered indicators and previous checks.

The software implementation of the precedent database is performed in the PostgreSQL database. The physical model of the database is shown in, Fig. 5.

The ontology of the RFP development process description contains a tree of production stages, operations included in these stages, and indicators for each operation with information about permissible values.

Based on the data stored in the ontology, a graphical operator interface is constructed in the form of a tree of RFP production stages and their operations. For each operation, input elements by indicators are displayed (a checkbox, if the indicator has a boolean type (yes/no), or a text field, if a numeric value should be entered for the indicator).

Deterministic automata S earlier were defined by the set of objects S(A, X, Y..), where A (1) is set of internal states of the automaton, X (2) set of input signals (input alphabet), x_1 is a letter of the input alphabet, Y (3) is set of output signals (output alphabet). The function of transitions, providing the unambiguous transition of the automaton to the state as from the state a_m under the action of the input signal x_F, i.e. a_s (4) is the function of outputs, characterizing the unambiguous value of the output signal yq depending on the state of the automaton am and the input signal x_f i.e. y_q (5).

$$A = \{a_0, a_1, a_2, \ldots, a_m\} \tag{1}$$

$$X = \{x_1,\ x_2 \ldots x_F\} \tag{2}$$

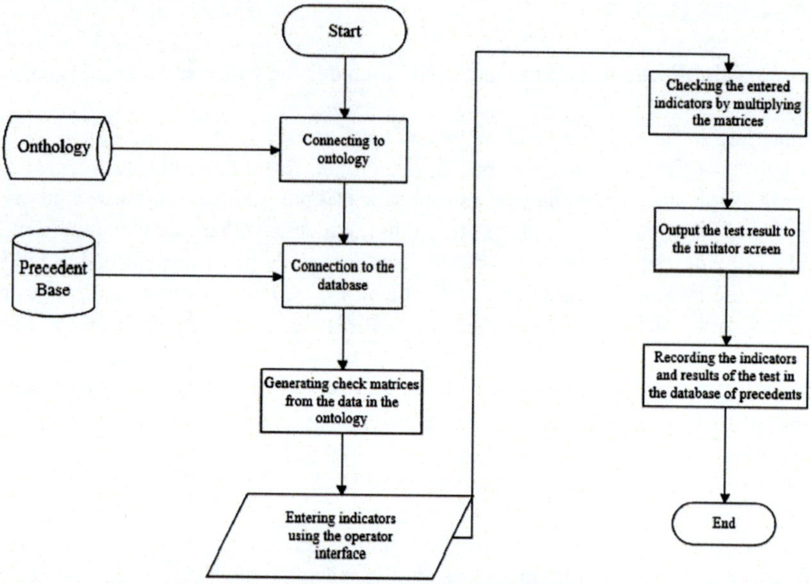

Fig. 3. The imitator's operation algorithm.

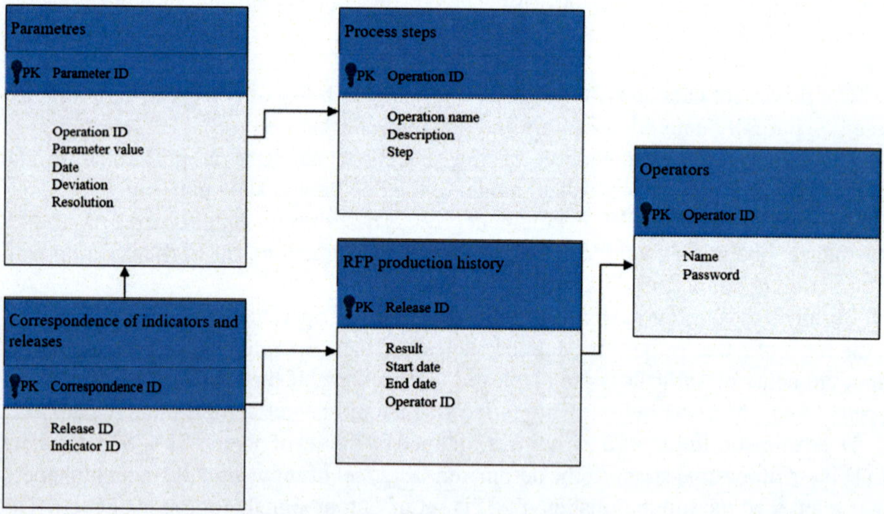

Fig. 4. Precedent database logic model.

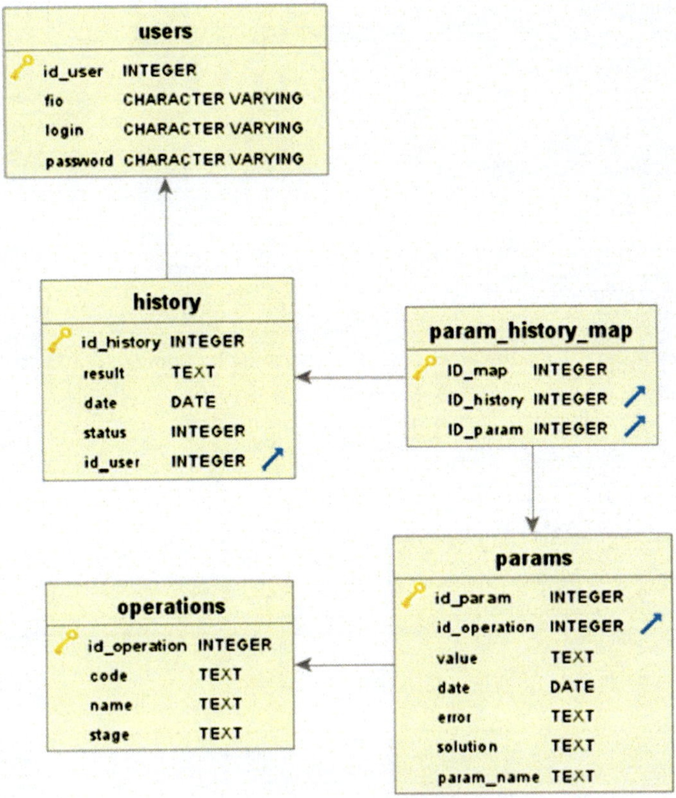

Fig. 5. Precedent database logi.

$$Y = \{y_1, y_2, \ldots, y_G\} \tag{3}$$

$$a_s = \left\{a_m, x_f\right\} \tag{4}$$

$$y_q = \left\{a_m, x_f\right\} \tag{5}$$

Let us consider the most general probabilistic model of the automaton, and specifically: knowing at the subsequent moment, as well as what output signal it produces in this case, Table 1, Table 2.

Automata in which knowing the state of the automaton a_m and the input signal x_f, we can only specify probabilities of transition to a new state and probabilities of appearance of in contrast to deterministic automata, where the functions of transitions and outputs are one-to-one (6) and (7), in PA these functions are probabilistic and specify probabilities of appearance of a state at time t + 1 and probabilities of appearance of an output letter. In a probabilistic automaton, the mechanism of randomness acts: the states of the automaton and the output letters appear randomly (8) This formula sets the

Table 1. State transition matrix.

a_x,x_f	a_0	a_1	….	a_m	….	a_M
a_0,x_1	P_{010}	P_{011}	….	P_{01m}	….	P_{01M}
a_0,x_2	P_{020}	P_{021}	….	P_{02m}	….	P_{02M}
….	….	……	….	….	….	…
a_m,x_F	P_{0F0}	P_{0F1}	….	P_{0Fm}	….	P_{0FM}
a_1,x_1	P_{110}	P_{111}	….	P_{11m}	….	P_{11M}
…	…	….	….	…	….	….
a_m,x_f	P_{mf0}	P_{mf1}	….	P_{mf1}	….	P_{mfM}
…	….	….	….	…	….	….
a_M,x_F	P_{MF0}	P_{MF1}	….	P_{MFm}	….	P_{MFM}

Table 2. Matrix of appearance of output signals.

a_m,x_f	y_1	y_2	….	y_y	….	a_M
a_0,x_1	q_{011}	q_{012}	….	q_{01y}	….	q_{01G}
a_0,x_2	q_{021}	q_{022}	….	q_{02y}	….	q_{02G}
….	….	……	….	….	….	…
a_m,x_F	q_{0F1}	q_{0F2}	….	q_{0Fy}	….	q_{0FG}
a_1,x_1	q_{111}	q_{112}	….	P_{11y}	….	P_{11G}
…	…	….	….	…	….	….
a_m,x_f	q_{mf1}	q_{mf2}	….	q_{mfy}	….	q_{mfG}
…	….	….	….	…	….	….
a_M,x_F	q_{MF1}	q_{MF2}	….	q_{MFy}	….	q_{MFG}

conditional probability that at time t + 1 the automaton will go to state a(t + 1) if at time t the automaton was in a state a(t), and x(t) has arrived.

$$a(t+1) = F[a(t), x(t)] \tag{6}$$

$$y(t) = \Phi[a(t), x(t)] \tag{7}$$

$$P\big[a(t+1)y(t)/a(t), x(t)\big] \tag{8}$$

This formula specifies the conditional probability that at time t + 1 the automaton will move to state a(t + 1) if at time t the automaton was in a state a(t) and x(t) was received.

Assume that the state of the automaton and the appearance of the input letter are independent. Under this condition, a distinction is made Mile's probabilistic automaton

(9) and Moore's probabilistic automaton (10)

$$P\big[a(t+1)y(t)/a(t),\ x(t)\big] = P[a(t+1)/a(t), x(t)] * P\big[y(t)/a(t), x(t)\big] \tag{9}$$

$$P\big[a(t+1)y(t)/a(t), x(t)\big] = P[a(t+1)/a(t), x(t)] * P\big[y(t)/a(t)\big] \tag{10}$$

After transforming the deterministic automata model of TP by keeping all its vertices and replacing signals at transitions and outputs by probabilities, let us interpret the probabilistic graph by a Markov chain, considering the sequence of vertices (TP operations) as a sequence of events with a countable number of outcomes and having the Markov property, which is shown by the fact that probability of appearance of this or that value at (K + 1) steps depends only on the value taken by this value at R-step, and does not depend on values of value at 1, 2,... k-steps. In other words, it is a sequence of random events in the vertices of the graph in which the probability of each event depends only on the state in which the TP is at the current step and does not depend on the earlier states.

Models of technological processes are described by the following probability distribution laws: binomial distribution, normal distribution law, Poisson's law, equal probability law, etc. These models refer to the simplest probability models of the distribution of a single quantity.

Knowing the number of operations and their probability values, which are used for approximation, we can calculate the number of vertices (steps-operations) that are included in the chain "yellow vertex - red vertex" by modeling with the Poisson distribution. Regarding the position of the yellow top, they are content related to those operations that are marked as risky. Their number is limited and in practice, they are easily defined. There can be many such nodes, and their location during TP is also well known.

The figure shows the simulation of the technological process using Poisson distribution (blue graph), normal distribution (blue graph), linear regression (purple graph), Fig. 6. By selecting the appropriate coefficients for each of the distributions, it is possible to construct graphs, as close as possible to the real values of failures of the technological process operations, obtained based on statistical data during the execution of the TP.

For the Poisson distribution (11), this coefficient is λ (mathematical expectation of a random variable), where x is the operation number. For the normal distribution (12), it is μ (mathematical expectation) σ2 (distribution variance). The linear regression is described by Eq. (13). Because Poisson distribution, normal distribution, and linear regression for different operations give different approximations to real values, and aggregating distribution (14) was constructed.

$$p(x) = \frac{\lambda^k}{k!} e^{-\lambda} \tag{11}$$

$$n(x) = \frac{1}{\sqrt{2\pi\sigma}} e^{\frac{-(x-\mu)^2}{2\sigma^2}} \tag{12}$$

$$l(x) = b \tag{13}$$

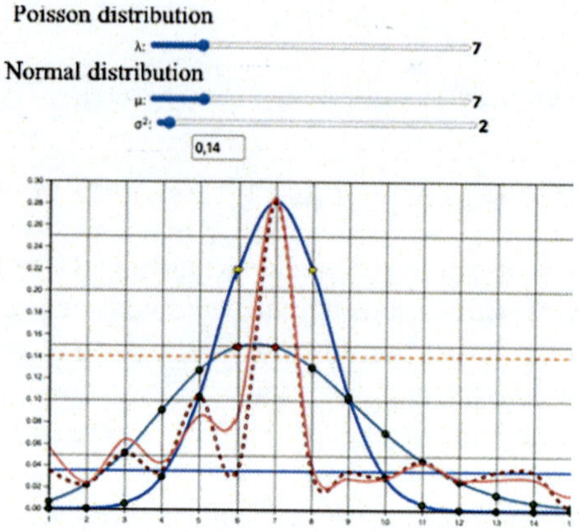

Fig. 6. Process modeling.

$$a(x) = \begin{cases} p(x),\ for\, x \in \{2, 3, 11, 12, 15, 20, 23, 30\} \\ n(x),\ for\, x \in \{5, 7, 10\} \\ l(x),\ for\, x \in \left\{ \begin{array}{l} 1, 4, 6, 8, 9, 13, 14, 16, 17, 18, 19, 21, 22, 24, 25, \\ 26, 27, 28, 29, 31, 32, 33, 34, 35, 36, 37, 38, 39, 40 \end{array} \right\} \end{cases} \quad (14)$$

4 Conclusion

Thus, a new information system has been developed that combines a centralized data storage - database server, a set of electronic computers with specialized software, and technological equipment involved directly in production. The system performs automated and manual recording of technological and control operations parameters, automated accounting of consumption and receipt documents, as well as stock balances of components and consumables, generates reports on stock balances for a given date, generation of outgoing accompanying documents, storage of incoming accompanying documents, stock accounting.

References

1. Uschold, M., Gruninger, M.: Ontologies: principles, methods and applications. Knowl. Eng. Rev. **11**(2), 93–136 (1996). https://doi.org/10.1017/s0269888900007797
2. Fernández-López, M., Gómez-Pérez, A., Juristo, N.: Methontology: from ontological art towards ontological engineering. In: Proceedings of the Ontological Engineering AAAI-97 Spring Symposium Series, pp. 33–40 (1997)
3. Sure, Y., Staab, S., Studer, R.: Ontology engineering methodology. In: Staab, S., Studer, R. (eds.) Handbook on Ontologies International Handbooks on Information Systems. IHIS, pp. 135–152. Springer, Heidelberg (2009). https://doi.org/10.1007/978-3-540-92673-3_6
4. Sowa, J.: Building large knowledge-based systems: representation and inference in the Cyc project. Artif. Intell. **61**(1), 95–104 (1993). https://doi.org/10.1016/0004-3702(93)90096-t

5. Schreiber, G., Wielinga, B., Jansweijer, W.: The KACTUS view on the 'O' word. In: Proceeding of the IJCAI Workshop on Basic Ontological Issues in Knowledge Sharing, pp. 159–168 (1995)
6. Jones, D., Bench-Capon, T., Visser, P.: Methodologies for ontology development. In: Proceeding of the 15th IFIP World Conference, pp. 20–35 (1998)
7. Vittikh, V., Sitnikov, P., Smirnov, S.: Ontological approach for the construction information-logical models in the processes of management of social systems. Herald Comput. Inf. Technol. **5**, 45–53 (2009)
8. Borgest, N., Simonova, E., Shustova, D.: Managing Project Tasks Using Ontological Systems. Samara Universty, Samara (2010)
9. Burdo, G., Palyukh, B., Vorobyeva, E.: Methodological bases of construction cad with developing the knowledge base. OSTIS-2014 (2014). https://libeldoc.bsuir.by/bitstream/123456789/26273/1/Burdo_SAPR.PDF. Accessed 27 May 2021
10. Burdo, G., Semenov, N., Isaev, A.: Intelligent procedures of technological processes design in integrated cad-cam systems. Search Int. Res. Pract. J. Softw. Syst. **1**(105), 60–64. Accessed 27 May 2021 (2014)
11. Ontology for Engineering Data. Open Systems (2013). https://www.osp.ru/os/2013/06/13036814. Accessed 27 May 2021
12. Andrich, O., Makyshkina, L.: Research evaluating quality ontological model methods. Contemp. Res. Innov. **3**(35), 11. Accessed 27 May 2021
13. Smirnov, S.: An Ontological Relativity and the Simulation Technology of Complex Systems, vol.1, pp. 66–71. Izvestiya of Samara Scientific Center of the Russian Academy of Sciences (2000)
14. Smirnov, S.: Ontological Analysis of Modeling Domains. Izvestiya of Samara Scientific Center of the Russian Academy of Sciences, vol. 1 (2001)
15. Smirnov, S.: The experience of semantic modeling and designing tools on the widely used platform. OSTIS-2015 (2015). https://libeldoc.bsuir.by/bitstream/123456789/4115/1/Smirnov_Opyt.PDF. Accessed 27 May 2021
16. Smirnov, S.: OSTIS-2013 (2013). https://libeldoc.bsuir.by/handle/123456789/4276. Accessed 27 May 2021
17. Sosnin, P.: Integrated cognitive assessment and optimal choice algorithm for ontological model. Researchgate (2014). https://www.researchgate.net/profile/Petr-Sosnin/publication/267509237_Ontologiceskaa_Podderzka_Konceptualnogo_Eksperimentirovania_v_Voprosno-Otvetnyh_Modeliruusih_Sredah/links/545118d80cf24884d886f805/Ontologiceskaa-Podderzka-Konceptualnogo-Eksperimentirovania-v-Voprosno-Otvetnyh-Modeliruusih-Sredah.pdf. Accessed 27 May 2021
18. Sosnin, P.: A personal ontology of professional experience. OSTIS-2014 (2014). https://libeldoc.bsuir.by/handle/123456789/26159. Accessed 27 May 2021

Structural Analysis of the Process Based on Extended Petri Nets with Semantic Relations

O. Kryukov(✉) and A. Voloshko

Tula State University, 92, Lenina Avenue, Tula 300012, Russia

Abstract. Process analysis and optimization is impossible without simulation. Models make it possible to represent and visualize the main interrelationships between the individual stages of the process, to investigate and predict the course of its development, to find bottlenecks and the ways of modernization. One of the important tasks in process optimization is the analysis of its structure, since various options for its organization impose corresponding restrictions on the possibility of its rebuilding. This paper discusses the issues of structural analysis of processes: the definition of linear structures, branches and cycles. Algorithms for the automatic analysis of the process represented by the extended Petri net with semantic relation are developed. An example of searching for such structures in the process of repairing a truck is considered.

Keywords: Petri nets · Manufacturing process · Simulation · Distributed process · Process structure · Analysis · Parallelization

1 Introduction

The current state of the industry is characterized by an increasing level of automation. Use of modern systems allows building and analyzing models based on process logs. However, the process of analyzing models remains a rather difficult task, since it depends on the goals and objectives of the analysis, as well as the modeling method.

Processes are a set of interrelated operations and have a complex structure. When analyzing and optimizing processes, it is often important to determine the type of structure to which the considered part of the process belongs. In this regard, it is necessary to develop methods for identifying process structures and, in the future, methods for their optimization.

2 Overview of Main Approaches to Structural Analysis

Issues related to the structure of the system can be resolved both at the stage of model creation by building special structural models, and at the stage of models analysis by carrying out structural analysis.

Structural modeling involves modeling the organizational structure of systems and subsystems, such as: informational, organizational, functional, management, i.e. modeling the composition and relations between the elements [1]. A detailed analysis of

A. A. Radionov and V. R. Gasiyarov (Eds.): RusAutoCon 2021, LNEE 857, pp. 224–236, 2022.
https://doi.org/10.1007/978-3-030-94202-1_22

possible methods for business processes modeling is presented in [2]. Special aspects of production processes within business processes and possible options for application of some models of production lines are considered in [3]. A large number of researchers suggest using the SADT methodology [4, 5] based on the IDEF [6–8] or BPMN notation and its extensions [9–11] for structural modeling. A comparative analysis of these approaches is presented in [12]. At the same time, methods have been developed to automate the creation of such models [13–15], which allows in the future also to automate the analysis of these models and the synthesis of new solutions.

Structural analysis can be applied for different types of models. Its main purpose is to search for static characteristics of the system related to the interconnections of its elements and subsystems at various levels, as well as to classify structures that are important for decision-making. Depending on the type of the studied models and on the selected classification parameter, the approaches to structural analysis may also differ [16–18]. It is shown in [19] that one of the promising methods for modeling production processes is mathematical apparatus of Petri nets and their extensions. Methods for structural analysis of Petri nets are discussed in [20–22]. However, these analyzes do not consider the possibility of restructuring the net to execute a series of steps in parallel. Therefore, it is necessary to develop your own algorithms that allow you to define the key structures that affect the possibility and the choice of the method of parallel organization of the process.

3 Extended Petri Nets with Semantic Relations as an Apparatus for Simulation of Distributed Manufacturing Processes

The simplest extended Petri net with semantic relations (ExPNSR) can be specified by the following set:

$$\Pi = \left\{A, \left\{Z^C, \tilde{R}^C, \hat{R}^C\right\}, \left\{Z^S, \tilde{R}^S, \hat{R}^S\right\}\right\},$$

where $A = \left\{a_{1(a)}, \ldots, a_{j(a)}, \ldots, a_{J(a)}\right\}$ – a finite set of places; $Z^C = \left\{z^C_{1(z^C)}, \ldots, z^C_{j(z^C)}, \ldots, z^C_{J(z^C)}\right\}$ – a finite set of transition by control relations; $\tilde{R}^C$ – an incidence matrix of size $J(a) \times J\left(z^C\right)$ that maps the set of places to a set of transitions by control relations; $\hat{R}^C$ – an incidence matrix of size $J\left(z^C\right) \times J(a)$ that maps the set of transitions by control relations to a set of places; $Z^S = \left\{z^S_{1(z^S)}, \ldots, z^S_{j(z^S)}, \ldots, z^S_{J(z^S)}\right\}$ – a finite set of transition by semantic relations; $\tilde{R}^S$ – an incidence matrix of size $J(a) \times J\left(z^S\right)$ that maps the set of places to a set of transitions by semantic relations for token; $\hat{R}^S$ – an incidence matrix of size $J\left(z^S\right) \times J(a)$ that maps the set of transitions by semantic relations to a set of places for token.

In addition, the following transition functions are set:

- input function of transitions by control relations:

$$I_A\left(Z^C\right) = \left\{I_A\left(z^C_{1(z^C)}\right), \ldots, I_A\left(z^C_{j(z^C)}\right), \ldots, I_A\left(z^C_{J(z^C)}\right)\right\};$$

- output function of transitions by control relations:

$$O_A\left(Z^C\right) = \left\{O_A\left(z^C_{1(z^C)}\right), \ldots, O_A\left(z^C_{j(z^C)}\right), \ldots, O_A\left(z^C_{J(z^C)}\right)\right\};$$

- input function of transitions by semantic relations:

$$I_A\left(Z^S\right) = \left\{I_A\left(z^S_{1(z^S)}\right), \ldots, I_A\left(z^S_{j(z^S)}\right), \ldots, I_A\left(z^S_{J(z^S)}\right)\right\};$$

- output function of transitions by semantic relations:

$$O_A\left(Z^S\right) = \left\{O_A\left(z^S_{1(z^S)}\right), \ldots, O_A\left(z^S_{j(z^S)}\right), \ldots, O_A\left(z^S_{J(z^S)}\right)\right\}.$$

More information about the mathematical apparatus of ExPNSR is presented in [23].

Between the places of the net, the control precedence relation is fulfilled. Place a_i is considered to be preceding by control relations for a_j (specified as $a_i <_C a_j$) if there is such a control path from the starting place of the process a_S to place a_j that it includes place a_i.

For the starting place of the process, it is true:

$$\exists z^C_{i(z^C)}\left(I_A\left(z^C_{i(z^C)}\right) = a_S\right) \wedge \nexists z^C_{j(z^C)}\left(O_A\left(z^C_{j(z^C)}\right) = a_S\right).$$

4 Structures of ExPNSR

In the process model, built in accordance with the definition of the ExPNSR, three types of structures can be distinguished, formed as a result of various combinations of places and control transitions:

- linear section (Ln);
- cycle (C);
- branching (Br).

The linear section (Fig. 1) is characterized by a sequence of linear control structures, implying the execution of some operation immediately after the completion of the previous one.

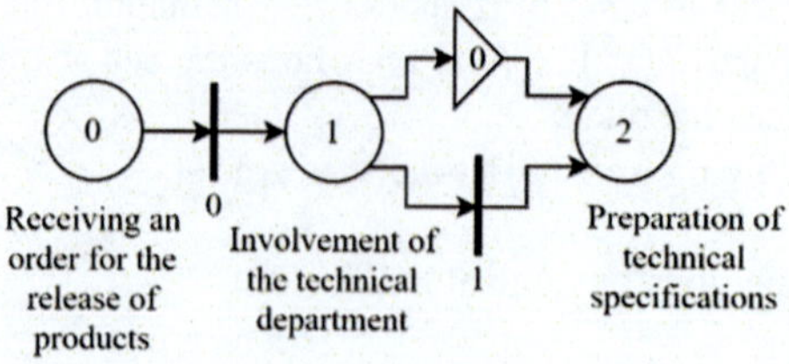

Fig. 1. An example of a linear section.

Cycles imply a return by control to the initial place a_{cB}, the presence of a condition (branch-place) a_{cond}, which takes control out of the cycle, and the body of the cycle (Cp), which contains a sequence of linear sections, branches and other cycles. Among the cycles, there are cycles with a post-condition (Fig. 2a) and cycles with a pre-condition (Fig. 2b).

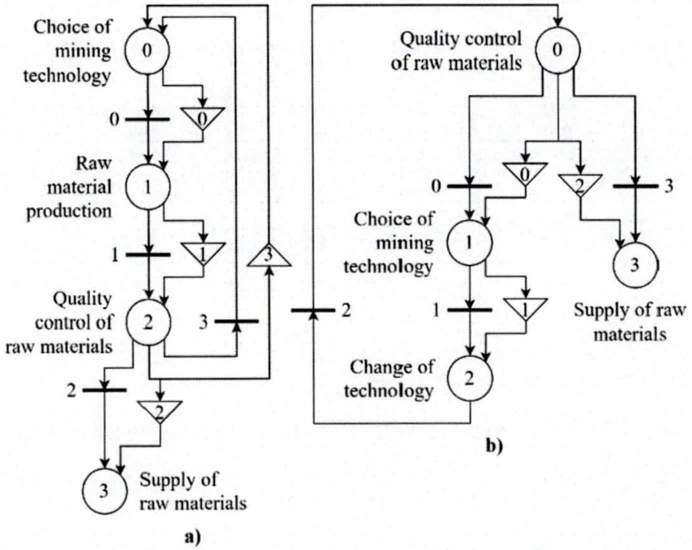

Fig. 2. Examples of cycles: a - with a post-condition; b - with a pre-condition.

Cycles are characterized by the following:

$$\exists z^C_{i(z^C)}\left(O_A\left(z^C_{i(z^C)}\right) = a_{cB} \wedge I_A\left(z^C_{i(z^C)}\right) = a_{i(a)}\right)|a_{cB} <_C a_{i(a)};$$

$$\left|Z^C_{Oi}\right| > 1\ \forall z^C_{i(z^C)} \in Z^C_{Oi}\left(I_A\left(z^C_{i(z^C)}\right) = a_{cond}\right) \wedge\ \nexists Br_i\ (a_{xB} = a_{cond});$$

$$\exists z^C_{i(z^C)}\left(I_A\left(z^C_{i(z^C)}\right) = a_{cond} \wedge O_A\left(z^C_{i(z^C)}\right) = a_{out}\right)|a_{out} \notin Cp$$

$$\wedge a_{out} \neq a_{cB}.$$

Branch (Fig. 3) is characterized by the presence of a condition describing options for actions according to its result, and the number of such options can vary from 2 to n, and contains the initial place the final place a_{xe}, as well as branches B_{xi}, $i = \overline{2, n}$. Each branch can contain a sequence of linear sections, other branches and cycles. Branches are characterized by the following:

$$\left|Z^C_{Oi}\right| > 1\ \forall z^C_{i(z^C)} \in Z^C_{Oi}\left(I_A\left(z^C_{i(z^C)}\right) = a_{xB}\right) \wedge .$$

$$\nexists C_i\ (a_{cB} = a_{xB}) \wedge\ \nexists z^C_{j(z^C)} \in Z^C_{Oi}\left(O_A\left(z^C_{j(z^C)}\right) = a_{i(a)}\right)|a_{i(a)} <_C a_{xB}.$$

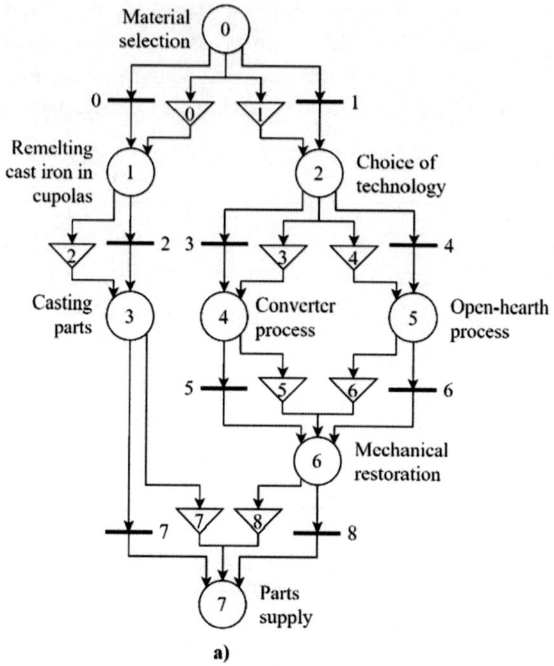

Fig. 3. Example of branch.

5 Structural Analysis of ExPNSR

To analyze the structure of the ExPNSR, it is necessary to build a model of the selected process. For this, each stage of the process is transformed into a place of the net, and the sequence of operations is specified by forming control transitions between places. The formation of semantic transitions, that is, the indication of material and informational dependencies between the elements of the process, is not required at this stage.

The constructed net is analyzed in the following order:

- Search for the starting place of the net.
- Search for cycles: determination of initial places.
- Finding branches: determining initial places and forming branches.
- Determination of conditions for cycles and nested cycles with the same initial place.
- Filling the contents of the branches of each of the branches.
- Filling the body of each of the cycles.

Of the described stages, the most difficult stage is the stage of determining the conditions of the cycles, since at this stage the nesting of cycles with the same initial place is also determined. Let's consider this stage in more detail.

Determination of conditions for cycles and nesting with the same initial place is performed as follows for each not yet considered loop $C_{i(C)}$ using the following notation:

- Conds – a set of places that are possible conditions for $C_{i(C)}$.;
- Z_O– a set of transitions to which $a_{cB_{i(C)}}$ leads;
- eC - the set of already considered cycles, the beginning of which coincides with $C_{i(C)}$;
- parC - the set of cycles in which can be nested.

1. Determine the set Z_O:

2. Determine the set *Conds:*

3. If $|Conds| = 1$, then there is no nesting with the same starting place, go to step 3, otherwise go to step 4.

4. If $|Z_O| > 1$, then a cycle with a precondition is found:

$$a_{cond_{i(C)}} = a_{cB_{i(C)}};$$

$$a_{out_{i(C)}} = a_{i(a)} | \exists z^C_{i(Z^C)} (I_A \left(z^C_{i(Z^C)} \right) = a_{cB_{i(C)}} \wedge$$

$$O_A \left(z^C_{i(Z^C)} \right) = a_{i(a)}) \wedge a_{i(a)} \notin Cp_{i(C)};$$

else, a cycle with a postcondition is found:

$$a_{cond_{i(C)}} = Conds_0;$$

$$a_{out_{i(C)}} = a_{i(a)} | \exists z^C_{i(Z^C)} (I_A \left(z^C_{i(Z^C)} \right) = a_{cB_{i(C)}} \wedge$$

$$O_A \left(z^C_{i(Z^C)} \right) = a_{i(a)}) \wedge a_{i(a)} \neq a_{cB_{i(C)}}.$$

complete the consideration of the cycle.

5. Determine the set *eC*:

$$eC = \left\{ C_{j(C)} \right\} | a_{cond_{j(C)}} = a_{cB_{i(C)}} \vee a_{cond_{j(C)}} \in Conds.$$

6. If $|eC| = 0$ and $|Z_O| > 1$, then the considered cycle is senior with the precondition:

$$a_{cond_{i(C)}} = a_{cB_{i(C)}};$$

$$a_{out_{i(C)}} = a_{i(a)} | \exists z^C_{i(Z^C)} (I_A \left(z^C_{i(Z^C)} \right) = a_{cB_{i(C)}} \wedge$$

$$O_A \left(z^C_{i(Z^C)} \right) = a_{i(a)}) \wedge a_{i(a)} \notin Cp_{i(C)};$$

complete the consideration of the cycle, else go to step 6.

7. Removing extra places from *Conds*:

$$Conds = Conds \backslash a_{i(a)} || Z_{Or}| = 1 \forall z^C_{i(Z^C)} \in Z_{Or} (I_A \left(z^C_{i(Z^C)} \right) = a_{i(a)})$$

$$Conds = Conds \backslash a_{i(a)} | \exists C_{j(C)} \left(a_{cond_{j(C)}} = a_{i(a)} \right)$$

8. Completing the formation of the cycle:

$$a_{cond_{i(C)}} = a_{i(a)} | a_{i(a)} \in Conds \wedge$$
$$\forall a_{j(a)} \in Conds\left(a_{j(a)} <_C a_{i(a)}\right);$$
$$a_{out_{i(C)}} = a_{i(a)} | \exists z^C_{i(Z^C)} I_A\left(z^C_{i(Z^C)}\right) = a_{cond_{i(C)}} \wedge$$
$$O_A\left(z^C_{i(Z^C)}\right) = a_{i(a)} \wedge a_{i(a)} \neq a_{cB_{i(C)}}.$$

9. If $|eC| = 1$, then this cycle is the parent for the considered one:

$$Cp_{eC_0} = Cp_{eC_0} \cup C_{i(C)};$$

else we search for the parent loop:

$$parC = \left\{C_{j(C)}\right\} | C_{j(C)} \in eC \wedge a_{cond_{j(C)}} \neq a_{cB_{j(C)}};$$
$$Cp_{j(C)} = Cp_{j(C)} \cup C_{i(C)} | C_{j(C)} \in parC \wedge$$
$$\forall C_{k(C)} \in parC\left(a_{cond_{j(C)}} <_C a_{cond_{k(C)}}\right).$$

Complete the consideration of the cycle.

As an algorithm for filling these structures, consider the filling of branches. Filling the body of cycles is done in a similar way.

To fill in the branching $Br_{i(Br)}$, each of its branches $Bx_{i(Br)}$ is analyzed according to the following algorithm with the notation:

- p – considered place;
- Ln – formed linear section;
- Z_I - set of transitions leading to p;
- np – the next place to be considered;
- Z_{In} - set of transitions leading to np.

1. Find the first place of the branch:

$$p = O_A\left(z^C_{i(Z^C)}\right) | I_A\left(z^C_{i(Z^C)}\right) = a_{xB_{i(Br)}}.$$

2. Form the current linear section:

$$Ln = \emptyset$$

3. Form the current linear section:

$$Z_I = \left\{z^C_{i(Z^C)}\right\} | O_A\left(z^C_{i(Z^C)}\right) = p \wedge \ \forall a_{i(a)} \in I_A\left(z^C_{i(Z^C)}\right)\left(a_{i(a)} <_C p\right).$$

4. If $|Z_I| > 1$, then branch completion is detected:

$$a_{xe_{i(Br)}} = p;$$
$$Bx_{i_{i(Br)}} = Bx_{i_{i(Br)}} \cup Ln | Ln \neq \emptyset;$$

complete the consideration of the branch.

5. If $\exists Br_{j(Br)} \left(a_{xB_{j(Br)}} = p \right)$, then a nested branch is detected:

$$Bx_{i_{i(Br)}} = Bx_{i_{i(Br)}} \cup Br_{j(Br)};$$

go to step 6, else go to step 11.If $a_{out_{j(Br)}} = \emptyset$, then fill the branching $Br_{j(Br)}$ in accordance with the algorithm.

6. Set the following place for consideration:

$$np = a_{xe_{j(Br)}}.$$

7. Save the linear section:

$$Bx_{i_{i(Br)}} = Bx_{i_{i(Br)}} \cup Ln | Ln \neq \emptyset.$$

8. Form a set Z_{In}:

$$Z_{In} = \left\{ z^C_{i(Z^C)} \right\} | O_A\left(z^C_{i(Z^C)} \right) = np \wedge \forall a_{i(a)} \in I_A\left(z^C_{i(Z^C)} \right) \left(a_{i(a)} <_C np \right).$$

9. If $|Z_{In}| > \left| Bx_{j(Br)} \right|$, then complete the consideration of the branch, else $p = np$; continue the consideration of the branch from step 2.
10. If $\exists C_{i(C)} \left(a_{cB_{i(C)}} = p \right)$, then nested cycle detected:

$$Bx_{i_{i(Br)}} = Bx_{i_{i(Br)}} \cup C_{i(C)};$$
$$p = a_{out_{i(C)}};$$
$$Bx_{i_{i(Br)}} = Bx_{i_{i(Br)}} \cup Ln | Ln \neq \emptyset;$$

continue the consideration of the branch from step 2.

11. Fill in the linear section:

$$Ln = Ln \cup p.$$

12. Set the next place for consideration:

$$p = a_{i(a)} | \exists z^C_{i(Z^C)} \left(I_A\left(z^C_{i(Z^C)} \right) = p \wedge O_A\left(z^C_{i(Z^C)} \right) = a_{i(a)} \right);$$

continue the consideration of the branch from step 3.

The described methodology will make it possible to form data structures that allow to unambiguously determine the presence and place of nonlinear structures in the process model, which, at the transformation stage, will facilitate the introduction of the hierarchy into the network by replacing the whole structure with one place.

6 The Example of Structural Analysis of the Process

Let's consider the process of structural analysis using the example of the technological process of repairing a truck. Figure 4 shows an ExPNSR that simulates this process. Structural analysis does not need the study of information and material relations, therefore the presented net includes transitions by only control relation in order to make the reading of the graph easier.

In accordance with the generalized structural analysis algorithm, the starting place of the network is the first determined. Since the place 0 fulfills the condition $\exists z^C_{i(z^C)}\left(I_A\left(z^C_{i(z^C)}\right) = a_S\right) \wedge \nexists z^C_{j(z^C)}\left(O_A\left(z^C_{j(z^C)}\right) = a_S\right)$, then $a_S = 0$.

At the next stage, the initial places of the cycles are determined. as a result of which structures of the following type are formed:

$$C_0 = \{a_{cB} = 3, a_{cond} = \emptyset, a_{out} = \emptyset, Cp = \emptyset\};$$

$$C_1 = \{a_{cB} = 10, a_{cond} = \emptyset, a_{out} = \emptyset, Cp = \emptyset\};$$

$$C_2 = \{a_{cB} = 20, a_{cond} = \emptyset, a_{out} = \emptyset, Cp = \emptyset\};$$

$$C_3 = \{a_{cB} = 32, a_{cond} = \emptyset, a_{out} = \emptyset, Cp = \emptyset\}.$$

Next, the initial places that allows to choose the further development of the process are determined and the branches are formed:

$$Br_0 = \{a_{xB} = 4, \{B_0 = \emptyset, B_1 = \emptyset\}, a_{xe} = \emptyset\};$$

$$Br_1 = \{a_{xB} = 11, \{B_0 = \emptyset, B_1 = \emptyset\}, a_{xe} = \emptyset\};$$

$$Br_2 = \{a_{xB} = 21, \{B_0 = \emptyset, B_1 = \emptyset\}, a_{xe} = \emptyset\}.$$

As a result, the definition of cycle conditions, previously formed structures are supplemented to the following form:

$$C_0 = \{a_{cB} = 3, a_{cond} = 3, a_{out} = 31, Cp = \emptyset\};$$

$$C_1 = \{a_{cB} = 10, a_{cond} = 10, a_{out} = 30, Cp = \emptyset\};$$

$$C_2 = \{a_{cB} = 20, a_{cond} = 20, a_{out} = 25, Cp = \emptyset\};$$

$$C_3 = \{a_{cB} = 32, a_{cond} = 32, a_{out} = 34, Cp = \emptyset\}.$$

When filling out the branches, linear sections are formed that belong to the every variant of the process development, and the nested cycles and branches are determined:

$$Br_0 = \{a_{xB} = 4, \{B_0 = \{Ln_0\}, B_1 = \{Ln_1, C_1\}\}, a_{xe} = 30\};$$

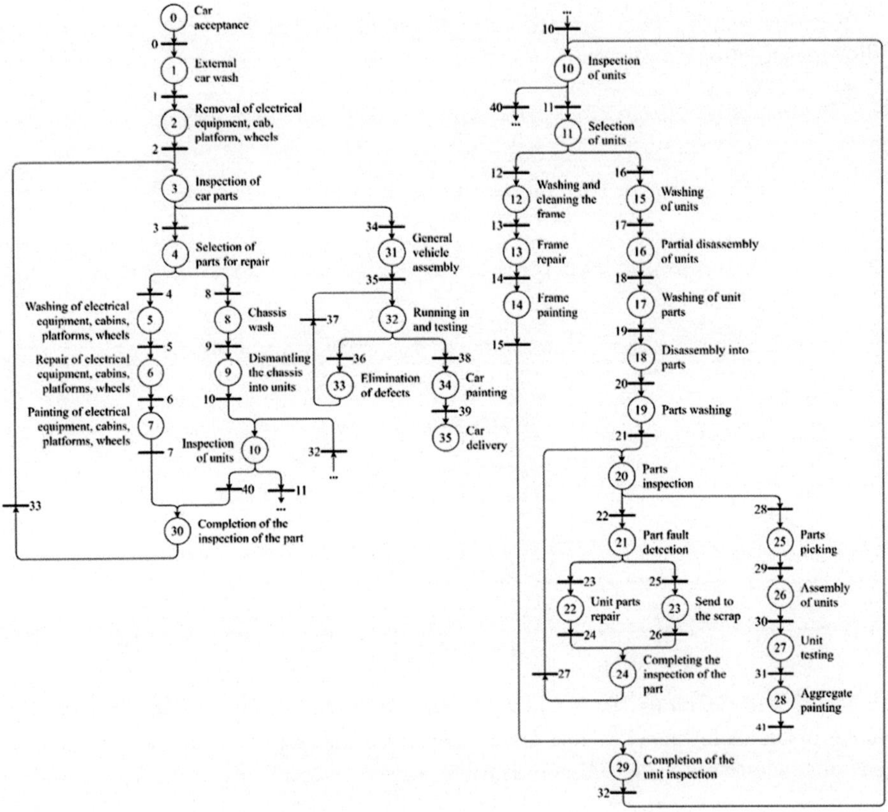

Fig. 4. Technological process diagram of a truck repair.

$$Br_1 = \{a_{xB} = 11, \{B_0 = \{Ln_2\}, B_1 = \{Ln_3, C_2, Ln_4\}\}, a_{xe} = 29\};$$

$$Ln_0 = \{5, 6, 7\};$$

$$Ln_1 = \{8, 9\};$$

$$Ln_2 = \{12, 13, 14\};$$

$$Ln_3 = \{15, 16, 17, 18, 19\};$$

$$Ln_4 = \{25, 26, 27, 28\};$$

$$Ln_5 = \{22\};$$

$$Ln_6 = \{23\}.$$

The same actions are performed when filling the body of the cycles. As a result the following structures are formed:

$$C_0 = \{a_{cB} = 3, a_{cond} = 3, a_{out} = 31, Cp = \{Br_0, Ln_7\}\};$$

$$C_1 = \{a_{cB} = 10, a_{cond} = 10, a_{out} = 30, Cp = \{Br_1, Ln_8\}\};$$

$$C_2 = \{a_{cB} = 20, a_{cond} = 20, a_{out} = 25, Cp = \{Br_2, Ln_9\}\};$$

$$C_3 = \{a_{cB} = 32, a_{cond} = 32, a_{out} = 34, Cp = \{Ln_{10}\}\};$$

$$Ln_7 = \{30\};$$

$$Ln_8 = \{29\};$$

$$Ln_9 = \{24\};$$

$$Ln_{10} = \{33\}.$$

The sets of positions {0, 1, 2}, {31} and {34, 35} are linear sections that do not belong to cycles or branches, and are not separated into separate structures. Thus, the general structure of the algorithm is as follows:

$$Alg = \{\{0, 1, 2\}, C_0, \{31\}, C_3, \{34, 35\}\}.$$

7 Conclusion

The paper describes the main nonlinear structures found in the model of a process based on ExPNSR. A method for analyzing the ExPNSR for the presence of nonlinear structures is proposed, which can be used during the optimization of production processes. The information obtained as a result of its application will make it possible to more efficiently distribute the resources available to the enterprise at the stage of planning the modernization of production or when developing new projects.

Acknowledgments. The research was partially supported by Grants of the President of the Russian Federation for state support of young Russian scientists – Ph.D. (project No. MK-1160.2020.9).

References

1. Danelyan, T.Ya.: Structural modeling. Stat. Econ. 6:166–169 (2014)
2. Trifonov, P.V.: Comparative analysis of models for describing business processes in an organization. Manag. Sci. Modern World **1**, 307 (2016)
3. Pignasty, O.M.: Review of models of controlled production processes of production lines production systems. Econ. Comput. Sci. **34**(7), 204 (2015)
4. Fedorova, O.V., Mamaeva, A.A., Yakunina, E.A.: Application of SADT and ARIS methodologies for modeling and management of business processes of information systems. Bull. Voronezh State Univ. Eng. Technol. **80**(1), 75 (2018)
5. Minnullina, A., et al.: Production strategy development for an energy company based on SADT. E3S Web Conf. EDP Sci. **217,** 07015 (2020)
6. Alontseva, E.N., Anokhin, A.N., Saakyan, S.P.: Structural Modeling of Processes and Systems. IATE NIYAU MIFI, Obninsk (2015)
7. Tsebrenko, K.N.: Functional-structural modeling of the system of automation of accounting of commodity-material values. J. Sci. **8**, 23 (2019)
8. Baghbani, M., Iranzadeh, S., Bagherzadeh Khajeh, M.: Extracting manufacturing process map in the form of the IDEF model prerequisite for the Implementation of PFMEA in the Sugar Industry. J. Mod. Process. Manuf. Prod. **7**(2), 79–93 (2018)
9. Aspridou, M.: Extending BPMN for modeling manufacturing processes. diss., Master thesis, Business Information Systems, TU/e (2017)
10. Erasmus, J., et al.: Using business process models for the specification of manufacturing operations. Comput. Ind. **123,** 103297(2020)
11. Yakimov, I.M., et al.: Structural modeling of business processes in BPMN EDITOR, ELMA, RUNAWFE systems. Bull. Kazan Technol. Univ. **17**(10) (2014)
12. Kelani, K.T.: Modelling techniques: comparison between BPMN2.0 and IDEF (0&3). Doctoral dissertation (2018)
13. Serysheva I.A., Chirkova, E.Yu.: Automation of analysis of technological schemes of production processes. Bull. Irkutsk State Tech. Univ. **12**(83) (2013)
14. Bocharov, E.P., Aleksentseva, O.N., Ermoshin, D.V.: Simulation model of the production process as an element of the management system of an industrial enterprise. Appl. Inform. **3** (2007)
15. Khan, R., et al.: A Framework for automated reengineering of BPMN models by excluding inefficient activities. In: Proceedings of the 2020 9th International Conference on Software and Computer Applications, pp. 147–151 (2020)
16. Honti, G., Dörgő, G., Abonyi, J.: Review and structural analysis of system dynamics models in sustainability science. J. Clean. Prod. **240,** 118015 (2019)
17. Jadhav, J.R., Mantha, S.S., Rane, S.B.: Analysis of interactions among the barriers to JIT production: interpretive structural modelling approach. J. Ind. Eng. Int. **11**(3), 331–352 (2014). https://doi.org/10.1007/s40092-014-0092-4
18. Lager, T.: A structural analysis of process development in process industry: a new classification system for strategic project selection and portfolio balancing. R&D Manag. **32**(1), 87–95 (2002)
19. Stepanenko, V.E., Frolov, D.N., Maryin, B.N.: Method of simulation modeling of the organization of production processes using extended Petri nets. Sci. Notes Komsomolsk-on-Amur State Tech. Univ. **1**(7), 71–78 (2011)
20. Abdul-Hussin, M.H.: A Structural analysis of petri nets-based siphons supervisors of flexible manufacturing systems. In: Proceedings of IEEE-UKSim-AMSS, 17th International Conference on Computer Modelling and Simulation, pp. 235–241 (2015)

21. Cabasino, M.P., Giua, A., Seatzu, C.: Structural analysis of Petri Nets. Control of Discrete-Event Systems, pp. 213–233 Springer, London (2013). https://doi.org/10.1007/978-1-4471-4276-8
22. Park, J.: Structural Analysis and control of Resource Allocation Systems using Petri Nets, Georgia Institute of Technology (2000)
23. Voloshko, A., Ivutin, A., Kryukov, O.: Petri nets based digital adaptive models for simulation of manufacturing processes. In: Advances in Automation II: Proceedings of the International Russian Automation Conference, RusAutoConf2020, September 6–12, 2020, pp. 223–234. Springer International Publishing, Sochi (2021)

Software for Modeling the Electron-Beam Welding in Steady State

V. Tynchenko(✉) and S. Kurashkin

Reshetnev Siberian State University of Science and Technology, 31, Krasnoyarsky Rabochy Avenue, Krasnoyarsk 660037, Russia

Abstract. At some of industrial enterprises, electron-beam welding technology is used in processes requiring the formation of permanent connections between equipment elements. However, its application is complicated by the need for accurate selection of the values of technological parameters. The software system proposed in this work allows simulating the temperature distribution over the volume of the products to be joined in the process of electron beam welding. The theory of welding processes is used as a mathematical apparatus. The software system has a modular structure and consists of six subsystems that implement specific functionality. All data, both input for mathematical models and results are stored in a single database, consisting of nine joined tables. The use of the proposed software allows both to reduce the cost and simplify the process of adjusting the technological parameters of electron beam welding, and to form the basis for the further implementation of effective control systems.

Keywords: Electron beam welding · Modeling · Technological parameters · Software · Energy distribution

1 Introduction

Currently, there is an active introduction of production process control technologies within the framework of the Industry 4.0 concept. For example, the authors of [1] consider models for the implementation of Industry 4.0 technologies in manufacturing companies. And the authors of [2] consider solutions for technological assistance to workers.

To implement an effective process control system, it is necessary to refine the parameters of such processes on digital twins [3, 4]. At the same time, mathematical modeling in conjunction with its software implementation is well suited to create digital twins [5, 6].

Within the framework of this study, the problem of the quality of control of electron beam welding (EBW) is considered, which can be solved by applying the concepts of Industry 4.0.

The process of electron beam welding is considered in a number of works [7–9], the authors of which propose conducting research on various metals and in various branches of mechanical engineering. In [10–12], the optimal options for the structural structure

A. A. Radionov and V. R. Gasiyarov (Eds.): RusAutoCon 2021, LNEE 857, pp. 237–246, 2022.
https://doi.org/10.1007/978-3-030-94202-1_23

of the welded joint are determined depending on the size of the allowance for machining in mathematical modeling.

At present, in order to further improve the quality of the technological process of electron beam welding, many authors have carried out mathematical modeling of this technological process in different modes and with different materials. For example, the authors of [13–15] considered the multicriteria optimization of the electron-beam welding process using experimental data obtained on the basis of real exact models of the electron-beam welding process, which describe the dependence of the geometry of welded joints on stainless steel on the parameters of the electron-beam welding mode... And the authors of works [16–18] investigated the processes of formation of the penetration channel in electron beam welding with full penetration of the material.

The implementation of the mathematical models of the EBW process previously applied by the authors in the steady state in the form of a software system will allow not only to work out the effective values of technological parameters, but also to form a software basis for the future EBW control system for thin-walled structures.

2 Materiats and Methods

2.1 Mathematical Modeling of Electron Beam Welding

The developed software makes it possible to simulate the heating of a thin-walled structure in the process of electron beam welding according to the following parameters of the technological process:

- Electron-beam current.
- Accelerating voltage.
- The speed at which the technological process is carried out.
- In addition, the applied model considers the physical parameters of the product.:
- Geometric dimensions.
- Thermophysical parameters of the product material.

The above parameters are used as inputs to the model, set by the user at the start of the simulation, and saved in the database. At the output of the model, the temperature values are calculated by the volume of the product in time.

The data obtained in the course of modeling can be used to optimize the parameters of the EBW process when working out the used or new modes. For this, the possibility of both data export and integration into the software system of the module for optimization is provided.

To describe the process of temperature distribution, we use the formulas [19] of an instantaneous point source over the surface of a semi-infinite body (1) and a linear source in an infinite plate (2):

$$T_1(V, q, v, t) = T_H + \frac{2q}{c\rho\sqrt{(4\pi a)^3}} e^{-\frac{vx}{2a}} \int_0^t e^{-\frac{v^2\tau}{4a} - \frac{R^2}{4a\tau}} \frac{d\tau}{\tau^{3/2}} \tag{1}$$

where $c\rho$ – heat capacity of steel, v – welding speed, a – thermal diffusivity, t – time counted from the moment the source passes through the section in which the point under consideration is located, R – distance from the direction of movement of the source to the point under consideration ($R^2 = x^2 + y^2 + z^2$), V – the volumetric coordinate of a point representing a vector (x, y, z).

$$T_2(V, q, v, t) = T_H + \frac{q}{4\pi\lambda\delta} e^{-\frac{vx}{2a}} \int_0^t e^{-\frac{v^2\tau}{4a} - \frac{2\lambda\tau}{c\rho\delta} - \frac{r^2}{4a\tau}} \frac{d\tau}{\tau} \quad (2)$$

where r – distance from the direction of movement of the source to the point under consideration ($r^2 = x^2 + y^2$), δ – thickness, λ – coefficient of thermal conductivity, t – heat propagation time.

In this work, a power model is used, which is function (3) written in the following form:

$$Q = I \cdot U \cdot \eta \cdot 0.24 \quad (3)$$

where U – accelerating voltage, I – beam current, η – Efficiency.

The complex fast-moving source was selected as the sum of two sources - point and linear, equivalent to the real ones described in the literature [19]. The calculation of the value of the functional is performed for an area whose dimensions are comparable to the dimensions of the penetration channel.

These formulas allow, when they are added (superposition of sources) in the calculation process, to describe the nature of the distribution of thermal energy after exposure to an electron beam.

2.2 Software Design

As a means of implementing the program:

- Programming system Embarcadero RAD Studio.
- Database management system MySQL.

The software system consists of 6th subsystems (Fig. 1).

- Subsystem of Mathematical modeling implements a mathematical model of the EBW process.
- Subsystem of field experiments data import allows you to download the data of the experiments carried out during the development of the technological process on the EBW installation.
- Subsystem of technological and modeling parameters configuration introduces technological and thermophysical parameters for modeling.
- The graphical user interface is the main window of the program.
- Subsystem of simulation data import allows you to load data and graphs obtained during the simulation.
- Subsystem of database interaction provides interaction between the GUI and the database.

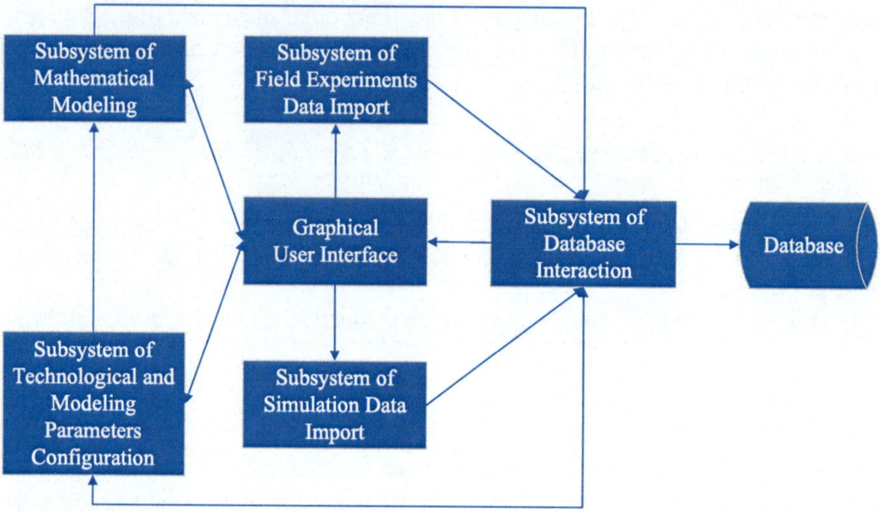

Fig. 1. Block diagram of the software system for mathematical modeling of the EBW process.

- Database that stores information on the settings of the technological process, the values of the thermophysical parameters of the materials being welded, as well as the simulation results.

The block diagram of the software system is shown in Fig. 2.

The first step is to check whether the current experiment exists in the database or not. If the experiment is new, technological and modeling parameters of experiment are entered and the values are saved to the database. If such an experiment already exists, then the program checks which parameters of the experiment are used. In case new parameters are used, technological and modeling parameters of experiment are entered by the user and the values are saved in the database. Otherwise, the experiment is loaded from the database. Next, the information about each of the types of experiment (mathematical modeling, a full-scale experiment, simulation experiment) is checked step by step. In the absence of information, one of two events can occur: mathematical modeling is carried out based on Eqs. (1), (2) and (3); import field experiment results or import simulation experiment results. At the last stage, the received and loaded data is saved to the database and displayed on the main form of the program.

The mathematical modeling algorithm is shown in Fig. 3.

At the initial stage of the algorithm, initializing of technological parameters (U, I, v) then input the mathematical model parameters: x_{min}, x_{max}, x_{step}, y_{min}, y_{max}, y_{step}, z_{min}, z_{max}, z_{step}, t_{min}, t_{max}, t_{step}. Further, the coordinates and time are assigned the minimum values ($x = x_{min}$; $y = y_{min}$; $z = z_{min}$; $t = t_{min}$).

The next stage is mathematical modeling calculation in point (x, y, z) in time t. This stage includes calculating the temperature field using Eqs. (1), (2) using the functional, which is understood as the arithmetic mean of the temperature, depending on the coordinate and time.

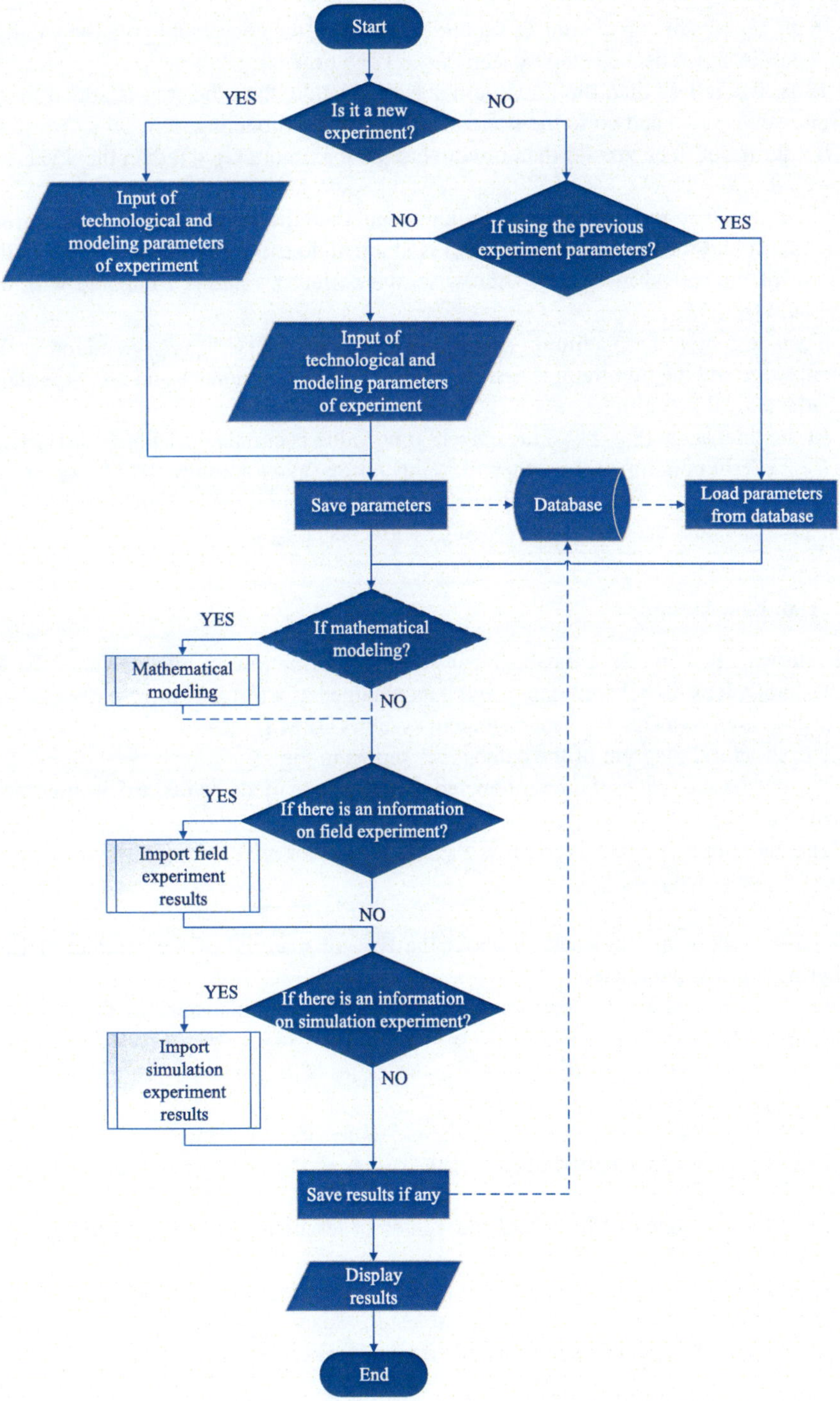

Fig. 2. Block diagram of the EBW modeling software.

In the process of calculation, to set and move along the coordinate grid, the value of each coordinate is checked step by step (x, y, z) and time (t).

Thus, if x is less than the maximum specified value, then the step is added to the current value (x_{step}) and go to the stage of mathematical modeling.

If x turns out to be greater than or equal to the maximum (x_{max}), then the y value is compared.

If y is less than the maximum specified value, then the step is added to the current value (y_{step}) and the minimum ($x = x_{min}$) is assigned to the x coordinate and go to the stage of mathematical modeling. Otherwise, the current z value is compared with the maximum (z_{max}).

If z is less than the maximum specified value, then the step (z_{step}) is added to the current value and the minimum is assigned to the x and y coordinates and go to the stage of mathematical modeling.

At the final stage ($z \geq z_{max}$) the current time value is compared with the maximum one (t_{max}). If the condition is met, then the temperature values are showed ($T(x,y,z,q,v,t)$). However, if $t < t_{max}$, then each coordinate is assigned a minimum value ($y = y_{min}$, $x = x_{min}$, $z = z_{min}$), and the time is increased by a given step t_{step}.

2.3 Database Design

The software system has a database for storing information on all conducted EBW studies, as well as model parameters and data obtained as a result of third-party studies. MySQL is used as a Database management system (DBMS) [20].

The structural diagram of the database is shown in Fig. 4.

The database contains 9 interconnected tables, and its model is in third normal form (3NF).

The main joining table is the experiment table. Tables are used to store information about the simulated product:

- Material – contains information about the type of material of the product and its thermophysical parameters.
- Workpiece – contains information about the geometric dimensions of the modeled product.
- Techprocess - contains information about the values of the parameters of the simulated technological process.
- To store information on a model experiment, tables are intended:
- Modeling – describes experiment metadata such as start time and date, comments, and simulation step.
- Data_model – contains the calculated values of temperatures in the volume of the product over time.

In addition, the database provides tables for loading the results of simulation (simulation and data_simul) and full-scale (practice) experiments.

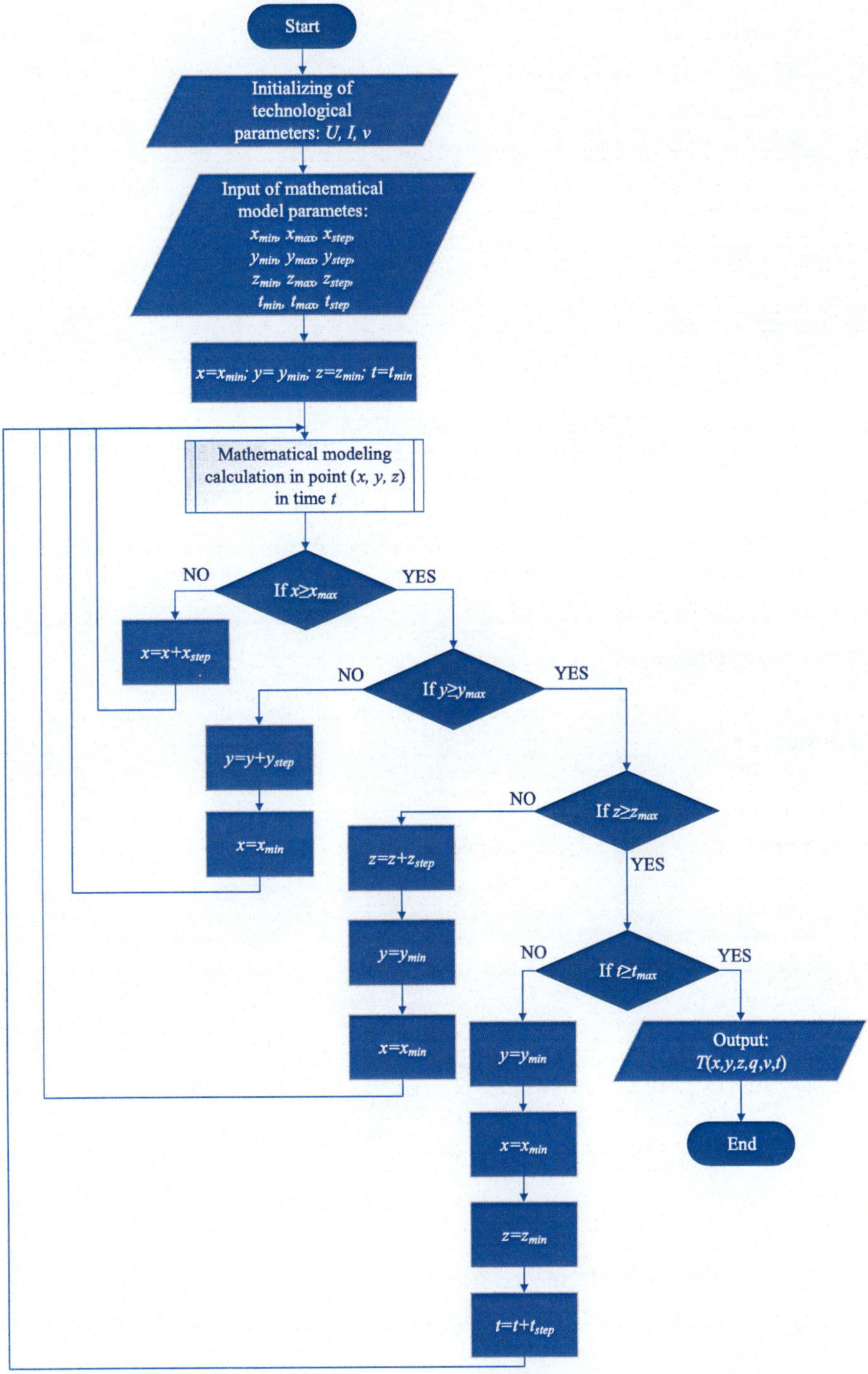

Fig. 3. Algorithm of EBW mathematical modeling.

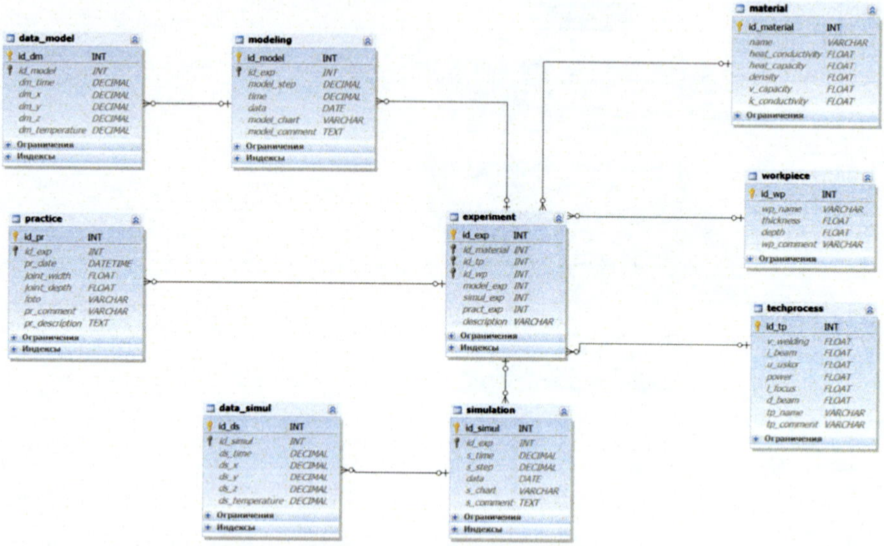

Fig. 4. Software data shema.

2.4 Software Operation

Figure 5 shows the main form of the software system for mathematical modeling of the EBW process.

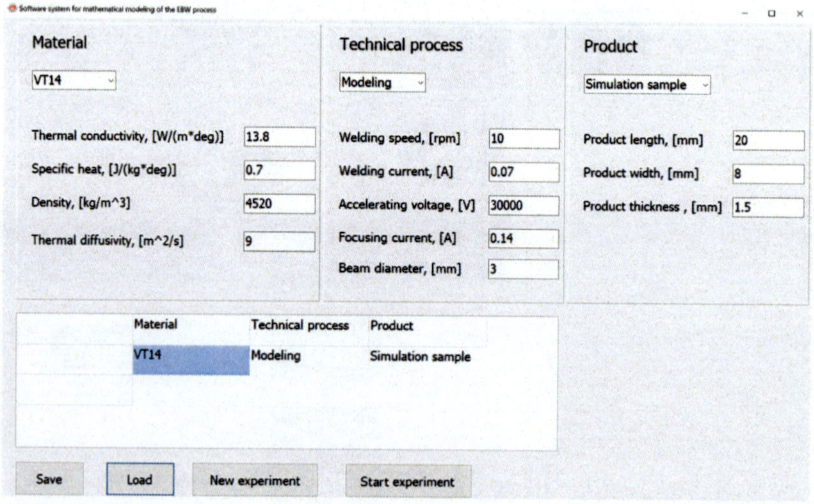

Fig. 5. The main form of the software system for mathematical modeling of EBW.

The main form contains three blocks:

- Material – it is possible to select a material and set thermophysical parameters.
- Technical process – technological parameters are indicated and the type of process is selected (field experiment, modeling).
- Product – select the type of sample (simulation disk, simulation sample and etc.) and set the geometric parameters.

At the bottom of the screen, there is a table that displays the saved experiments in the database. With the help of the corresponding buttons it is possible to save, load, create a new experiment and start modeling.

3 Conclusion

The proposed software system can be used to simulate the distribution of energy in the process of electron beam welding of thin-walled structures, considering the specified values of the technological parameters of the process and the thermophysical parameters of the workpiece. The results of such simulations are stored in a single database along with the results of third-party studies such as simulation modeling and field experiments.

The information collected in a single database can be used to select the effective values of the EBW parameters in order to improve the quality of welded joints in thin-walled structures.

In the future, on the basis of these results, it is possible to create and program implementation of control systems for technological processes of forming permanent joints, which make it possible to increase the accuracy of control.

References

1. Frank, A.G., Dalenogare, L.S., Ayala, N.F.: Industry 4.0 technologies: Implementation patterns in manufacturing companies. Int. J. Prod. Econ. **210**, 15–26 (2019)
2. Gorecky, D., Schmitt, M., Loskyll, M., Zühlke, D.: Human-machine-interaction in the industry 4.0 era. In: 2014 12th IEEE International Conference on Industrial Informatics (INDIN), pp. 289–294 (2014)
3. Bai, C., Dallasega, P., Orzes, G., Sarkis, J.: Industry 4.0 technologies assessment: a sustainability perspective. Int. J. Prod. Econ. **229**, 107776 (2020)
4. Saucedo-Martínez, J.A., Pérez-Lara, M., Marmolejo-Saucedo, J.A., Salais-Fierro, T.E., Vasant, P.: Industry 4.0 framework for management and operations: a review. J. Amb. Intell. Hum. Comput. **9**(3), 789–801 (2018)
5. Ghobakhloo, M.: Industry 4.0, digitization, and opportunities for sustainability. J. Clean. Prod. **252**,119869 (2020)
6. Zheng, T., Ardolino, M., Bacchetti, A., Perona, M.: The applications of Industry 4.0 technologies in manufacturing context: a systematic literature review. Int. J. Prod. Res. **59**(6), 1922–1954 (2021)
7. Yunlian, Q., Ju, D., Quan, H., Liying, Z.: Electron beam welding, laser beam welding and gas tungsten arc welding of titanium sheet. Mater. Sci. Eng. A280, **280**(1), 177–181 (2000)
8. Salomatova, E.S.: Electron beam welding - from invention to the present day. Bull. Perm Natl. Res. Polytech. Univ. Mech. Eng. Mater. Sci. **1**, 74–87(2013)

9. Permyakov, G.L., Olshanskaya, T.V., Belenkiy, V.Ya., Trushnikov, D.: Simulation of electron beam welding to determine the parameters of welded joints of dissimilar materials. Bull. Perm Natl. Res. Polytech. Univ. Mech. Eng. Mater. Sci. **1**(4), 48–58 (2013)
10. Yang, Z., Fang, Y., He, J.: Numerical investigation on molten pool dynamics and defect formation in electron beam welding of aluminum alloy. J. Mater. Eng. Perform. **29**(10), 6570–6580 (2020). https://doi.org/10.1007/s11665-020-05111-2
11. Yang, Z., Fang, Y., He, J.: Numerical simulation of heat transfer and fluid flow during vacuum electron beam welding of 2219 aluminium girth joints. Vacuum **175,** 109256 (2020)
12. Kaisheva, D., Angelov, V., Petrov, P.: Simulation of heat transfer at welding with oscillating electron beam. Can. J. Phys. **97**(10), 1140–1146 (2019)
13. Mladenov, G., Koleva, E., Belenky, V.Ya., Trushnikov, D.N.: Modeling and optimization of electron beam welding of steels. Bull. Perm Natl. Res. Polytech. Univ. Mech. Eng. Mater. Sci. **16**(4), 7–21 (2014)
14. Kanigalpula, P.K.C., Jaypuria, S., Pratihar, D.K., Jha, M.N.: Experimental investigations, input-output modeling, and optimization of spiking phenomenon in electron beam welding of ETP copper plates. Measurement **129**(1), 302–318 (2018)
15. Luo, M., Hu, R., Liu, T., Wu, B., Pang, S.: Optimization possibility of beam scanning for electron beam welding: physics understanding and parameters selection criteria. Int. J. Heat Mass Transf. **127**(1), 1313–1326 (2018)
16. Ignat'eva, M.A., Kadyrov, R.F., Mazo, A.B.: Calculation of the temperature field of a plate when electron-beam welding. Uchenye Zapiski Kazanskogo Universiteta. Seriya Fiziko-Matematicheskie Nauki **148**(4), 23–34 (2006)
17. Das, D., Das, A.K., Pratihar, D.K., Roy, G.G.: Prediction of residual stress in electron beam welding of stainless steel from process parameters and natural frequency of vibrations using machine-learning algorithms. Proc. Inst. Mech. Eng. C J. Mech. Eng. Sci. **235**, 2008–2021 (2020)
18. Luo, M., Hu, R., Liu, T., Wu, B., Pang, S.: Optimization possibility of beam scanning for electron beam welding: physics understanding and parameters selection criteria. Int. J. Heat Mass Transf. **127**, 1313–1326 (2018)
19. Konovalov, A.V.: Theory of Welding Processes. Izd-vo MGTU im. N.E. Bauman, Moscow (2007)
20. MySQL. https://www.mysql.com/. Accessed 5 Jan 2021

Approaches to Energy Systems Digital Twins Development and Application

D. Zolin(✉) and E. Ryzhkova

National Research University "Moscow Power Engineering Institute", 14, Krasnokazarmennaya Street, Moscow 111250, Russia

Abstract. This article highlights the power systems digital twins development and application, and also proposes digital twin architecture along with the digital twin software implementation prototype. The digital twin is a key building block of a high-tech control system. The energy system digital twin is considered as the main tool for high-tech distributed energy infrastructure intelligent management. As an example of the digital twin application, it is given the hybrid power supply system optimal configuration calculation. Also, a power system active low voltage consumer software implementation prototype digital twin based on the products Nrjpack, Matlab Simulink and Homer PRO was performed.

Keywords: Digital twin · Intelligent management · Software implementation prototype · Internet of things

1 Introduction

The digital twin concept is fundamental in the fourth industrial revolution (Industrie 4.0) context. A digital twin (DT) is a technical object virtual copy that faithfully reproduces and sets the structure, state and behavior of the original in real time [1]. As an intelligent superstructure on top of the Internet of Things (IoT) environment, the digital twin is a high-tech control system key building block. According to Gartner, by 2021, almost half of large industrial companies will use digital twin technology to improve the assessing product performance and technical risks accuracy, while achieving an increase in the products operational efficiency by about 10% [2].

Among the high-tech facilities, which management is advisable to organize on the digital twins basis, are modern distributed energy systems, including a power receivers variety, local generating equipment based on renewable sources and energy storage. However, it is natural that technologies for constructing digital twins borrowed from the engineering industries have characteristic drawbacks, such as the need for bulky expensive software tools and highly qualified personnel, which is clearly visible in a few examples of power systems digital twins [3]. There is an unambiguously interpreted reliable data in standard machine-readable formats, adequate mathematical models, and instrumentation lack. It is not clear how to automatically assemble an integral large power system digital twin from components twins, taking into account their connection rules. Technologies such as Generative Design, which automatically find energy supply optimal design solutions, are developing very slowly [4]. These shortcomings are especially acute in the low voltage mass small consumers power systems life cycle.

A. A. Radionov and V. R. Gasiyarov (Eds.): RusAutoCon 2021, LNEE 857, pp. 247–255, 2022.
https://doi.org/10.1007/978-3-030-94202-1_24

The article proposes approaches to overcoming these shortcomings in distributed energy facilities control systems design and operation. These approaches development will make it possible to effectively organize energy resources generation, consumption, storage, transfer assessment and forecasting on the digital twins basis in all aspects, as well as mode control, equipment condition predictive monitoring, models and algorithms verification, design solutions virtual approbation and optimization, facility personnel training.

2 Energy System Digital Twin Architecture

By its structure, Digital twin is interconnected computer models complex capable to reliably display the original object, its state and behavior under various environmental conditions and control actions. Models form an object entire life cycle representation, which allows you to detect, analyze, predict and prevent undesirable situations during the object operation. In particular, the models are used to solve the following tasks:

- original computer representation reconciliation with real object data;
- personnel notification and decision support;
- predicting changes in the original over time;
- identification of new possibilities for using the original object;
- original object operation economic effects for various options calculation.

Table 1. Energy system digital twin architecture.

Enterprise Resource Planning (ERP)	Electronic Document Management System (EDMS)	Geographic Information System (GIS)	Product Lifecycle Management (PLM)	Regulatory and Reference Information Unified System (RRIUS)	Information Security Tools (IST)
Energy system digital twin					
Mathematical and simulation models					
Digital circuits and maps	Electronic documentation		Information models	Operative information	
Ontological model					
Article Processing Charges Systems (APCS)	Manufacturing Execution System (MES)	Automated Measuring and Information System for Electric Power Fiscal Accounting (AMIS EPFA)	Automated Energy Resources Accounting System (AERAS)	Automatic Dispatcher Control System (ADCS)	Internet of Things (IoT)
Measuring instruments and automation					

These tasks are highly relevant for power systems as well. Models with real world data reconciliation implies the automatic supply of digital twin mathematical and simulation models with structured up-to-date initial data from basic information components that describe the power system in various aspects and are filled from adjacent software systems in real time as they arise. To eliminate discrepancies in the concepts naming and interpretation digital twin operates with, an energy infrastructure ontological model is placed at the basis of its information support, for its formation a significant backlog has been accumulated [5]. The following information components of the digital twin of the power system are built over it, as shown in Table 1:

- digital diagrams and maps (primarily a single-line power supply diagram);
- electronic documentation (design estimates, operational, etc.);
- information models (master data – information about subjects, about objects, about the equipment composition and characteristics, related reference books, etc.);
- operational information (the consumption instrumental measurements results and the equipment technical condition primary characteristics).

For large multicomponent objects digital twins, such as power systems, the problem is that it is difficult to combine the constituent components working models (twins) into a single coherent whole. In fact, the connection requires virtually reproducing the power system building process on information and mathematical models, with the correct interconnections between them. A promising approach to solving this problem is proposed on the theory of categories mathematical apparatus basis [6].

3 Information Modeling of Active Consumer Infrastructure

The digital twin forming process is shown in large block in Fig. 1. It can be seen that ontological modeling constitutes the design stage foundation in it and serves as the basis for the formation of information support.

For consumers energy infrastructure, ontological modeling is complicated by the fact that the common information models (CIMs), which serve as terms and relations source, are mainly focused on large energy facilities: power plants, power lines, substations.

Low voltage active energy consumption objects information modeling is a difficult task, including due to the high variability of their generating and consuming energy equipment. The missing terms and relationships are borrowed from ISO 17800 (microgrid models), IEC 61850 (smart electrical device models), OASIS EMIX (market information exchange models), and observation and weather forecasting description models, etc.

Low voltage active energy consumption objects information modeling is a difficult task, including due to the high variability of their generating and consuming energy equipment. The missing terms and relationships are borrowed from ISO 17800 (microgrid models), IEC 61850 (smart electrical device models), OASIS EMIX (market information exchange models), and observation and weather forecasting description models, etc.

Information models, the structure of which is formed on the basis of open ontologies and standards, are called open (OIM) and are naturally the preferred basis for integrating

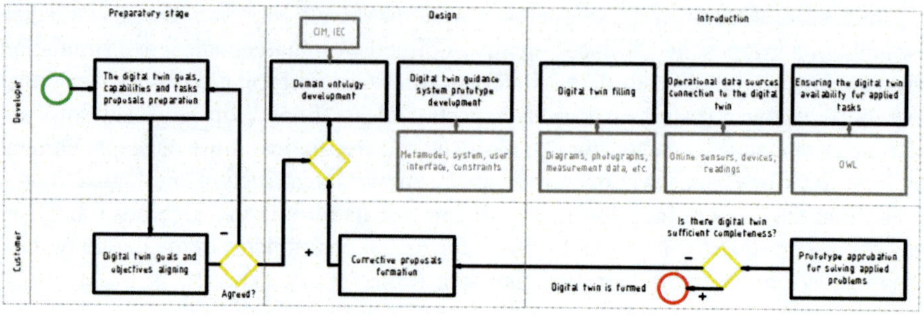

Fig. 1. The digital twin formation process.

information management and market applications with a digital twin. However, as shown in the architecture diagram above, the digital twin must be integrated with a wide range of specialized tools that are used in the design and operation of energy facilities, including CAD, APCS, etc. Information models implemented in such tools are often have a closed character (CIM), and for integration based on them, it is required to introduce connecting elements - conjugation models, as shown in Fig. 2. For example, such an approach is appropriate for connecting to the support system digital twin for the design of small power systems based on CAD AutoCAD, as an information model number of fragments interactive "editor". Another example is integration with gateways for collecting operational data from sensors and issuing commands to actuators developed on the basis of the Eclipse Kura IoT platform.

To maintain information models and other pilot facility basic energy system digital twin components, the domestic software Nrjpack can be used. Complex screens examples intended for information model displaying and entering elements are shown in Fig. 3.

Complex architectural basis is a metamodel - information entities description, characteristics, connections between them, obtained from the ontology. This design solution allows you to automate the execution of routine programming tasks:

- user interface forms layout for performing creating, viewing, updating and deleting data operations (CRUD operations);
- database structure and content formation;
- unloading the information model for transmission to the input to mathematical and simulation models.

This significantly reduces the time and labor spent on iterative filling and updating of the twin, in comparison with the traditional approach (domain driven design, DDD [7]). The fact is that traditionally ontology elements are represented in the form of classes and objects described directly in the source code of programs. In such a situation, when the ontology is changed, it is not enough to reconfigure the metamodel through the user interface: you need to change the source code, recompile, test and install a new software version.

The Nrjpack software package is built on the CQRS (command query request segregation) architecture, which involves dividing the set of user calls to the system into two independent threads: requests and commands. When making changes to the information model or metamodel, the user actually generates a change request, to which the system reacts by changing the data structure or filling it by generating appropriate events.

It should be emphasized that the CQRS architecture is implemented not only for the information model entities (data), but also for the metamodel entities (types), which made it possible to separate into a separate stream events aimed at changing the information model structure, and not just filling. One way to handle events is to represent the information model in a relational database. This approach, which is traditional for highly loaded systems, is of particular interest in the tasks of creating and maintaining a

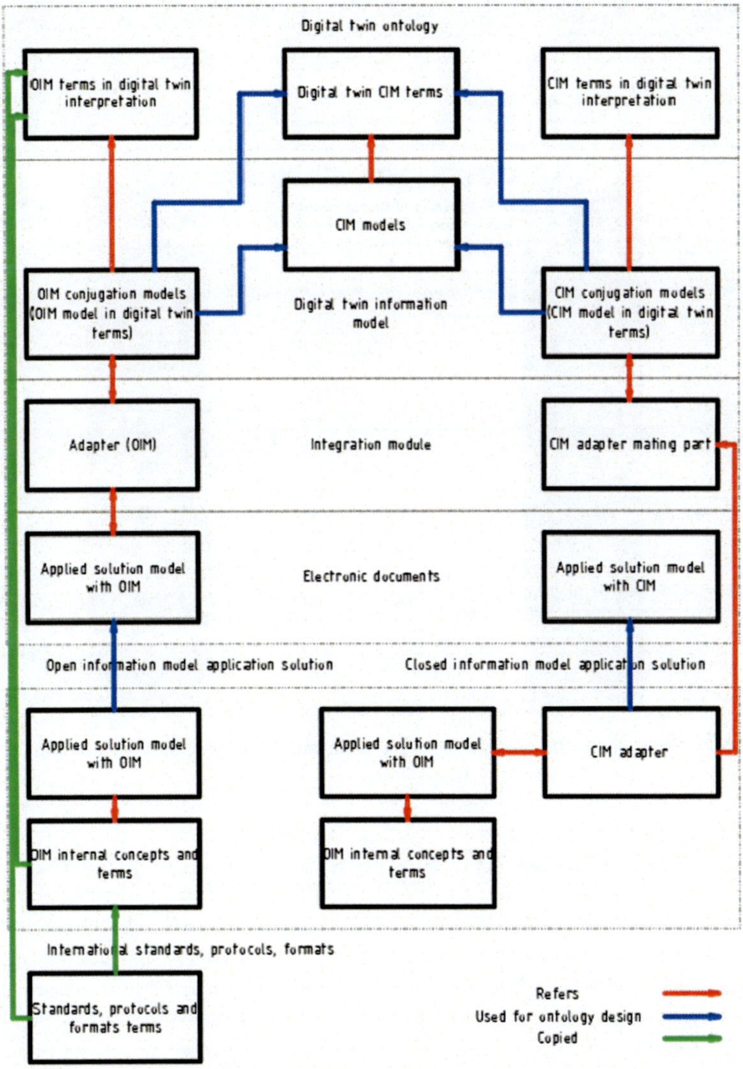

Fig. 2. Digital twin integration with applications.

digital twin, since it allows the formation of different relational projections, optimized for different twin uses.

On the metamodel basis, the software package implements complete information set automatic unloading about an object into a file in the ontological modeling OWL (Ontology Web Language) standard language. The upload tools are based on the metamodel and therefore automatically adapt to changes in the ontology.

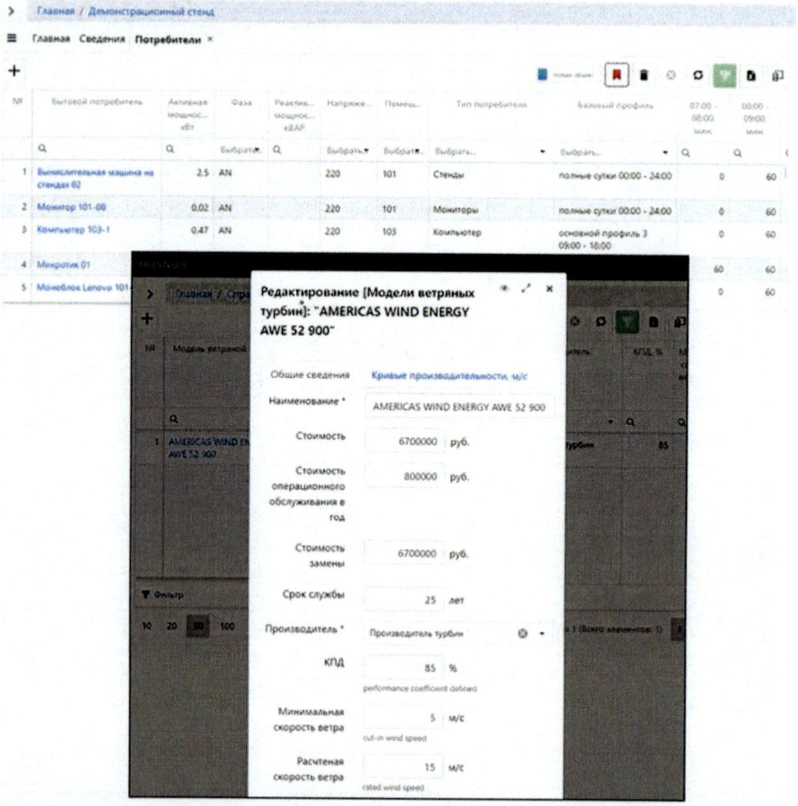

Fig. 3. Screens example for displaying and entering information model elements.

4 Finding the Optimal Hybrid Power System Configuration

As an applied problem for the digital twin practical application, let us consider an automatic search for the Hybrid Renewable Energy System (HRES) optimal configuration for a given facility, for which the consumption profile and weather conditions are known. Such a system contains local generating equipment (including renewable energy sources) and energy storage units, which, instead of or in addition to the centralized power grid, supply the facility with electricity. The optimal configuration assumes such a combination of generation and storage components, which gives the greatest effect in terms of operation in comparison with passive power supply from an external network. Optimization is performed for three target functions: the annual cost of the system, the probability of loss of power supply and the amount of harmful emissions into the atmosphere. The design variables set the number of devices of each type and settings; with restrictions in the form of ranges of variable values. An example is the solution of this problem using an evolutionary algorithm [8]. However, most of these algorithms require significant computational resources; therefore, in practice, it is possible to use lighter weight HRES optimizers. An example is the HOMER Pro package, designed specifically to quickly find economically suboptimal configurations of distributed energy systems. The model in HOMER Pro for finding the optimal configuration is included in the mathematical model of the digital twin of the power system.

The most important digital twin functions include checking the found configuration operation in simulation mode (simulation). This function can be implemented by adapting the well-known typical electrical model of the power system in the Matlab Simulink software package [9]. The model contains elements of distributed energy resources: an electric network, elements of renewable energy sources, a diesel generator, a storage device, and the means for connecting them into a given configuration are software implemented. The configuration resulting from the HOMER Pro model is automatically recreated in the dynamic Matlab Simulink model. In this way, in practice, an automatic connection of component models into a single complex is achieved. Further, the user of the digital twin gets the opportunity to execute the resulting simulation model under various parameters and conditions of interest to him, for example, to evaluate the power system reaction to one or another control action, as shown in Fig. 4.

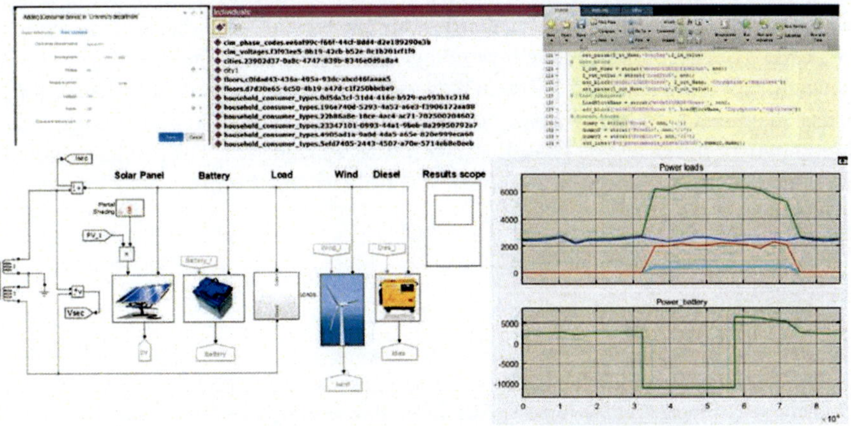

Fig. 4. Power system functioning simulation model example.

In turn, the consumer object description (information model) generated in the Nrjpack software package is passed to the input to the HOMER Pro package. The transfer occurs automatically, by transforming and loading the information model representation into HOMER Pro by an OWL file generated in NrjPack. At the same time, for each power receiver, according to the parameters specified in the information model, such as nominal power and operating mode, an estimated hourly consumption profile is formed. If actual consumption data is available from metering devices accessible through IoT gateways, the evaluation profile is calibrated. In a situation where one metering device covers several power receivers, the total profile measured by it is programmatically "decomposed" into their profiles, by means of so-called disaggregation algorithms [10]. The power receivers generated and calibrated profiles are transmitted to the Matlab Simulink model input for simulation, and the combined load profile from them is fed to the HOMER Pro model input to select the HRES configuration suboptimal for this consumption profile.

As an alternative to electrical models written in Matlab Simulink, models built by machine learning of deep neural networks are considered. Neural networks architectures and examples, including recurrent and convolutional ones, are proposed for solving a number of intelligent control key problems in the power industry, such as load forecasting, electricity price forecasting, load distribution optimization between available generating equipment, assessment and power equipment technical state prediction, diagnostics failures and disasters. It can be expected that in the future, neural network models, supported by other computer mathematical statistics means, will be able to take a leading position in the power system digital twin.

5 Conclusion

The approaches to the construction of a digital twin proposed in this work make it possible, without unnecessary labor and time, to form an easy-to-use virtual copy of the power system that can reproduce the structure, state and behavior of the original with a degree of completeness, reliability and efficiency sufficient for a number of practical purposes. This was confirmed in the development cycle of a digital twin prototype for power systems of pilot active low voltage consumers [11]. The model allows you to calculate suboptimal configurations of the power system and perform a realistic imitation of its behavior, including in transient modes (switching between sources). Thus, a classic digital twin of the base type (Baseline Twin) was realized [12]. The results obtained are fundamental for the design of intelligent control systems for the energy infrastructure of the future.

References

1. Madni, A.M., Madni, C.C., Lucero, S.D.: Leveraging digital twin technology in model-based systems engineering. Systems **7**(1), 7 (2019). https://www.mdpi.com/2079-8954/7/1/7
2. Pettey, C.: Prepare for the Impact of Digital Twins. Gartne, Stamford (2017)
3. Brosinsky, C., Westermann, D., Krebs, R.: Recent and prospective developments in power system control centers: Adapting the digital twin technology for application in power system control centers. In: Proceedings of the 2018 IEEE International Energy Conference ENERGYCON. IEEE, pp. 1–6 (2018)
4. Kovalyov, S.P.: An approach to develop a generative design technology for power systems. In: Proceedings VI International Workshop "Critical Infrastructures: Contingency Management, Intelligent, Agent-Based, Cloud Computing and Cyber Security" (IWCI 2019). Advances in Intelligent Systems Research, vol. 169, pp. 79–82 (2019). https://www.atlantis-press.com/proceedings/iwci-19/125917306

5. Kovalev, S.P.: Application of ontologies in the development of distributed automated information-measuring systems. Avtometriya **44**(2), 41–49 (2008)
6. Nolan, J.S., Pollard, B.S., Breiner, S., Anand, D., Subrahmanian, E.: Compositional models for power systems. In: Proceedings of the Applied Category Theory Conference ACT 2019. NIST (2019). https://www.nist.gov/publications/compositional-models-power-systems
7. Evans, E.: Domain-Driven Design - Tackling Complexity in the Heart of Software, p. 529. Addison-Wesley, New York (2004)
8. Ming, M., Wang, R., Zha, Y., Zhang, T.: Multi-objective optimization of hybrid renewable energy system using an enhanced multi-objective evolutionary algorithm. Energies **10,** 674 (2017). https://www.mdpi.com/1996-1073/10/5/674
9. Hiroumi, M.: Simplified model of a small scale micro-grid. https://se.mathworks.com/help/physmod/sps/examples/simplified-model-of-a-small-scale-micro-grid.html
10. Faustine, A., Mvungi, N.H., Kaijage, S., Michael, K.: A survey on non-intrusive load monitoring methodies and techniques for energy disaggregation problem (2017). arXiv. https://arxiv.org/abs/1703.00785
11. Andryushkevich, S.K., Kovalyov, S.P., Nefedov, E.: Composition and application of power system digital twins based on ontological modeling. In: Proceedings of the 17th IEEE International Conference Industrial Informatics INDIN'19. Helsinki-Espoo, pp. 1536–1542. IEEE (2019)
12. Erikstad, S.: Design patterns for digital twin solutions in marine systems design and operations. In: Proceedings of the 17th International Conference Computer and IT Applications in the Maritime Industries COMPIT' 2018, pp. 354–363. Technische Universität Hamburg, Hamburg (2018)

Automated Information System for Control and Diagnostics of the Blast Furnace Slag Mode

I. Gurin(✉), N. Spirin, and V. Lavrov

Ural Federal University, 51, Avenue, Lenin, Yekaterinburg 620075, Russia
ivan.gurin@urfu.ru

Abstract. The numerical simulation model of the slag mode control and diagnostics system contains the following calculation blocks: determination of yield, composition and polytherm viscosity of final slag; calculation of slag desulfurizing ability; slag mode diagnostics. The model is applicable to slags with permissible variations of Al2O3 content from 3 to 17%, MgO - from 5 to 20%, SiO2 content of 35% and more in slags, CaO content of 35% and more in slags. The developed automated information system calculates the following slag properties: slag temperature, slag viscosity at slag discharge temperature, slag basicity (CaO/SiO2), (CaO + MgO)/SiO2 and (CaO + MgO)/(SiO2 + Al2O3); temperatures at which slag viscosity is 7 and 25 poise; slag viscosity gradients in the range of 7–25 poise and in the range of 1400–1500 °C. The system can be used by a process team both during blast furnace smelting to control and diagnose the slag mode for a selected time frame, and during simulation of slag properties with regard to the specified rates and properties of the loaded burden.

Keywords: Slag-forming processes · Mathematical model · Blast furnace · System · Control · Diagnostics

1 Introduction

The slag mode (slag-forming processes, composition, and properties of slag) of blast-furnace smelting determines in a great measure the key technical and economic indicators of a blast furnace - the specific coke consumption and production capacity. The slag composition, specific yield and properties affect the gas-dynamic, thermal and reduction processes occurring in a blast furnace.

Slag mode numerical simulation is performed to tackle the following process tasks [1–5]:

- Production of slag with required properties during the entire slag formation process, starting with the softening stage, the subsequent filtration of the primary slag melt in the coke packing and ending with obtaining the final slag melt.
- Production of pig iron of the required chemical composition with the permissible sulfur content therein.

A. A. Radionov and V. R. Gasiyarov (Eds.): RusAutoCon 2021, LNEE 857, pp. 256–264, 2022.
https://doi.org/10.1007/978-3-030-94202-1_25

- Production of slag, which provides sufficient gas permeability of the slag-forming zone, the even flow of the burden in the furnace, as well as a stable thermal condition of a blast furnace.
- Numerical simulation of slag properties when changing its composition.

2 Numerical Simulation Model of Slag Mode Control and Diagnostics

Slag mode numerical simulation is performed to tackle the following process tasks [1–5]:

Numerical simulation model of the system for control and diagnostics of the slag mode is shown in Fig. 1. It contains the following calculation blocks: determination of yield, composition and polytherm viscosity of final slag; calculation of slag desulfurizing ability; slag mode diagnostics.

The calculation of slag yield and composition (the content of CaO, SiO_2, Al_2O_3, MnO, MgO, TiO_2 in the slag) is based on the equations of material balances of the main chemical constituents and their compounds. For instance, one can calculate the slag yield from the slag-forming constituents' balance, calcium oxide balance [1–4]. The chemical compositions of all iron ore materials, fluxes and coke ash are considered in the calculation.

Three-component CaO–SiO_2–Al_2O_3 and four-component CaO–SiO_2–Al_2O_3–MgO slag systems are the most used models for calculating viscosity of the blast furnace slag [7–10]. In this paper, the analytical estimation of the slag viscosity is based on the mathematical treatment of the four-component slag system CaO-SiO_2-Al_2O_3-MgO at temperatures of 1400 and 1500 °C in the field of real values of blast-furnace slags and the known dependence of viscosity of homogeneous slag melts on temperature. The influence of other oxides FeO, MnO, TiO_2 and others on the slag viscosity is considered [11–20].

In Fig. 2 one shows the slag diagram of the four-component system CaO-SiO_2-Al_2O_3-MgO at the 10% MgO content in the slag. Viscosity of the slag at 1500 °C and the content of the following constituents in the slag: CaO – 40%, SiO_2 – 40%, Al_2O_3 – 10%, MgO – 10% according to the experimental data is 2.8 poise, and 2.6 poise - according to the computational algorithm. Viscosity of the slag at 1500 °C and the content of the following constituents in the slag: CaO – 35%, SiO_2 – 40%, Al_2O_3 – 15%, MgO – 10% according to the experimental data is 4.4 poise, and 4.46 poise - according to the computational algorithm [21–25].

In Table 1 presents comparative data on slag viscosity from experimental studies and estimated values at different temperatures [15–18].

Comparison of the viscosity of slags of different compositions and at different temperatures showed that the experimental data correspond to the calculated values with a relative error of 4–7%.

The content of elements of the four-component system CaO-SiO_2-Al_2O_3-MgO in the blast furnace slag according to observations in 2015–2020 is 95–97%. Table 2 shows the composition of blast furnace slag for different periods at different blast furnaces.

The suggested numerical simulation model for control and diagnostics of the slag mode allows taking into account the complex influence of slag constituents on its properties by calculating the viscosity polytherms in the range of temperatures present in

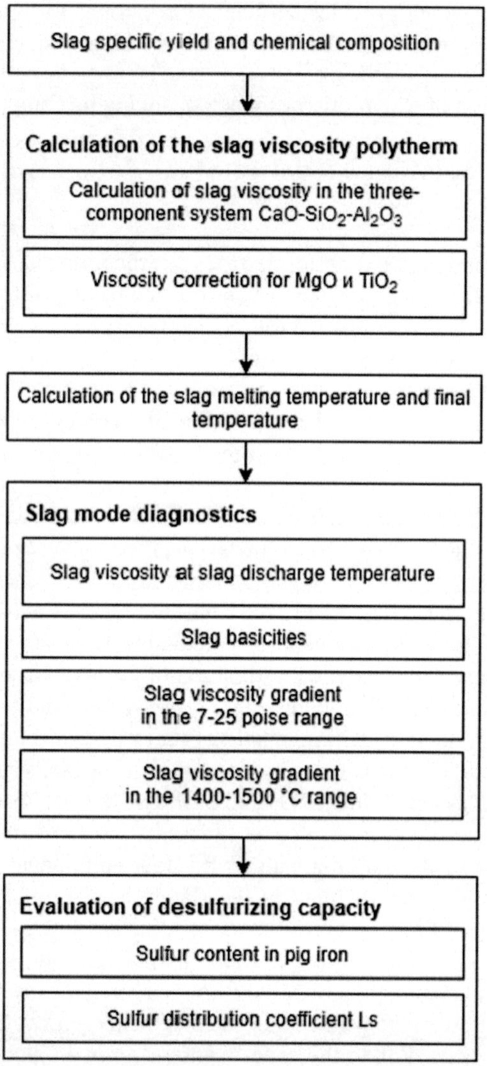

Fig. 1. Numerical simulation model of slag mode control and diagnostics.

the blast furnace. The model is applicable to slags with allowable variations in their composition:

- Al2O3 between 3 and 17%;
- MgO between 5 to 20%;
- SiO2 of 35% or more in the slags;
- CaO of 35% or more in the slags;
- CaO/SiO2 slag basicity between 0.8 and 1.20

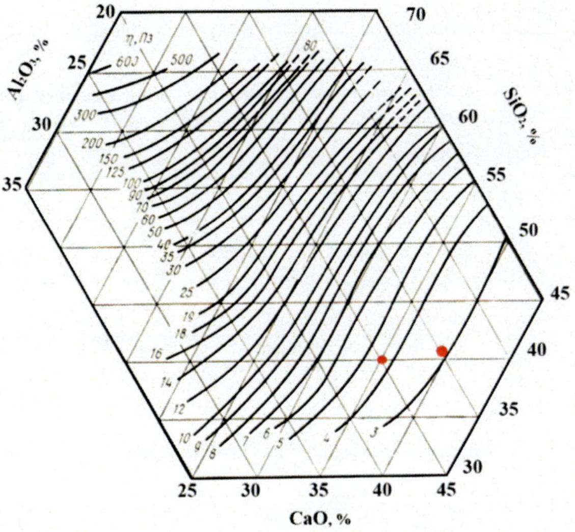

Fig. 2. Viscosity of melts (in poises) of the system CaO-SiO_2-Al_2O_3-MgO at 1500 °C and 10% MgO in the slag [24].

One of the key characteristics of the slag mode is the stability of slag systems ('short' and 'long' slags). Changes in the processes of slag formation cause fluctuations in the basicity of the slag, which in turn lead to changes in the temperature interval of the viscoplastic state.

Slag viscosity gradients are used to estimate slag properties.

One determines the viscosity gradient (gradient-1), which shows how fast the viscosity of the slag changes from 7 to 25 poise when the slag temperature changes by 1 °C, poise/°C:

$$\Delta\eta_{07}^{25} = \frac{25 - 7}{t_{\text{slag}}^{7} - t_{\text{slag}}^{25}} \tag{1}$$

One determines the viscosity gradient (gradient-2), which shows the rate of increase in the viscosity of the slag when the slag temperature changes by 1 °C in the range from 1400 to 1500 °C, poise/°C:

$$\Delta\eta_{1400}^{1500} = \frac{\eta_{1400} - \eta_{1500}}{1500 - 1400} \tag{2}$$

According to gradient-1 one can estimate how quickly the slag loses mobility in the area of lower temperatures ('short' or 'long' slag).

According to the gradient-2 one can estimate how stable the slag is in the range of operating temperatures (1400–1500 °C).

The 'shorter' the slag, the higher the gradient-1 and lower the gradient-2, that is, in the area of operating temperatures, the 'short' slag is more mobile and stable.

The slag viscosity gradient along with the allowable slag viscosity ranges at different slag temperatures is used in slag mode modeling as a limiting factor for slag mode

Table 1. Comparative data on the slag viscosity from experimental studies and estimated values.

Slag composition, %				Slag basicity (CaO/SiO_2), units	Temperature, °C	Slag viscosity, poise		Difference, poise/%
CaO	*SiO_2*	*Al_2O_3*	*MgO*			*Exper*	*Calculation*	
40	40	10	10	1.0	1500	2.5	2.6	0.1/4.0
40	40	10	10	1.0	1450	4.2	4.0	0.2/4.8
40	40	10	10	1.0	1400	7.2	7.5	0.3/4.2
37.5	37.5	15	10	1.0	1500	4.0	4.2	0.2/5.0
37.5	37.5	15	10	1.0	1450	5.6	5.9	0.3/5.4

Table 2. Chemical composition of PAO MMK blast furnace slag, %.

Blast furnace No.	CaO	SiO_2	Al_2O_3	MgO	TiO_2	MnO	FeO	S	Slag basicity (CaO/SiO_2), units	Content of CaO-SiO_2-Al_2O_3-MgO system constituents, %
1	39.6	39.1	9.8	7.9	0.7	0.3	0.3	0.7	1.01	96.4
2	36.4	36.8	12.8	9.3	1.3	0.4	0.3	0.7	0.99	95.3
4	39.5	39.3	10.2	7.7	0.9	0.2	0.3	0.7	1.01	96.7
6	37.3	36.4	12.3	8.7	1.1	0.5	0.3	0.8	1.02	94.7
7	38.1	38.1	11.2	8.4	0.9	0.5	0.3	0.7	1.00	95.8
8	39.0	38.7	11.3	7.5	0.9	0.3	0.3	0.7	1.01	96.5
9	38.4	38.3	11.3	7.8	0.8	0.2	0.3	0.7	1.00	95.8
10	38.8	38.5	10.7	7.5	0.8	0.2	0.3	0.7	1.01	95.5

diagnostics. The choice of the limit values of each range and viscosity gradient is made by expert assessment.

For one, for the conditions of PAO MMK blast furnaces, the following limiting values are considered most often:

- If the viscosity of the slag at 1450 °C is in the range of 5.5–6.5 poise, the slag has increased viscosity in the temperature range of 1400–1500 °C;
- If the slag viscosity at 1450 °C exceeds 6.5 poise, then the viscosity of the slag in the range of operating temperatures exceeds the allowable limits for the normal smelting mode;
- If the viscosity of the slag at 1450 °C is less than 2 poise, then the slag has low viscosity, is highly aggressive to the refractory work and destroys the slag lining;
- If the viscosity gradient $\geq \Delta\eta_{07}^{25} 0.35$, then the slag is of the 'short' type. The blast furnace hearth may be clogged. The operation (operation mode) of the furnace is hindered;
- If the viscosity gradient $\Delta\eta_{1400}^{1500} > 0.085$ the slag is unstable in the range of operating temperatures (1400–1500 °C). Slag can solidify in the spouts.

3 Specification of the Slag Mode Control and Diagnostics System

To control and diagnose the blast furnace slag mode at blast furnaces of PAO MMK one has developed an automated information system (AIS). In Fig. 3 one shows the architecture of the developed system.

The system contains two modules: a module for continuous data collection on the process parameters (composition and properties of the burden materials, composition and properties of liquid smelting products etc.) and a data presentation module.

The information, which continuous collection module receives, is averaged by hours, shifts, days, weeks, months, and taps. The data presentation module is designed as a web application.

In Fig. 4 one shows the window for displaying information on process parameters over time. The window with the AIS results is shown in Fig. 5.

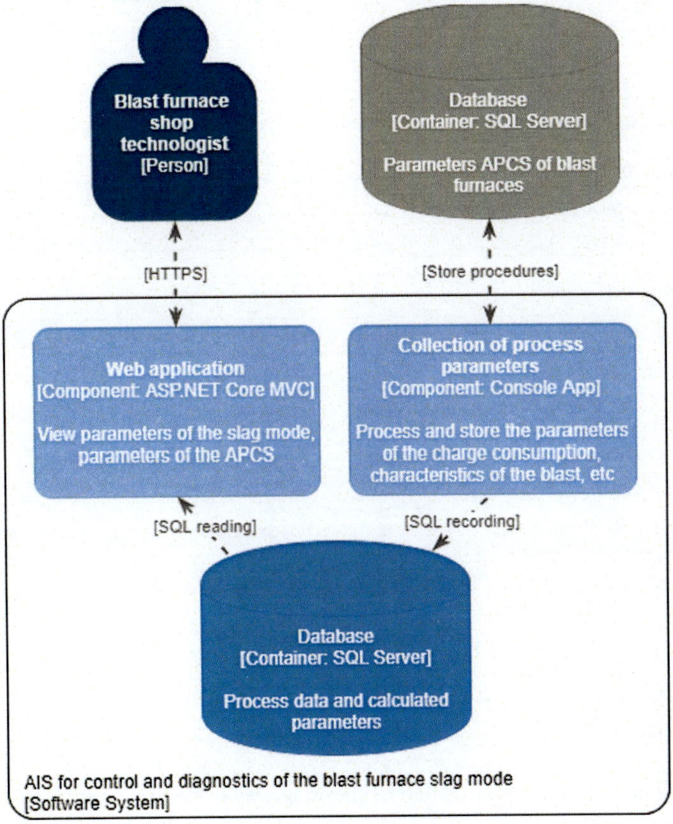

Fig. 3. Architecture of the automated information system for control and diagnostics of the blast furnace slag mode.

In the web application, one can view some mode and calculation parameters of the slag control and diagnostics system:

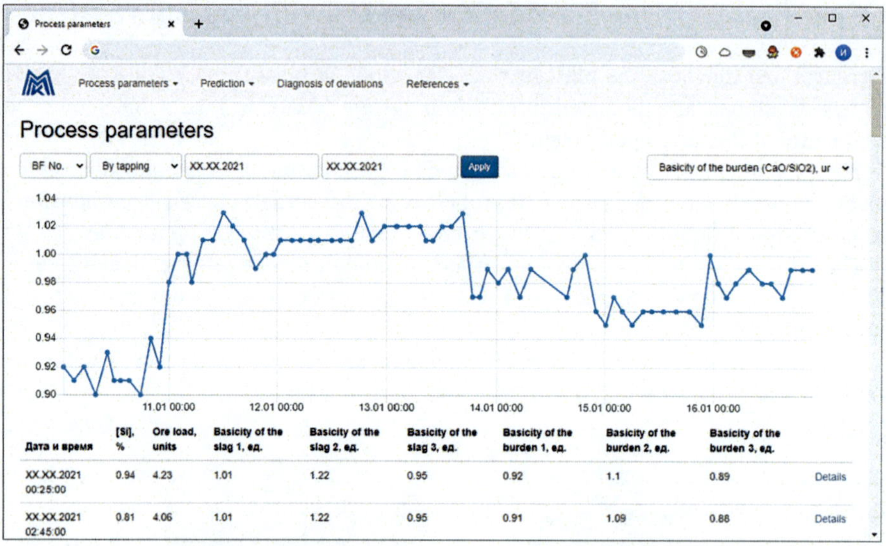

Fig. 4. Display window for information on process parameters over time.

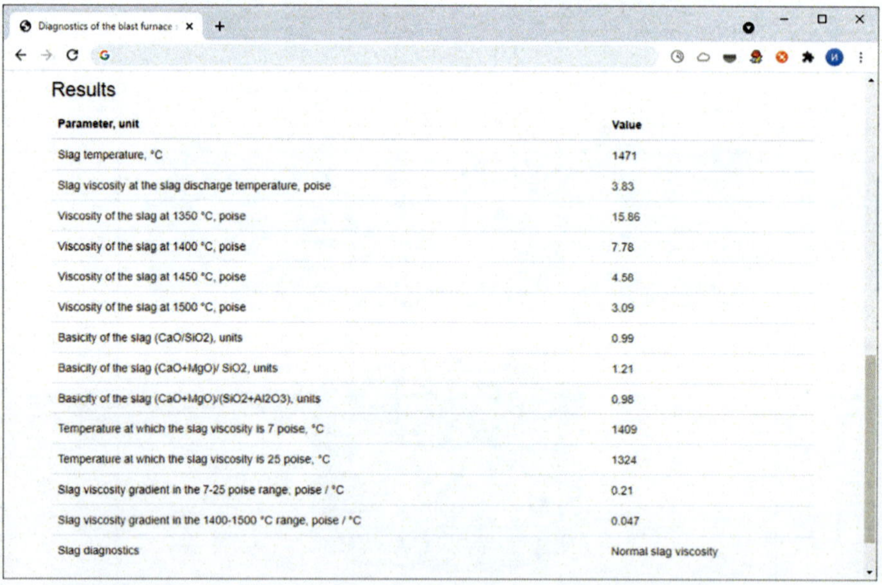

Parameter, unit	Value
Slag temperature, °C	1471
Slag viscosity at the slag discharge temperature, poise	3.83
Viscosity of the slag at 1350 °C, poise	15.86
Viscosity of the slag at 1400 °C, poise	7.78
Viscosity of the slag at 1450 °C, poise	4.58
Viscosity of the slag at 1500 °C, poise	3.09
Basicity of the slag (CaO/SiO2), units	0.99
Basicity of the slag (CaO+MgO)/ SiO2, units	1.21
Basicity of the slag (CaO+MgO)/(SiO2+Al2O3), units	0.98
Temperature at which the slag viscosity is 7 poise, °C	1409
Temperature at which the slag viscosity is 25 poise, °C	1324
Slag viscosity gradient in the 7-25 poise range, poise / °C	0.21
Slag viscosity gradient in the 1400-1500 °C range, poise / °C	0.047
Slag diagnostics	Normal slag viscosity

Fig. 5. Window of calculation results of the slag mode properties in AIS.

- Ore load, units;
- Content [Si] in the pig iron, %;
- Content (CaO) in the slag, %;
- Content (SiO2) in the slag, %;

- Basicity of the slag (CaO/SiO_2), ($CaO + MgO)/SiO_2$ and ($CaO + MgO)/(SiO_2 + Al_2O_3)$, units;
- Basicity of the iron ore materials (IOM) (CaO/SiO_2), ($CaO + MgO)/SiO_2$ and ($CaO + MgO)/(SiO_2 + Al_2O_3)$, units;
- Basicity of the burden (CaO/SiO_2), ($CaO + MgO)/SiO_2$, ($CaO + MgO)/(SiO_2 + Al_2O_3)$, units.

4 Conclusion

The developed AIS calculates the following slag properties: slag temperature, °C; slag viscosity at the slag discharge temperature, poise; slag basicity (CaO/SiO_2), ($CaO + MgO)/SiO_2$ and ($CaO + MgO)/(SiO2 + Al_2O_3)$, units; temperatures at which the slag viscosity is 7 and 25 poise, °C; the slag viscosity gradient in the range of 7–25 poise and the slag viscosity gradient in the range of 1400–1500 °C, poise/°C. Based on the calculated data, a diagnostic message on the estimation of slag mode parameters is generated.

The system can be used by a process team both during blast furnace smelting to control and diagnose the slag mode for a selected time frame, and during simulation of slag properties, considering the specified rates and properties of the loaded burden.

References

1. Spirin, N.A., Onorin, O.P., Rybolovlev, V.Y., Perminov, A.I., Shchipanov, K.A.: Simulation of slag processes in blast-furnace smelting. Steel in Transl. **35**(8), 7–11 (2005)
2. Oporin, O.P., Spirin, N.A., Lavrov, V.V.: Computer methods of modelling the gas-dynamic and slag conditions of blast furnace heat. Stal **6**, 55–58 (2005)
3. Polinov, A.A., Pavlov, A.V., Pishnograev, S.N., Logachev, G.N., Spirin, N.A.: Effect of slag regime on alkaline compound behavior in a blast furnace. Metallurgist **61**(3–4), 193–197 (2007)
4. Spirin, N.A., Lavrov, V.V., Rybolovlev, V.Y., et al.: Model systems for the support of decision making in automatic systems of control over the technological process of blast-furnace smelting in metallurgy. In: UrFU, Ekaterinburg, p. 462 (2011)
5. Spirin, N.A., Lavrov, V.V., Rybolovlev, V.Y., et al.: Mathematical modeling of metallurgical processes in automated process control systems. In: Spirin, N.A. (Ed.) UrFU, Ekaterinburg, p. 558 (2014)
6. Gordon, Y., Izumskiy, N.: Mathematical model and stabilization system for slag mode of blast furnace operation. In: AISTech 2017 Iron and Steel Technology Conference, Nashville, United States, AISTech - Iron and Steel Technology Conference Proceedings, vol. 1, pp. 797–805 (2017)
7. Gan, L., Lai, C.: A General Viscosity Model for Molten Blast Furnace Slag. Metal. Mater. Trans. B. **45**(3), 875–888 (2013). https://doi.org/10.1007/s11663-013-9983-9
8. Iida, T., Sakai, H., Kita, Y., Shigeno, K.: An equation for accurate prediction of the viscosities of blast furnace type slags from chemical composition. ISIJ Int. **40**, 110–114 (2000). https://doi.org/10.2355/isijinternational.40.suppl_s110
9. Shu, Q.: A viscosity estimation model for molten slags in Al2O 3-CaO-MgO-SiO2 system. Steel Res. Int. **80**(2), 107–113 (2009). https://doi.org/10.2374/SRI08SP085

10. Jiang, D., Zhang, J., Wang, Z., Feng, C., Jiao, K., Xu, R.: A Prediction model of blast furnace slag viscosity based on principal component analysis and k-nearest neighbor regression. JOM **72**(11), 3908–3916 (2020). https://doi.org/10.1007/s11837-020-04360-9
11. Jia, R., Deng, L., Yun, F., Li, H., Zhang, X., Jia, X.: Effects of SiO2/CaO ratio on viscosity, structure, and mechanical properties of blast furnace slag glass ceramics. Mater. Chem. Phys. **233**, 155–162 (2019). https://doi.org/10.1016/j.matchemphys.2019.05.065
12. Shen, X., Chen, M., Wang, N., Wang, D.: Viscosity property and melt structure of CaO-MgO-SiO2-Al2O3-FeO slag system. ISIJ Int. **59**, 9–15 (2019). https://doi.org/10.2355/isijinternational.ISIJINT-2018-479
13. Li, T., Sun, C., Song, S., Wang, Q.: Influences of Al2O3 and TiO2 content on viscosity and structure of CaO–8%MgO–Al2O3–SiO2–TiO2–5%FeO blast furnace primary slag. Metals **9**(7), 743 (2019). https://doi.org/10.3390/met9070743
14. Xu, R.Z., Zhang, J.L., Han, W.X., Chang, Z.Y., Jiao, K.X.: Effect of BaO and Na2O on the viscosity and structure of blast furnace slag. Ironmaking Steelmaking **47**(2), 168–172 (2020). https://doi.org/10.1080/03019233.2018.1498761
15. Xing, X., Pang, Z., Mo, C., Wang, S., Ju, J.: Effect of MgO and BaO on viscosity and structure of blast furnace slag. J. Non-Cryst. Solids **530**, 119801 (2020). https://doi.org/10.1016/j.jnoncrysol.2019.119801
16. Zheng, H., et al.: Viscosity prediction model for blast furnace slag with high Al2O3. Steel Res. Int. **92**(1), 1900635 (2021). https://doi.org/10.1002/srin.201900635
17. Zhang, Y., Wang, D., Chen, S., Liu, Z., Pan, W., Zhao, Z.: Effects of Basicity, FeO, and TiO2 on Phase Transformation and Viscosity of TiO2-Bearing Primary Slag in Blast Furnace. In: Li, J., et al. (eds.) Characterization of Minerals, Metals, and Materials 2021. TMMMS, pp. 187–199. Springer, Cham (2021). https://doi.org/10.1007/978-3-030-65493-1_18
18. Yang, D., Zhou, H., Wang, J., Pang, Z., et al.: Influence of TiO2 on viscosity, phase composition and structure of chromium-containing high-titanium blast furnace slag. J. Market. Res. **12**, 1615–1622 (2021). https://doi.org/10.1016/j.jmrt.2021.03.069
19. Chang, Z.Y., Jiao, K.X., Zhang, J.L., Ning, X.J., Liu, Z.Q.: Effect of TiO2 and MnO on viscosity of blast furnace slag and thermodynamic analysis. ISIJ Int. **58**(12), 2173–2179 (2018). https://doi.org/10.2355/isijinternational.ISIJINT-2018-379
20. Jiao, K., Zhang, J., Liu, Z., Chen, C.: Effect of MgO/Al2O3 ratio on viscosity of blast furnace primary slag. High Temp. Mater. Processes (London) **38**, 354–361 (2019). https://doi.org/10.1515/htmp-2018-0019
21. Zhilo, N.L.: Formirovanie i svoistva domennykh shlakov (Formation and Properties of Blast Furnace Slag). Metallurgiya, Moscow (1974)
22. Voskoboinikov, V.G., Dunaev, N.E., Mikhalevich, A.G.: Svoistva zhidkikh domennykh shlakov (Properties of Liquid Blast Furnace Slags). Metallurgiya, Moscow (1975)
23. Vegman, E.F.: Domennoe proizvodstvo, Tom 1. Podgotovka rud i domennyi protsess (Blast Furnace Production, Vol. 1: Preparation of Ores and Blast Furnace Processes), Metallurgiya. Moscow (1989)
24. Slag Atlas. 2nd Edition, Ed. VDEh, Ed, Verlag Stahleisen GmbH: Düsseldorf, p 636 (1995)
25. Shalimov, A.G., Kuklev, V.G.: Izvestiya Akademii nauk SSSR, OTN. Metallurgija I Toplivo **5**, 43–51 (1962)

Adaptive Weather Forecasting Based on Local Characteristics of the Territory

R. V. Sharapov(✉)

Vladimir State University, 23, Orlovskaya Street, Murom 602264, Russia

Abstract. At present, global models describing the dynamics of the atmosphere, such as GFS, NAM, ECMWF, UkMet and others, are actively used to compile weather forecasts. Modern global models make it possible to build weather forecasts of a sufficiently high accuracy. Nevertheless, for local territories, errors often arise related to the peculiarities of a particular area. It is rather difficult to provide the proper level of detail for global models - it is necessary to set an extended description of the territory for the entire model, which is quite difficult (especially for poorly explored areas). At the same time, the volumes of the initial data increase and the complexity of the calculation algorithms increases. For this reason, the use of adaptive models that allow adjusting forecasts from global models for specific territories seems quite attractive.

Keywords: Forecast · Weather · Local characteristics · Weather model · Adaptive weather forecast

1 Introduction

People have always been interested in weather forecasts. What is the weather going to be like tomorrow? Will it rain or freeze? No one will give an exact answer to these questions. Nevertheless, modern forecasts can give an idea of the most likely weather in the coming days.

In recent years, weather observation and forecasting have been actively developing. A large number of meteorological stations have appeared that are capable of automatically collecting and transmitting weather data to unified meteorological centers. If earlier weather data were collected by meteorological stations at airports, now anyone can purchase an inexpensive meteorological station and share its readings via the Internet [1]. Gigabytes and terabytes of data are being accumulated on weather conditions in various parts of the world. The use of these data allow move to a new level in the preparation of weather forecasts, improving their quality, detailing, as well as improving forecast models by identifying their inaccuracies.

At present, a lot of research has been devoted to the problem of making so-called mesoscale forecasts. Many researchers propose to use the data of several global models and, on their basis, make adaptive forecasts.

A. A. Radionov and V. R. Gasiyarov (Eds.): RusAutoCon 2021, LNEE 857, pp. 265–273, 2022.
https://doi.org/10.1007/978-3-030-94202-1_26

2 Weather Forecast Models

Any modern weather forecasting model is based on a mathematical model of the atmosphere proposed by Wilhelm Bjerknes [2]. For forecasting, Bjerknes suggested using a system of equations: one equation for each dependent variable describing the atmosphere. In particular, he identified seven main variables: pressure, temperature, density, humidity, and three components of air flow velocity (x, y, z velocity).

To obtain correct forecasts, a significant number of primary data from meteorological satellites and meteorological stations is required. The highest resolution of the best modern forecasting models is up to 3 km (in most cases 25 km) when using spatial interpolation algorithms.

There are three main centers for collecting meteorological data in the world, where forecasting models are implemented, which are different approaches to solving the equations of atmospheric dynamics, the core of which contains the Richardson equations [3]:

1. Global Forecasting System (GFS) - is operated by NCEP (National Centers for Environmental Prediction), which is a division of NOAA (National Oceanic and Atmospheric Administration, National Oceanic and Atmospheric Administration), NWS (National Weather Service, National Weather Service), USA [4–7].

The GFS model is updated four times a day (00:00, 06:00, 12:00 and 18:00 UTC) and provides forecasts up to 16 days (384 h). The model is executed in two parts: the first part has a higher resolution and runs up to 192 h (8 days), the second part runs from 192 to 384 h (16 days).

NAM (North American Mesoscale) regional mesoscale model with enhanced elevation data and improved surface and prediction process parameters.

The Weather Research and Forecasting (WRF) model is a next generation mesoscale numerical weather prediction model designed to meet the needs of both operational forecasting and atmospheric research. The WRF is updated 4 times a day and generates forecasts 78 h ahead in 1-h increments. Forecasts include wind speed and direction, wind gusts, temperature, total cloud cover and precipitation [8] (Fig. 1).

2. European Center for Medium-Range Weather Forecasts (ECMWF), Reading, UK. To date, this model is the newest, is rapidly developing and is the leader in the accuracy of long-term forecasts [9] (Fig. 2)

3. The UkMet model is the brainchild of the world's oldest meteorological service, MetOffice, the British Government Meteorological Office. It is believed that it is richer than others in terms of scientific potential and gives the most accurate short-term forecast.

In Russia, the main organization performing weather forecasting is the Hydrometeorological Center of Russia, which also uses hydrodynamic models of the atmosphere,

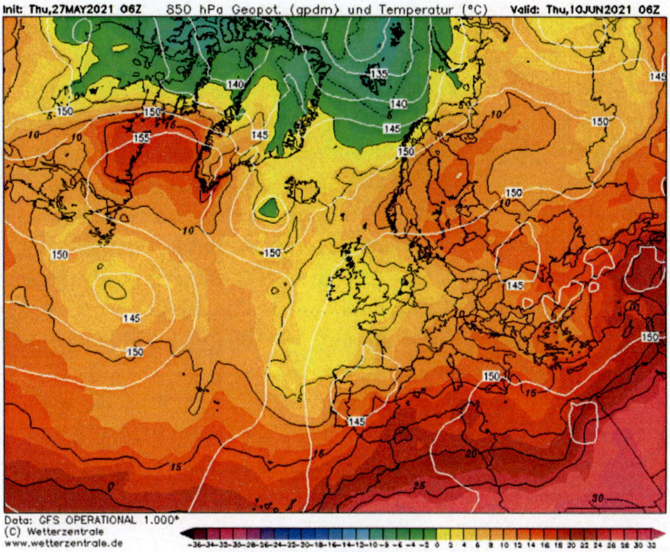

Fig. 1. GFS model.

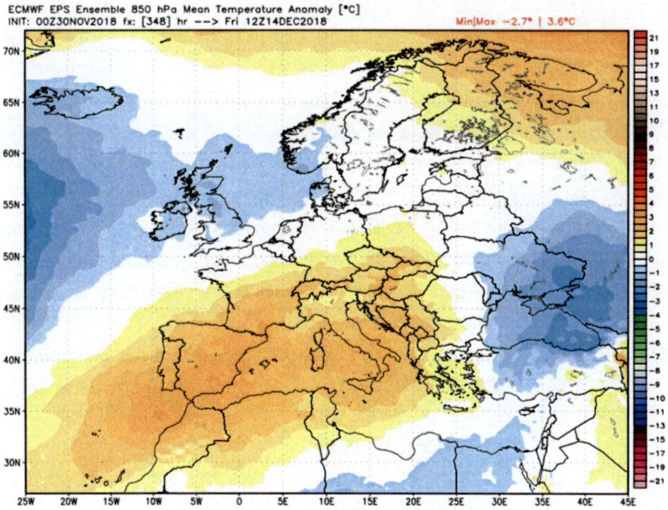

Fig. 2. ECMWF model.

in which the main weather-forming mechanisms are presented: cloud-radiation interactions, moisture phase transitions, turbulence in the boundary layer, heat transformations, etc. moisture in the upper soil layer, interaction with vegetation, etc. However, some of the physical processes are deliberately not taken into account or roughened due to limited resources.

Hydrodynamic models predict not the point, but the characteristics averaged over the cells of the computational coordinate grid. All the variety of properties of the atmosphere and underlying surface inside the cell is represented by spatially averaged grid values.

Spatio-temporal sampling and smoothing affect the ability of models to reproduce local features of meteorological fields and, first of all, extreme characteristics and sharp changes in weather, as a rule, of greatest interest to forecast consumers.

All of these models have significant limitations. Systematic errors arise primarily due to the lack of initial data. For example, in Europe the average distance between meteorological stations is 25 km, in Russia it is 100 km at best for the European part (for Siberia, as you yourself understand, there may be 1000 km between stations). In addition, all these models have a resolution of 50–100 km, poorly take into account the relief, water areas, and local features of the territory.

For example, forecasts are made for the Vladimir region for the main settlements. The data from several meteorological stations in Vladimir (Vladimir, Gus-Khrustalny) and neighboring (Vyksa, Pavlovsky Posad, Volzhskaya GMO) regions are taken as a basis. Nevertheless, the forecasts obtained are rather approximate: they are poorly tied to settlements and are not accurate enough.

3 Adaptive Weather Forecasting

Existing weather forecasting models use huge amounts of data, require significant computing power, and make it possible to predict the weather in regions with sufficient accuracy [10–12]. However, the lack of a sufficient number of observation stations (or prompt access to their data), insufficient consideration of the local characteristics of the territory (terrain, plantations, water bodies, technogenic impact), does not allow making accurate forecasts for individual settlements and points of the terrain [13–15].

On the other hand, the development of our own models without the use of global data (movement of atmospheric layers, changes in meteorological parameters, including in remote areas) will not allow obtaining a more accurate picture of forecasts.

To solve this problem, it is proposed to use the data of existing weather forecasting models with their further correction. The use of existing global models makes it possible to solve the problem of taking into account global changes in the atmosphere without attracting large computing power.

The adjustment affects three main aspects:

1. Systematic errors in forecasting for specific territories and their correction [16].
2. Accounting for data from local observational meteorological stations and other sources.
3. Taking into account the local characteristics and features of the territory that affect weather conditions.

At present, there is a fairly large amount of data on weather forecasts for various territories and the actual values of meteorological parameters [17]. Comparison of predicted and actual values reveals forecast errors and inaccuracies. If by themselves, errors in past forecasts are not a sufficient basis for correcting future forecasts, then identifying the causes of errors, associated factors (including meteorological), local features of the terrain, allow to find systematic errors and correct them.

Due to the fact that the volumes of meteorological data and forecasts are large [18] (for example, WRF and GFS forecasts occupy more than 60 GB and are updated every minute), it is possible to effectively identify their relationships and correct forecasts using machine learning methods [19–23]. For these purposes, neural networks, support vectors, etc. can be successfully used [24–28]. The use of factor analysis makes it possible to identify the relationship between environmental parameters and their influence on weather changes.

As a set of parameters for machine learning, it is proposed to use a wide set of indicators of both the current weather state and unsuccessful forecasts:

- Temperature in 3 h increments,
- Average temperature,
- The greatest temperature difference,
- Average temperatures over the past week,
- Atmosphere pressure,
- Change in atmospheric pressure,
- Wind speed and direction,
- Relative humidity,
- Weather indicators in the vicinity (the area is divided into zones with a set step, for example, 1×1, 5×5, 10×10 km),
- The average temperature for the current day based on long-term statistics,
- The difference between the current temperature value and the average temperature value based on long-term statistics,
- Cloudiness,
- Height above sea level,
- Elevation difference with the surroundings,
- The share of water bodies at the forecast point
- Etc.

Due to the fact that the traditional network of meteorological stations does not give a detailed idea of the weather conditions at specific points in the region (where there are no meteorological stations), it is of interest to use additional sources of meteorological information that can provide information about barometric pressure, temperature, humidity, air velocity. Such sources can be temperature and humidity sensors, precipitation sensors, automatic and compact weather stations, etc. Since these data can have a different structure and form of presentation, and also not uniformly cover the territory of the Vladimir region, the problem arises of integrating and processing such heterogeneous and incomplete meteorological data.

4 Structure of Adaptive Weather Forecasting System

Sources of data for compiling adaptive weather forecasts are:

- Data of current weather observations coming from the ground network of meteorological stations, meteorological satellites, radiosondes, as well as local meteorological stations of users.
- Archival data of weather observations. Currently, extensive archives of weather observation data are collected in the National Climatic Data Center (NCDC), National Oceanic and Atmospheric Administration (NOAA), European Climate Assessment & Dataset, Hydrometeorological Center of Russia, etc. Weather data available for the last 100 years. In our country, the most complete archives of weather observations have begun to be collected over the past 10 years. Much data is collected automatically and is available online.
- Archived data of weather forecasts compiled on the basis of existing models of GFS, WRF, ECMWF, etc.
- A dataset of local characteristics of territories, allowing to take into account their individual characteristics (for example, changes in the relief, the presence of water bodies, green spaces, etc.)
- Forecast data from existing global forecasting models. Models such as the GFS and WRF give a fairly good idea of the changing weather conditions on Earth, but sometimes have inaccuracies for specific points in the terrain. For this reason, the predictions of such models are valuable sources of data.

Modern global models used to make weather forecasts are quite complex and require very large computing power. Computational clusters are often used for calculations. Making changes to such models to adapt them to specific conditions is an extremely complex process. For this reason, it seems to be a more efficient solution to use the data obtained from the operation of such models (WRF, GFS, ECMWF, etc.) in large scientific centers. In this case, it becomes possible to use ready-made forecasts from different models without the cost of computing power.

For making adaptive forecasts, one of the most valuable data sources is information on correspondence of forecasts to actual weather values. For these purposes, the system contains a module that compares forecasts and observational data and finds stable errors in forecasts. Information about such errors and their numerical expressions (differences between predicted and actual temperature, humidity, air velocity) are an additional source of data for compiling adaptive weather forecasts.

The core of the system is a data processing module using machine learning. The module receives sets of initial data and weather forecasts, processes them, and corrects forecasts based on previously identified stable errors. For this, multilayer neural networks are used.

Multilayer neural network is a neural network consisting of input, output and one or more hidden layers of neurons located between them (Fig. 3).

The system is constantly learning based on the data of the detected persistent errors. In this case, the following difficulties arise:

Global forecasting models are constantly evolving and revising. For this reason, in new versions of models, errors can appear or disappear. For this reason, the most significant errors are those that have arisen in the most recent forecasts.

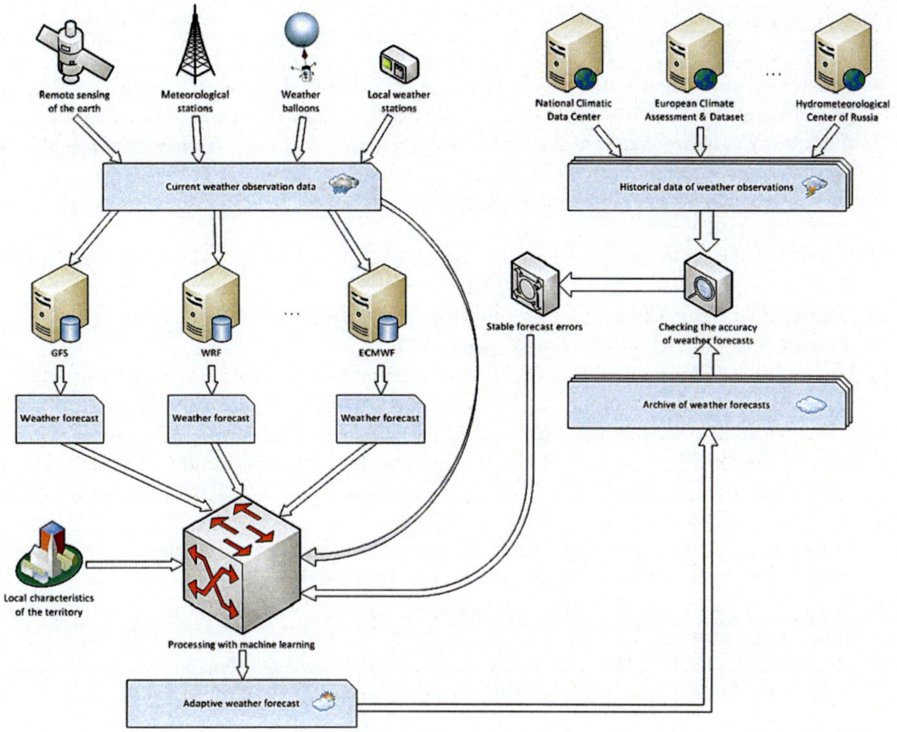

Fig. 3. Adaptive weather forecasting.

The adaptive forecasting system can also make mistakes. For this reason, it is necessary to monitor the convergence of the forecasts issued by the system and real observational data. Some predictions can be so unfortunate that the number of errors will be large. And this, in turn, can affect the quality of future forecasts.

Errors sometimes occur in the observational and forecast data streams. For example, data from some weather station is not received, or the values of these data do not correspond to reality (for example, extremely large or small values, which in fact are not). For this reason, careful verification of the incoming data should be carried out and clearly inconsistent data should be excluded from consideration.

5 Conclusion

Modern global models make it possible to build weather forecasts of sufficiently high accuracy. Nevertheless, for local territories, errors often arise related to the peculiarities of a particular area. It is rather difficult to provide the proper level of detail for global models - it is necessary to set an extended description of the territory for the entire model, which is quite difficult (especially for poorly explored areas). At the same time, the volumes of the initial data increase and the complexity of the computational algorithms increases. For this reason, the use of adaptive models that allow correcting forecasts from global models for specific territories seems to be quite attractive.

References

1. Local Weather Forecast. News and Conditions, Weather Underground. https://www.wunderground.com
2. Bjerknes, V.: The problem of weather prediction, considered from the viewpoints of mechanics and physics. Meteorol. Z. **21**, 1–7 (1904)
3. Richardson, L.: Weather Prediction by Numerical Process, p. 236. The University Press, Cambridge (1922)
4. Skamarock, W.C.: A description of the Advanced Research WRF version 3. NCAR Tech. Note NCAR/TN-475+STR, p. 113 (2008)
5. Hamill, T.M., Whitaker, J.S.: Increasing NOAA's computational capacity to improve global forecast modeling. A NOAA White Paper (2010)
6. Deshpande, M., Johny, C.J., Kanase, R.: Implementation of Global Ensemble Forecast System (GEFS) at 12km Resolution. Contribution from IITM Technical Report No.TR-06 ESSO/IITM/MM/TR/ 02(2020)/200
7. Liu, Q., Marchok, T., Pan, H.L., Bender, M., Lord, S.: Improvements in hurricane initialization and forecasting at NCEP with global and regional (GFDL) models. TPB 472, National Weather Service, US Department of Commerce, p. 7 (2000)
8. Andersson, E., Hollingsworth, A.: Typhoon bogus observations in the ECMWF data assimilation system. ECMWF Tech. Memo. 148, ECMWF, Reading, UK (1988)
9. Arakawa, A., Mintz, Y.: The UCLA atmospheric circulation model. Department of Meteorology, California (1974)
10. Tracton, M.S., Mo, K., Chen, W., Kalnay, E., Kistler, R., White, G.: Dynamical extended range forecasting (DERF) at the national meteorological center. Mon. Wea. Rev. **117**, 1604–1635 (1989)
11. Yu, T.-W., Iredell, M., Keyser, D.: Global data assimilation and forecast experiments using SSM/I wind speed data derived from a neural network application. Wea. and Fcst. **12**, 859–865 (1997)
12. Pegion, P., Whitaker, J., Hamill, T., Bates, G., Gehne, M., Kolcynski, W.: Stochastic parameterization development in the NOAA/NCEP Global Forecast System. In: ECMWF, Reading, UK (2016)
13. LeMarshall, J., et al.: Improving global analysis and forecasting with AIRS. Bull. Amer. Met. Soc. **87**, 891–894 (2006)
14. Kiehl, J.T., Hack, J.J., Bonan, G.B., Boville, B.A., Williamson, D.L., Rasch, P.J.: The national center for atmospheric research community climate model CCM3. J. Climate **11**, 1131–1149 (1998)
15. Ji, M., Kumar, A., Leetma, A.: A multiseason climate forecast system at the national meteorological center. Bull. Amer. Meteor. Soc. **75**, 569–577 (1994)
16. Admassu, A., Teshome, A., Wondifraw, D., Scher, S., van der Burgt, F., de Wit, A.: Validation of the European Centre for Medium-range Weather Forecasting (ECMWF) short-range forecasts over Ethiopia. National Meteorological Agency of Ethiopia and CommonSense project (2017)
17. Rogers, D., Tsirkunov, V.V.: Weather and Climate Resilience: Effective Preparedness through National Meteorological and Hydrological Services. World Bank (2013)
18. Yoshida, H., Terai, T.: Modeling of weather data by time series analysis for air conditioning load calculations. ASHRAE Trans. **98**, 328–345 (1992)
19. Kawashima, M., Dorgan, C.E., Mitchell, J.W.: Hourly thermal load prediction for the next 24 hours by ARIMA, EWMA, LR, and an Artificial neural network. ASHRAE Trans. **101**, 186–200 (1995)

20. Bartok, J., Habala, O., Bednar, P., Gazak, M., Hluchy, L.: Data mining and integration for predicting significant meteorological phenomena. Procedia Computer Science **1**, 37–46 (2010)
21. Henze, G.P., Kalz, D.E., Felsmann, C., Knabe, G.: Impact of forecasting accuracy on predictive optimal control of active and passive building thermal storage inventory. HVAC & R Res. **10**, 153–178 (2004)
22. Chen, T.Y., Athienitis, A.K.: Ambient temperature and solar radiation prediction for predictive control of HVAC systems and a methodology for optimal building heating dynamic operation. ASHRAE Trans. **102**, 26–36 (1996)
23. Allen, G., LeMarchall, J.: An evaluation of neural networks and discriminant analysis methods for application in operational rain forecasting. Aust. Meteorol. Mag. **43**, 17–28 (1994)
24. McGullagh, J., Choi, B., Bluff, K.: Genetic evolution of a neural networks input vector for meteorological estimations. In: ICONIP 1997, New Zealand, pp. 1046–1049 (1997)
25. Solomatine, D., Dulal, K.N.: Model trees as an alternative to neural networks in rainfall-runoff modelling. Hydrol. Sci. J. **48**, 399–411 (2003)
26. Jareanpon, C., Pensuwon, W., Frank, R.J.: An adaptive RBF network optimised using a genetic algorithm applied to rainfall forecasting. In: International Symposium on Communications and Information Technologies 2004 (ISCK 2004), pp. 1005–1010 (2004)
27. Kemmoku, Y., Orita, S., Nakagawa, S., Sakakibara, T.: Daily insolation forecasting using a multi-stage neural network. Sol. Energy **66**, 193–199 (1999)
28. Pan, X., Wu, J.: Bayesian neural network ensemble model based on partial least squares regression and its application in rainfall forecasting. In: 2009 International Joint Conference on Computational Sciences and Optimization, Chine, pp. 49–52 (2009)

Choosing a Rational Design of the Engine Fastening Span

V. A. Tereshonkov[1], D. S. Shavelkin[1], and I. V. Pocebneva[2](✉)

[1] Moscow Aviation Institute (National Research University), 4, Volokolamskoe Shosse, Moscow 4125993, Russia
[2] Voronezh State Technical University, 14, Moskovsky Prospect, Voronezh 394000, Russia
ipocebneva@vgasu.vrn.ru

Abstract. The work is devoted to assessing the reliability of the attachment to the fuselage of aircraft engines, which directly determines the safety and reliability of the aircraft. To do this, the choice was made of the rational design of the front span. Comparison of design options and the selection of the most preferred of them were based on the chosen criterion. The choice of the variant of the scheme was carried out in order to minimize the weight of the optimized unit. The criterion function value was calculated using the End Elements Method (ICE) in the COSMOS program in SolidWorks (CAE). Tensions, loads and deformations were calculated. The results of calculations for the spangot were also verified using COSMOS. As a result, the divergence of the results of theoretical calculations and calculations COSMOS was no more than 1.2%.

Keywords: Attachment · Engine attachment reliability · Thin-walled construction stability · Spangot model

1 Introduction

Today, the typical version of the location of engines on an administrative class aircraft is their attachment in the tail part of the fuselage [1]. Naturally, this part of the fuselage carries an additional load and is a place that requires increased attention of the designer and designer at all stages of the creation of the new L.A. Consequently, the reliable mounting of engines directly affects the safety and reliability of the L.A. development [2]. This issue therefore requires careful further study and study.

The point loads experienced by the power set of the tail part of the fuselage determine the difficult conditions of its functioning.

It was in this regard that increased attention was paid to the design of the engine gondola attachment sites to the corresponding spans.

As a research work, the rational design of the front engine attachment was chosen. Comparison of design options and the selection of the most preferred of them were based on the chosen criterion. The choice of the variant of the scheme was carried out in order to minimize the weight of the optimized unit. The criterion function value was calculated

A. A. Radionov and V. R. Gasiyarov (Eds.): RusAutoCon 2021, LNEE 857, pp. 274–289, 2022.
https://doi.org/10.1007/978-3-030-94202-1_27

using the End Elements Method (ICE) in the COSMOS program in SolidWorks (CAE) [3, 4].

Of course, it should be noted that in real-world conditions the designer has to take into account a much larger number of requirements for the created design. In particular, we have to take into account the requirements for the technology of the assembly, the technology of the assembly, special requirements, for example, related to the layout of the cabin (in the case when the inner part of the spangot should be hollow). Taking into account all these requirements leads to the need to develop and account for an entire system of criteria (indicators) and limitations [5, 6].

The front span gout has been selected as it is more difficult to load and functions in more difficult conditions.

2 Basic Raw Data

Further analysis was carried out on the scheme of active loads based on the design of the engine pylon and its attachment to the spangots.

The main load that experiences the first spangot fixing the pylon is the weight, and the load from the thrust of the engine. The engine pylon is attached to the first spangot by the pylon suspension node lying in its plane, as well as a sloping beam that takes the thrust of the engine and passes it to the second spangoute [7].

This mount converts the thrust of the engines into a pair of forces that work to compress the front spangot. In this case, the weight of the engines is transferred to the front spangot in the form of a combination of the moment and cutting force. The load definition is shown on Fig. 1.

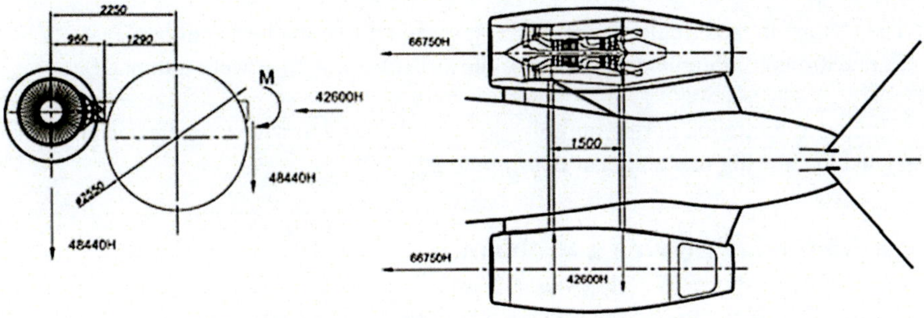

Fig. 1. Identify the loads acting on the researched spangot.

The shape of the engine pylons, their size and location during the optimization were not used. In addition, the shape of the spangot is limited: the local diameter of the fuselage, the constant value of the thickness of the spangot, the values of the characteristics of the material, chosen among the set in COSMOS.

An allluminal alloy of 7050-was chosen as the material of the spangot.
T7651 with the following physical characteristics:
$E = 7200$ daN/mm^2.
$G = 2700$ daN/mm^2.

3 Assumptions

In the calculations of power spangots usually used ready-made dependencies derived from the solution of static uncertainty of closed rings.

In the course of optimization, the following assumptions were made:

- It's a non-natural design, a very idealized version;
- The ratio of the transverse size of the beam to the curvature radius is small;
- Local loads on the spangout are balanced by tangent efforts in the shell;
- The rigidity of the shell on the bend compared to the rigidity of the spangot is small;
- Local or general loss of stability of the thin-walled structure is not taken into account;
- The design of the pylon suspension nodes is fixed and is not considered during optimization (the pylon attachment nodes are considered to be integrated into the span putt belt);
- The gyroscopic moment of the engine from the rotation of its moving parts in the force scheme was not taken into account;
- Engine vibrations were not taken into account;
- The temperature extension of the structure material was not taken into account;
- No effort and deformations are taken into account, which are transmitted to the spangout on other elements of the structure (panels of cladding, beams, reinforced stringers).
- The change in the circular shape of the spangout under load was minor;
- Calculations for fatigue destruction of the structure and wear were not made.

The main limitation during optimization was the need to withstand the design of the assigned loads without significant deformations.

4 A Method of Solving a Problem

To solve the problem, the design scheme was optimized by busting a number of options [8]. For each variant of the loading scheme, COSMOS programs were calculated with the help of COSMOS ICS. The results of calculations for the spangot were also verified using COSMOS [9, 10]. As a result, the divergence of the results of theoretical calculations and calculations COSMOS was no more than 1.2%.

Von Mises' voltage or equivalent voltage is a quantitative characteristic of the voltage component. COSMOS uses the von Mises formula to calculate the quantitative characteristics of voltages. Although the von Mises formula does not fully describe the voltages at the point, it provides information adequate enough to assess the safety of the project for a variety of plastic materials.

Unlike the voltage component, von Mises' voltages have no direction. They are fully determined by the amount of voltages. The van Mises formula is used as a criterion for assessing damage to plastic materials.

The voltage is calculated from six voltage components according to the following formula:

$$\sigma_{VON} = \left\{ \begin{array}{l} 0.5\left[(\sigma_X - \sigma_Y)^2 + (\sigma_X - \sigma_Z)^2 + (\sigma_Y - \sigma_Z)^2\right] + \\ +3\left(\tau_{XY}^2 + \tau_{XZ}^2 + \tau_{YZ}^2\right) \end{array} \right\}^{\frac{1}{2}} \quad (1)$$

The optimization was based on the specified geometric characteristics of the design.

Weight was chosen as the target function, which had to be minimized during the research. The parameters of the spangout section and the geometric characteristics of the shape of the spangout (e.g., the cut-out radius) were selected as variables. Details of the composition and rational values of the groups reviewed are presented in the next chapter. The limitation in optimization was the strength limit for the material you selected [11].

5 Steps to Solve a Problem

During the simulation, it was necessary to bring the model closer to the real conditions of operation of the design during the operation of the L.A. [5]. To do this, the Spangot was first designed in SolidWorks CAD. To reduce the calculation time and load on the computer, the symmetry function in COSMOS was used.

The model was configured only to calculate and optimize one of the two symmetrical halves of the unit design. Because COSMOS's built-in features were used, this simplification did not affect the accuracy and reliability of the calculation results [12, 13]. Then, as the point of fixing the structure plays a decisive role in the calculations and optimization, as well as for taking into account the real conditions of the structure, the model to the spangot was attached to the cladding, which was fixed half a meter from the spangot (which corresponds to the real distance between the span). The spangot was fixed in the symmetry sphere, provided that the symmetry function was used in COSMOS to simulate the functioning of an entire spangoute. Then, the loads according to the accepted power scheme were applied at the points lying in the plane of the spangot, at the points of the pylon suspension nodes [14–16].

An illustration of an example of a model pinning and the loads that are working on it is shown in Fig. 2.

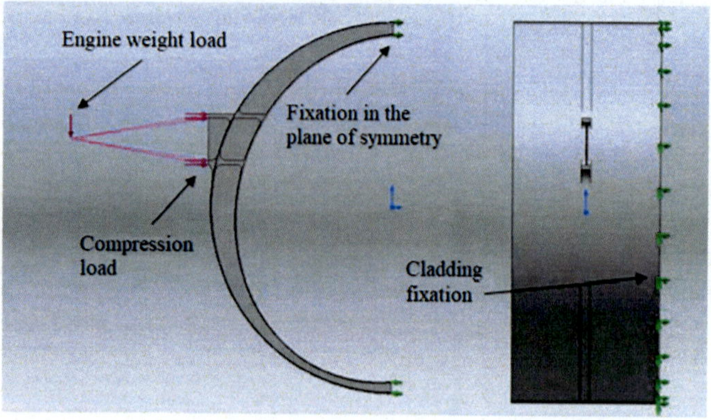

Fig. 2. How to attach and distribute loads in SOSMOS.

6 Results

We compared five different proposed versions of the spangout scheme by weight.

Possibility I
Round spangout with a round neckline, inside which is a beam (Fig. 3).

Variable options:

- The thickness of the spaniel belt
- The thickness of the spangoat wall
- The height of the spangot
- The thickness of the beam belt
- The thickness of the beam wall

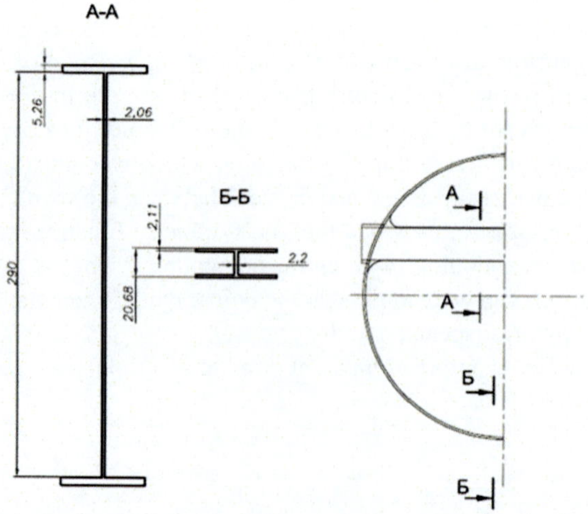

Fig. 3. Round spangout with a round neckline, inside which is a beam.

Possibility II
Round spangout, with a cut-out in the form of an oval double curvature inside (Fig. 4).

Variable options:

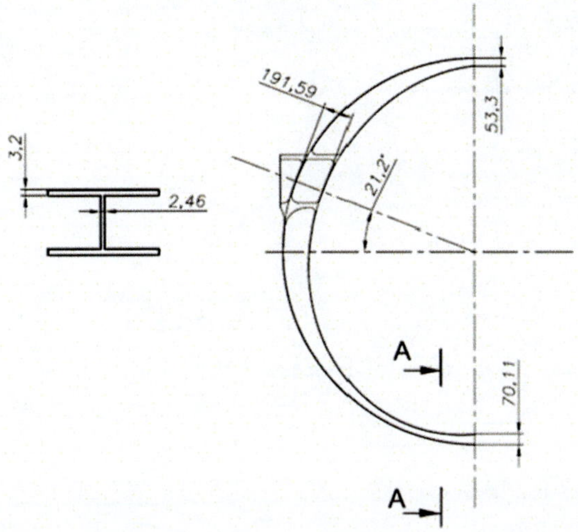

Fig. 4. Round spangout, with a cut-out in the form of an oval double curvature inside.

Possibility III
Round spangout with a solid thickened wall (a frame with a blank wall) on which the tides are located (Fig. 5) [17].

Variable options:

- The thickness of the spaniel belt
- The thickness of the spangoat wall
- Tide length

Possibility IV
Round span with a round neckline in the middle (Fig. 6) [18].

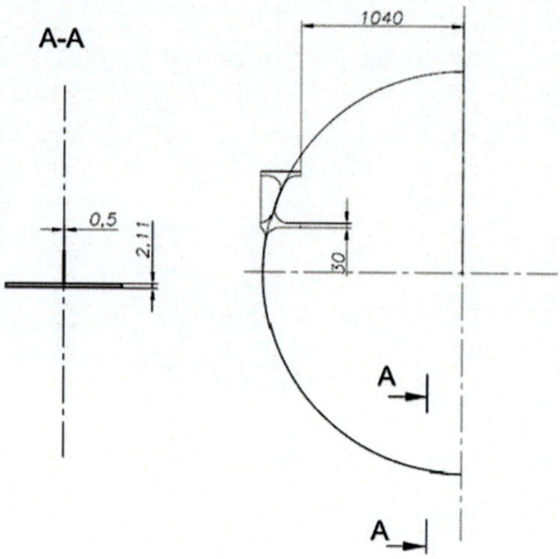

Fig. 5. Round spangout with a solid thickened wall on which the tides are located.

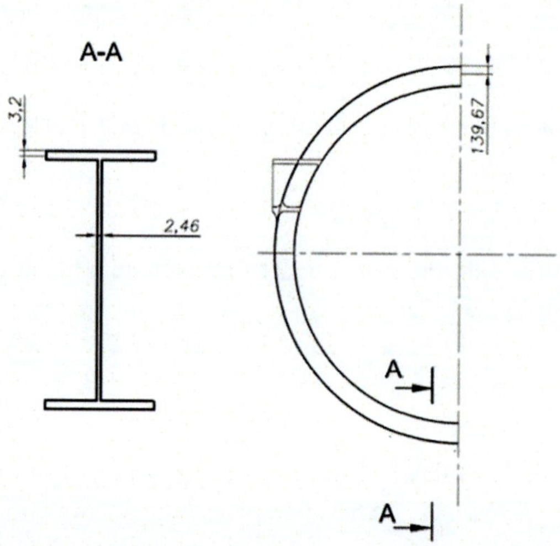

Fig. 6. Round span with a round neckline in the middle.

Variable options:

- The thickness of the spaniel belt
- The thickness of the spangoat wall
- The height of the spangoat wall

Possibility V

Round spangout with a thin wall with two oval cutouts (Fig. 7) [19].

Variable options:

- Width and height of clippings
- Tide length

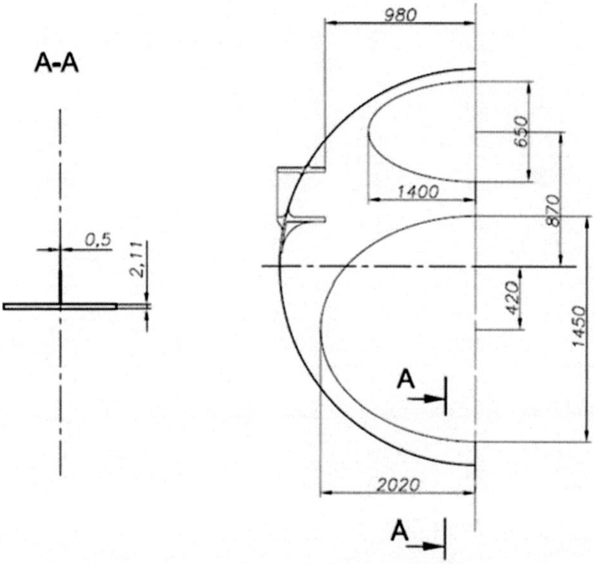

Fig. 7. Round spangout with a thin wall with two oval cutouts.

A comparison of the results is shown in the Table 1.

Table 1. Comparisons calculated variant.

Variant	I	II	III	IV	V
The number of administrative aircraft in which this type of spank is designed (the total number of planes - 25)	15	7	3	-	-

(*continued*)

Table 1. (*continued*)

Variant	I	II	III	IV	V
Massive parameters					
Total spangout weight after optimization (kg)	21,030	23,594	36,584	26,936	27,568
Weight of the spangout wall (kg)	0,976	6,064	17,184	10,904	6,920
Weight of the spank belt (kg)	5,226	7,574	5,484	7,942	5,472
Weight of the beam (kg)	10,442	-	7,874	-	9,136
Weight of pylon suspension nodes (kg)	4,386	9,956	6,042	8,090	6,040
Deformation parameters					
Maximum displacement (mm)	23,3	6,8	2,2	6,5	14,0
Displacement of the top node of the pylon suspension (mm)	6,3	1,9	1,3	2,1	1,5
Displacement of the lower suspension node pylon (mm)	8,0	5,5	1,8	5,7	4,6
Optimization options					
Number of variables	5	5	3	3	5
Number of iterations	23	23	15	15	23
Number of items	58942	55344	69422	55458	54807
The size of one item	22.6 mm × 1.1 mm	23.1 mm × 1.2 mm	24.2 mm × 1.2 mm	23.4 mm × 1.2 mm	24.1 mm × 1.2 mm
Number of nodes	118921	111289	138840	111289	110323

The distribution of loads and deformations for each design version is shown in the following figures (Figs. 8, 9, 10, 11, 12, 13, 14, 15, 16, and 17).

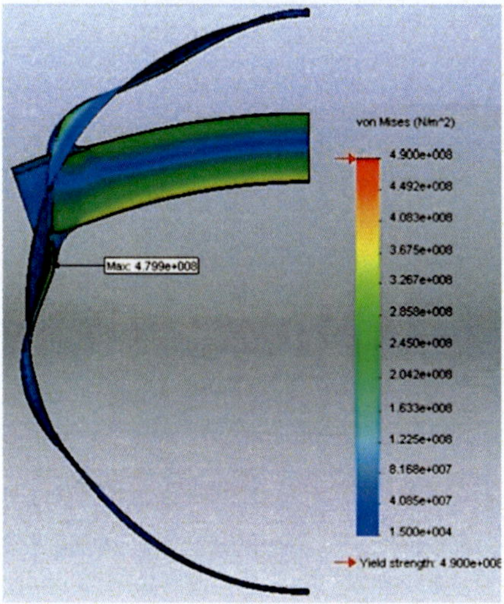

Fig. 8. Illustration of the distribution of loads (for the visuality of the deformation shown with a scale of 1:10) - Possibility I.

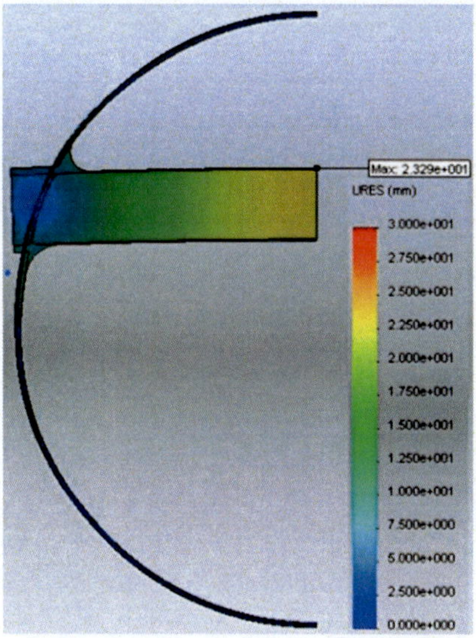

Fig. 9. The deformation illustration is Possibility I.

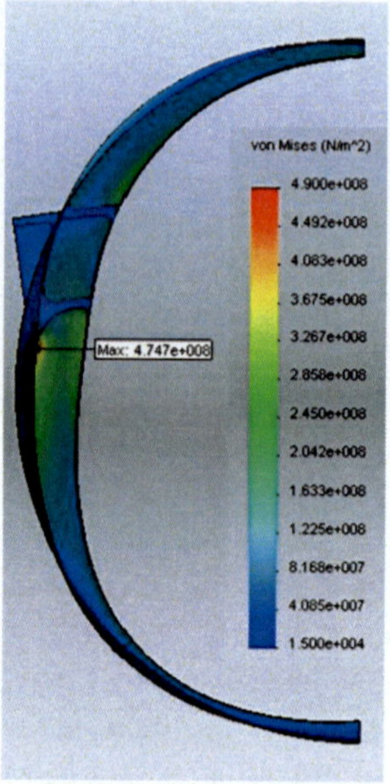

Fig. 10. Illustration of the distribution of loads (for the visuality of the deformation shown with a scale of 1:10) - Possibility II.

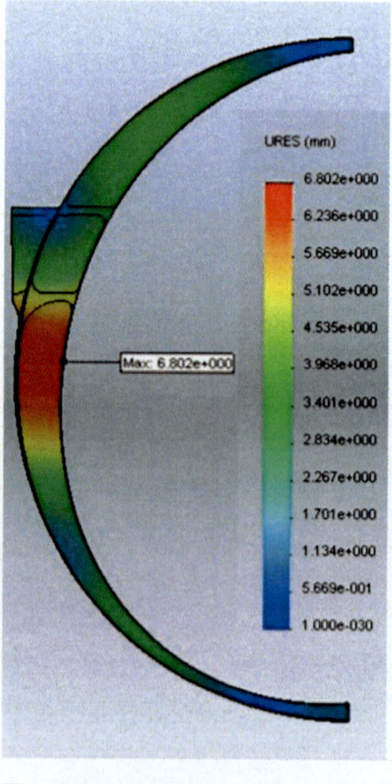

Fig. 11. The deformation illustration is Possibility II.

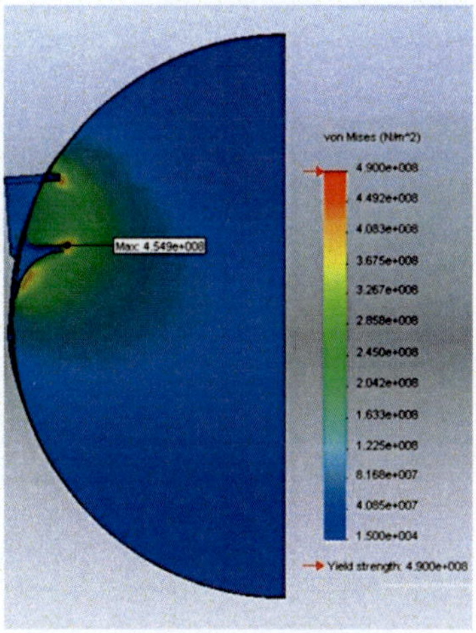

Fig. 12. Illustration of the distribution of loads (for the visuality of the deformation shown with a scale of 1:10) - Possibility III.

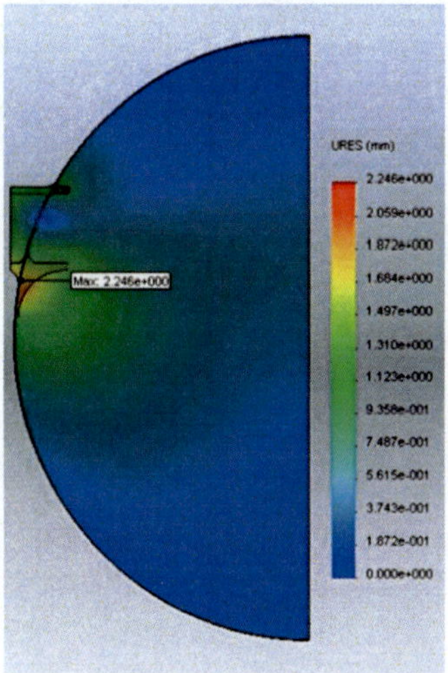

Fig. 13. The deformation illustration is Possibility III.

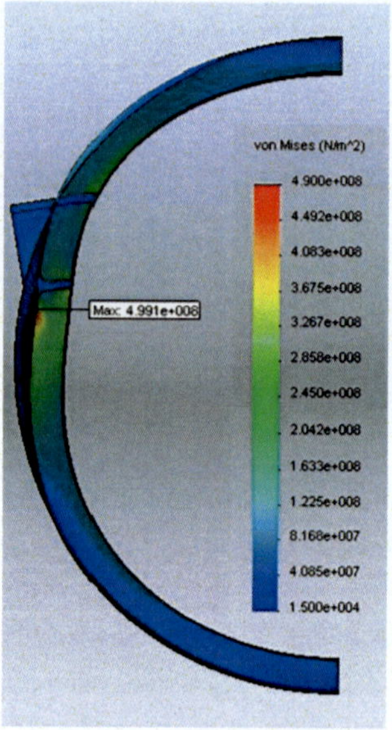

Fig. 14. The load distribution illustration (shown with a scale of 1:10) is Possibility IV for the perstence of the deformation.

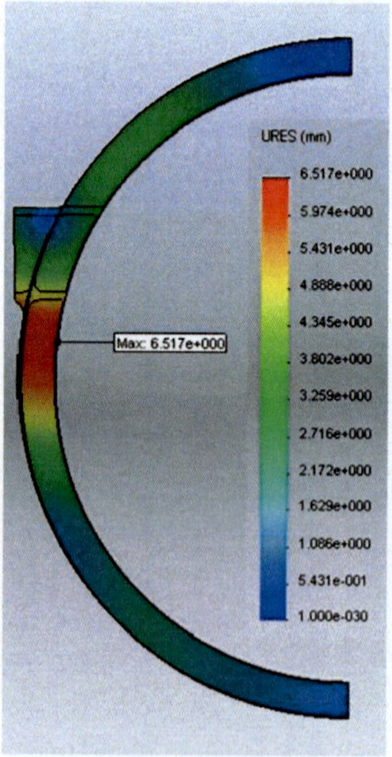

Fig. 15. The deformation illustration is Possibility IV.

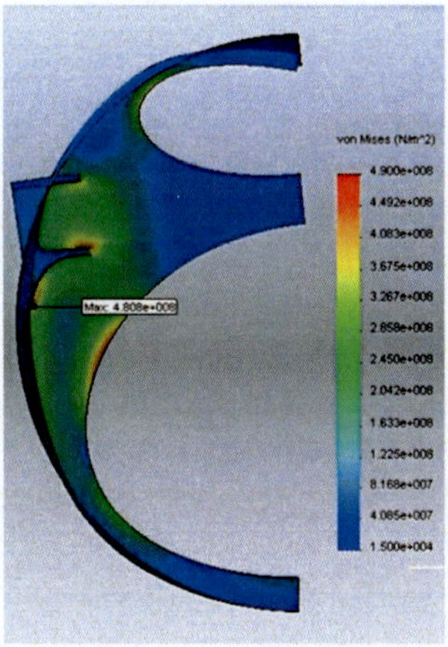

Fig. 16. An illustration of the distribution of loads (for the visibility of the deformation shown with a scale of 1:10) is Possibility V.

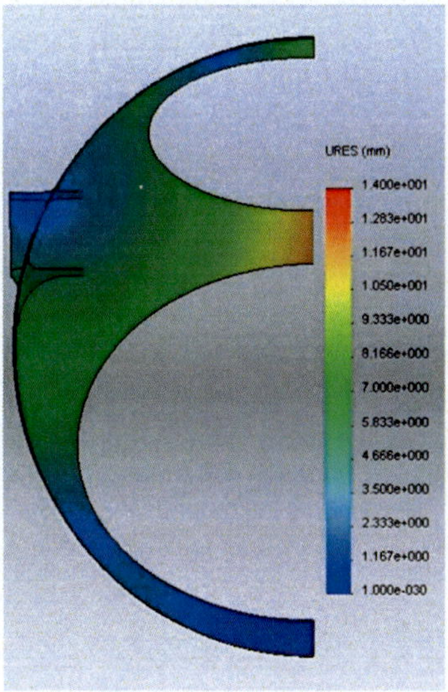

Fig. 17. The deformation illustration is Possibility V.

7 Conclusions

Based on the data in the table, it is clear that the flyweight is the spank as a result of optimization turned out option I (21.03 kg), in the cut of which passes the beam. It is 12% lighter than Option II and 73% lighter than Option III.

However, there is much more bias for option I than for everyone else. The maximum displacement occurs at the top of the central part of the beam (23.3 mm). The suspension node shift is also as high as the first option for Option I and is 8.0 mm. Option III, although heavier in weight, but very well keeps the shape under load. The maximum offset for this option is only 2.2 mm. Option II is a good choice for the designer, because its weight is only 2,564 kg heavier than for option I, but the bias in the suspension nodes is small (from 1.9 to 5.5 mm).

The difference in weight between a spangout with an oval-shaped neckline and a circle-shaped one is about 14%. At the same time, the biases for these two design variants are practically no different. It is obvious that in terms of ease of manufacture, the cut-out in the shape of a circle is much easier than the cut-out in the form of a double curvature oval. Thus, the transition to a simpler manufacturing process requires a weight gain of 3.3 kg.

Option V is actually a further development of Option III. In this way we get a reduction in weight to 32%. On the other hand, the thin wall begins to lose its stability and there are a number of large shifts with values up to 14 mm. Although the displacement in the anchorage of the engines is still not large (up to 4.6 mm) in the lower mount.

It should be noted that this study is only an indicative approximation in the preliminary design of the spangout. The purpose of the study is to provide a brief and clear picture of the weight and deformities for each option as a result of the award. As a result of real design, after taking into account the impact of production requirements and other important factors, the end results of weight and strength calculations may differ significantly from the results of the studies.

One possible way to deepen the research is to take into account the created model of fatigue, vibrations from engines (resonance) during optimization and taking into account the thermal effect of the engine on the structure. In-depth studies can also be carried out through the COSMO programme.

References

1. Smolyaninov, A.V., Pocebneva, I.V., Chernenkaya, L.V.: Mathematical model of asynchronous motor with frequency-cascade regulation. In: Paper Presented at the Proceedings of the 2019 International Russian Automation Conference, RusAutoCon 2019 (2019). https://doi.org/10.1109/RUSAUTOCON.2019.8867604
2. Dolgov, O., Bibikov, S., Pocebneva, I.: Elements of the synthesis method for the layout of a front-line aircraft. In: Paper presented at the E3S Web of Conferences, vol. 110 (2019). https://doi.org/10.1051/e3sconf/201911001068
3. Desyatirikova, E.N., Ivanov, S.A., Sergeeva, S.I., Zuev, S.A.: Creation of digital archives of design documents. In: Paper presented at the Proceedings of the 2020 IEEE International Conference "Quality Management, Transport and Information Security, Information Technologies", IT and QM and IS 2020, pp. 357–359. (2020). https://doi.org/10.1109/ITQMIS51053.2020.9322944
4. Schlichtmann, U., Das, S., Lin, I.C., Lin, M.P.H.: Overview of 2019 CAD contest at ICCAD. In: 2019 IEEE/ACM International Conference on Computer-Aided Design (ICCAD), pp. 1–2. (2019). https://doi.org/10.1109/ICCAD45719.2019.8942133
5. Pocebneva, I., Deniskin, Y., Yerokhin, A., Artiukh, V., Vershinin, V.: Simulation of an aerodynamic profile with sections of ad hoc concavity. In: Paper Presented at the E3S Web of Conferences, vol. 110 (2019). https://doi.org/10.1051/e3sconf/201911001074
6. Deniskin, Y., Miroshnichenko, P., Smolyaninov, A.: Geometric modeling of surfaces dependent cross sections in the tasks of spinning and laying. In: Paper Presented at the E3S Web of Conferences, vol. 110 (2019). https://doi.org/10.1051/e3sconf/201911001057

7. Deniskin, Y., Deniskina, A., Pocebneva, I., Revunova, S.: Application of complex information objects in industry management systems. In: Paper Presented at the E3S Web of Conferences, vol. 164 (2020). https://doi.org/10.1051/e3sconf/202016410042
8. Bityukov, Y., Deniskin, Y., Deniskina, G., Pocebneva, I.: Application of wavelets and conformal reflections to finding optimal scheme of fiber placement at 3d printing constructions from composition materials. In: Paper Presented at the E3S Web of Conferences, vol. 244 (2021). https://doi.org/10.1051/e3sconf/202124405004
9. Smolyaninov, A., Pocebneva, I., Fateeva, I., Singur, K.: Software implementation of a virtual laboratory bench for distance learning. In: Paper Presented at the E3S Web of Conferences, vol. 244 (2021). https://doi.org/10.1051/e3sconf/202124411009
10. Desyatirikova, E.N., Myshovskaya, L.P., Lutin, V.I., Mager, V.E., Khripunov, Y.V.: Algorithmization of the evaluation of decisions by the neumann-pearson criterion. In: Paper Presented at the Proceedings of 2020 23rd International Conference on Soft Computing and Measurements, SCM 2020, pp. 178–181 (2020). https://doi.org/10.1109/SCM50615.2020.9198783
11. Yurin, D., Deniskina, A., Boytsov, B., Karpovich, M.: Quality 4.0. time of revolutionary changes in the QMS. In: Paper Presented at the E3S Web of Conferences, vol. 244 (2021). https://doi.org/10.1051/e3sconf/202124411010
12. Aruvelli, S.V., Dolgov, O.S.: A method of increasing the electric aircraft flight range by reducing weight during flight. Russ. Aeronaut. **63**(3), 405–412 (2020). https://doi.org/10.3103/S1068799820030058
13. Desyatirikova, E.N., Efimova, O.E., Polukazakov, A.V., Akimov, V.I., Chernenkaya, L.V.: Methods of assessing the factors significance, influencing the engineering network pipelines technical condition. In: Paper Presented at the Proceedings of the 2021 IEEE Conference of Russian Young Researchers in Electrical and Electronic Engineering, ElConRus 2021, pp. 861–865 (2021). https://doi.org/10.1109/ElConRus51938.2021.9396455
14. Desyatirikova, E.N., Polukazakov, A.V., Akimov, V.I., Tzaregorodtceva, O.V., Khripunov, Y.V.: Software infrastructure management systems for smart territories. In: Paper Presented at the Proceedings of the 2020 IEEE International Conference "Quality Management, Transport and Information Security, Information Technologies, IT and QM and IS 2020, pp. 202–205, (2020). https://doi.org/10.1109/ITQMIS51053.2020.9322880
15. Desyatirikova, E.N., Chernenkaya, L.V., Mager, V.E.: Subsystem for on-line diagnostics of cutting process in flexible manufacturing. In: Paper Presented at the Proceedings of the 2019 International Russian Automation Conference, RusAutoCon 2019 (2019). https://doi.org/10.1109/RUSAUTOCON.2019.8867814
16. Deniskina, G.Y., Deniskin, Y.I., Bityukov, Y.I.: About some computational algorithms for locally approximation splines, based on the wavelet transformation and convolution. In: Radionov, A.A., Gasiyarov, V.R. (eds.) RusAutoCon 2020. LNEE, vol. 729, pp. 182–191. Springer, Cham (2021). https://doi.org/10.1007/978-3-030-71119-1_19
17. Desyatirikova, E.N., Akimov, V.I., Polukazakov, A.V., Polyakov, S.I., Mager, V.E.: Development, modeling and research of automation systems for "smart" heating of a residential building. In: Paper Presented at the Proceedings of the 2021 IEEE Conference of Russian Young Researchers in Electrical and Electronic Engineering, ElConRus 2021, pp. 849–854 (2021). https://doi.org/10.1109/ElConRus51938.2021.9396401
18. Dolgov, O., Safoklov, B., Sergeeva, S., Ivanova, A.: Application of automated systems for quality control of ground anti-icing treatment of aircraft. In: Paper Presented at the E3S Web of Conferences, vol. 244. (2021). https://doi.org/10.1051/e3sconf/202124408001
19. Soare, V., Varzaru, G., Zarnescu, A., Ungurelu, R., Ionescu, C.: Considerations Regarding Computer Aided Design of a Structure to be Manufactured in Occam Technology. In: 2018 IEEE 24th International Symposium for Design and Technology in Electronic Packaging (SIITME), pp. 124–127 (2018). https://doi.org/10.1109/SIITME.2018.8599201

VERTICAL CAD in the Design of Efficient Technologies for Making Aircraft Glider Parts

V. I. Bekhmetyev[1](✉), V. A. Tereshonkov[1], and V. Lepeshkin[2]

[1] Moscow Aviation Institute (National Research University), 4, Volokolamskoe Shosse, Moscow 125993, Russia

[2] Voronezh Institute of High Technologies, 73A, St. Lenin, Voronezh 394043, Russia

Abstract. In the modern aerospace industry, there is an intensive development of computer technology. This paper discusses the effectiveness of the automated design system of the Russian firm ASCON's automated design system for the technical design of aviation parts.

Keywords: CNC · Electronic model · Rig · Detail · Automation

1 Introduction

VERTICAL CAD is a system of automated process design that solves most of the problems of automation of manufacturing preparation processes (CCI), namely:

- to design the manufacturing and assembly processes of structures in several automated modes,
- calculate the material and labor costs of the production of products,
- calculate cutting, welding and other technological parameters,
- automatically form all necessary sets of technological documentation in accordance with the RUSSIAN GOST and the standards used in the enterprise (STP), for which there are necessary additional interface settings),
- to carry out parallel design of complex and end-to-end process by a group of technologists, and in real time,
- to carry out data verification in the process (on the relevance of reference data, as well as norm control),
- to form orders for the design of special technology equipment and the creation of management programs for equipment with numerical software control (CNC),
- to maintain the relevance of technological information through change management processes
- to support the process of building a single information space in the enterprise to manage the product's lifecycle from development to recycling [1].

VERTICAL CAD supports all business processes of electronic engineering paperwork, including the management of technological changes, and can also form blocks of

A. A. Radionov and V. R. Gasiyarov (Eds.): RusAutoCon 2021, LNEE 857, pp. 290–299, 2022.
https://doi.org/10.1007/978-3-030-94202-1_28

management programs in route technology. The system takes a qualitatively new approach to the organization of process data, based on the object model of presentation and processing of information.

Thanks to these capabilities and, as practice has shown, the system can be successfully applied to set and solve the problems of technological design of aviation equipment and its structural elements, but has its own features.

In the SYSTEM "VERTICAL" technologist can create technical procedures of three kinds:

- The process of making a part,
- The process of manufacturing an assembly unit,
- A typical (group) process [2].

2 Vertical SAPR Structure

Design and technology information are in the window of oneprogram.

At the heart of the system is a tree of constructive-technological elements of the part (CTE), or simply CE, as shown in the Fig. 1, and the tree of the process itself (the tree of TP or detailed-structural units - DSE) (Fig. 1). In the workspace there are also windows for the output of a 3D model and 2D drawing of the part, which gives a number of advantages for the technologist [3].

If you specify any technological transition in the process tree, the CTE tree will show the element to which it belongs, indicating all the parameters of the element, and the 3D models will be illuminated by the treated surfaces. Thus, on the screen we will see not only all the data about the item, but also its location on the details. And it is worth noting that the communication is two-way - if you choose the surface on the 3D model, the CTE will be displayed and the current technological transition in the TP tree, through which the processing of this surface is carried out [4].

Automated design of the process is carried out using a library of design and technological elements, made in the form of a universal technological directory UTS. THE UTS handbooks are an extensive database, but it does not always contain the required equipment, tool or rig [5].

It contains data on all materials used in the industry.

The system also provides for a mechanism of collective work on the design of the process. The technologist, working on the project, can send the task for design, for example, welding operation to a welding technologist, who in turn, after design, returns it to the process. All of this can be implemented through department heads who will appoint performers with simple deadlines and design priorities. This is recommended through the Lotzman Technologist app [6].

The relationship between the Vertical CAD and the main applications that form a single information space of the enterprise (EIP) is shown in the Fig. 2.

The formation of technological documentation in the system is implemented through the "Vertical Reports" application. This application can be used without the Vertical system and does not require a license to use it. You can save the set of formed documentation in various formats independent of CAD from SAPR (.pdf, xls, tif, zff) [7].

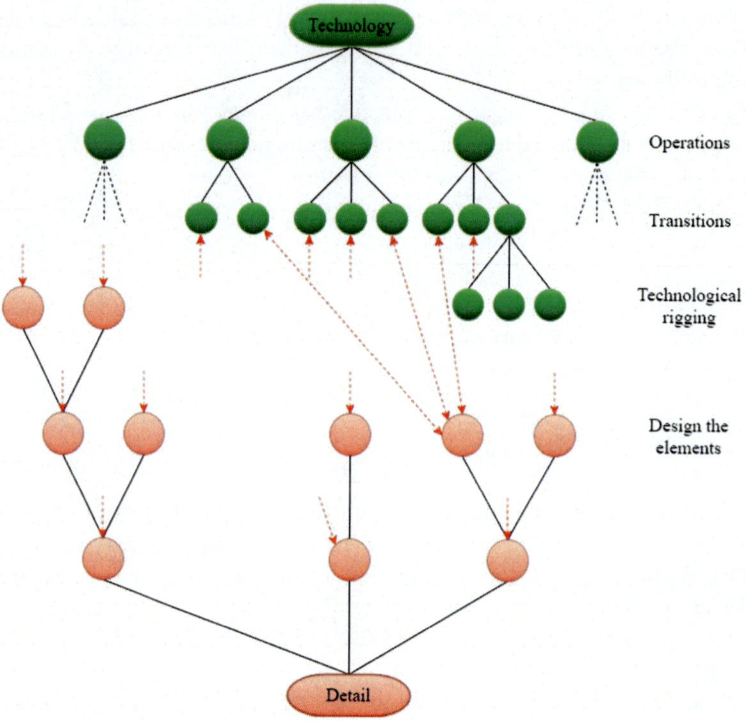

Fig. 1. The relationship between technological and design elements in the design process.

3 Process of Technological Design

Now we will carry out the development of the process of technological design of milled aviation part - bracket of the upper node of the steering wheel direction of the multipurpose maneuverable fighter (Fig. 3).

Before you start choosing a blank, you need to analyze the design feature: the material, configuration and dimensions of the part [8].

The bracket of the upper node of the steering wheel of the direction is a milled mechanical detail (Fig. 3).

Detail size: 98 × 1 × 58.

There are 2 10 o'1 in the details. It's a 3 mm and one non-squisms. 9 mm.

The roughness of the treated surfaces is Ra 6.3, except for the interior surfaces, where the roughness is from Ra 0.8 to Ra 1.6.

This part is made of the material D19chat GOST17232-79, the strength limit $\sigma_B \geq$ 40 kgs/mm^2, control group onOST1 00021-78 (Table 1).

The outline is milled according to the program on a milling machine with CNC under the management program, according to the recommendations.

Use Ahn coating. 8,000, EP-140 grey enamel, horizontal drying on OST 190055-85. After mechanical processing, the part undergoes galvanic treatment of the surface followed by control on the absence of cracks. The weight of the part is 0.420 kg.

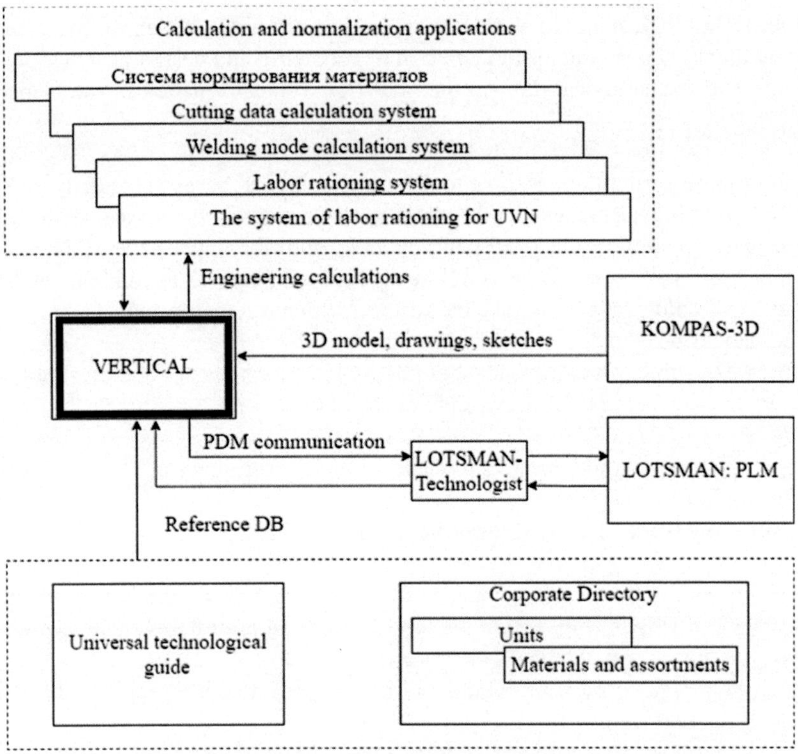

Fig. 2. Building ane-digital information space of the enterprise (EIP).

Fig. 3. The electronic model of the bracket of the upper node of the steering wheel direction of the maneuverable fighter.

Table 1. Physical and mechanical properties D19chat.

Temporary resistance to destruction	Yield point	Material density	NSW's hardness
$\sigma_B = 405$ MPa	$\sigma_{0,2} = 320$ MPa	$\rho = 7{,}20\ \tau/\text{cm}^3$	No more than 185

In the technological design of the bracket based on its CAD model, we will create an operating blank, which contains, in addition to geometric and design information, also geometric and technological data on the operational transformation of the blank into detail [9]:

- when installing parts in a machine tool for technological bases necessarily take real surfaces, directly in contact with the installation elements of the device. When selecting technological bases for processing the bracket of the upper node of the steering wheel direction, it is necessary to follow the following rules: - to combine the bases, i.e. as a technological base to take the surface, which is a measuring and design base at the same time;
- to observe the principle of permanence of bases. If the permanence of the technological base cannot be achieved, the new technological base must necessarily select the treated surfaces;
- technology bases should ensure sufficient stability and rigidity of the installation of the blank;
- to deprive the blank of all degrees of freedom [10, 11].

Thus, in relation to the process under study, we break down the base process into two installations (Fig. 4, Fig. 5). On the first set we have draft bases (unprocessed still surfaces), on the second, the final - clean.

Using Vertical's corporate machine tooling database, the most appropriate options for making the part were:

1. Horizontal milling machine 6P82G. It is used for various milling of steel, cast iron and other alloys.
2. Vertically - milling machine FP-17. On the machine, you can mill parts bounded by flat surfaces or shaped contours with a constant angle of tilt forming, such as beams, ribs, spars, brackets and others. Drilling, zenkering, pre-sharpening of holes. The machines can be treated with convex and concave surfaces of the double curvature with the help of shaped cutters with a ball end.
3. DMU Processing Center - 60. High-performance milling processing center DMU 60 with Heidenhein CNC system with 3D modeling function.
4. Vertical drilling machine 2N135. Designed for drilling, drilling, zing, zenkering, deployment.
5. Machine fixtures. They are the largest group and make up 70 to 80% of the total number of devices.

Aircraft factories of mass production are equipped mainly with universal metal cutting machines. Each machine is designed to perform a certain job with a given precision. For such machines use special devices that expand the technological capabilities of the equipment. With the help of such devices on the machine do work, for the implementation of which requires a machine of a completely different type.

To attach the blank of the study part to the machine with CNC is recommended to use the grip (Fig. 4b) and a special machine tool (Fig. 5).

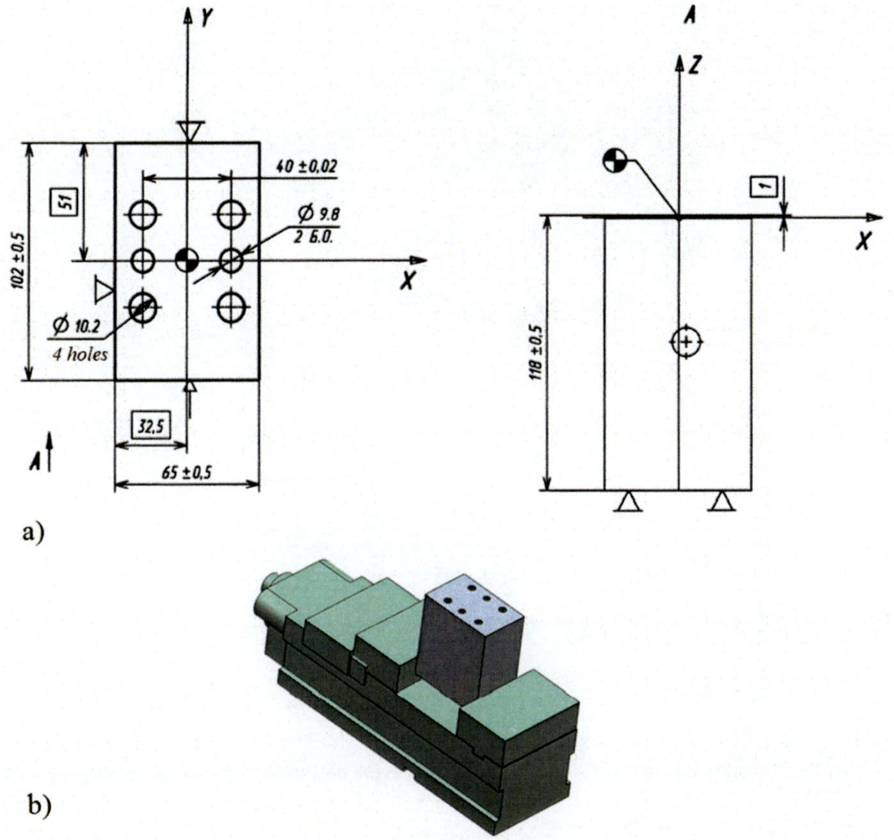

Fig. 4. The blank (a - the scheme of the base), b - fixing in the stagnant grip on the first installation.

The choice of cutting tool is carried out by the type of technological operation (sharpening, milling, deployment), the size of the treated surface, the properties of the material being processed, the accuracy of processing and the necessary roughness of the surface [12, 13].

Depending on the roughness parameter, a method of processing a given surface is chosen, which corresponds to its specific cutting tool. The choice of tool material is important for processing. For thin (finishing) methods of processing materials with high cutting speeds (over 500 m/min) the use of super-hard instrumental materials is recommended.

The swirls for the processing of materials such as D19chAT GOST17232-79 are made from the fast-cutting steel P6M5.

Cutting tool is recommended to choose with:

- The use of a normalized and standard tool
- Processing methods
- The sizes of the surfaces being processed,

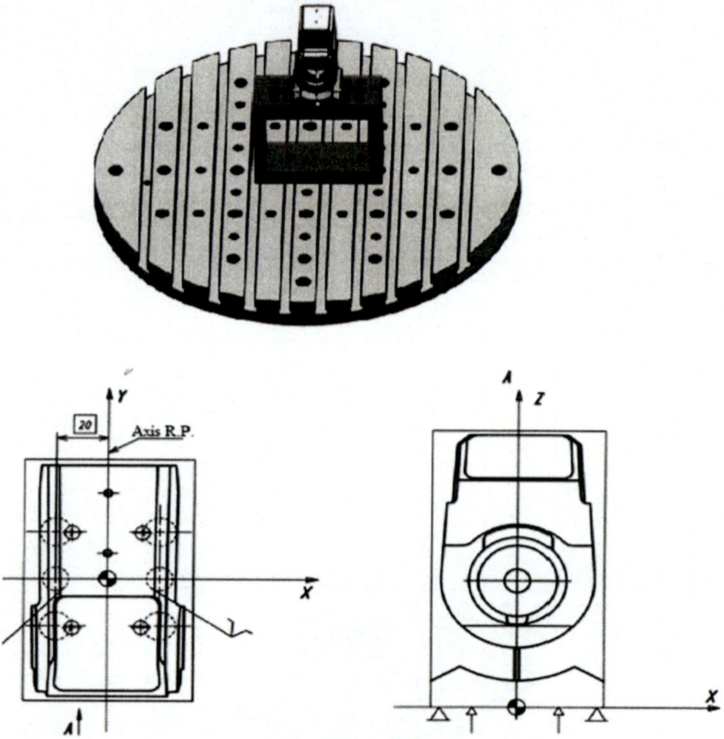

Fig. 5. The preparation and its base in a special device on the second (final) installation.

- Precision processing and surface quality,
- The type of material being processed,
- the durability of the instrument, its cutting properties and durability,
- Processing stages (black, clean, finishing),
- Type of production.

Based on the above, a cutting toolis chosen for the serial production of the aircraft part considered in thiswork.

During the operation of the vertical system, it was necessary to work with the functionality of other systems, applications and modules (LATMAN: PLM, Universal Technology Handbook, Corporate Directory "Materials and Sortaments" etc.).

To better visualize the details and subsequent data imports, a 3D model file process and an electronic detail drawing are connected to the process. The greatest effect with the use of high-level systems can be achieved using a complex system of automated design (design CAD-system) and a system of technological design (SAPR TP) [14, 15].

The design of the process was carried out by the method of forming a tree TP. The selection of equipment for TP operations, cutting tool, calculation of bracket surface processing modes, time standards for the manufacture of the part was made using the relevant vertical databases, UTS tabs and embedded algorithms [16].

The calculation of processing modes took into account the type and geometry of the processed structural element, the rigidity of the "Machine-adaptation-tool-detail" system, the physical and mechanical properties of the material and the state of the surface layer of the blank, the passport data of the machine and the parameters of the cutting tool. The System provides an opportunity to calculate cutting modes for cases of multi-instrumental mechanical processing, including in the context of automatic turning operations [17, 18].

After creating all operations in the Vertical system, a ready-made or created a new technological sketch can be connected to each TP operation to a more visual representation of the sizes performed on a particular operation, the quality of surfaces, spatial deviations, the base of the part, etc. [19]. To create working process documentation that is transferred directly to work sites, the Map Shaper feature is used on the Program toolbar. The master of the formation of technological documentation determines the desired kind of technological document, according to the settings appointed by the technologist [20]. The technologist is given the opportunity to choose the optimal combination of design modes.

4 Conclusions

Thus, the main possibilities of automated design of technological processes are considered, recommendations are given on the structure of technological design of special parts and processes in the SAPR TSRT, related to the design and technological features of the aviation industry products.

References

1. Deniskin, Y., Deniskina, A., Pocebneva, I., Revunova, S.: Application of complex information objects in industry management systems. Paper presented at the E3S web of conferences, p.164 (2020). https://doi.org/10.1051/e3sconf/202016410042
2. Bityukov, Y., Deniskin, Y., Deniskina, G., Pocebneva, I.: Application of wavelets and conformal reflections to finding optimal scheme of fiber placement at 3D printing constructions from composition materials. Paper presented at the E3S web of conferences, p. 244 (2021). https://doi.org/10.1051/e3sconf/202124405004
3. Smolyaninov, A., Pocebneva, I., Fateeva, I., Singur, K.: Software implementation of a virtual laboratory bench for distance learning. Paper presented at the E3S web of conferences, p. 244 (2021). https://doi.org/10.1051/e3sconf/202124411009
4. Desyatirikova, E.N., Myshovskaya, L.P., Lutin, V.I., Mager, V.E., Khripunov, Y.V.: Algorithmization of the evaluation of decisions by the Neumann-Pearson criterion. Paper presented at the proceedings of 2020 23rd international conference on soft computing and measurements, SCM 2020, pp. 178–181 (2020). https://doi.org/10.1109/SCM50615.2020.9198783
5. Dolgov, O., Bibikov, S., Pocebneva, I.: Elements of the synthesis method for the layout of a front-line aircraft. Paper presented at the E3S web of conferences, p. 110 (2019). https://doi.org/10.1051/e3sconf/201911001068
6. Desyatirikova, E.N., Ivanov, S.A., Sergeeva, S.I., Zuev, S.A.: Creation of digital archives of design documents. Paper presented at the proceedings of the 2020 IEEE international conference "quality management, transport and information security, information technologies", IT and QM and IS 2020, p 357–359 (2020). https://doi.org/10.1109/ITQMIS51053.2020.9322944

7. Schlichtmann, U., Das, S., Lin, I.-C., Lin, M.P.-H.: Overview of 2019 CAD contest at ICCAD. In: 2019 IEEE/ACM International Conference on Computer-Aided Design (ICCAD), pp. 1–2 (2019), https://doi.org/10.1109/ICCAD45719.2019.8942133
8. Pocebneva, I., Deniskin, Y., Yerokhin, A., Artiukh, V., Vershinin, V.: Simulation of an aerodynamic profile with sections of ad hoc concavity. Paper presented at the E3S web of conferences, p. 110 (2019). https://doi.org/10.1051/e3sconf/201911001074
9. Deniskin, Y., Miroshnichenko, P., Smolyaninov, A.: Geometric modeling of surfaces dependent cross sections in the tasks of spinning and laying. Paper presented at the E3S web of conferences, p. 110 (2019). https://doi.org/10.1051/e3sconf/201911001057
10. Desyatirikova, E.N., Akimov, V.I., Polukazakov, A.V., Polyakov, S.I., Mager, V.E.: Development, modeling and research of automation systems for “smart” heating of a residential building. Paper presented at the proceedings of the 2021 IEEE conference of Russian young researchers in electrical and electronic engineering, ElConRus 2021, pp. 849–854 (2021). https://doi.org/10.1109/ElConRus51938.2021.9396401
11. Yurin, D., Deniskina, A., Boytsov, B., Karpovich, M.: Quality 4.0. Time of revolutionary changes in the QMS. Paper presented at the E3S web of conferences, p. 244 (2021). https://doi.org/10.1051/e3sconf/202124411010
12. Desyatirikova, E.N., Efimova, O.E., Polukazakov, A.V., Akimov, V.I., Chernenkaya, L.V.: Methods of assessing the factors significance, influencing the engineering network pipelines technical condition. Paper presented at the proceedings of the 2021 IEEE conference of Russian young researchers in electrical and electronic engineering, ElConRus 2021, pp. 861–865 (2021). https://doi.org/10.1109/ElConRus51938.2021.9396455
13. Desvatirikova, E.N., Polukazakov, A.V., Akimov, V.I., Tzaregorodtceva, O.V., Khrinunov, Y.V.: Software infrastructure management systems for smart territories. Paper presented at the proceedings of the 2020 IEEE international conference “quality management, transport and information security, information technologies”, IT and QM and IS 2020, pp. 202–205 (2020). doi:https://doi.org/10.1109/ITQMIS51053.2020.9322880
14. Desyatirikova, E.N., Chernenkaya, L.V., Mager, V.E.: Subsystem for on-line diagnostics of cutting process in flexible manufacturing. Paper presented at the proceedings - 2019 international Russian automation conference, RusAutoCon 2019 (2019). https://doi.org/10.1109/RUSAUTOCON.2019.8867814
15. Deniskina, G.Y., Deniskin, Y.I., Bityukov, Y.I.: About some computational algorithms for locally approximation splines, based on the wavelet transformation and convolution. In: Radionov, A.A., Gasiyarov, V.R. (eds.) RusAutoCon 2020. LNEE, vol. 729, pp. 182–191. Springer, Cham (2021). https://doi.org/10.1007/978-3-030-71119-1_19
16. Dolgov, O., Safoklov, B., Sergeeva, S., Ivanova, A.: Application of automated systems for quality control of ground anti-icing treatment of aircraft. Paper presented at the E3S web of conferences, p. 244 (2021). https://doi.org/10.1051/e3sconf/202124408001
17. Soare, V., Varzaru, G., Zarnescu, A., Ungurelu, R., Ionescu, C.: Considerations regarding computer aided design of a structure to be manufactured in Occam technology. In: IEEE 24th International Symposium for Design and Technology in Electronic Packaging (SIITME), pp. 124–127 (2018). https://doi.org/10.1109/SIITME.2018.8599201
18. Kuliev, E., Kureichik, V., Kureichik, V.: Mechanisms of swarm intelligence and evolutionary adaptation for solving PCB design tasks. In: 2019 International Seminar on Electron Devices Design and Production (SED), pp. 1–5 (2019). https://doi.org/10.1109/SED.2019.8798449

19. Fedorov, E., Ferenetz, A.: Computer-Aided design of vehicle electrical harnesses. In: 2017 International Conference on Industrial Engineering, Applications and Manufacturing (ICIEAM), pp. 1–4 (2017). https://doi.org/10.1109/ICIEAM.2017.8076382
20. Marinova, G., Chikov, O., Rodič, B.: E-Content and tool selection in the cloud-based online-CADCOM platform for computer-aided design in communications. In: 15th International Conference on Telecommunications (ConTEL), pp. 1–5 (2019). https://doi.org/10.1109/ConTEL.2019.8848533

Research of the Emission of Electromagnetic Interference from a Secondary Power Supply

A. V. Kirsha(✉) and S. F. Chermoshentsev

Kazan National Research Technical University named after A. N. Tupolev – KAI, 10, K.Marx Street, Kazan 420111, Russia

Abstract. In this work, studies are carried out the emission of electromagnetic interference from the secondary power supply of the unmanned aerial vehicle power system. An approach to predicting electromagnetic emission from a secondary supply and its influence on the operation of onboard equipment is proposed. A practical example of predicting the emission of electromagnetic interference from a secondary power supply, which is located on board an unmanned aerial vehicle, is discussed.

Keywords: Onboard equipment · Electromagnetic compatibility · Forecasting · Electromagnetic interference · Electromagnetic environment · Aircraft

1 Introduction

The development of aircraft design is now becoming more and more complex. Polymer composite materials are widely used to reduce weight and dimensions and improve aerodynamic properties. But as an inevitable consequence, the use of composite materials significantly degrades the electromagnetic characteristics. The advantages of composite materials lead to their widespread use in the design of unmanned aerial vehicles (UAVs) [1–3].

Degradation of electromagnetic performance requires special measures to ensure electromagnetic compatibility (EMC). The reliability, functional safety and performance of UAVs are directly dependent on EMC. EMC of onboard equipment is its ability to function in a real electromagnetic environment in accordance with the established requirements. The electromagnetic environment is a combination of external electromagnetic influences and intersystem interactions with other onboard equipment [4, 5].

One of the most important problems of EMC is the prediction of the emission of electromagnetic interference from power lines of power supply. A great contribution to the solution of these issues was made by Russian scientists and specialists [6–8]. Other authors, that must be mentioned are next [9–11]. From the analysis of publications, it can be concluded that the problem of emission of electromagnetic interference radiation from secondary power sources (SPS) and circuits formed by direct and return conductors of the power supply system has been insufficiently studied.

A. A. Radionov and V. R. Gasiyarov (Eds.): RusAutoCon 2021, LNEE 857, pp. 300–309, 2022.
https://doi.org/10.1007/978-3-030-94202-1_29

The requirements for the quality of electricity onboard the aircraft are becoming more stringent due to the increase in the number of electricity consumers with various characteristics. Power quality requirements are established by the Russian standard GOST R 54073-2010 “Power supply systems for aircraft and helicopters. General requirements and standards of power quality” [12] and the American standard MIL-STD-704“ Aircraft electric power characteristics” [13]. Conducted interference of various origins makes a significant contribution to the quality of electricity in the UAV’s power supply system. Conducted interference can be induced by an electromagnetic field of artificial or natural origin, and can also be generated by secondary power converters and the supplied loads themselves [14]. The SPS is EMI generator in a wide frequency range, contributing to all types of interference presented in Fig. 1. The main generators of EMI are power elements that have parasitic parameters that have a significant impact on the interference range above 30 MHz.

A complex electromagnetic environment can occur during the operation of the aircraft power supply system. It has a consequence of a violation of the quality of the onboard equipment functioning [15]. The used method for predicting the electromagnetic environment is an integral part of the general methodology for designing a power supply system for UAVs, taking into account the EMC requirements. The electromagnetic environment is created by electromagnetic fields from the UAV’s power supply system.

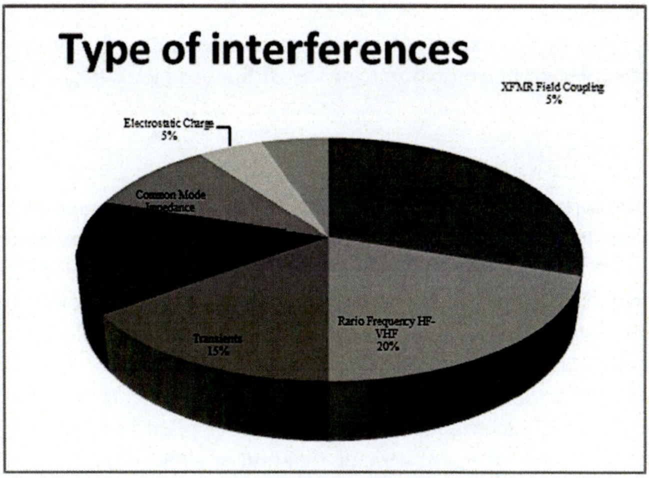

Fig. 1. Diagram of types of electromagnetic interference in a UAV.

The loops, which are formed by the conductors of the power supply system and the powered devices, emit a low-frequency magnetic field. This is due to the high currents flowing from the SPS. Therefore, to reduce the induction of low-frequency magnetic fields, it is necessary to reduce the area of the current circuits. For this, the grounding system is designed rationally and a reference potential equalization system is created [15, 16].

The aim of this work is to study the emission of electromagnetic interference when it is emitted from power lines of SPS.

2 Research Methods and Models

2.1 Research Methods

There are three levels of detail in the study and evaluation of the EMC of the UAV onboard equipment. This takes into account both the intrasystem electromagnetic interaction and external electromagnetic influences [17, 18]. This study examines the first level of detail. It is a study of the electromagnetic environment in the intra-fuselage space of an aircraft. In accordance with the immunity requirements of a particular type of equipment, a comparison is made between the maximum permissible value of the electromagnetic field strength and the value of the electromagnetic field strength obtained as a result of simulation.

To assess the level of emission of electromagnetic interference from SPS that are radiated by the field, the following three approaches are used: analytical calculations, experimental studies and simulation.

The analytical approach does not allow for accurate forecasting, because it is difficult to take into account the parameters of the aircraft material, equipment layout, etc. Experimental studies can be applied when a prototype or construction of a device and an aircraft already exist.

The work uses the program of electrodynamic modeling to study the emission of electromagnetic interference radiation from the SPS loops [19, 20].

2.2 Research Models

For research and forecasting, the influence of radiation from SPS, a simulation model of the SPS has been developed. To solve this problem, a mixed model was compiled, which combines the SPS circuit in the SPICE format (point 4 of block diagram on the Fig. 2) and the parameters of the printed conductors obtained as a result of the application of the method, presented in the form of a 3D model. Figure 2 shows a block diagram of the design of the physical topology of the SPS.

The proposed genetic algorithm for the placement of elements of a SPS, taking into account the criterion of electromagnetic radiation, is applied to a test example, a SPS circuit with a power of 0.5 kW, a schematic diagram of which contains 27 elements. At this stage, point 1 of block diagram on the Fig. 2 was completed. Figure 3 shows the results of the using this algorithm.

A heuristic routing algorithm was applied to the placement result, which minimizes the area of the current loops between the power elements of the power supply when developing the topology. At this stage, point 2 of block diagram on the Fig. 2 was completed. Figure 4 shows the result of the using this algorithm.

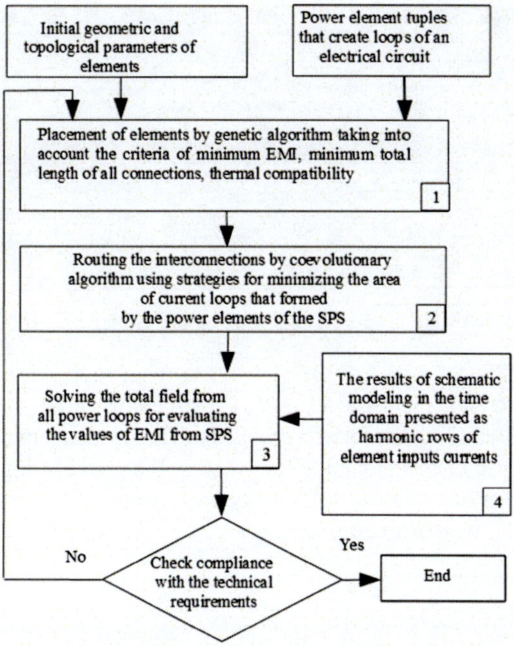

Fig. 2. Block diagram of SPS physical topology creation.

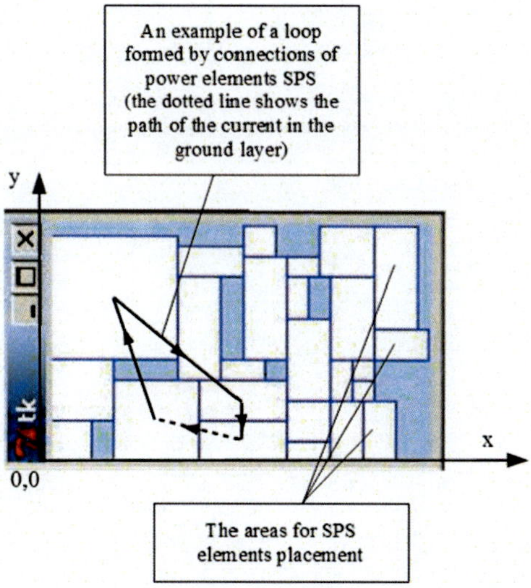

Fig. 3. Elements of SPS placed by the proposed algorithm.

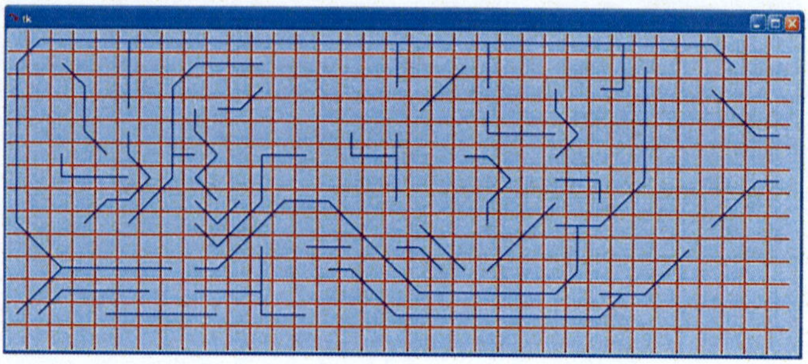

Fig. 4. Traces made by the proposed algorithm on the printed circuit board of the SPS.

The combined model is shown in Fig. 5. Co-simulation allows obtaining the exact value of the signal, including the interference signal, at the inputs of the printed conductors obtained in step 2 of the sequence for creating the physical topology of the SPS. Printed conductors serve as emitters of the electromagnetic field. Thus, point 3 of block diagram on the Fig. 2 is performed.

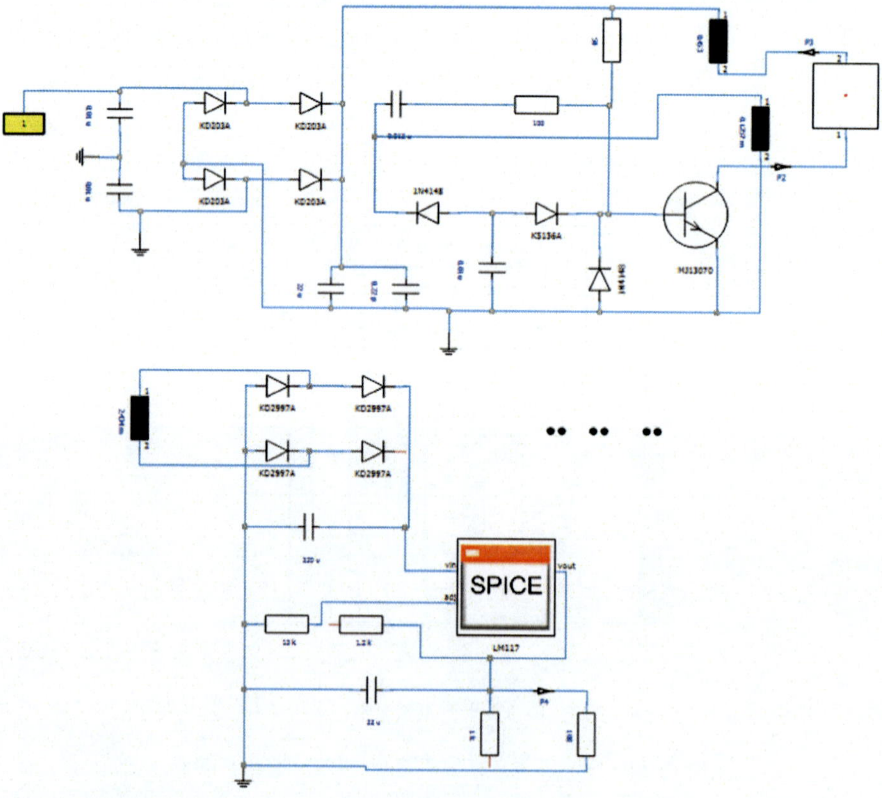

Fig. 5. Combined model of the investigated secondary power supply, consisting of electrodynamic and circuitry parts.

The distribution of the strength of the electric and magnetic fields is calculated for the analysis of the electromagnetic environment. The exact electrophysical, geometric and structural parameters of the research object are taken into account when modeling the emission of electromagnetic interference from the power lines of the power supply.

3 Research Results

The energy 0.5 kW (with a frequency of 400 Hz) supplies the SPS, when the electromagnetic field from this SPS are examined.

The observation points are located with a step of 100 mm up to a distance of 1 m, respectively, a total of 10 observation points exist.

The common view of the distribution of the electric and magnetic fields is shown in Fig. 6, Fig. 7 and Fig. 8.

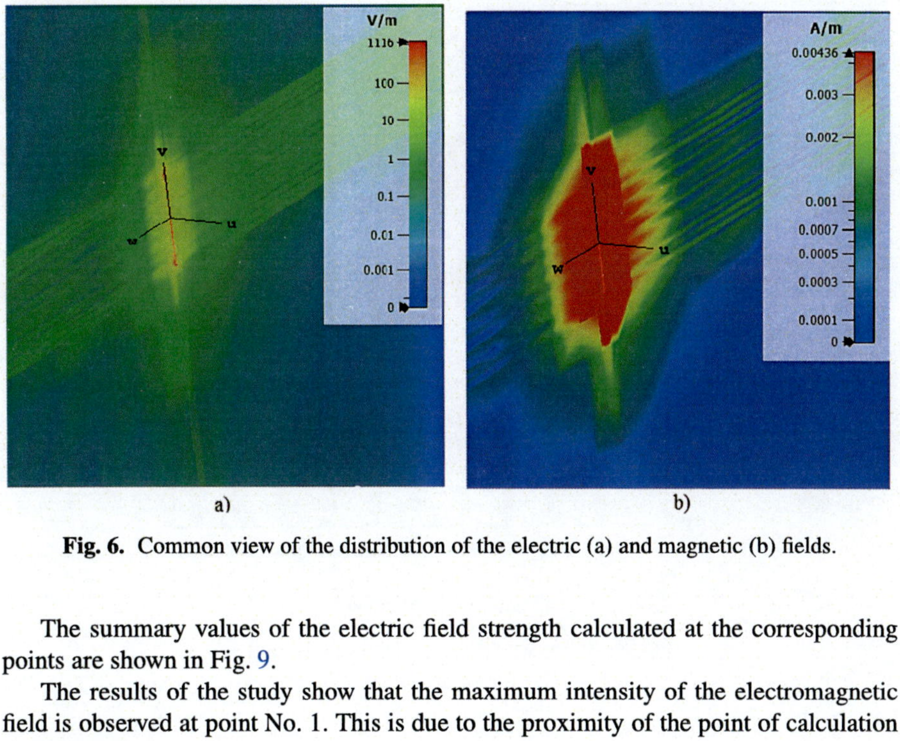

Fig. 6. Common view of the distribution of the electric (a) and magnetic (b) fields.

The summary values of the electric field strength calculated at the corresponding points are shown in Fig. 9.

The results of the study show that the maximum intensity of the electromagnetic field is observed at point No. 1. This is due to the proximity of the point of calculation to the SPS.

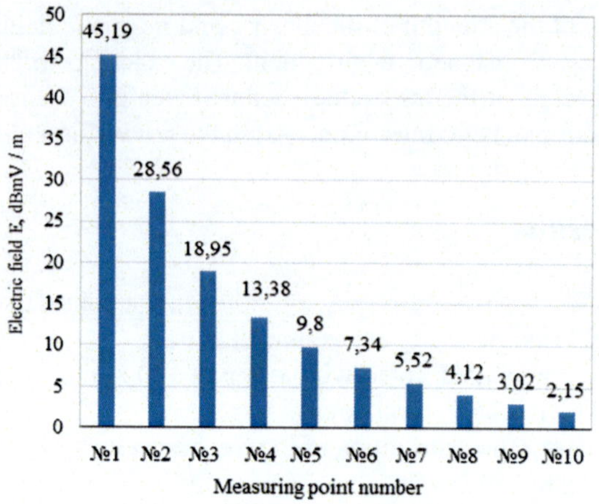

Fig. 7. Summary values of the electric field.

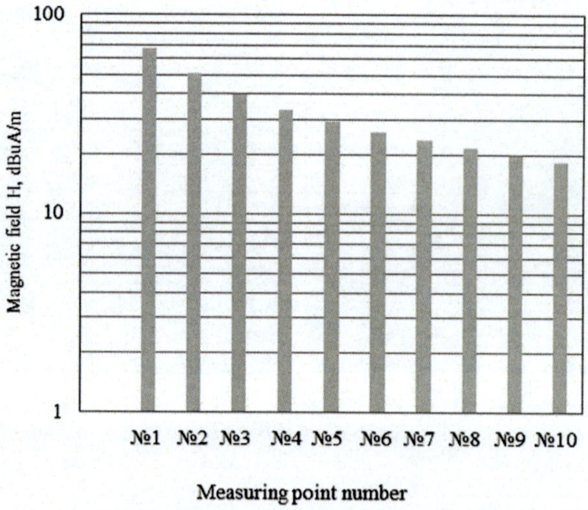

Fig. 8. Summary values of the magnetic field.

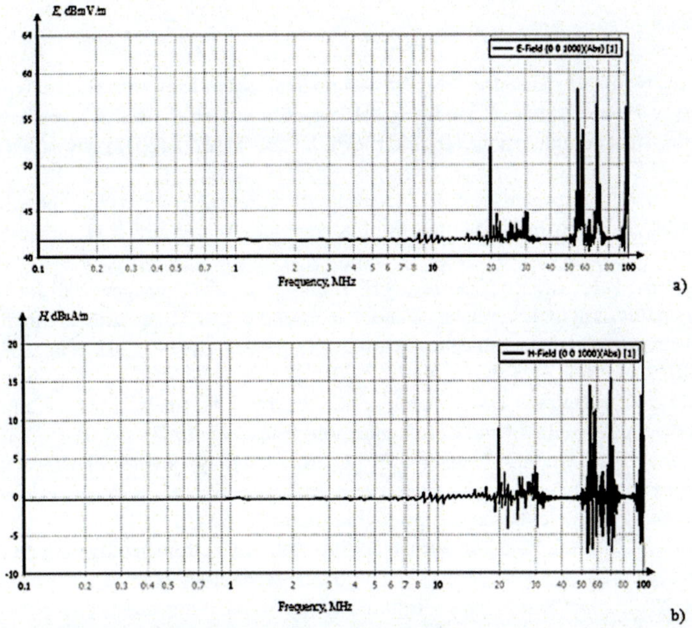

Fig. 9. The value of the electric (a) and magnetic (b) fields at a distance of 1 m from the SPS.

4 Discussion of Research Results

The following generalizations can be made from the simulation results: the distribution of the electric field is smoothness; the maximum electric field strength are observed near the SPS; the magnetic field strength at a distance of 1 m from the SPS reaches 18.25 dB mA/m; comparison of the requirements of the KT-160D standard (clause 19.3.3) to the maximum permissible value of the magnetic field strength of 120 A/m at a test frequency of 400 Hz, with the obtained result shows that the equipment of the class CC will meet the requirement [21–25].

5 Conclusions

1. An approach to the study of the emission of electromagnetic interference when it is emitted from the power lines of the SPS is proposed.
2. The approach is based on a combination of the schematic and electrodynamic model of the power supply, which allows obtaining the distribution of the electromagnetic field for a specific operating mode of the power supply and a specific topology of its power conductors.
3. Application of the approach makes it possible to predict emission levels from power lines of different types of SPS at the design stage.
4. A practical example of the application of the proposed approach for assessing emissions from SPS placed on UAV board is considered.

References

1. Eason, G., Noble, B., Sneddon, I.N.: On certain integrals of Lipschitz-Hankel type involving products of Bessel functions. Modeling the impact of powerful electromagnetic interference on the electrical complex of the aircraft, Electr. J. Proc. MAI **71** (2013). www.mai.ru/science/trudy/
2. Kirsha, A.V., Chermoshentsev, S.F.: Modeling of conducted noise in the onboard network of an aircraft in a composite-based body. Bull. Kazan State Tech. Univ. A.N Tupolev **3**, 183–189 (2014)
3. Gaynutdinov, R.R., Chermoshentsev, S.F.: Emission of electromagnetic disturbances from coupling paths of avionics unmanned aerial vehicles. In: 2017 International Siberian Conference on Control and Communications (SIBCON), Astana, pp. 1–5 (2017). https://doi.org/10.1109/SIBCON.2017.7998580
4. Kravchenko, V.I., Bolotov, E.A., Loginova, N.: Electronic Means and Strong Electromagnetic Interference, p. 256. Radio and Communications, Moscow (1987)
5. Gaynutdinov, R.R., Chermoshentsev, S.F.: Elektromagnitnaya sovmestimost 'perspektivnykh aviatsionnykh kompleksov (Electromagnetic compatibility of prospective aviation complexes) Tekhnologii elektromagnitnoy sovmestimosti **2,** 62–78 (2018)
6. Kechiev, L.N., Lemeshko, N.V.: Virtual certification of electronic equipment on the level of interference emission formulation of the problem. Technology EMC **2**, 3–15 (2010)
7. Kirillov, V.U., Marchenko, M.V., Tomilin, M.M.: EMC Onboard Cable Network Aerial Vehicles, p. 172. Publishing House of the Moscow Aviation Institute, Moscow (2014)
8. Gaynutdinov, R.R., Chermoshentsev, S.F.: Electromagnetic Interference emission from communication lines of onboard equipment of an unmanned aerial vehicle. J. Commun. Technol. Electron. **65**, 221–227 (2020). https://doi.org/10.1134/S1064226920020059
9. Paul, C.R.: Analysis of Multiconductor Transmission Lines, p. 623. John Wiley & Sons Inc., Hoboken (2007)
10. Sorensen, M., Hubing, T.H., Jensen, K.: Study of the Impact of board orientation on radiated emissions due to common-mode currents on attached cables. In: Proceedings of the 2016 IEEE International Symposium on Electromagnetic Compatibility (Ottava, July 25–29 2016), Ottava, pp. 36–39 (2016)
11. Baklezos, A.T.: Electromagnetic emission modeling in case of shielded cabling with respect to the ground dielectric properties. IEEE Trans. Electromagn. Compat. **58**(6), 1694–1700 (2016)
12. GOST R 54073–2010 Aircraft and Helicopter Power Supply Systems. General Requirements and Standards for the Quality of Electricity. Standartinform, Moscow, p. 33 (2011)
13. MIL-STD-461F Department of Defense Interface Standard Requirements for the Control of Electromagnetic Interference Characteristics of Subsystems and Equipment, December 10, p. 255 (2007)
14. Tooley, M., Wyatt, D.: Aircraft Electrical and Electronic Systems. Elsever., Amsterdam. p. 424 (2009)
15. Williams, T., Armstrong, K.: EMC for Systems and Installations, p. 508. Publishing House "Technologies", Moscow (2004)
16. Gaynutdinov, R.R., Chermoshentsev, S.F.: Stalemate. No. 157194RF, MPK H 01R 4/66, Device of alignment of basic potential; applicant and patent holder of KNITU-KAI. -2014147477/07; zayavl. 25.11.2014; opubl. 11/27/2015 Bulletin No. 33 (2015). (in Russian)
17. Gainutdinov, R.R., Chermoshentsev, S.F.: Methodology to ensure the intrasystem electromagnetic compatibility of UAV avionics. Russian Aeron. **59**(4), 613–618 (2016). https://doi.org/10.3103/S1068799816040279

18. Gaynutdinov, R., Chermoshentsev, S.: Study radiation from radio transmitters antennas influence on the UAV onboard equipment. In: 2019 International Conference on Electrotechnical Complexes and Systems (ICOECS), Ufa, Russia, pp. 1–4 (2019). https://doi.org/10.1109/ICOECS46375.2019.8949988
19. Chermoshencev, S.F., Gaynutdinov, R.R.: Modeling the external electromagnetic influences on the complex electronic equipment. In: 2015 XVIII International Conference on Soft Computing and Measurements (SCM), St. Petersburg, pp. 90–92 (2015). https://doi.org/10.1109/SCM.2015.7190420
20. Gaynutdinov, R.R., Chermoshentsev, S.F.: Immunity research of the electronic systems elements at the influence of intentional ultrashort electromagnetic pulses. In: 2016 17th International Conference of Young Specialists on Micro/Nanotechnologies and Electron Devices (EDM), Erlagol, pp. 214–218 (2016). https://doi.org/10.1109/EDM.2016.7538727
21. RTCA D0–160E: Environmental Conditions and Test Procedures for Airborne Equipment (2005)
22. Gaynutdinov, R., Chermoshencev, S.: Virtual testing of electronic systems susceptibility by electromagnetic compatibility requirements. In: 2018 International Russian Automation Conference (RusAutoCon), Sochi, pp. 1–5 (2018). https://doi.org/10.1109/RUSAUTOCON.2018.8501726
23. Gaynutdinov, R.R., Chermoshentsev, S.F.: Virtual testing of electronic systems by electromagnetic compatibility requirements. In: 2018 XIV International Scientific-Technical Conference on Actual Problems of Electronics Instrument Engineering (APEIE), Novosibirsk, pp. 320–323 (2018). https://doi.org/10.1109/APEIE.2018.8545765
24. Clarke, C.A., Larsen, W.E.: Aircraft electromagnetic compatibility (USA-CR-181051) Aircraft Electromagnetic Compatibility Final Report, September June, p. 146 (1985)
25. Chermoshencev, S., Kirsha, A.: Placement of elements and trace interconnection power supply taking into account the criterion of electromagnetic radiation. In: 2015 XVIII International Conference on Soft Computing and Measurements (SCM), pp. 144–146 (2015). https://doi.org/10.1109/SCM.2015.7190437

Control Theory

Building an Aggregate Rating of Popular SaaS Services Based on Organization of Customer Support Channels

S. V. Razumnikov(✉)

Yurga Institute of Technology, Tomsk Polytechnic University,
26, ul. Leningradskaya, Yurga 652055, Russia
demolove7@inbox.ru

Abstract. The article analyzes popular SaaS services based on the organization of customer support channels. The main "strengths" of the customer service for these SaaS projects are shown, and several positions are noted that should be improved. Also, based on the analysis, a rating of popular SAAS services was built based on the used customer support channels using the non-compensatory threshold aggregation method. The calculation was carried out in the developed software.

Keywords: SAAS services · Cloud technologies · Rating · Threshold aggregation · Communication support channels

1 Introduction

Software as a service (SaaS) is a relatively young trend even within the IT market. At the same time, the dynamics of the development of SaaS services in Russia, according to experts, exceeds the average rate of development of the information technology market.

This profitable business model allows both parties to optimize costs. Indeed, by accessing the software via the Internet, the customer saves on purchases of physical infrastructure, and also spends much less money on "renting" the software than on the classic licensing scheme. For a growing business, customer service is becoming one of the most important strategic issues. Dynamically evolving SaaS services, without a doubt, use the most advanced technologies to support their users. Let's analyze the quality of online customer support in this most promising part of the IT sector [1–4].

To do this, based on various ratings (CNews Analytics: the largest SaaS providers in Russia, Parallels: research of the cloud services market, ratings of startups, etc.) and key data on the number of registered users, revenue from services and project growth rates, we will single out 20 popular SaaS services in Russia.

These include: CallbackHunter, amoCRM, InSales, DaOffice, JivoSite, Jimdo, LiveTex, MANGO OFFICE, LPgenerator, SeoPult, UMI, Webinar.ru, YouScan, Bitrix24, Kontur-Extern, Megaplan, «My business», «My stock», «Telfin», «Elba». As part of this analysis, experiments were carried out on channels that are in the public

A. A. Radionov and V. R. Gasiyarov (Eds.): RusAutoCon 2021, LNEE 857, pp. 313–323, 2022.
https://doi.org/10.1007/978-3-030-94202-1_30

domain, mentioned on the websites of companies and are intended for both existing and potential customers. The most popular channels were thoroughly tested and experimented with: e-mail, online chat, FAQ and Knowledge Base, forum, social networks, customer community. It is these channels that will be analyzed as criteria [5–7].

2 Support Channels: Trends

One of the basic indicators of the quality of customer service can be considered the number of declared support channels. We checked which tools are mentioned on the companies' websites as channels for supporting users of SaaS services.

All presented projects use several different tools to communicate with their users. Along with the classic communication channels, such as telephone and e-mail, most of the services have their own knowledge base for independent search for information and groups in social networks.

On average, these resources provide 6–7 communication channels to their customers. The leaders in the use of communication tools are LPgenerator (9 points), Insales and Kontur-Extern (8 points of contact).

Cloud service providers try to educate, help, solve problems, listen, and try to help. The technical support division is completely based on this approach, hence the large number of contact channels with the "cloud" team of the provider. The customer service problem is evident in most businesses, especially in Russia. That is why this approach to working with people is seen as a strong competitive advantage [8–11].

The most popular channels of communication with users among the listed services are FAQ and Knowledge Base, as well as Facebook. However, this social network is currently used in most cases as an information source, and not as a channel for customer support.

When comparing the popularity of service channels among SaaS services and TOP-40 Internet providers, two new trends can be noted.

- Classic support channels are much less popular with SaaS services. The phone, which was a communication channel for absolutely all Internet operators, was only in 4th place for SaaS services. The traditional forum is not very popular either. Only 15% of companies provide such support, compared with 42.5% for telecom operators. Besides, ICQ and guestbook are not used at all. Some telecom providers still use these "out of fashion" tools, although their number tends to zero.
- Self-service channels are preferred. In the first place in popularity among SaaS providers were FAQ and Knowledge Base (95%). More companies use the customer support community (25% versus 5% for telecom operators). The popularity of social networks, where you can also see a ready-made answer or start a discussion with more experienced users, also reaches record highs (95% - Facebook, 85% - Vkontakte). While for telecom providers this figure was 60%.

Thus, the audience of SaaS services, being more advanced in Internet technologies, clearly prefers self-service and efficiency in customer service. Software-as-a-service providers try to meet the requirements of their users in everything, thus shifting the focus of customer service to more advanced channels.

The most popular online communication channels of SaaS services were considered and the quality of user service in them was analyzed. Integration with a broader corporate strategy in the field of IT and business.

2.1 E-mail

As part of testing this user service channel, requests were written to the e-mail indicated on the website to 18 services. Depending on the specifics of the product provided, questions related to the functionality of the service, integration with a CRM system, rules for accessing an account, etc.

The speed of response to such a request is an important indicator of the quality of customer service. According to Lightspeed Research, a user is willing to wait no more than 24 h for an email response. The speed of reaction to treatment was divided into several time intervals. Table 1 shows the distribution of SAAS services according to the speed of response to a request.

An additional advantage of 6 of the presented services (amoCRM, InSales, JivoSite, LPgenerator, MANGO OFFICE, "MoyKklad") is the presence of an auto-response, which helps the user to understand that his request has been accepted and processed. A good example of an auto-reply came from LPgenerator. It contained information about the processing time of the request, about other ways of contacting the company and links to the knowledge base, where the user can independently view the answer.

As for the answers to the questions asked, three companies (MANGO OFFICE, Kontur-Extern and MoySklad) received an incomplete answer, which forced them to write additional questions, go to the site to independently search for the necessary information, or contact another division of the company. For example, Kontur-Extern was contacted with the following questions: "How many users can have access to one account? Is it possible to restrict an employee's access to certain sections and documents?" In response, we received a hyperlink that redirected to another service channel.

The other services provided the correct answer to the question. From two of them, LPgenerator and Livetex, the answers were the most complete and understandable.

Note that only two companies (amoCRM and InSales) requested feedback on the support service.

It should be noted that the effectiveness of this support channel for SaaS resources is higher than that of Internet providers in Moscow and the region. According to the analysis carried out among Internet service providers [12], one third of operators did not respond to a question asked by e-mail within 24 h. The simple and convenient auto-reply function is used by only 3 Internet providers who support clients by e-mail. While in 95% of the studied SaaS services, they respond to users within an acceptable time frame, and 6 out of 18 have connected an auto-response.

2.2 Online Chat

Only 8 out of 20 SaaS resources use an online consultant as a support channel. Two of them - Jivosite and Livetex - are services in which online chat is a product of the company.

To analyze the quality of this support channel, we contacted online consultants and asked non-trivial questions about their product. The response time to contact in the online chat of all the studied resources was high - for all of them less than 10 min, and on average it takes 1–2 min. Table 1 shows the response time to a call.

In two services, Kontur-Extern and Elba, support consultants turned out to be less competent, both in matters of the product itself and in the level of communication and support. In the Kontur-Extern service, the operator did not answer the question raised and recommended contacting the service center. When communicating with the Elba online consultant, I had to literally "pull" the necessary answers from the specialist.

2.3 FAQ and Knowledge Base

According to studies of Western vendors of customer self-service systems, the effectiveness of the Knowledge Base is determined by 5 criteria [13]:

- accessibility from any page of the site;
- Convenience of structure and information retrieval;
- content and clarity of articles;
- usefulness and its assessment by clients;
- relevance of articles.

This support channel has been tested for SaaS services. The 19 presented solutions have a section that helps the user to independently find the necessary information. A number of companies have a Knowledge Base presented with a large number of answers to questions, almost all of them have a search bar that makes it much easier to find the information you need.

For 18 services, the Knowledge Base is accessible from any page of the site, which increases the speed of navigation and the convenience of finding a question of interest for the user. Table 1 shows the number of questions and answers in the FAQ for SAAS services.

By the convenience of the structure of information placement, the content and clarity of the materials provided, the following SaaS services can be distinguished: amoCRM, CallbackHunter, InSales, JivoSite, LPgenerator, UMI, Webinar.ru, Kontur-Extern, Megaplan, MoyCklad.

The knowledge bases of these companies are distinguished by a clear categorization of sections, a convenient hierarchical structure, the ability to go to other questions and sections from the last page, a visual presentation of information due to different types of content: text, images, video. Other services in the Knowledge Base have the following shortcomings that complicate the search and perception of the necessary information: incomplete relevance of articles to sections; lack of a search string; confusing menu structure; duplication of content; the presence of sections without information; long references with monotonous content (text only); no FAQ, only the Knowledge Base.

10 companies out of 20 have an assessment of the usefulness of the article by users. This tool allows you to receive feedback on the importance and appropriateness of placing this material in the Knowledge Base. For two of these 10 companies, the utility

can be assessed only through "reposts" to social networks, which significantly reduces the likelihood of getting a response from the user.

2.4 Forum

This tool is the least popular among SaaS services. Of the 20 companies, only three companies have an open forum on their websites: SeoPult, Kontur-Extern and Megaplan. The response rate of support specialists can vary greatly.

In terms of the quality of the channel support, the work of Kontur-Extern and SeoPult specialists can be noted: high involvement of administrators (all user requests receive answers) and the relevance of topics to forum sections. Less stability is observed on the Megaplana forum: quick answers from administrators can give way to long expectations of users, some questions are left unattended, there are examples of irrelevance of the request and the topic of the section, very often support specialists redirect the user to another channel.

2.5 Social Networks

All SaaS services have pages in at least one of the popular social networks (Facebook, Vkontakte, and Twitter). We do not include Telfin in this analysis, as there is no integration of social networks with the site. Thus, 95% of services have representation on Facebook, Vkontakte - 85%, Twitter - 75%. However, not everyone consults users on social networks.

On Facebook, only two services provide support to users: MANGO OFFICE and YouScan. On Vkontakte, the number of projects with support increases to 9. In other cases, social networks are used as an information channel. However, even if the company does not have a special support section on the social network page, users always have the opportunity to ask a question and get advice from the company in the general account feed.

An indicator of high-quality customer service in social media is a quick response to a request. According to Edison Research, 42% of consumers who ask a company a question on social media expect a response within an hour.

The response time in the dedicated support channel and the responses of company representatives to requests in the general feed were separately checked.

The analysis of the timing showed a strong fragmentation in time intervals. Even within the framework of one account, they could range from several minutes to 3 days. Only two services, Kontur-Extern and YouScan, usually respond within an hour. Most SaaS resources do not seek to quickly close questions. This could indicate a lack of attention from companies to customer support in this channel, or a lack of a dedicated specialist for this channel.

It is believed that it is necessary to respond to user inquiries on social media as soon as possible. The faster the answer, the higher the client rates the company's business reputation. But often it is not the speed of the response that is important, but its quality.

In addition, common mistakes were noted in the accounts of SaaS services on Facebook and Vkontakte, affecting the quality of customer support in social media. There are examples when user requests and negative reviews remained unanswered or a comment appeared after a long time. There are a lot of examples when company representatives redirect a user to contact another department, forcing him to perform unnecessary manipulations [14].

2.6 Client Communities

Client communities are represented by five services: InSales, Jimdo, JivoSite, LPgenerator, Bitrix24. However, this indicator for Saas services is, in percentage terms, higher than that of the TOP-40 telecom providers (25% and 5%, respectively). The online customer support community is a specialized tool and has a number of advantages over forums and other channels:

- contains a structured "Knowledge Base" with an assessment of the benefits of articles;
- provides "smart search" when creating a new case. Right during typing, the system searches for similar calls in previously created topics and the knowledge base, offering solutions to the user automatically;
- contains an analytics module that allows you to evaluate the work of the support service for all important KPIs.

The Insales community is distinguished by a structured posting of information, but it has very low user activity and the community itself is difficult to find on the site.

The LPgenerator and Bitrix24 communities turned out to be the least successful in implementing and providing support. They noted: the difficulty of finding the community itself; lack of a search string (Bitrix24); irrelevance of discussion to topics and sections of the community; redirection to another division of the company; in many cases, a long response time (it can take up to a month) and a lack of responses to user requests [15–17].

3 Building an Aggregate Rating

Based on the analysis, data were collected (Table 1) and a four-grade assessment scale was formed (Table 2). Using this scale and the collected data, gradation scores were distributed across 6 criteria for 20 assessed SaaS services (Table 3).

The rule of threshold aggregation is that first the numbers of units are compared, that is, the number of estimates "unsatisfactory", ("bad") in vectors x and y. If they are not equal, then the option (vector) that will have fewer units (worst estimates) is preferable. If there are equal numbers of ones in x and y, then the number of average grades is compared (the number of twos). The option that has fewer of them will be considered more preferable, etc. If the number of all estimates is equal (and ones, and twos, and triples, etc.), then such vectors are considered equal and incomparable. That is, if there are equal numbers of bad assessments, then the best option will be the one with less than the average [18–20].

Using the threshold aggregation model (1), preference indices were calculated and a rating of SaaS services was formed based on the used customer support channels (Fig. 1).

Table 1. SaaS customer support channel data.

No	SAAS services	Criteria (customer support channels)					
		Email	Online chat	FAQ and knowledge base	Forum	Social networks	Client communities
1	amoCRM	30 min to 3 h	–	69		+	
2	CallbackHunter	<30 min	–	33		+	
3	DaOffice	<30 min	–	0		+	
4	InSales	From 3 h to a day	1 min	277		+	+
5	Jimdo	From 3 h to a day	–	157		+	+
6	JivoSite	30 min to 3 h	2 min	56		+	+
7	LiveTex	30 min to 3 h	2 min	37		+	
8	LPgenerator	<30 min	6 min	165		+	+
9	MANGO OFFICE	<30 min	–	290		++	
10	SeoPult	From 3 h to a day	–	269	+	+	
11	UMI	>days	–	82		+	
12	Webinar.ru	–	–	183		+	
13	YouScan	30 min to 3 h	–	7		+ +	
14	«Bitriks24»	–	–	113		+	+
15	«Contour-Extern»	–	2 min	269	+	+ +	
16	«Megaplan»	30 min to 3 h	–	214	+	+	
17	"My business"	30 min to 3 h	2 min	574		+	
18	"MySklad"	<30 min	–	137		+	
19	Telfin	<30 min	3 min	70		–	
20	"Elbe"	–	1 min	106		+	

Table 2. Four-grade scale for evaluating SaaS services on six indicators

Data ranges for five metrics				Score (gradation)	Description
Email (Time intervals of the speed of reaction to requests)	Online chat (response time to requests)	FAQ and knowledge base (number of "questions and answers")	Forum, social networks, Client communities		
No answer for more than a day	No	0–75	+ available, but not actively used; ++ available and actively used; – is absent	1	The minimum indicator/availability of this channel for the service
From 3 h to a day	6 min	76–150		2	Indicator above minimum values/available and used
30 min to 3 h	3 min	151–250		3	The value of the indicator is at a good level/well served
<30 min	1–2 min	251–300 and more		4	Excellent level of use

The calculation was carried out in the developed C# program "Formation of Aggregate Ratings".

$$F(x) = \sum_{j=1}^{m} C_{n-V_j(x)+m-j-1}^{m-j} = \sum_{j=1}^{m} C_{n-(\eta(1)+\eta(2)+\ldots+\eta(j))+m-j-1}^{m-j} \quad (1)$$

Within the framework of our cloud services rating model, here n is the number of criteria for assessment, m = 4 is the number of gradations (ratings). In our case, 1, 2, 3, 4. $V_j(x)$– is the number of ratings from the provider x. F - preference index.

Based on the results of calculations, the best SaaS services for organizing customer support channels turned out to be: LPgenerator, InSales, Kontur-Extern.

Table 3. Gradation diversity.

No	SAAS services	Criteria (customer support channels)					
		Email	*Online chat*	*FAQ and knowledge base*	*Forum*	*Social networks*	*Client communities*
1	amoCRM	3	1	1	1	2	1
2	CallbackHunter	4	1	1	1	2	1
3	DaOffice	4	1	1	1	2	1
4	InSales	2	4	4	1	2	4
5	Jimdo	2	1	3	1	2	4
6	JivoSite	3	4	1	1	2	4
7	LiveTex	3	4	1	1	2	1
8	LPgenerator	4	3	3	1	2	3
9	MANGO OFFICE	4	1	4	1	4	1
10	SeoPult	2	1	4	4	2	1
11	UMI	1	1	2	1	2	1
12	Webinar.ru	1	1	3	1	2	1
13	YouScan	3	1	1	1	4	1
14	«Bitriks24»	1	1	2	1	2	3
15	«Contour-Extern»	1	4	4	4	4	1
16	«Megaplan»	3	1	3	3	2	1
17	"My business"	3	4	4	1	2	1
18	"MySklad"	4	1	2	1	2	1
19	Telfin	4	3	1	1	1	1
20	"Elbe"	1	2	2	1	2	1

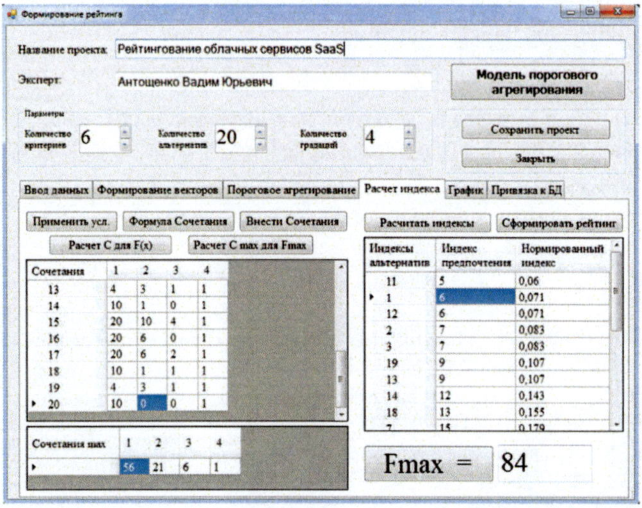

Fig. 1. Tab "Index calculation" of the "Formation of aggregated ratings" program

4 Conclusion

SaaS services provide truly advanced customer support with a focus on online channels. This situation is largely dictated by the expectations of the audience of such resources, which not only refers to advanced Internet users, but also made the World Wide Web an important working tool.

Among the indisputable advantages of supporting customers with SaaS services, one can single out the presence of a large number of communication channels with users; the work of competent specialists in the support service; high speed of response to requests by e-mail and online chat; focus on more modern channels of communication with customers, which increases user self-service capabilities and reduces the costs of companies for customer service.

However, there are still many tasks to be worked on in order to improve the quality of customer service for SaaS projects. Among them:

- organization of systematic support for users in social networks;
- increasing the level of work with negative reviews;
- fight against redirecting users to other channels to answer a question;
- increasing the initiative of support specialists: independent redistribution of requests within the company;
- optimization of the Knowledge Base: improving the search convenience and visibility of the posted material, introducing a tool for evaluating the usefulness of an article.

The article analyzes popular SaaS services based on the organization of customer support channels. The main "strengths" of the customer service for these SaaS projects are shown, and several positions are noted that should be improved. Also, based on the analysis, a rating of popular SAAS services was built based on the used customer support channels using the non-compensatory threshold aggregation method. The calculation was carried out in the developed software.

Acknowledgment. This work was supported by the Russian Science Foundation grant № 21-71-00041 "Development of methods, algorithms and diagnostic models using binary relations and collective solutions for the analysis and optimization of cloud computing management in organizations."

References

1. Liyanage, M., Gurtov, A., Ylianttila, M.: EPC in the cloud. In: Software Defined Mobile Networks (SDMN): Beyond LTE Network Architecture, pp. 107–128. Wiley (2015)
2. Reynolds, P., Yetton, P.: Aligning business and IT strategies in multi-business organization. J. Inf. Technol. **30**(2), 101–118 (2015)
3. Murugesan, S., Bojanova, I.: Compliance in Clouds. Encyclopedia of Cloud Computing, pp. 267–2733. IEEE (2016)
4. Hinchey, M.: A model-centric approach to the design of resource-aware cloud applications. In: Software Technology: 10 Years of Innovation in IEEE Computer, pp. 315–326. IEEE (2018)

5. Elamir, A., Jailani, N.: Framework and architecture for programming education environment as cloud computing service. Procedia Technol. **11**, 1299–1308 (2013)
6. Sultan, N.: Knowledge management in the age of cloud computing and Web 2.0. Int. J. Inf. Manage. **33**(1), 160–165 (2013)
7. Breedveld, S., Craft, D., Haveren, R., Heijmen, B.: Multi-criteria optimization and decision-making in radiotherapy. Eur. J. Oper. Res. **277**(1), 1–19 (2019)
8. Razumnikov, S.V., Kurmanbay, A.K.: Models of evaluating efficiency and risks on integration of cloud-base IT-services of the machine-building enterprise: a system approach. IOP Conf. Ser. Mater. Sci. Eng. **124**(1), 012089 (2016)
9. Razumnikov, S.V., Prankevich, D.A.: Integrated model to assess cloud deployment effectiveness when developing an IT-strategy. IOP Conf. Ser. Mater. Sci. Eng. **127**(1), 012018 (2016)
10. Razumnikov, S.V.: Automation of selecting cloud it services for machine-building enterprise. In: 2019 International Russian Automation Conference (RusAutoCon), Sochi, Russia, pp. 1–5 (2019)
11. Razumnikov, S.V.: Application of the STEM method in the multi-criteria task of linear programming in the designt of a cloud technology development strategy. In: Proceedings - 2020 International Russian Automation Conference, RusAutoCon, pp. 98–103 (2020)
12. Amini, A., Jamil, N.: A comprehensive review of existing risk assessment models in cloud computing. J. Phys. Conf. Ser. **1018**(1), 012004 (2018)
13. de Gasperis, G.: Building an AIML chatter bot knowledge-base starting from a FAQ and a glossary. J. E-Learn. Knowl. Soc. **6**(2), 75–83 (2010)
14. Wanda, P., Jie, H.J.: DeepFriend: finding abnormal nodes in online social networks using dynamic deep learning **11**(1), 34 (2021)
15. Armbrust, M., et al.: A view of cloud computing. Commun. ACM **53**(4), 50–58 (2010)
16. Benmerzoug, D.: Towards AiP as a service: an agent based approach for outsourcing business processes to cloud computing services. Int. J. Inf. Syst. Serv. Sect. **7**(2), 1–17 (2015)
17. Álvarez-Miranda, E., Garcia-Gonzalo, J., Pais, C., Weintraub, A.: A multicriteria stochastic optimization framework for sustainable forest decision making under uncertainty. Forest Policy Econ. **103**, 112–122 (2019)
18. Aleskerov, F., Chistyakov, V.V., Kalyagin, V.A.: The threshold aggregation. Econ. lett. **107**(2), 161–162 (2010)
19. Aleskerov, F., Yakuba, V., Yuzbashev, D.: A threshold aggregation of three-graded rankings. Math. Soc. Sci. **53**, 106–110 (2007)
20. Aleskerov, F., Chistyakov, V., Kalyagin, V.: Social threshold aggregations. Soc. Choice Welfare **35**(4), 627–646 (2010)

Method of Forming an Updated List of Technical Products Fuzzy Quality Indicators Based on Fuzzy Clustering

G. T. Pipiay, L. V. Chernenkaya, and V. E. Mager(✉)

Peter the Great St.Petersburg Polytechnic University, 29, Polytechnicheskaya Street, St.Petersburg 195251, Russia
mv@qmd.spbstu.ru

Abstract. At the stage of the technical products' production planning, the key point of the success for meeting the customer's expectations is to determine customer's needs and to convert these needs into product characteristics, which are eventually reflected in the formed product quality indicators list. To monitor quality indicators at the production stage manufacturers often use the theory of fuzzy sets to combine several groups of quality indicators into one indicator, which simplifies the process of monitoring and evaluating product quality, but complicates the process of interpreting these indicators on the principle of "acceptable – not acceptable". To solve the problem of interpretation, an updated list based on fuzzy sets should contain not only information about the types of quality indicators, but also information on evaluation or measurement scales for each product quality indicator. In this paper, we propose a method for forming a gradations scale for product quality indicators based on fuzzy clustering. A method for forming a rank scale used for particular indicators in a two-level product quality optimization model is developed and justified. The proposed method of a fuzzy term set constructing allows to solve the problem of determining an updated list of fuzzy quality indicators.

Keywords: Fuzzy sets · Measure of uncertainty · Two-level optimization model · Fuzzy clustering · Graph

1 Introduction

Formed indicators of technical products quality at the planning stage are necessary to meet the needs of all interested parties. At the same time, product quality management at the production stage often requires the definition of evaluation or measurement scales, not only for the product technical characteristics, but also for economic or organizational indicators. The combination of heterogeneous product quality indicators and the formation of fuzzy quality indicators allow to cover several aspects of production at once. Thus, the introduction of the established Updated List of Fuzzy Quality Indicators (ULFQI) with the use of evaluation or measurement scales allows you to get answers to questions:

A. A. Radionov and V. R. Gasiyarov (Eds.): RusAutoCon 2021, LNEE 857, pp. 324–336, 2022.
https://doi.org/10.1007/978-3-030-94202-1_31

- is the quality optimal taking into account the investments?
- is it needed to develop corrective or preventive measures regarding the product quality?

It should be noted that the use of a multi-criteria or multi-level evaluations based on modern mathematical methods, in particular, optimization theory [1], complicates the problem of ULFQI forming, since it requires to consider a larger amount of information, which leads to the growth of fuzzy rules and the emergence of the so-called dimension problem. In this paper, we will propose a method for forming a rank scale used for partial indicators of a two-level product quality optimization model.

2 Description of the Product Quality Assessment Model

The evaluation of product quality is carried out using a two-level optimization model, which is based on the principle of decentralized management. A decentralized approach from the management standpoint refers to an approach in which the top management transfers part of its functions to subordinates. In such conditions, the low level units get a degree of freedom in their activities, but at the same time they are still under the upper level control.

Concerning the product quality, the decentralized approach provides independent management of the enterprise structural departments, related to product development, economic support, logistics, material supply and production. At the same time, assessment of the effectiveness and efficiency of the listed structural elements activities is provided by the Quality management department. One of the ways to assess the level of quality when implementing a decentralized approach is the theory of optimization, in particular the theory of two-level optimization.

The two-level optimization problem was described in 1973–1974 by J. Bracken and J. McGill [2, 3]. The model for assessing the quality level in these works is formulated through the two-level decentralized model. The solution to the problem of assessing the quality level, based on a given model, is found by solving the problem of two-level optimization, where the upper level is the leader, and the sublevel is the follower. In this setting, the follower makes the decision first, taking into account the wishes of the leader, and after the follower found a decision, the leader considers follower wishes and determines his own optimal solution. The main advantages of the two-level optimization model in solving of quality assessing problems, related to the instrument-making products, are given in authors' papers [4–6].

A detailed description of the objective functions, the partial summands of the objective functions, the system of constraints, and the algorithm for finding the optimal point will be given in the following sections.

The assessment of the instrument-making products quality level will be determined by the set-theoretic model $\langle Q, X, F_i, Y_i \rangle$, where Q is the target function of the main level (the leader's function), X is the area for the numerical values determining of the main level function, F_i are the i-th target functions of sublevels (the followers' functions), Y_i is the area for determining of the i-th target function values.

Target function of the main level Q is determined by a continuous sequence of numbers $a_{11}, a_{12}, ..., a_{ij}$ $(j = 1, 2, ..., n)$, $x_1, x_2, ..., x_i$ $(i = 1, 2, ..., m)$. In this case a_j : $0 \leq a_j \leq 1$ (for a fixed i-th value of x), and $x_i : \sum_{i=1}^{m} x_i = 1$.

The hierarchy of quality indicators sets additional restrictions on the scope of the Q definition; therefore, the main TF will take the form:

$$Q\ [X, F_i(Y_i)], \tag{1}$$

where $x \in X : Ax \leq d,\ |a_{ij}| = A$, sublevels $y \in Y : B_i \cdot y \leq d,\ |b_{ij}| = B_i,\ 0 \leq b_j^k \leq 1$ (for a fixed i-th value of y), and $0 \leq \sum_{i=1}^{m} y_i \leq 1$.

The search for the optimal value of the function (1) is performed from the bottom to top: first, the optimal value of sublevels $F_i(Y^i)$ is determined, after which the founded values of Y^i are substituted in (1), and the function value for the main level is searched.

The optimization problem for the upper level, taking into account the constraints imposed by sublevels, will look like [7]:

$$\min_x \{Q[y(x), x] : G[y(x), x] \leq 0, H[y(x), x] = 0, y(x) \in \psi(x)\}, \tag{2}$$

where $y(x) = F_i,\ F_i : y(x) \in \psi(x),\ \psi(x)$ is a polyhedron, the domain of its constraints is such, that $Q : R^n \times R^m \to R,\ G : R^n \times R^m \to R^k,\ H : R^n \times R^m \to R^l$ (for the k-th indices there are constraints with the sign "≤", and for the l-th indices – with the sign "="). Then the optimization problem for the lower level is represented as follows:

$$\min_y \{f(x, y) : g(x, y) \leq 0, h(x, y) = 0\}. \tag{3}$$

3 Description of the Problem Area

As mentioned above, the ULFQI is designed to solve the problem of monitoring and evaluating the technical products quality, covered quality indicators of which are based on fuzzy sets, and the evaluation itself is carried out using a multi-level optimization model. In this case, the quality indicators' forming is carried out using a fuzzy inference system based on the Tagaki – Sugeno algorithm [8], in which the fuzzy rules are set as:

$$R_j : u_1(x_1) = a_{1j}AND, ..., ANDu_i(x_i) = a_{ij} \to y_j = b_j, \tag{4}$$

where R_j is the singular inference rule, $j = 1, 2, \ldots, n$ (n is the total number of inference rules); $u_i(x_i)$ is the membership function of the fuzzy variable x_i, $i = 1, 2, \ldots, m$ (m is the number of antecedents in the j-th rule); a_{1j} is the fuzzy term evaluating the membership function $u_i(x_i)$; y_j is the fuzzy inference variable; b_j is the fuzzy term evaluating the fuzzy variable y_i; $y_i = b_j$ is the consequent of the j-th rule.

The output linear variables are set as follows:

$$y_j = a_{1j}x_1 + a_{2j}x_2 + ... + a_{ij}x_i + a_0, \tag{5}$$

where a_{1j} is a fuzzy term (in our case, this is a fuzzy term of a linguistic variable); x_i are variables that define the scale of the fuzzy term and its contribution to the resulting value of y_j compared to other fuzzy terms of the j-th rule.

Fuzzy rules allow to determine the output value of a fuzzy system by applying combinations of judgments, where the antecedents act as a combination of judgments, and the output value is the consequents. The content of antecedents and consequents is determined out of the task being solved. Based on the analysis of the existing literature and scientific publications on the topic of fuzzy inference systems modeling, it can be concluded that the accuracy of the fuzzy inference system depends not only on the inference algorithm, but also on the power of the term set and the number of fuzzy rules. The last two problems in the theory of fuzzy modeling are called the "curse of dimensionality", i.e., when the accuracy of the output of a fuzzy system is affected by the dimensionality of the fuzzy rules set and the power of the term-set. For example, in [9] authors are modeling a fuzzy control system for electronic devices, where they reduce 2500 rules of the standard fuzzy system to 500 rules, using a top-down hierarchical training approach. In [10], new approaches are proposed to solve the dimension problem by measuring the dimension, which is carried out using the particle swarm optimization and differential evolution.

This problem (the dimension problem) is related to the system of quality indicators fuzzy inference as follows. To determine the quality criteria required for regulation, the quantitative scale should contain reference points (divisions), for example, "very high" – "high" – "moderately high" – "medium" – "low" – "very low". Since the decision-making is grounded on 9 partial criteria, and each criterion can contain more than 10 divisions on a quantitative scale, top-manager will have more than 100 different combinations in the decision-making process, where the columns are private criteria, and the rows are their formalized verbal-numerical values.

The described situation resembles the formation of a fuzzy inference rule base and, as a result, the emergence of a dimension problem when forming a combination matrix containing more than 100 different combinations. To solve this problem, it is needed to:

- determine the cardinality of an extended fuzzy rules set;
- develop a way to reduce the dimension;
- define a refined version of the fuzzy rules set.

Based on the fuzzy sets description, by extended fuzzy set we mean a fuzzy set with linguistic variables that have not three, but more gradations.

It is known that the decision accuracy depends on the result of the evaluation, which, in turn, is the result of comparing the obtained numerical value with so-called quantitative scale. At the same time, depending on the received point position on the quantitative scale (estimated value), one or the other measures are produced. Since the recorded gradations on the quantitative scale carry not only a quantitative expression, but also, as a rule, a description of the obtained numerical value influence on the overall product quality assessment, i.e., some physical meaning. At the same time, with the increasing of gradations' levels, the degree of a given value specification impact on the overall quality assessment result increases, and, as a sequence, the accuracy of the measures developed to improve or ensure the current quality increases.

To determine factors that affect the change of target functions individual criteria and, as a result, the assessment of product quality, it is necessary to describe the operation of the product quality monitoring and evaluation model.

4 Description of the Product Quality Monitoring and Evaluation Model Operation

To present the approach to the extended fuzzy set formalization, it is needed to define the content of fuzzy sets, namely, to describe linguistic variables. As an example, we consider the content of the quality indicator "The level of usable products' output", given in the authors' work [11] (Table 1).

From Table 1, it can be concluded that linguistic variables are related to the following departments activities: economic, technological, production, quality office. Indicators listed in the Table 1 in the form of linguistic variables are the result of the listed departments activities (services or departments), which means that these linguistic variables express the state of not only the products quality, but also the quality of the production and structural divisions functioning.

Table 1. Description of the indicator "the level of usable products' output" content.

Partial criterion	Name of linguistic variable
The level of usable products' output x1	Percentage of component parts with deviation permits
	Percentage of component parts with acts on defects
	Percentage of component parts with acts on non-conformities
	Percentage of purchased items included in the product

To determine the relationship between the two-level optimization model, linguistic variables, and the state of structural elements (departments), it is necessary to identify their relationships. The structural relationship between listed departments is shown in Fig. 1.

The functions shown by arrows in Fig. 1 implement the following actions:

- function I: coordination of production and output goals and objectives;
- function II: transfer of templates for the accumulation and storage of primary information by departments and structural divisions;
- function III: structuring, grouping and distribution of primary information that characterizes production and manufactured products;
- function IV: distribution of the processed information by structural divisions and departments for the purpose of transmission to the input of the fuzzy inference system;

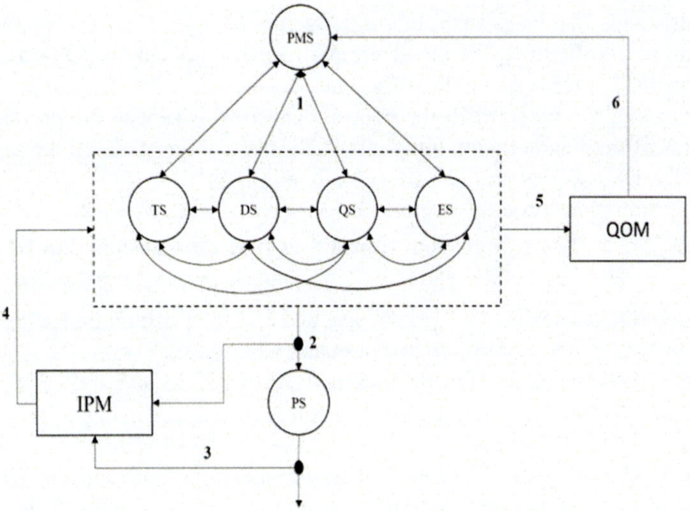

Fig. 1. Structural relationship between departments (PMS - top management, TS - technological department, DS - design department, QS - quality service, ES - economic department, PS - production process, QOM – quality assessment methodology based on two-level optimization, IPM - forms of incoming information).

- function V: transmitting the output values of the fuzzy inference system;
- function VI: providing information that characterizes production and manufactured products in terms of quality.

The work of production structures or departments (according to Fig. 1) it is carried out as follows:

1. In the PMS, vectors $x = (x_1, ..., x_n)$, $y = (y_1, ..., y_n)$, $z = (z_1, ..., z_n)$, $i = 1, 2, ...n$ are formed that determine the target impact $x_i \subset x^t : x^t \Rightarrow x^t = \left[x^-, x^+\right]$, where x, y, z are singular criteria in accordance with Table 1, x^t is the target impact of singular criteria, x^- and x^+ are the tolerance lower and upper limits, respectively.
2. Through the function I, the target vector (target influence) $x_i \subset x^t : x^t \Rightarrow x^t = \left[x^-, x^+\right]$ is transmitted from the PMS to TS, DS, QS, ES and PS, and under its influence, target vectors $y_i \subset y^t : y^t \Rightarrow y^t = \left[y^-, y^+\right]$ and $z_i \subset z^t : z^t \Rightarrow z^t = \left[z^-, z^+\right]$ are formed in TS, DS, QS, ES and PS, which set the tolerance limits of target functions consisting of three groups of variables X, Y, Z, where y and z are singular criteria, y^t and z^t are target values of singular criteria, $\left[y^-, y^+\right]$ and $\left[z^-, z^+\right]$, respectively, the lower and upper tolerance limits. In that case of the optimization problem existing in a convex or concave set, the target constraints will be defined using Jensen's inequalities [12].

   In addition to the target vectors formulating, PMS together with TS, DS, QS, ES and PS form the partial values of target functions $c = (c_1, ..., c_m)$, $d_{(1)} = \left[d_{(1)1}, ..., d_{(1)m}\right]$, $d_{(2)} = \left[d_{(2)1}, ..., d_{(2)m}\right]$ as normals to the hyper planes, composing the direction of the target function growth and decline in the area specified by

the matrices A, $B_{(1)}$, $B_{(2)}$ such, that $a_{ij}x_j \in Ax$, $b_{(1)ij}y_j \in B_{(1)}y$, $b_{(2)ij}z_j \in B_{(2)}z$. It should be recalled that the target vectors formation occurs together between the departments TS, DS, QS and ES (Fig. 1).
3. Then, through the function II, the control command is sent to the production, and forms for filling in and maintaining information about the status of the main blocks (PMS, TS, DS, QS, ES and PS) are submitted to IPM.
4. In the production process vectors $x = (x_1, ..., x_n)$, $y = (y_1, ..., y_n)$, $z = (z_1, ..., z_n)$, $i = 1, 2, ...n$ are sent from PS to IPM through function III, such that $x_i \subset x^r$, $y_i \subset y^t$, $z_i \subset z^t$. The change in numerical values of vectors x, y and z within the boundaries $[x^-, x^+]$, $[y^-, y^+]$ and $[z^-, z^+]$ occurs under the influence of two factors' types: controlled and unmanaged.
5. The processed information is sent from the IPM to TS, DS, QS and ES via the function IV, and further the incoming information is formalized using a fuzzy inference system.
6. The obtained numerical values (partial quality indicators) are sent to QOM via the function V to solve the next optimization problem, the optimal solution of which exists if the conditions [13, 14]:
- there are permissible solutions on the set

$$S = \left\{ (x, y) \in X \times Y_i : Ax + \sum_{i=1}^{k} B_i y \geq d,\ A_i x + B_i y \geq d_j,\ i = 1, 2, ..., k,\ j = 1, 2, ..., n \right\};$$

- there are permissible solutions for the lower level under constraints imposed by the upper level.
7. Through the function VI, user information that reflects the main problem areas and product quality assessment is transmitted to PMS for decision-making on production and manufactured products. Then, from PMS, the TS, DS, QS, and ES receive control impacts to eliminate problem areas.

From the presented description of the relations between structural units and elements of the fuzzy inference system, it can be seen that the information supplied to the input of functions II and III depends on the work of the structural units and production sites, i.e. these functions depend on arguments reflecting the state of work these structural units. However, to define an extended set, it is necessary to determine how the state of the structural unit will affect both the products quality and the quality of the interrelated departments work. Hence the following problem arises: how to determine the importance of the structural unit work and the necessary number of gradations on the evaluation scale, so that it'll become possible to regulate the products quality flexibly.

As it is known from TQM [15], the quality management system consists of interrelated processes, divided into the main, auxiliary, providing and managerial. At the same time, each process contributes to the consumer product value. Besides, the contribution of the process can be both positive and negative, for example, tighter control at the stage of the production process leads to an increase in the products cost, which reduces attractiveness to potential customers. However, there could be not only external, but also internal customers of the process result. At the same time, the internal customers' satisfaction is directly related to the satisfaction of external ones, for example, the occurrence of defects at the production stage leads to an increase in the production time of finished products, which affects the delivery time of products to the customer. Thus, to determine the evaluation scale, it is necessary to distinguish the final functions of the process, the complexity of the process and customers of the process. Then, based on the selected process features, it is necessary to determine the quantitative measure of the process information, and, as the numerical measure value is greater, the more important this process is. After that, based on the process importance, we must determine the number of gradations on the quantitative scale.

The most effective tool for describing the states of physical and non-physical processes is the theory on the amount of information based on B. Hotling and (or) C. Shannon measure [16]. The need to apply this theory is caused by the following:

a. the state of the quality management system processes, which is difficult for modeling from the perspective of probability theory. For a probability-theoretic processes functioning states' description, it is necessary to have a large amount of statistical information. It is important that the processes functioning depends on such factors as the number of personnel and their qualifications, which is almost impossible to describe using normal, Poisson, exponential and other distribution laws;
b. taking into accounting the heterogeneous information, such as staff qualifications, technologies used, labor intensity;
c. the presence of structural elements that determine the process functioning and its state. With this information, it is possible to assess the process complexity, and, consequently, its impact, since the complexity of the process depends not only on its structure, but also on the number of output nodes (Fig. 1);
d. the ability of a process to be in multiple states at the same time;
e. disorganization of the process, which means that any process is not ideal, thus it is difficult to apply improvements;
f. permanent changes in external conditions.

Based on these points, the need appears to develop a methodology for measuring the structural information measure of described processes grounded on the information theory [17].

5 Development of a Method for Finding an Extended Fuzzy Set of Fuzzy Rules

This section defines the information measure of the forming partial quality indicators process (see Table 1). To define the processes information measure and an extended fuzzy set of fuzzy rules, it is necessary to take into account the following process features:

- process customers;
- scope of the process;
- resources and tasks submitted to the process input;
- process functions;
- process execution technologies;
- outputs of the process.

The process of obtaining information on partial quality indicators should consist the following operations:

a. transfer of resources and basic materials to perform functions of structural units involved in the process of generating information on partial quality indicators;
b. performing the functions of the structural units involved in the process of forming information on partial quality indicators;
c. control points for checking the results of structural divisions activities;
d. making a decision on the results obtained;
e. elimination of comments.

To define an extended fuzzy set, we describe the process of forming information about quality indicators through a directed graph *G*(*V*, *E*) [18]. Let's define this graph with the following features:

1. Vertices of the graph are the vector $V = (v_1, v_2, ..., v_n)$, where $i = \overline{1, n}$ are the process operations, and the edge is the vector $E = (e_1, e_2, ..., e_m)$, where $j = \overline{1, m}$ is the number of the edge [19].
2. When moving from one operation to another within the scope of the process definition, resources are spent, defined as a vector $S = (s_1, s_2, ..., s_m)$, where $e_j^+ = (v_i, v_{i+1})$ are planned costs, and $e_j^- = (v_i - 1, v)$ are expenses.
3. The amount of expenses depends on the method of comments elimination and the stage of comments detection.
4. Each expense is characterized by its own entropy, therefore, the content of entropy is the volume of negative factors that lead to expenses:

$$H(s_i) = (-k)\log_2\left(1 - \frac{s_i}{SP}\right), \tag{6}$$

where *SP* are the planned costs; *k* is the number of operation consumers, on which the comments were noticed; the total entropy of the graph by cost is $H(S) = \sum_{i=1}^{n} H(s_i)$.

To determine the method for constructing a rank scale, we use statistical methods for constructing histograms, namely, finding the number of intervals for a quantitative scale. There are many ways to do this. We will focus on the method used for the equally probable distribution law, since the chosen method for finding entropy (6) is based on the so-called structural entropy (Hartley entropy), the peculiarity of which is the equal probability of occurrence of all given events [20]:

$$k = 4\lg(n), \tag{7}$$

where *n* is the sample size, and *k* is the number of partitioning intervals.

To determine expenses, a matrix of the incident $v_n \times s_m$ is preformed, in which v_n are vertices (process operations), s_m are edges (costs/expenses).

Only the negative values of the edges $-s_j$ are involved in the calculation of the partition intervals number (7). After determining the number of intervals, a rank scale is constructed, where 0 is the absence of entropy (no comments), and *n* is the maximum rank of the scale (the maximum entropy value). The value of n is calculated by (7). An example of the correspondence of the entropy scale and the rank scale at $H(s_1) \approx 11$ and $k(8.04) \approx 14$ is shown in Table 2. The following conditions are applied to the scale construction:

$$n = \begin{cases} 1 \; if \; k < 0, \\ 0 \; if \; k = 0, \\ 2 \; if \; 0.85 \le k < 1 \end{cases} . \tag{8}$$

6 Method for Reducing the Fuzzy Set Dimension and Finding an Updated List of Fuzzy Quality Indicators

To reduce the dimension of the fuzzy set or the ranks of the rank scale, an approach based on the fuzzy clustering of c–means is used [21, 22]. We define the matrix of observations for applying the clustering algorithm in the form $N = \{x_{ij}\}$, where the scale ranks go by indices i ($i = \overline{1, n}$), and attributes of each rank go by indices j ($j = \overline{1, m}$). The feature vector x_j will consist of the following quantitative factors:

1. expenses on eliminating comments without returning to the previous operation and using resources;
2. expenses on eliminating comments without returning to the previous operation, but using resources;
3. expenses on eliminating comments with a return to the previous operation and using resources;
4. the number of quality management system documents used;
5. the number of design and technological documentation used;
6. the number of process operation functions;
7. the number of potential consumers of the process operations.

After determining the observation matrix, a random set of clusters is formed, according to which a fuzzy cluster matrix $M = \{c_{ij}\}$, $i = \overline{1, T}$, $j = \overline{1, m}$ is constructed, where the clusters go by indices i, and by the indices j – the degree of the rank, belonging to a certain cluster of the rank scale. The fuzzy cluster matrix satisfies conditions.

$c_{ij} \in [0, 1]$, $\sum_{i=1}^{T} c_{ij} = 1 \; if \; j = \overline{1, m}$, and $0 < \sum_{j=1}^{m} c_{ij} < m \; if \; j = \overline{1, m}$. Quality assessment of the rank scale division into clusters is determined by the degree of belonging [23]:

$$J = \sum_{i=1}^{T} \sum_{j=1}^{m} \left(c_{ij}\right)^{w} d\left(l_i x_j\right), \tag{9}$$

where $d(l_i x_j)$ is the Euclidean distance between the j–th object and the i–th center of the cluster l_i, $w \in (1, \infty)$ is the exponential weight that determines the blurriness of the cluster.

Table 2. Rank scale.

Edges of the incident matrix s_i	Entropy H	Number of partitioning intervals k	Number of operations n
0.2	11.60964	14.14901	14
0.25	10.0	13.28771	13
0.3	8.684828	12.47399	12
0.35	7.572866	11.68336	11
0.4	6.60964	10.89829	10
0.45	5.760015	10.10429	10
0.5	5.0	9.287712	9
0.55	4.312482	8.434074	8
0.6	3.684828	7.526389	7
0.65	3.107442	6.54291	6
0.7	2.572866	5.453505	5
0.75	2.075187	4.212967	4
0.8	1.60964	2.746954	3
0.85	1.172326	0.917497	2
0.9	0.760015	−1.5836	1
0.95	0.370003	−5.73757	1
1.0	0	0	0

The cluster centers make up a matrix $V = \{v_{ik}\}$, whose components are calculated by the formula:

$$v_{ik} = \frac{\sum_{j=1}^{m} (c_{ij})^{w} x_{ik}}{\sum_{j=1}^{m} (c_{ij})^{w}}, \quad k = \overline{1, n}. \tag{10}$$

Our task is to find a fuzzy clusters matrix, at which j is minimal. In subsequent iterations, elements c_{ij} are calculated as follows:

$$\begin{gathered} c_{ij} = \frac{1}{(d_{ij})^{\frac{2}{w-1}} \sum_{k=1}^{T} \frac{1}{(d_{kj})^{\frac{2}{w-1}}}} \quad if \ d_{ij} > 0; \\ c_{ij} = \begin{cases} 1, \ k = i, \\ 0, \ k \neq i \end{cases} \quad if \ d_{ij} = 0. \end{gathered} \tag{11}$$

Calculations should be continued until the difference $\|M - M^{*}\|$ becomes minimal (M^{*} are matrices in the previous iteration). The proof of this algorithm convergence is presented in [24].

7 Conclusions

The proposed method for constructing a fuzzy term-set allows to solve the problem of determining an updated list of fuzzy 1uality indicators, for their further application in a two-level model for optimizing of product quality assessment. This method is also capable of processing large numerical values of scale division intervals.

Acknowledgment. The reported study was funded by RFBR, project number 20-37-90012.

References

1. Cococcioni, M., Foschini, L., Lazzerini, B., Marcelloni, F.: Complexity reduction of Mamdani Fuzzy Systems through multi-valued logic minimization. In: 2008 IEEE International Conference on Systems, Man and Cybernetics, pp. 1782–1787 (2008). https://doi.org/10.1109/ICSMC.2008.4811547
2. Bracken, J., McGill, J.T.: Mathematical programs with optimization problems in the constraintsю. Oper. Res. **21**(1), 37–44 (1973). https://doi.org/10.1287/opre.21.1.37
3. Bracken, J., McGill, J.T.: Defense applications of mathematical programs with optimization problems in the constraints. Oper. Res. **22**(5), 1086–1096 (1974). https://socionet.ru/d/repec:inm:oropre:v:22:y:1974:i:5:p:1086-1096. http://dx.doi.org/10.1287/opre.22.5.1086
4. Pipiya, G.: A model for monitoring quality parameters in a multiple criterion environment. Stand. Qual. **3**, 108–110 (2019)
5. Pipiya, G.T., Chernenkaya, L.V.: Method for formalization of single quality criteria for instrumentation products in a two-level model. Part II. Formalization of single upper- and lower-level criteria. J. Instrum. Eng. **63**(8), 749–755 (2020). https://doi.org/10.17586/0021-3454-2020-63-7-650-656
6. Pipiay, G., Chernenkaya, L., Mager, V.: Fuzzy formalization of individual quality criteria for quality level evaluation by using two-level optimization model. In: Radionov, A.A., Gasiyarov, V.R. (eds.) ICIE 2021. LNME, pp. 557–565. Springer, Cham (2021). https://doi.org/10.1007/978-3-030-54817-9_65
7. Zhang, G., Lu, J., Gao, Y.: Multi-Level Decision-Making, p. 377. Springer, Heidelberg (2015). https://doi.org/10.1007/978-3-662-46059-7
8. Hodashinsky, I.A., Gorbunov, I.V.: Algorithms of the tradeoff between accuracy and complexity in the design of fuzzy approximators. Optoelectron. Instrum. Data Process. **49**(6), 569–577 (2013). https://doi.org/10.3103/S875669901306006X
9. Dovydaitis, J., Jasinevicius, R., Petrauskas, V., Vrubliauskas, A.: Training, retraining, and self-training procedures for the fuzzy logic-based intellectualization of IoT&S environments. Int. J. Fuzzy Syst. **17**(2), 133–143 (2015). https://doi.org/10.1007/s40815-015-0035-2
10. Chen, S., Montgomery, J., Bolufé-Röhler, A.: Measuring the curse of dimensionality and its effects on particle swarm optimization and differential evolution. Appl. Intell. **42**(3), 514–526 (2014). https://doi.org/10.1007/s10489-014-0613-2
11. Pipiya, G.T., Chernenkaya, L.V.: Method for formalization of single quality criteria for instrumentation products in a two-level model. Part I. Single criteria for target quality functions. J. Instrum. Eng. **63**(7), 650–656 (2020). https://doi.org/10.17586/0021-3454-2020-63-7-650-656
12. Miyamoto, S., Ichihashi, H., Honda, K.: Algorithms for Fuzzy Clustering: Methods in c-Means Clustering with Applications, p. 247. Springer, Heidelberg (2008). https://doi.org/10.1007/978-3-540-78737-2

13. Kim, J.H.: Further improvement of Jensen inequality and application to stability of time-delayed systems. Automatica **64**, 121–125 (2016)
14. Pop, P.C., Fuksz, L., Marc, A.H., Sabo, C.: A novel two-level optimization approach for clustered vehicle routing problem. Comput. Ind. Eng. **115**, 304–318 (2018). https://doi.org/10.1016/j.cie.2017.11.018
15. Mager, V.E.: Quality Management: Handbook, p. 176. INFRA-M, Moscow (2015)
16. Dehmer, M., Emmert-Streib, F., Chen, Z., Li, X., Shi, Y. (eds.) Mathematical Foundations and Applications of Graph Entropy, p. 296. Wiley-VCH, New York (2017)
17. Ghosh, S., Yadav, V.K., Mukherjee, V., Yadav, P.: Evaluation of relative impact of aerosols on photovoltaic cells through combined Shannon's entropy and Data Envelopment Analysis (DEA). Renew. Energy **105**, 344–353 (2017). https://doi.org/10.1016/j.renene.2016.12.062
18. Xie, X.L., Beni, G.: A validity measure for fuzzy clustering. IEEE Trans. Pattern Anal. Mach. Intell. (PAMI) **13**(8), 841–847 (1991)
19. Höppner, F., Klawonn, F., Kruse, R., Runkler, T.: Fuzzy Cluster Analysis, p. 300. Wiley, Chichester (1999)
20. Saraswat, S.K., Digalwar, A.K.: Evaluation of energy alternatives for sustainable development of energy sector in India: an integrated Shannon's entropy fuzzy multi-criteria decision approach. Renew. Energy **171**(C), 58–74 (2021). https://doi.org/10.1016/j.renene.2021.02.068
21. Bezdek, J.C.: Pattern recognition with fuzzy objective functional algorithms. In: Advanced Applications in Pattern Recognition, p. 272. Plenum Press, New York (1981). https://doi.org/10.1007/978-1-4757-0450-1
22. Ruspini, E.H., Bezdek, J.C., Keller, J.M.: Fuzzy clustering: a historical perspective. IEEE Comput. Intell. Mag. **14**(1), 45–55 (2019). https://doi.org/10.1109/MCI.2018.2881643
23. Li, C., et al.: A comparison of fuzzy clustering algorithms for bearing fault diagnosis. J. Intell. Fuzzy Syst. **34**(6), 3565–3580 (2018). https://doi.org/10.3233/JIFS-169534
24. Goldberg, D.E.: Genetic Algorithms in Search, Optimization, and Machine Learning, p. 432. Addison-Wesley, Reading (1989). https://doi.org/10.5860/choice.27-0936

Combined Control of Angular Velocities of an Aircraft Based on a Predictive Model

V. N. Trofimenko[1,2](✉) and A. A. Volkova[1]

[1] Don State Technical University, 1, Gagarin Sq., Rostov-on-Don 344000, Russia
[2] Rostov State Transport University, 2, Rostovskogo Strelkovogo Polka Narodnogo Opolcheniya Sq., Rostov-on-Don 344038, Russia

Abstract. Unmanned aerial vehicles (UAVs) are widely used in various sectors of the national economy. Therefore, improving the efficiency of algorithms for controlling the angular position of aircrafts is relevant. Predictive model algorithms based on the optimization of the generalized work functional are characterized by good computational efficiency. The paper presents a simulation of angular velocity control of an axisymmetric aircraft based on the predictive model. It is shown that disturbing moments lower the quality of control – there appear errors that can not completely eliminated by the control. The structure of combined control is proposed. It includes the main control loop using the predictive model method and the perturbation moment compensation loop. The compensation is determined based on the estimates of the disturbing moments. The estimates are obtained using predictive model algorithm. The results of modeling are presented, confirming the increase in the efficiency of control in the conditions of disturbances – the accuracy of regulation increases.

Keywords: Aircraft · Angular velocity control · Predictive model · Combined control · Perturbing moments estimation · Compensation

1 Introduction

Increasing the efficiency of algorithms for controlling the angular position of aircrafts is relevant since UAVs are widely used in various areas of economy [1, 2]. Optimal control algorithms allow minimizing cost and duration of the control.

Predictive model (PM) algorithms based on minimization of the generalized work function result in better computational efficiency [3, 4] comparing to algorithms based on minimization of classic functionals [5]. The joint maximum principle algorithms (JMP) also reveal good computational efficiency [6].

We will consider an axisymmetric UAV, which dynamics is described by nonlinear differential equations. For such an UAV there were derived control laws in analytical form based on the PM [7–10] and JMP [11] algorithms. JMP-based algorithms allows considering of control actions constraints, but the discrete nature of control actions leads to the sliding mode. PM-based algorithms do not result in sliding mode; consequently, it is more precise [12].

A. A. Radionov and V. R. Gasiyarov (Eds.): RusAutoCon 2021, LNEE 857, pp. 337–345, 2022.
https://doi.org/10.1007/978-3-030-94202-1_32

However, modeling of PM-based algorithm mentioned above did not consider perturbation moments. Perturbing conditions lead to degradation of the control quality [13] because of control errors. The reason is that deviation control lowers influence of perturbations, but does not completely eliminates it. Excessive control errors can result in mission failure of the UAV. Estimating and compensating of the disturbing perturbation will decrease static control error, influenced by the perturbation.

The paper presents results of simulation of combined angular velocity control of an axisymmetric aircraft including the perturbation moment compensation loop, based on the predictive model.

2 Influence of the Perturbation Moments on the Control Quality

Let us consider algorithm for angular velocities control of an axisymmetric UAV based on the predictive model [7].

Angular movement of an axisymmetric UAV is described by following equations [14]:

$$\begin{aligned} \dot{\omega}_1 + A\omega_2\omega_3 = u_1,\ \omega_1(t)\Big|_{t=t_0} &= \omega_1(t_0), \\ \dot{\omega}_2 - A\omega_1\omega_3 = u_2,\ \omega_2(t)\Big|_{t=t_0} &= \omega_2(t_0), \\ \dot{\omega}_3 = u_3,\ \omega_3(t)\Big|_{t=t_0} &= \omega_3(t_0), \end{aligned} \tag{1}$$

where A – reduced inertia moment; $\omega_1, \omega_2, \omega_3$ and u_1, u_2, u_3 – angular velocities and normalized control moments as time dependent functions which constitute vectors ω and u respectively.

The Krasovsky function for synthesis of the control law for object (1) is determined by the following expression:

$$J = \int_{t_0}^{t_1} \left(\omega^T \mathbf{Q}\,\omega\right)dt + \frac{1}{2}\int_{t_0}^{t_1} \left(\mathbf{u}^T\mathbf{K}^{-1}\mathbf{u} + \mathbf{u}_O^T\mathbf{K}^{-1}\mathbf{u}_O\right)dt, \tag{2}$$

where K and Q – positively defined diagonal matrices; ω – angular velocity vector; u, $\mathbf{u}_O$ – variable and optimal control moments vectors; ^{T} – the transposing symbol.

According to the PM algorithm, optimal control for object (1) is determined by the following expression [7]:

$$\mathbf{u}_O(t) = -K\int_{t}^{t_1} G^T(s,\ t)\,\frac{\partial}{\partial\omega_m}\left(\omega_m^T\mathbf{Q}\,\omega_m\right)ds, \tag{3}$$

where s – accelerated prediction time; ω_m – vector for predicted velocities of object (1) free movement; $G(s,\ t)$ – fundamental matrix, which is determined by free rotation of system (1).

Free rotation of an axisymmetric UAV is determined by the following system:

$$\begin{aligned} &\dot{\omega}_1 + A\omega_{m2}\omega_{m3} = 0,\ \omega_{m1}(s)\Big|_{s=t} = \omega_1(t), \\ &\dot{\omega}_2 - A\omega_{m1}\omega_{m3} = 0,\ \omega_{m2}(s)\Big|_{s=t} = \omega_2(t), \\ &\dot{\omega}_{m3} = 0,\ \omega_{m3}(s)\Big|_{s=t} = \omega_3(t). \end{aligned} \tag{4}$$

Fundamental matrix $G(s,\ t)$ for system (4) is determined by following equation:

$$\frac{\partial G(s,t)}{\partial s} = F_\omega \cdot G(s,t),\ G(s,t)|_{s=t} = \begin{pmatrix} 1\,0\,0 \\ 0\,1\,0 \\ 0\,0\,1 \end{pmatrix}, \tag{5}$$

where F_ω – the Jacobi matrix for system (4).

The Jacobi matrix F_ω is determined by expression:

$$F_\omega = A\begin{pmatrix} 0 & -\omega_{m3} & -\omega_{m2} \\ \omega_{m3} & 0 & \omega_{m1} \\ 0 & 0 & 0 \end{pmatrix}. \tag{6}$$

Analytical solution for Eq. (3), derived by symbolic integration of (4) and (5), is represented in paper [7].

The optimal control $\mathbf{u}_O$, which minimizes the generalized work function and stops rotation of the UAV, is determined by expressions:

$$u_{O1}(t) = -k_1\left\{\omega_1(q_1+q_2)(t_1-t) - \frac{q_1-q_2}{2\alpha}[\omega_1 \sin 2\beta - \omega_2(\cos 2\beta) - 1]\right\}, \tag{7}$$

$$u_{O2}(t) = -k_2\left\{\omega_2(q_1+q_2)(t_1-t) + \frac{q_1-q_2}{2\alpha}[\omega_2 \sin 2\beta - \omega_1(\cos 2\beta) - 1]\right\}, \tag{8}$$

$$\begin{aligned} u_{O3}(t) = &-k_3\{2\omega_3 q_3\,(t_1-t) + 0,25\alpha^{-2}(q_1-q_2) \\ &\times \left[\left(\omega_2^2 - \omega_1^2\right)(\sin 2\beta - 2\beta \cos 2\beta) - 4\omega_1\omega_2 \right. \\ &\times (\cos 2\beta + 2\beta\ \sin 2\beta - 1)\}, \end{aligned} \tag{9}$$

where $\beta = A\omega_3(t)\cdot(t-t_1)$; $k_1,\ k_2,\ k_3$ – diagonal elements of matrix; q_1, q_2, q_3 – diagonal elements of Q.

In previous papers [5, 7, 10, 12] results of simulations according the algorithm (7)–(9) were derived without considering influence of the disturbing moments.

Results of simulation according the algorithm (7)–(9), which consider the disturbing moments will be represented below.

Initial values of angular velocities for simulation are: $\omega_1(0) = 0.5$ rad/s, $\omega_2(0) = -0.5$ rad/s, $\omega_3(0) = 0.25$ rad/s. Perturbation moments were simulated as 3-channel rectangular impulses (Fig. 1) having following values: $-2\ \text{s}^{-2}$ on time interval [0.3 s, 0.5 s]; $2\ \text{s}^{-2}$, on time interval [0.1 s, 0.2 s]; $1.5\ \text{s}^{-2}$ on time interval [0.1 s, 0.4 s].

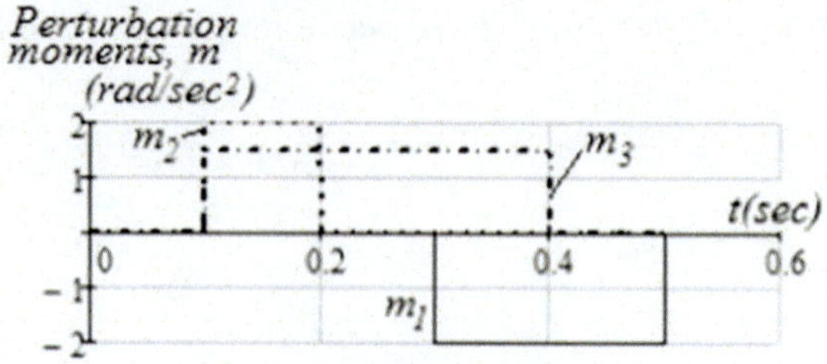

Fig. 1. Perturbation moments, *m*.

Figure 2, are presents angular velocities graph obtained through control simulation without perturbation moments applied. (Fig. 2b) represents angular velocities graph obtained through control simulation with perturbation moments (Fig. 1) applied.

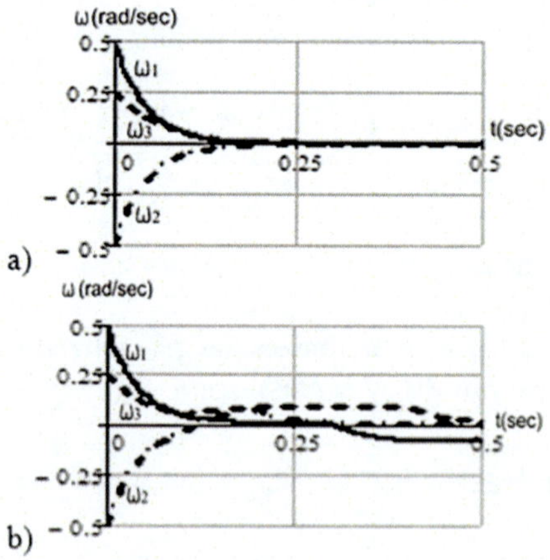

Fig. 2. Angular velocities dynamics without (a) and with (b) perturbation moments applied.

Analysis of graphs reveals that in eliminable angular velocities deviations appear during perturbation. Deviations achieve as much as 0.1 s^{-1}. As shown by the simulation, the control system, implementing algorithm (7)–(9), eliminates these deviations after the perturbation period.

Consideration and compensation of perturbations allows decreasing the value of angular velocities deviations and time, necessary for control.

3 Combined Control with Perturbation Compensation

Automatic system, invariant to perturbations, assume presence of perturbation meters. A disadvantage of invariant systems is compensating of influence only such perturbations, which can be measured [15]. The measurement of perturbation influence is not always available. In this case, a method for indirect estimates might be applied.

Such problems are usually solved using methods, based on probabilistic representation of the process being explored, – estimation of a system state is determined by observation results and known statistical properties of the system and measurement channels. These methods include many algorithms for linear and non-linear estimation, for example, Bayesian filter, maximum plausibility algorithms, the Kalman filter, etc. [16]. For non-repeating processes, estimates should be 'good' for a single realization, not only for the set average. In such cases a dynamical system condition is being estimated concerning a single realization observation by minimizing the residual function [17].

However, optimization problem for estimation based on the residual function is ill-posed in the sense of Hadamard [18], since it comes down to solving a Fredholm equation of the first kind [19, 20]. In terms of problem of estimation of dynamical system condition, the generalized work function (Krasovskii functional) relates to Tikhonov regularized functions. This is due to the fact, that the generalized work function optimization comes down to solving a Fredholm equation of the second kind, which is well-posed in the sense of Hadamard [20].

The PM algorithms, based on the generalized work function optimization, show decent computational performance. The efficiency of implementing the PM algorithms for non-linear dynamical systems estimation problems was demonstrated in [20–25].

Therefore we will use the estimation algorithm based on generalized work function optimization based on the PM. The estimations obtained will be combined with controls, derived using the PM algorithm (7)–(9).

To state a problem for estimating perturbation effect on an axisymmetric UAV we will add unknown perturbation moments to the angular movement Eq. (1).

In this case, initial dynamics equations for the estimation problem will be as follows:

$$\begin{aligned} \dot{\omega}_1 + A\omega_2\omega_3 - u_{O1} &= m_1, \ \omega_1(t)|_{t=t_0} = \omega_1(t_0), \\ \dot{\omega}_2 - A\omega_1\omega_3 - u_{O2} &= m_2, \ \omega_2(t)|_{t=t_0} = \omega_2(t_0), \\ \dot{\omega}_3 - u_{O3} &= m_3, \ \omega_3(t)|_{t=t_0} = \omega_3(t_0), \end{aligned} \tag{10}$$

where u_{O1}, u_{O2}, u_{O3} – reduced control moments being determined by expressions (7)–(9); m_1, m_2, m_3 – unknown perturbation moments on respective axes as components of vector m; $\omega_1(t_0), \omega_2(t_0), \omega_3(t_0)$ – initial angular velocity values for estimation time t_0.

It should be noted, that perturbation vector m is generalized perturbation vector, in the sense that it represent not only external unknown perturbation moments, but also parametrical perturbations caused by deviation of the mathematical model (1) from the real object, for example, UAV degradation.

According to the PM-based algorithm for perturbing moments estimation, $\hat{m}_{O1}, \hat{m}_{O2}, \hat{m}_{O3}$ are determined by optimizing the residual function regularized as the generalized work function:

$$\begin{aligned} J[\hat{\omega}] &= \int_{t_0}^{t_1} \Psi(\hat{\omega})dt + \frac{1}{2}\int_{t_0}^{t_1} \hat{m}^T R^{-1}\hat{m}\,dt + \frac{1}{2}\int_{t_0}^{t_1} \hat{m}_O^T R^{-1}\hat{m}_O dt, \\ \Psi(\hat{\omega}) &= (\omega - \hat{\omega})^T P(\omega - \hat{\omega}), \end{aligned} \tag{11}$$

where $\hat{\omega}$ – angular velocities estimation; $\hat{m}_O$ – desired optimal perturbation estimation; R, P – positively defined diagonal matrices of weight coefficients.

The estimation problem comes down to determination perturbation moments $\hat{m}_O$, which would minimize regularized residual function (11). In fact, this is problem definition for optimal control of the following model:

$$\begin{aligned} \dot{\hat{\omega}}_1 + A\hat{\omega}_2\hat{\omega}_3 - u_{O1} = \hat{m}_1,\ \hat{\omega}_1(t)\big|_{t=t_0} = \omega_1(t_0), \\ \dot{\hat{\omega}}_2 - A\hat{\omega}_1\hat{\omega}_3 - u_{O2} = \hat{m}_2,\ \hat{\omega}_2(t)\big|_{t=t_0} = \omega_2(t_0), \\ \dot{\hat{\omega}}_3 - u_{O3} = \hat{m}_3,\ \hat{\omega}_3(t)\big|_{t=t_0} = \omega_3(t_0). \end{aligned} \tag{12}$$

Or we can represent it as matrix

$$\dot{\hat{\omega}} + f(\hat{\omega}) = \hat{m},\ \hat{\omega}(t)\bigg|_{t=t_0} = \omega(t_0), \tag{13}$$

where $f(\hat{\omega})$ – generating function of the system (12).

According to the PM algorithm, optimal estimation $\hat{m}_O$ for problem (10)–(11) is determined by expression

$$\begin{aligned} \hat{m}_O(t) &= -R \cdot \int_t^{t+t_W} \Phi^T(s,t) \cdot \frac{\partial}{\partial\hat{\omega}}(\Psi(\hat{\omega}_m, s)) \cdot ds \\ &= -2R \cdot \int_t^{t+t_W} \Phi^T(s,t) \cdot P \cdot (\omega - \hat{\omega}_m) \cdot ds, \end{aligned} \tag{14}$$

where t_W – estimation interval; $\hat{\omega}_m \equiv \hat{\omega}_m(s)$ – predicted dynamics of the object (12) assuming zero perturbation moments (free movement of the object in accelerated time s) with the following initial conditions

$$\hat{\omega}_m(s)\big|_{s=t} = \omega(t); \tag{15}$$

$\Phi(s,t)$ – fundamental matrix of the system, which is determined by equation:

$$\frac{\partial\Phi(s,t)}{\partial s} = \mathrm{E}_{\hat{\omega}} \cdot \Phi(s,t),\ \Phi(s,t)|_{s=t} = I, \tag{16}$$

where $\mathrm{E}_{\hat{\omega}} = \frac{\partial f}{\partial\hat{\omega}}$ – Jacobi matrix of equation for object dynamics (12).

Numerical integration of systems for fundamental matrix (16) and free movement of system (12) is the base for computing control according to expression (14) [26].

Figure 3 represents structure of the combined control. The controller includes two control channels: one for determining angular velocities control according to algorithm (7)–(9) and another for determining compensation moments according to algorithm (14). In fact, the perturbation moment compensation loop is an observation device [15].

Results of combined control simulation considering perturbation moments are represented below. Difference between angular velocities of perturbed (Fig. 2b) and non-perturbed (Fig. 2a) trajectories without compensation control channel is represented on (Fig. 4a). Difference between angular velocities of perturbed and non-perturbed trajectories with perturbation moments compensation control channel is represented on (Fig. 4b).

Analysis of the graphs above shows that combined control decreases angular velocities deviation for several times (in the example above – by four times) during perturbation moments action. Deviation peaks at (Fig. 3b) result from delay in estimations obtaining. The delay depends on size of the estimation "window" t_W [20]. Since angular velocities deviation decreases, control duration decreases also.

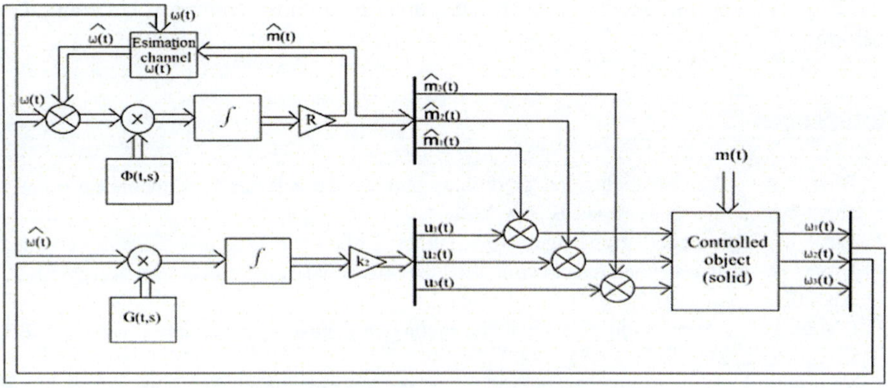

Fig. 3. Structure of the combined control.

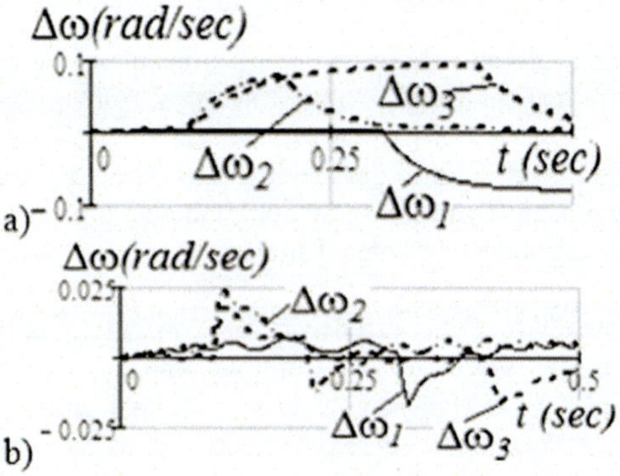

Fig. 4. Difference between angular velocities without (a) and with (b) compensation of perturbation moments.

4 Conclusion

Obtained results show that:

- lowering angular velocities deviation in the problem of angular position of UAV control in perturbing condition is possible through producing appropriate compensating forces;

- evaluation of the compensating forces is possible through perturbation moments estimation based on the generalized work function optimization according the predictive model algorithm;
- combined control decreases angular velocities deviation and control duration decreases during perturbations.

The presented combined control structure may be used for control of UAV angular position.

References

1. Boyko, A.: Areas of application of drones. http://robotrends.ru/robopedia/oblasti-primeneniya-bespilotnikov. Accessed 10 Apr 2021
2. Faro, U.: The second drone age: How Turkey Defied the U.S. and Became a Killer Drone Power. https://theintercept.com/2019/05/14/turkey-second-drone-age/. Accessed 11 Apr 2020
3. Clarke, D.W.: Application of generalized predictive control to industrial processes. IEEE Control Syst. Mag. **8**(2), 49–55 (1988)
4. Yunusova, S., Izmailova, R., Mamyrov, U.: Algorithm of synthesis of predictive control of an electromechanical object. Young Sci. **3**(107), 238–241 (2016). https://moluch.ru/archive/107/25623. Accessed 06 Mar 2020
5. Vasiliev, F.P.: Lectures on methods for solving extreme tasks. Moscow University Press, Moscow (1974)
6. Kostoglotov, A.A., Kostoglotov, A.I., Lazarenko, S.: Joint maximum principle in the problem of synthesizing an optimal control of nonlinear systems. Autom. Control Comput. Sci. **41**, 274–281 (2007)
7. Taran, V.N., Trofimenko, V.N.: Sintez optimalnogo algoritma uglovoy stabilizatsii metodom prognoziruyushchey modeli (Synthesis of an optimum algorithm of angular stabilization by method of the predicting model) Avtomatikaitelemechanica (Autom. Equip. Telemech.) **5**, 82–85 (1997)
8. Taran, V.N., Trofimenko, V.N.: Generalized work functional in regularization of the estimation of the state of a dynamical system. Autom. Control Comput. Sci. **33**(4), 30–39 (1999)
9. Taran, V.N., Trofimenko, V.N.: Transport systems intellectualization based on analytical control synthesis of angular velocities for the axisymmetric spacecraft. Adv. Intell. Syst. Comput. **680**, 154–160 (2018)
10. Volkova, A., Agapov, A., Trofimenko, V.: Synthesis of the spacecraft control algorithm based on the functional of the generalized work using the predictive model method. In: Current Problems and Prospects for the Development of Transport, Industry and the Russian Economy ("TransPromEk-2019"), pp. 25–29. Rostov State Transport University, Rostov-on-Don (2019)
11. Kostoglotov, A.: Method for synthesis of optimal attitude stabilization algorithm based on joint maximum principle. Autom. Control Comput. Sci. **5**(36), 21–28 (2002)
12. Kostoglotov, A., Taran, V., Trofimenko, V.: Fuzzy topological approach to a solid control task. Adv. Intell. Syst. Comput. **874**, 373–381 (2019)
13. Ankhimyuk, V., Opeyko, O., Mikheev, N.: Theory of Automatic Control. Design PRO, Minsk (2000)
14. Tikhonravov, M.K., Yatsunsky, I.M., Maximov, G., Bazhinov, I.K., Gurko, O.V.: Bases of the Theory of Flight and Elements of Design of Artificial Earth Satellites. Mashinostroeniye, Moscow (1967)

15. Ankhimyuk, V.L., Opeyko, O.F., Mikheev, N.N.: Theory of Automatic Control. Design PRO. Minsk (2000)
16. Farina, A., Studer, F.: Radar Data Processing. Volume 1 – Introduction and Tracking. Wiley, Letchworth (1985)
17. Cirlin, A.M., Balakirev, V.S., Dudnikov, E.G.: Variational Methods of Optimization of Controlled Processes. Energiya, Moscow (1975)
18. Lapedes, D.N.: McGraw-Hill Dictionary of Scientific and Technical Terms. McGraw-Hill Education, New York (1989)
19. Krylov, V.I., Bobkov, V.V., Monastyrsky, P.I.: Computational Methods, vol. 2. Nauka, Moscow (1977)
20. Taran, V.N., Trofimenko, V.N., Trofimenko, I.V.: Generalized work functional in regularization of the estimation of the state of a dynamical system. Autom. Control Comput. Sci. **33**(4), 30–39 (1999)
21. Kostoglotov, A.A., Taran, V.N.: The use of the predictive-model method to solve inverse problems of gas dynamics. J. Appl. Math. Mech. **58**(5), 865–871 (1994)
22. Trofimenko, V.N.: Regularizing effects of derivative measurement in dynamic-system state estimation. Autom. Control Comput. Sci. **34**(6), 37–44 (2000)
23. Trofimenko, V., Trofimenko, K.: Algorithm for air density estimation by measuring parameters of test body movement based on the criteria of minimum of the generalized work functional. In: Proceedings of SPIE - The International Society for Optical Engineering. Proceedings of the 1999 6th International Symposium Atmospheric and Ocean Optics. Sponsors: Russian Academy of Sciences, Russian Foundation for Basic Research, SPIE Russia Chapter, Optical Society of America, International Center for Fundamental Physics in Moscow, Tomsk, pp. 321–327 (1999)
24. Trofimenko, V., Trofimenko, E., Taran, V.: Determination of the parameters of the motion of an aircraft by indirect measurements based on the Krasovsky functional. In: Theory, Methods and Means of Measurement, Control and Diagnostics. Materials of the XI International Scientific and Practical Conference, Computer Technologies in Science, Production, Social and Economic Processes: Materials of the XII International Scientific and Practical Conference, Novocherkassk, pp. 37–41 (2012)
25. Trofimenko, V., Polovinchuk, N., Ivanov, S.: An algorithm for estimating the parameters of the trajectory of an aircraft. Dualtechnologies **1**(66), 30–33 (2014)
26. Lazarenko, S.V., Trofimenko, V.N., Volkova, A.A., Kuchugura, N.O.: Synthesis of control of the aircraft angular velocities based on algorithm with a predictive model. In: IOP Conference Series: Materials Science and Engineering, Dynamics of Technical Systems (DTS 2020), Rostov-on-Don, 11–13 September 2020, vol. 1029 (2020)

Machine Learning, Big Data, Internet of Things

Development of On-Board Electronic Communication Devices for Intelligent Transportation Systems

D. Chkalova(✉)

Vladimir State University named after Alexander and Nikolay Stoletovs, 87, Gorky Street, Vladimir 600000, Russia

Abstract. The work is devoted to the task of developing means of operational communication for intelligent transportation systems under unstable cellular coverage. A study of current trends in the development of intelligent transport systems field is carried out, an example of a prototype of the developed on-board radio-electronic device is shown. The manufactured device is installed on vehicles of civil and special purposes (including emergency, rescue and utilities), and is intended to provide an online communications service by organizing a dynamic communication network between moving cars in the presence of a partial or complete absence of traditional radio communications. Areas and scenarios of the possible application of development are considered, the most promising functions related to road safety, operational interaction of vehicles with special services, prevention of traffic jams are described.

Keywords: V2X · DSRC · Mesh network · VANET · Road safety

1 Introduction

The development of operational communication means for intelligent information systems in transport under unstable cellular coverage is an urgent research task in the context of improving road safety and, in the long term, developing autonomous driving systems [1–3].

The rapid increase in the number of vehicles leads to an increase in the number of road accidents. In this regard, there is a need for intelligent information systems that can autonomously (without driver) quickly respond to the traffic situation and immediately inform road users about possible collisions, adverse weather conditions, traffic jams, etc. Results of the development of intelligent transportation systems (ITS) are also the basis for a promising segment of autonomous (unmanned) vehicles, where the need for such communications is felt most acute [4–6].

Currently, the problem of ensuring road safety is one of the most important tasks worldwide. In order to solve this problem, a digital electronic device has been developed, designed to provide an operational communication service by organizing a dynamic communication network between moving cars. Among the tasks of the developed device:

A. A. Radionov and V. R. Gasiyarov (Eds.): RusAutoCon 2021, LNEE 857, pp. 349–358, 2022.
https://doi.org/10.1007/978-3-030-94202-1_33

reducing the concentration of traffic accidents on the road network of urban agglomerations, optimizing traffic flows, creating an automated monitoring system focused on interaction with roads users.

2 Cooperative Intelligent Transport Systems

An intelligent transport system is a system that uses innovative developments in the modeling of transport systems and regulation of traffic flows, which provides end users with greater information and security, as well as a qualitatively higher level of interaction between traffic participants compared to conventional transport systems. In recent years, the emphasis in ITS has turned specifically towards the new generation - Cooperative Intelligent Transport Systems (C-ITS) [7–9], in which vehicles interact with each other and/or with infrastructure. It is cooperative ITS that can significantly increase the quality and reliability of information available about vehicles, their location and road environment. The serious potential is created for vehicle movement in real transport conditions without human intervention, since a human-driven vehicle is a vehicle of increased danger.

Currently, the most actively developing customer-oriented approach to vehicle driving: ITS technologies aimed at interacting with an individual driver (car), and not with a traffic stream. Nevertheless, the modern organization of traffic requires a completely different level of awareness. It is necessary to control not only the activity of the car itself, but also the surrounding traffic situation. This trend is clearly seen in many reports and speeches at the latest major international events in the field of ITS [10–14].

Cooperative ITSs are built on the basis of vehicle-to-vehicle (V2V), vehicle-to-infrastructure (V2I, I2V), infrastructure-to-infrastructure (I2I) and vehicle-to-pedestrian (V2P) communication systems [15–17]. Together, V2V and V2I technologies are usually referred to as V2X. In other words, intelligent systems provide us with information about cars and their location, about road conditions, allow us to optimize and secure road network traffic, as well as accelerate the response to traffic incidents and accidents. Intelligent road network adapts to actual changes in real-time. Messages about traffic intensity, incidents and accidents become available throughout the network. To ensure safe driving, the driver needs to evaluate and control traffic situation. The purpose of cooperative ITSs is to help drivers maintain a safe speed and distance, make lanes, avoid overtaking in critical situations and safely pass road intersections. In addition, with the help of DSRC-based (Dedicated Short-Range Communication) hardware, it is possible to determine optimal parking location, transfer data on parking spaces availability to the information resource, deliver this information over the network to an appropriate vehicle, lay the route, helping the driver to reach the final destination faster [18–21]. But the most important thing in cooperative ITS is the ability to identify potential risks in real time. This, of course, will have a positive impact on road safety and traffic management. Services provided by cooperative ITS can be divided into six categories:

- assistance for safe driving;
- traffic flow control, smoothing traffic flow;
- improving driving comfort through the use of information technology;

- response to incidents and emergency traffic situations;
- supporting economic activity, creating a private services market;
- support in road services activities.

Successful implementation of tasks set for cooperative ITS is based on two components. The first component is a reliable channel for multiservice communication between cars, as well as between a car and infrastructure. Such a channel should at least satisfy the following requirements:

- to be standardized and OSI compliant;
- to have sufficient bandwidth, low latency, high speed of establishing a connection with an object moving at maximum speed (to solve the entire spectrum of problems);
- to be protected;
- to be weatherproof.

The second component is various mathematical models focused on solving a particular problem. A common property of all these models is the need to represent the reflection of a specific real road network or to represent a limited small section of such a network.

3 On-Board Communication System for Unstable Cellular Coverage Conditions

The system of operational information data exchange under conditions of unstable coverage of cellular networks is intended both to transmit information to drivers about the accident that has occurred, and about the possibility of its occurrence, with the aim of improving traffic safety [22].

The main element of the system is a specialized communication device installed in the car (Fig. 1). The device is integrated with several sensors and has the ability to exchange data via the following wireless channels: GSM (3G/4G), Wi-Fi, V2X (Fig. 2).

The exchange of information between vehicles, in conditions of a poor level or lack of a signal, is carried out by transmitting it via a V2X network through vehicles located within a radius of 1.5 km (using the nearest vehicles as a signal repeater) from a vehicle transmitting the necessary information.

The on-board automotive electronic device has the following characteristics:

- powered from onboard power supply network 12/24 V;
- power consumption in active mode, no more than 10 W;
- the maximum range of radio communication between two onboard devices, not less than 1000 m.
- maximum number of simultaneously serviced connections with the nearest on-board devices is at least 10;
- the maximum relative speed of vehicles with interacting on-board devices, not less than 200 km/h;
- operating frequency range is 5.9 GHz (5855–5925 MHz);

- the maximum bandwidth of the air interface between two on-board devices, at least 20 Mbps;
- providing packet data functions between on-board device applications;
- transit of messages through intermediate on-board devices with support for priority service functions and "on network ready" delayed delivery system;
- ability to integrate with sensors of GPS/GLONASS/BeiDou satellite positioning systems;
- ability to integrate with personal mobile devices using Bluetooth or USB technology;
- ability to integrate with GPRS/EDGE/3G/LTE networks;
- built-in sound and optical indication.

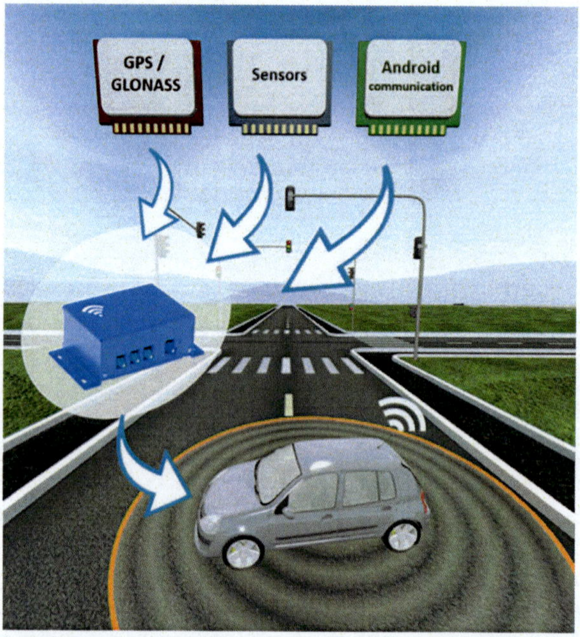

Fig. 1. On-board electronic communication device.

The system operates on the basis of the developed software. Embedded software is an integral part of the on-board device and is distributed in a pre-installed form or as a binary image of non-volatile memory, which includes all the necessary system and application software, in the particular operating system. The control software is designed to implement target functions of the onboard DSRC device of smart car network, it is responsible for starting and initializing the operating system service, within which the device software modules that control its hardware components function: DSRC-modem, LTE-modem, GPS/GLONASS receiver, accelerometer and gyroscope, Bluetooth and Wi-Fi network interfaces. The program contains means for tuning modules parameters using configuration files, control tools in the form of a system protocol, and also means of modules interaction for data exchange.

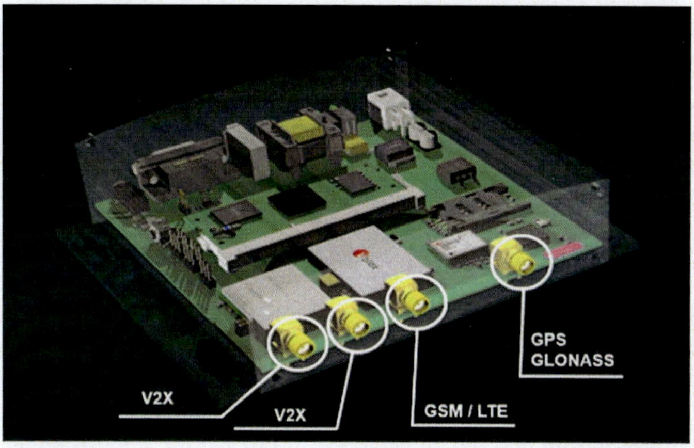

Fig. 2. Wireless communication channels of the developed communication device.

The developed solution is based on a principle of VANET (Vehicle Ad-hoc Networks) networks, i.e. specialized networks of "machine-to-machine" type [23]. This type of network is a case of MESH networks that have been actively developed in recent years in various application areas (other representatives are, for example, MANET - a network of mobile devices [24, 25] and FANET - a network of unmanned aerial vehicles [26, 27]).

The principal differences between MESH networks and traditional information and communication networks are: peer-to-peer structure; high variability of topology; the use of elements of "self-organization", and the intellectual nature of infrastructure algorithms at the network level. In relation to VANET networks and the scenarios of their use described above, it is possible to highlight additional specific features that were considered during the work on the project:

- transmission of short informational messages with high priority; implementation of large data files transfer functions, in particular multimedia;
- the concentration of network nodes along known routes (roads), a significant length of topology graph with a small cross section;
- the bilateral nature of nodes movement, the stability of their relative positions when moving in one direction;
- a large range of node density: from thousands of units/sq. km in conditions of urban congestion to single cars in conditions of country roads;
- hybrid nature, providing for cooperative interaction with conventional mobile networks through nodes located in the access zone of their base stations.

Protocol stack algorithms of network and transport levels should ensure the fastest delivery of datagram messages, considering the possible partial loss of network connectivity. For this, in particular, a dynamic routing algorithm, switching algorithms with priority queues and the possibility of deferred delivery at the time of connectivity restoration, dynamic load balancing algorithms were applied.

Particular attention should be paid to providing flexibility in the device configuration in order to increase ease of use and ensuring further expansion of the functions performed. For this, it is possible to optionally combine the device under development with equipment for various purposes, for example: car PC, personal mobile device, satellite positioning device, cellular communication device.

4 Application Areas

The main scenarios for application of the development imply conditions for a partial or complete absence of traditional radio communications, including broadcasting. Such conditions may develop on road sections located at a considerable distance from settlements or passing in rough terrain. The surroundings of some popular tourist destinations with mountainous terrain and underdeveloped infrastructure are also characterized by insufficient cellular coverage. In addition, such conditions can occur situationally due to weather disasters and other emergencies, as well as due to failure or overload of cellular stations equipment.

Among the most important scenarios for the use of developed means, the following can be distinguished:

- broadcast urgent alerts [28]: about emergencies, about worsening weather conditions, about the road situation, about carrying out repair work, about blocking traffic due to mass events, etc.;
- transmission of emergency alarms from motor vehicles to specialized services (Fig. 3): requesting medical assistance, requesting technical assistance, reporting of accidents (cooperation with the ERA-GLONASS/eCall system), robbery, etc.;
- automatic monitoring of the traffic situation in order to collect statistical data, quickly detect traffic jams and further inform both drivers and traffic control services about them (Fig. 4);
- collision avoidance between vehicles [29], including unmanned vehicles, by directly exchanging data with environmental sensors, positioning devices, or measuring the power of a radio signal (Fig. 5);
- telemetry information transfer from autonomous cars and special vehicles to data centers (Fig. 6);
- redundancy of automatic means of speed control to reduce an errors probability and ensure operation in difficult weather conditions.

The application is designed to connect to the telematics services of the information network of the car and serves to assist the driver when using the vehicle, as well as to prevent accidents and traffic accidents in conditions of limited visibility, heavy traffic, adverse weather events, complex topographic features of the area by visualizing the current situation on the road on the screen of a portable device (Fig. 7). The software facilitates the driver's orientation in a dense traffic flow, warns the driver in advance of potentially dangerous maneuvers by the nearest road users by means of information and sound alerts, and broadcasts emergency messages from special and road services (traffic police, ambulance, road transport weather services).

Fig. 3. Alarm signal transmission.

Fig. 4. Route control in unfamiliar territory.

Fig. 5. Vehicle collision avoidance.

Fig. 6. Warning on the passage of special vehicles.

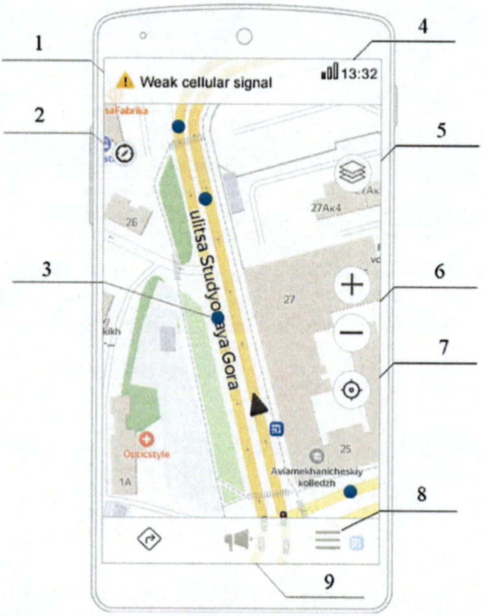

Fig. 7. The graphical interface of the software: 1 - messages and system alerts, 2 - reset the orientation of the map, 3 - vehicles, 4 - mobile network status, 5 - map display mode, 6 - map scaling, 7 - camera linking, 8 - settings, 9 - broadcast alerts.

5 Conclusion

Currently, the field of intelligent transport systems is a promising, rapidly developing and popular investment target. An assessment of international markets showed that leading countries are actively developing not only regulatory documents and standards, but also provide government support for various projects in the field of autonomous driving. State structures and major private companies are developing both in direction of assistance systems for piloting cars, and automating the interaction of vehicles and infrastructure. At the same time, almost all leading car manufacturers not only develop autonomous driving means, but also actively test various technologies. A number of companies have already implemented high-level automation solutions in the premium segment.

Despite the widespread popularity of intelligent systems, the use of such innovative technologies is often relevant only for modern cars of the latest generations, while the percentage of new cars among all vehicles in individual countries is quite low, which is a complication for the dissemination of technology. The world's leading markets focus on the use of technologies with the transfer of information mainly through cell towers and, in most cases, use expensive mobile Internet to transfer data to processing centers [30]. Full implementation of such systems can cause widespread dissatisfaction of drivers in view of the increase in vehicle operating cost. A number of legal restrictions, including the uncertain status of unmanned or combined control vehicles, introduces limitations on the implementation of developed technological devices.

Thus, integrated ITS have the greatest potential for rapid mass implementation due to lower cost and lack of a monthly fee. An analysis of the market for telematic operative communication systems for intelligent information systems under unstable cellular coverage showed rather high development prospects in the context of solving problems of autonomous driving and improving road safety.

References

1. Janušová, L., Čičmancová, S.: Improving safety of transportation by using intelligent transport systems. Procedia Eng. **134**, 14–22 (2016). https://doi.org/10.1016/j.proeng.2016.01.031
2. Baskar, L.D., et al.: Traffic control and intelligent vehicle highway systems: a survey. IET Intell. Transport Syst. **5**, 38–52 (2011). https://doi.org/10.1049/iet-its.2009.0001
3. Vasilchenkova, D.G.: Solution of problems to ensure road safety by means of intelligent transport systems. IOP Conf. Ser. EES **666**(2), 022003 (2021). https://doi.org/10.1088/1755-1315/666/2/022003
4. Zhang, T., et al.: Current trends in the development of intelligent unmanned autonomous systems. Front. Inf. Technol. Electron. Eng. **18**, 68–85 (2017). https://doi.org/10.1631/FITEE.1601650
5. Finn, A., Scheding, S.: Developments and challenges for autonomous unmanned vehicles: a compendium. Intell. Syst. Ref. Libr. **3**, 128–154 (2010)
6. Nonami, K., et al.: Autonomous control systems and vehicles – Intelligent unmanned systems. Intell. Syst. Control Autom. Sci. Eng. **65** (2013). https://doi.org/10.1007/978-4-431-54276-6
7. Alexander, P., Haley, D., Grant, A.: Cooperative intelligent transport systems: 5.9-GHz Field Trials. Proc. IEEE **99**, 1213–1225 (2011). https://doi.org/10.1109/JPROC.2011.2105230
8. Sjöberg, K., et al.: Cooperative intelligent transport systems in Europe: current deployment status and outlook. IEEE Veh. Technol. Mag. **12**, 89–97 (2017). https://doi.org/10.1109/MVT.2017.2670018
9. Festag, A.: Cooperative intelligent transport systems standards in Europe. IEEE Commun. Mag. **52**, 166–172 (2014). https://doi.org/10.1109/MCOM.2014.6979970
10. Chandra, Y.R.V.S., Shiva Harun, M., Reshma, T.: Intelligent transport system. Int. J. Civ. Eng. Technol. **8**(4), 2230–2237 (2017)
11. Qureshi, K.N., Abdullah, A.H.: A survey on intelligent transportation systems. Middle East J. Sci. Res. **15**(5), 629–642 (2013). https://doi.org/10.5829/idosi.mejsr.2013.15.5.11215
12. Perallos, A., et al.: Intelligent Transport Systems: Technologies and Applications. John Wiley & Sons, New York (2015)

13. Singh, B., Gupta, A.: Recent trends in intelligent transportation systems: a review. J. Transp. Lit. **9**, 30–34 (2015). https://doi.org/10.1590/2238-1031.jtl.v9n2a6
14. Grant-Muller, S., Usher, M.: Intelligent transport systems: the propensity for environmental and economic benefits. Technol. Forecast Soc. Change **82**, 149–166 (2014). https://doi.org/10.1016/j.techfore.2013.06.010
15. Liu, C., et al.: Opportunities and challenges of vehicle-to-home, vehicle-to-vehicle and vehicle-to-grid technologies. Proc. IEEE **101**(11), 2409–2427 (2013). https://doi.org/10.1109/JPROC.2013.2271951
16. Wang, J., et al.: A survey of vehicle to everything (V2X) testing. Sensors **19**(2), 334 (2019). https://doi.org/10.3390/s19020334
17. Schmidt, R.K., Leinmüller, T., Böddeker, B.: V2X communication. In: Proceedings of 17th Aachener Kolloquium (2008)
18. Ma, X., Chen, X., Refai, H.H.: Performance and reliability of DSRC vehicular safety communication: a formal analysis. EURASIP J. Wirel. Commun. Netw. **2009**(1), 1–13 (2009). https://doi.org/10.1155/2009/969164
19. Yin, J., et al.: Performance evaluation of safety applications over DSRC vehicular ad hoc networks. In: VANET – Proceedings of the First ACM International Workshop on Vehicular Ad Hoc Networks, pp. 1–9 (2004). https://doi.org/10.1145/1023875.1023877
20. Xu, Q., et al.: Vehicle-to-vehicle safety messaging in DSRC. In: VANET – Proceedings of the First ACM International Workshop on Vehicular Ad Hoc Networks, pp. 19–28 (2004). https://doi.org/10.1145/1023875.1023879
21. Bai, F., Krishnan, H.: Reliability analysis of DSRC wireless communication for vehicle safety applications. In: IEEE Conference on Intelligent Transportation Systems, pp. 355–362 (2006). https://doi.org/10.1109/ITSC.2006.1706767
22. Kochuev, D., Chkalov, R., Chernikov, A.: Problems of operative communication means for intelligent transport information systems under unstable cellular coverage. In: Proceedings 2019 International Russian Automation Conference, pp. 1–5 (2019). https://doi.org/10.1109/RUSAUTOCON.2019.8867781
23. Zeadally, S., et al.: Vehicular ad hoc networks (VANETS): status, results and challenges. Telecommun. Syst. **50**(4), 217–241 (2012). https://doi.org/10.1007/s11235-010-9400-5
24. Conti, M., Giordano, S.: Mobile ad hoc networking: Milestones, challenges and new research directions. IEEE Commun. Mag. **52**(1), 85–96 (2014). https://doi.org/10.1109/MCOM.2014.6710069
25. Carlsson Redell, P., Redell, P.C.: Foucault, Art, and Radical Theology. Routledge, London (2018)
26. Chriki, A., et al.: FANET: communication, mobility models and security issues. Comput. Networks **163**, 106877 (2019). https://doi.org/10.1016/j.comnet.2019.106877
27. Bekmezci, I., Sahingoz, O.K., Temel, S.: Flying ad-hoc networks (FANETs): a survey. Ad Hoc Netw. **11**(3), 1254–1270 (2013). https://doi.org/10.1016/j.adhoc.2012.12.004
28. Vasilchenkova, D.: Message relay between short-range radio devices in traffic flow. In: Proceedings 2020 International Russian Automation Conference, pp. 443–447 (2020). https://doi.org/10.1109/RusAutoCon49822.2020.9208117
29. Vasilchenkova, D.: Computational models of vehicle collision prediction by means of DSRC technology. In: Proceedings 2020 International Russian Automation Conference, pp. 170–174 (2020). https://doi.org/10.1109/RusAutoCon49822.2020.9208229
30. Abboud, K., Omar, H.A., Zhuang, W.: Interworking of DSRC and cellular network technologies for V2X communications: a survey. IEEE Trans. Veh. Technol. **65**(12), 9457–9470 (2016). https://doi.org/10.1109/TVT.2016.2591558

Using Decision Trees to Determine the Important Characteristics of Ice Hockey Players

M. Gliznitsa(✉) and N. Silkina

South Ural State University, 76, Lenin Avenue, Chelyabinsk 454080, Russia

Abstract. This study uses the machine learning algorithm of decision tree classification to determine the features most useful for predicting overall productivity of an ice hockey player in an NHL (National Hockey League) season. While most existing studies use detailed data on a small number of players, this analysis is performed using a publicly accessible dataset consisting of data from 44 NHL seasons. The simplicity of the used algorithm allows to directly determine the most important features used in prediction. The results indicate that the number of assists per game has a significantly higher impact on the player's productivity during the season and that the player's experience and body weight are useful factors in predicting his productivity level. Our analysis shows that limited publicly accessible data contains features that can be useful in predicting a player's productivity with average accuracy. However, more detailed data, which is usually only collected during specific trials, would be necessary to provide more practical application.

Keywords: Data analysis · Machine learning · NHL · Prediction system · Competitive sports · Decision tree · Classification

1 Introduction

Predictive modeling and machine learning have received notable attention in the academic literature. In recent years, machine learning algorithms have been improving from relatively simple models to powerful deep neural networks.

However, simpler machine learning models are still being used in various tasks today. Different classification and regression algorithms have found numerous applications in areas such as healthcare, where support vector machines (SVM) were used for tasks related to drug classification [1], and ensembles models, like boosting, found an application in predicting different drug properties [2].

Recently, deep neural network models have gained a lot of popularity. For example, they were proposed for tasks such as advertisement click-through rate prediction, where they showed good performance [3].

One of the topics that have always attracted great interest is sports, and many ways were found to apply machine learning algorithms to it.

A. A. Radionov and V. R. Gasiyarov (Eds.): RusAutoCon 2021, LNEE 857, pp. 359–369, 2022.
https://doi.org/10.1007/978-3-030-94202-1_34

2 Literature Review

Association football, also known as soccer, is the most popular sport in the world. This fact makes it an extremely popular target for building various predictive models. One of the most obvious goals of modeling is to predict match results, and many different methods were used to reach that goal. For example, polynomial classification found an application in predicting football match results [4], as did random forest, which was used to predict scores of football matches in international tournaments [5]. Gradient boosting was also useful when compared to different models in predicting match results [6].

Similar research has been done for other sports, not necessarily involving common machine learning models. For example, Boulier-Stekler methodology was successfully used to predict the winners in NCAA basketball games [7], while for American football, a wide range of machine learning models was used with the goal of comparing their performance [8].

In a more unusual approach, a Hidden Markov Model (HMM) methodology was used to recognize specific behaviors of a swimmer in a swimming pool, and it was able to achieve high accuracy [9].

Simpler machine learning methods have also been applied to human activity classification, where some models, namely SVM and logistic regression, showed good results in predicting type of human activity and their speed using data from wearable sensors [10]. A similar, but more complex research focused on using support vector regression (SVR) for gait analysis, where they were able to extract estimates of fundamental gait parameters, such as stride length and velocity [11].

Experts performing athlete selection have also found machine learning techniques useful in their decision-making processes, where a model was able to provide them with supporting information to assist with their decision [12]. The same authors used an unsupervised machine learning technique together with statistical approaches to acquire additional information from analyzing multiple events [13].

In baseball, machine learning was used to predict the next-season injury risk and injury location for players, where more advanced machine learning models were compared to logistic regression and usually outperformed it. In particular, these models were capable of predicting injuries, including the anatomic region of injury, for position players, but not for pitchers [14].

One other popular sport for predictions is hockey, both on ice and in a field. Prediction are often based on statistical models, not involving machine learning. In one study, various statistical models were used to predict the results of an ice hockey match [15]. Many other interesting applications of modeling have been found, including predicting the performance of individual players based on the data on their physical capabilities. Examples of such studies exist in both field hockey and ice hockey. In field hockey, a model built on various data from 14 and 15-year-old female players was used to distinguish between successful and less successful of them [16]. In a similar study for ice hockey, off-ice testing was used to acquire data on women's ice hockey players, which was then used to select the strongest predictors of player's game performance. For example, 40-yd dash time of a player was found to be useful to predict their skating speed [17].

A particularly interesting application of modeling is to use the results to identify the most important characteristics that drive the outcome of a specific situation that can occur during the match. In particular, speed of the players was useful in predicting the outcome of one-vs-one player interactions in women's ice hockey. In that particular study, decision trees were the model used for prediction [18].

Most existing studies focus on collecting a high number of parameters from a limited number of participants. For this purpose, a group of players can be recruited to complete skill tests, which results can then be used for research. For example, a number of players can be recruited for a specific purpose of completing tests, the results of which can then be used to identify their talent. In junior Australian football, this approach was used in classifying players into elite and subelite groups [19].

One of the possible reasons for this approach is the lack of detailed statistics. While datasets that provide data on hundreds and thousands of players exist, they rarely provide detailed information on physical characteristics of the players, instead focusing on the scores acquired in the game itself. However, while limited, the data provided by these datasets might still be useful in determining some of the important features for predicting the performance of the players.

3 Setting of the Problem

This study aims to determine the characteristics that affect the performance of an individual player in an NHL (National Hockey League) season. For this purpose, publicly accessible datasets that provide data on a high number of players can be used.

Most of the features publicly accessible datasets contain are the ones that provide information on the player's performance in matches, for example, individual score. However, some general individual characteristics of the players can also be present in these datasets. In this study, we hypothesize that the data provided in publicly accessible datasets can be detailed enough to determine some of the important parameters that affect the player's performance in matches.

One of the popular algorithms for data analysis is the machine learning algorithm known as decision tree classification. It is a type of machine learning algorithm, which builds a tree-like model of decisions to make predictions in regards to the target value. The tree consists of internal nodes, or conditions, which split the tree into branches, or edges. The end of every branch is called a leaf.

One of the most important features of decision trees is their simplicity, particularly the fact that actual feature used to make the split and its impact on the prediction accuracy can be easily seen from the tree visualization. One consequence of this is the fact that a computer is not necessary to use a trained decision tree model. Another consequence is that the decision tree can be used as a tool for determining the most important features of a dataset or a part of a dataset.

This study is an attempt to advance knowledge on the significance of individual player's characteristics for their performance in games, while trying to avoid the limitations brought by the limited number of players participating in a survey.

4 Materials and Methods

Analysis was performed using the publicly available NHL players' statistics, compiled into a dataset [20]. The dataset provides various data on NHL players and spans from season'75 –'76 to season'19–'20. There are 40 features provided for every player, including individual characteristics, position on the field, and scores acquired in games. Features used in the analysis are listed in Table 1.

Every player who appeared in more than one season or more than one team has multiple entries. Since these duplicate entries are unlikely to be useful for analysis, they were removed with preserving one entry per player. The preserved entry was the one with most games played, which is likely to be the most informative.

The data present in the dataset was then used to create additional features: GoalsPG and AssistsPG, which correspond to the number of goals and assists divided by the number of games played.

Players were then split into four groups based on their position on the field. The groups and the percentage of players in them can be seen in a pie chart form on Fig. 1. Total number of players remaining in the dataset is 2013.

Aside from the listed four groups, three more were created: right wing and left wing combined (as these positions are likely to have similarities), right wing, left wing and center combined (as all non-defense groups), and all groups combined.

The productivity, or total time on ice divided by points, was used to split the players into two groups: players with above and below average productivity. These groups were then used as the target variable for the decision tree.

Table 1. Features used in the analysis.

Feature	Meaning
Games_played	Number of games the player participated in during the season
Goals	Number of goals scored by the player
Assists	Number of assists earned by the player
Production	Productivity of the player, or the time on ice divided by points scored
Position	Position of the player on the field (center, left wing, right wing or defense)
Height	Height of the player
Weight	Weight of the player
Body_mass_index	Body mass index of the player
Age	Age of the player
Experience	Number of seasons passed since the player's first appearance in NHL

To prevent overfitting, the data for every group was further divided into two sections: first one for training the model and the second one to test the trained model. If the error on the training set is much lower than the one on the test set, the model suffers from overfitting and needs adjustment.

To build the decision tree classifier, scikit-learn was used. Scikit-learn is a Python module integrating a wide range of machine learning algorithms for medium-scale supervised and unsupervised problems [21]. Supported algorithms contain the decision tree classifier, which was used to build the models in our research.

Finally, a common decision tree classifier for all four groups and separate decision tree classifiers for every group of players were built. The depth and other parameters of decision trees were adjusted in the way most likely to avoid overfitting based on the error percentages on the training set and the test set. Entropy was selected as the function used to measure the quality of the split.

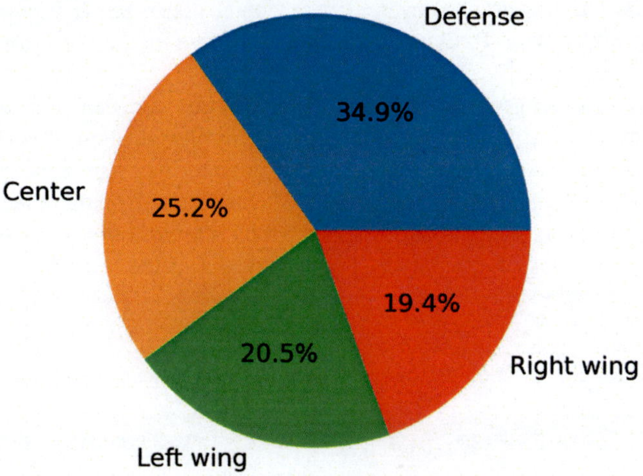

Fig. 1. Percentage of players in all positions

The fact that the decision tree builds a very simple model, which clearly shows the features used for classification, allows to easily determine both the most important feature found by the algorithm and the actual importance of the feature for classification.

In order to validate the results, it was important to score the tree's performance, which would not only allow preventing overfitting, but also determining whether the built model is useful in predicting the player's performance. Mean accuracy on the training and test sets was used as the score.

In the first test, the models were built on features, which included goals per game and assists per game. They were used to determine the most prevalent way of raising productivity relative to the player's position on the field. These values are directly linked to productivity, so the models built are likely to have high accuracy. Then, these features were deleted from the used dataset, and the models were built using the remaining individual player's attributes, which were no longer directly linked to their productivity.

5 Results

The aim of the study was to determine the most useful features for predicting productivity. First, the models were built to determine whether goals per game or assists per game were more important for the task. The accuracy of the built models and the most important feature determined are listed in Table 2.

As expected, built models have high accuracy (usually over 80%) on both training and test sets. For most groups the most important feature was determined to be assists per game. For example, one of the groups where the first decision classified the players with higher amounts of assists per game as having higher productivity was the defense.

Still, the players with the highest amounts of assists per game were classified (on the second level of the decision tree) into a group with lower productivity. The decision tree for the defense group is shown on Fig. 2.

The rest of the trees showed a common trend: players with high amounts of assists per game had lower productivity; however, players with no assists per game were still likely to have lower productivity. To illustrate this point, the tree for the left wing productivity is shown on Fig. 3.

Then, the assists per game and goals per game features were deleted from the dataset and the models were built on the remaining data. This change allowed the built models to contain only the characteristics of the player, such as their age and experience, but not the statistical information on their scores. The models built on these features will be more useful for the original goal of this study, which is to determine the most important personal characteristics of the player. The accuracy and the most important feature determined are listed in Table 3.

Table 2. Results of building decision trees on goals per game and assists per game.

Group	Size (both sets)	Train set accuracy	Test set accuracy	Most important feature
All	2013	0.80	0.76	Goals per game
Center	507	0.80	0.82	Assists per game
Defense	703	0.90	0.89	Assists per game
Right wing	391	0.89	0.93	Assists per game
Left wing	412	0.81	0.85	Assists per game
Both wings	803	0.88	0.85	Assists per game
All non-defense	1310	0.82	0.78	Assists per game

Built models have significantly lower accuracy than the ones built in the previous test. For most groups, experience was chosen as the most important attribute in classification.

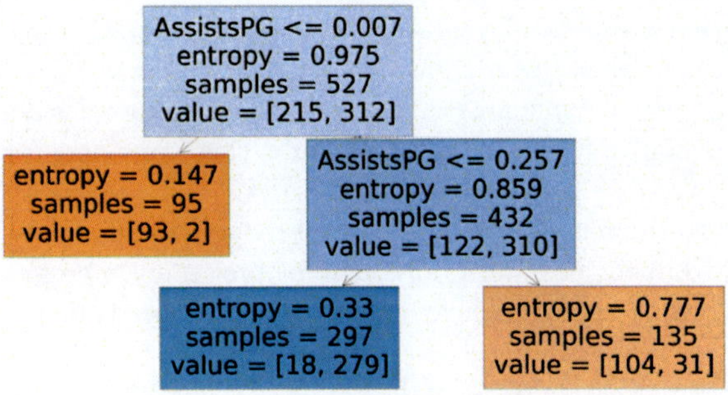

Fig. 2. Decision tree for the defense group.

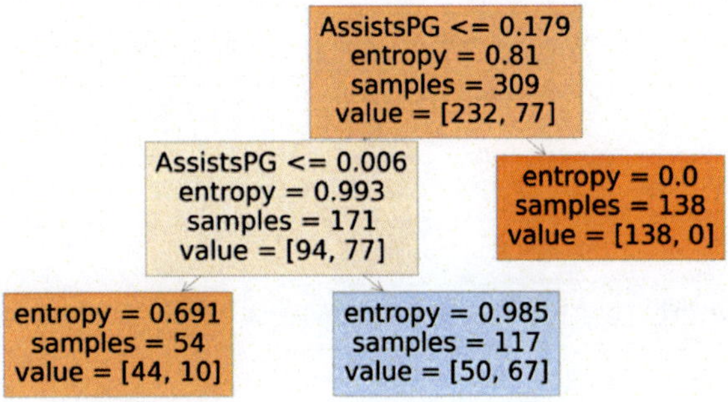

Fig. 3. Decision tree for the left wing group.

Table 3. Results of building decision trees on individual player characteristics

Group	Size (both sets)	Train set accuracy	Test set accuracy	Most important feature
All	2013	0.64	0.59	Weight
Center	507	0.76	0.72	Age
Defense	703	0.69	0.70	Experience
Right wing	391	0.80	0.71	Weight
Left wing	412	0.72	0.81	Experience
Both wings	803	0.75	0.78	Experience
All non-defense	1310	0.75	0.77	Experience

The important fact is that for most groups, players with the highest and the lowest experience were classified as less productive, while players with closer to average experience were classified as more productive. The tree for the left wing group is a good illustration for this trend, because it classifies both the players with a very high (8 and more years) experience and the players with low (1 year) experience into less productive groups. The aforementioned tree is shown on Fig. 4.

The number of nodes containing a certain feature among all trees can be used to demonstrate the overall significance of the feature. These numbers for all features are listed in Table 4.

The feature most often used to determine the productivity group of the player seems to be experience, while the next most important feature seems to be weight. Height, age and body mass index all appear to have less importance in prediction.

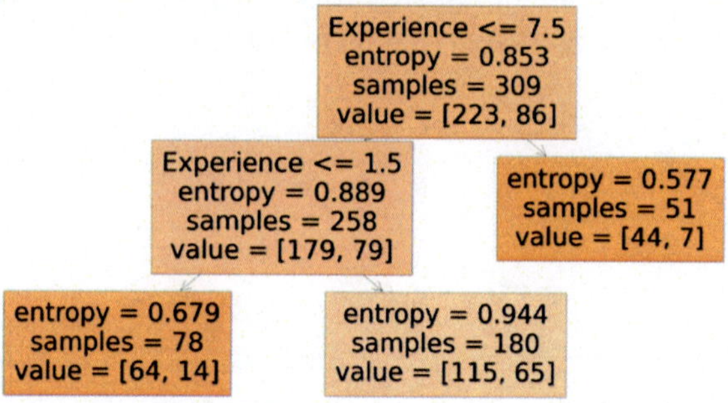

Fig. 4. Decision tree for the left wing group (for individual characteristics).

Table 4. The number of nodes using a specific feature for classification.

Feature	Number of nodes
Height	2
Weight	5
Body mass index	1
Age	2
Experience	8

6 Discussion

The study aimed at determining parameters that affect the individual player's productivity using publicly accessible data. Most of the publicly accessible data is focused on the player's scores, time on ice and similar features, while the data on player's individual characteristics is scarce. For this reason, the first built models contained predictions based on the data on player's scores.

The ways to achieve points for a player in hockey are goals and assists. The results of the study indicate that assists per game were noticeably more important for the player's overall productivity than goals per game regardless of the player's position on the field. Of the parameters which had no direct connection to productivity, experience, which is defined as the number of seasons since the player's first appearance, appeared the most in classification.

In line with the hypothesis, some of the parameters appear to be more important than others for determining the player's productivity. Built models indicate that goals per game are not as useful for determining the player's overall productivity as assists per game, which allow classifying the player into the more productive or less productive group with an accuracy nearing 80%. Experience of the player and the player's weight also seem to be important features, although models built on the group of parameters which cannot be directly linked to productivity have a noticeably lower accuracy of 70–75%.

As for the values of parameters, the highest productivity is shown to be linked to more moderate values of both assists per game and experience, while extreme values (both the highest and lowest) are associated with lower productivity.

While previous research has focused on collecting data from a limited number of players, which allowed them to collect a high number of features, these results demonstrate that the limited number of features present in publicly accessible databases with high amounts of players can also be used to predict the player's productivity.

Due to the lack of data on personal attributes of the players, only the general physical parameters like weight or height could be used in the models. For that reason, the built models are unlikely to be particularly useful in determining the expected productivity of a new player based on the results they could show in a physical test.

Future studies on more detailed data could produce results with more practical applications; for example, establish links between certain physical test results and productivity of the player.

7 Conclusion

The aim of the present research was to determine the most important parameters that affect the player's performance in an ice hockey game using the publicly accessible datasets that include information on a large quantity of players. These results of this study indicate that some of the present parameters can actually be used to predict whether a certain player had lower or higher productivity than average during the season. A major limitation of this study is the fact that publicly accessible datasets for the most part consist of statistical data, which is directly linked to productivity, and only include limited information on the player's individual physical attributes. Collection of more detailed data on the players will help to produce results with more practical application.

References

1. Wale, N., Watson, I.A., Karypis, G.: Comparison of descriptor spaces for chemical compound retrieval and classification. Knowl. Inf. Syst. **14**(3), 347–375 (2008). https://doi.org/10.1007/s10115-007-0103-5
2. Arodź, T., Yuen, D.A., Dudek, A.Z.: Ensemble of linear models for predicting drug properties. J. Chem. Inf. Model. **46**(1), 416–423 (2006). https://doi.org/10.1021/ci050375+

3. Zhang, S., Liu, Z., Xiao, W.: A hierarchical extreme learning machine algorithm for advertisement click-through rate prediction. IEEE Access **6**, 50641–50647 (2018). https://doi.org/10.1109/ACCESS.2018.2868998
4. Martins, R.G., et al.: Exploring polynomial classifier to predict match results in football championships. Expert Syst. Appl. **83**, 79–93 (2017). https://doi.org/10.1016/j.eswa.2017.04.040
5. Groll, A., Ley, C., Schauberger, G., Van Eetvelde, H.: A hybrid random forest to predict soccer matches in international tournaments. J. Quant. Anal. Sport. **15**(4), 271–287 (2019). https://doi.org/10.1515/jqas-2018-0060
6. Hubáček, O., Šourek, G., Železný, F.: Learning to predict soccer results from relational data with gradient boosted trees. Mach. Learn. **108**(1), 29–47 (2018). https://doi.org/10.1007/s10994-018-5704-6
7. Stekler, H.O., Klein, A.: Predicting the outcomes of NCAA basketball championship games. J. Quant. Anal. Sport. **8**(2) (2012). https://doi.org/10.1515/1559-0410.1373
8. Hsu, Y.C.: Using machine learning and candlestick patterns to predict the outcomes of American football games. Appl. Sci. **10**(13), 4484 (2020). https://doi.org/10.3390/app10134484
9. Chen, H.L., Tsai, M.J., Chan, C.C.: A Hidden Markov Model-based approach for recognizing swimmer's behaviors in swimming pool (2010). https://doi.org/10.1109/ICMLC.2010.5580797
10. Mannini, A., Sabatini, A.M.: Automatic machine learning methods for analysis of signals from accelerometers: classification of human activity and walking–running speed estimation. Gait Posture **33**, S24 (2011). https://doi.org/10.1016/j.gaitpost.2010.10.031
11. Zhang, H., Guo, Y., Zanotto, D.: Accurate ambulatory gait analysis in walking and running using machine learning models. IEEE Trans. Neural Syst. Rehabil. Eng. **28**(1), 191–202 (2020). https://doi.org/10.1109/TNSRE.2019.2958679
12. Ofoghi, B., Zeleznikow, J., Macmahon, C., Dwyer, D.: Supporting athlete selection and strategic planning in track cycling omnium: a statistical and machine learning approach. Inf. Sci. **233**, 200–213 (2013). https://doi.org/10.1016/j.ins.2012.12.050
13. Ofoghi, B., Zeleznikow, J., Dwyer, D., Macmahon, C.: Modelling and analysing track cycling Omnium performances using statistical and machine learning techniques. J. Sports Sci. **31**(9), 954–962 (2013). https://doi.org/10.1080/02640414.2012.757344
14. Karnuta, J.M., et al.: Machine learning outperforms regression analysis to predict next-season major league baseball player injuries: epidemiology and validation of 13,982 player-years from performance and injury profile trends, 2000–2017. Orthop. J. Sport. Med. **8**(11), 232196 (2020). https://doi.org/10.1177/2325967120963046
15. Marek, P., Šedivá, B., Ťoupal, T.: Modeling and prediction of ice hockey match results. J. Quant. Anal. Sport. **10**(3), 357–365 (2014). https://doi.org/10.1515/jqas-2013-0129
16. Nieuwenhuis, C.F., Spamer, E.J., Van Rossum, J.H.A.: Prediction function for identifying talent in 14- to 15-year-old female field hockey players. High Abil. Stud. **13**(1), 21–33 (2002). https://doi.org/10.1080/13598130220132280
17. Bracko, M.R., Georgem, J.D.: Prediction of ice skating performance with off-ice testing in women's ice hockey players. J. Strength Cond. Res. **15**(1), 116–120 (2001). https://doi.org/10.1519/1533-4287(2001)015%3c0116:POISPW%3e2.0.CO;2
18. Morgan, S., Williams, M.D., Barnes, C.: Applying decision tree induction for identification of important attributes in one-versus-one player interactions: a hockey exemplar. J. Sports Sci. **31**(10), 1031–1037 (2013). https://doi.org/10.1080/02640414.2013.770906
19. Woods, T.E.C., Raynor, J.A., Bruce, L., McDonald, Z.: The use of skill tests to predict status in junior Australian football. J. Sports Sci. **33**(11), 1132–1140 (2015). https://doi.org/10.1080/02640414.2014.986501

20. Kaggle: NHL Players Statistics (2020). https://www.kaggle.com, https://www.kaggle.com/alexbenzik/nhl-players-statistics/metadata. Accessed 14 Dec 2020
21. Pedregosa, F., et al.: Scikit-learn: machine learning in Python. J. Mach. Learn. Res. **12**, 2825–2830 (2011)

Optimizing the Daily Energy Consumption of an Enterprise

O. Yu. Maryasin and A. I. Lukashov(✉)

Yaroslavl State Technical University, 88, Moskovsky Prospect, Yaroslavl 150023, Russia
lukashovai@ystu.ru

Abstract. The paper considers the problem of determining the optimal daily energy consumption of an enterprise. By solving this problem, a reduction of electricity costs can be achieved by reducing energy consumption during periods of high prices, increasing energy consumption during hours when electricity prices are lower, as well as by cardinally reducing energy consumption during peak load hours. To determine the optimal daily energy consumption, three different optimization methods are proposed – linear programming method, genetic algorithm, and particle swarm optimization algorithm. The results are provided regarding the solution of the problem using the specified methods and specific input data, as well as data on actual and forecasted electricity prices. The study includes an estimation of savings obtained using various optimization methods. The effect of forecasted prices and price volatility on the size of savings is investigated as well. The methods for determining the optimal daily energy consumption discussed in this study can be useful for enterprises that pay for electricity at market prices.

Keywords: Energy optimization · Optimal consumption profile · Forecasting · Linear programming · Genetic algorithm · Particle swarm optimization

1 Introduction

One of the ways to improve energy efficiency and reduce electricity costs is energy consumption optimization. This topic is reflected in a large number of studies in the world scientific literature [1–3]. All energy optimization problems can be divided into optimal energy management problems and optimal energy scheduling problems [4]. The first type corresponds to a closed-loop system, where the values of optimized parameters are determined from the current state of the system. An example of this problem type is a problem of optimal control of the microclimate and energy consumption of a building [5]. One of the most popular methods for optimal energy consumption control is model predictive control (MPC) [5, 6].

Optimal energy scheduling problems correspond to the open-loop system when the values of the optimized parameters are determined in advance for the entire scheduling period based on the prediction of the most important system indicators. Such problems are common in the demand side management (DSM) systems of smart grids (SG) [7]. In

A. A. Radionov and V. R. Gasiyarov (Eds.): RusAutoCon 2021, LNEE 857, pp. 370–382, 2022.
https://doi.org/10.1007/978-3-030-94202-1_35

particular, these include the recently popular problems of energy consumption scheduling for household appliances [8, 9].

This study considers the problem of determining the optimal daily energy consumption profile of an enterprise, which also relates to the problems of optimal energy consumption scheduling. In the context of the Russian Federation, this problem is relevant for companies being energy consumers of 3, 4, 5 and 6 price categories. These enterprises can save on electricity costs by controlling energy consumption with the consideration of:

- changes in hourly market electricity prices;
- the power payment during peak load hours.

In the first case, it is possible to reduce energy consumption during periods of high prices and increasing the load during hours when the electricity prices become lower. In the second case, the energy consumption shall be cardinally decreased during the peak load hours. At the same time, in all cases, it is necessary to take into account the limitations imposed on the minimum and maximum energy consumption associated with the enterprise's main area of competence. These can include limitations associated with the production equipment operation, production safety, activity of personnel, etc. Therefore, the problem of determining the optimal energy consumption profile of an enterprise can be defined taking into account peak load hours and other limitations on energy consumption. Reducing energy consumption during high price hours and peak load hours is possible by changing the enterprise's operation schedule and optimizing the equipment operation modes, as well as by utilizing own energy sources including renewable (solar panels, wind generators) or other local generators and energy storage devices [10].

The retail energy market of the Russian Federation regions related to the price zones has a particular aspect – the data on market electricity prices and the peak load hours of the current month becomes available only after the tenth day of the next month. Such input data specifics determine the relevance of predicting market electricity prices and peak load hours a month ahead. Forecasting allows solving problems of optimal energy scheduling and obtaining significant economic effects even with a relatively low accuracy of forecasting results.

The main issues that are considered in this study are the follows:

- How can the optimal energy consumption profile of an enterprise be determined taking into account the peak load hours and other limitations on energy consumption?
- What size of savings can be obtained by optimal energy scheduling?
- How may the prediction accuracy for market electricity prices and peak load hours affect the optimization results and savings?
- How can the price volatility affect the savings resulting from optimization?

Furthermore, the paper shows the methods that allow solving the problem of determining the optimal energy consumption profile of an enterprise. The results are provided regarding the solution of this problem for the specific input data, data on the energy consumption, as well as data on actual and forecasted electricity prices.

2 Optimization Problem Definition

Let us suppose that the enterprise continuously consumes electricity during the day. Let us take x_i as the energy consumption by the enterprise per hour i, $i = 1,\ldots,24$, and c_i retail electricity price per hour i. Hourly market electricity prices in the current month can be determined by forecasting [11]. Then the optimality criterion for the problem of optimizing the daily energy consumption profile can be described as follows:

$$f(x) = c_1x_1 + c_2x_2 + \ldots + c_{24}x_{24} \quad (1)$$

Let us denote the daily energy consumption required for the enterprise operation as X_d. Then the following condition is fair:

$$x_1 + x_2 + \ldots + x_{24} = X_d. \quad (2)$$

Let us define lower x_{il} and upper x_{iu} limits of hourly energy consumption:

$$x_{il} \leq x_i \leq x_{iu}, i = 1, \ldots, 24, i \neq h_{pp}, \quad (3)$$

where h_{pp} – peak load hour. These limits are determined by the enterprise's ability to reduce and increase its energy consumption without any significant consequences for the enterprise's operation. In addition, the energy consumption shall be reduced to the lowest possible value of X_{pph} at the peak load hour h_{pp}:

$$x_i = X_{pph}, i = h_{pp}. \quad (4)$$

Thus, the problem of optimizing the daily energy consumption profile of an enterprise consists of minimizing the criterion (1) if the conditions (2)–(4) are met. The advantages of such problem definition are its simplicity and versatility, while disadvantages – a relatively large problem order and lack of conditions for connecting energy consumption with the main indicators of the enterprise's economic activity.

3 Optimization Methods

We will use three different optimization methods to solve the problem of determining the optimal energy consumption profile. Since the problem (1)–(4) is linear in variable x_i, linear programming (LP) is a logical choice for the main method of its solution. An example of LP application for solving the problem of energy consumption optimization is described in [12].

The genetic algorithm (GA) will be used as one of the alternative solution methods. GA is one of the most frequently used algorithms for solving energy scheduling problems. In [3] specifically, GA is used together with an artificial neural network to optimize the energy consumption of an office building. In [9], GA is used to schedule the energy consumption of all automatic household appliances in a residential building in order to minimize electricity costs. GA is widely used in DSM systems of SG [13, 14]. The main GA parameters are population size N_{ps}, mutation probability P_m, crossover probability

P_c, a ratio of elite individuals R_e, a ratio of parents in population R_p, selection type, mutation type, crossover type.

Another popular method for solving energy scheduling problems is the particle swarm optimization (PSO) algorithm. The utilization of PSO in a DSM system to schedule energy consumption of household appliances is considered in [15]. Other PSO applications for energy optimization are described in [10, 16]. The main PSO parameters are number of particles N_p, individual memory coefficient k_{im}, collective memory coefficient k_{cm}, inertia coefficient k_i.

4 Daily Energy Consumption Profile Optimization

4.1 Input Data

The input data to solve the energy consumption profile optimization problem are data on energy consumption by one of the enterprises of the Yaroslavl region of the Russian Federation. This enterprise is the energy consumer of the third price category. For definitiveness, we will use data on energy consumption and electricity prices for February 2020. The enterprise's energy consumption schedule for February 2020 is shown in Fig. 1. Let us take 06.02.2020 as a specific date to define the optimal energy consumption profile of the enterprise.

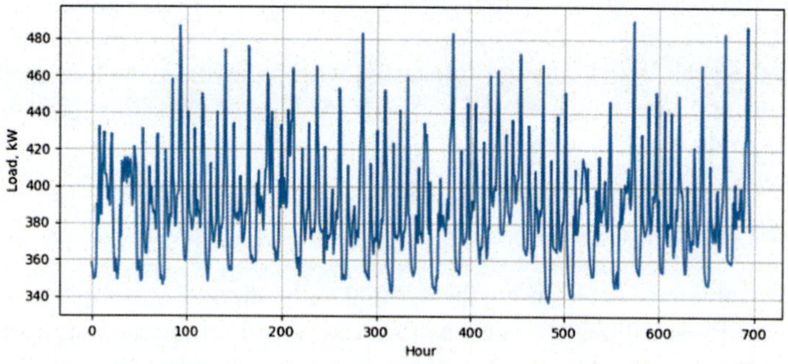

Fig. 1. Enterprise's energy consumption schedule for February 2020.

Data on market electricity prices in the Yaroslavl region are published on the website of the energy provider, TNS Energo Yaroslavl [17]. As mentioned earlier, data on actual electricity prices for the current month become available only after the tenth day of the next month. This requires using forecasted electricity prices for the energy consumption profile optimization. A month-long forecast can be obtained using the application previously developed by the authors for forecasting market electricity prices described in [18]. The hourly market electricity prices forecasting results for February 6, 2020 are shown in Fig. 2. The forecasting accuracy amounted to 1.7 MAPE (Mean Absolute Percentage Error). The standard deviation for actual and forecasted electricity prices was approximately 0.2. Figure 2 shows that the electricity price is significantly lower at night and in the morning hours while reaching its maximum at 12–16 o'clock.

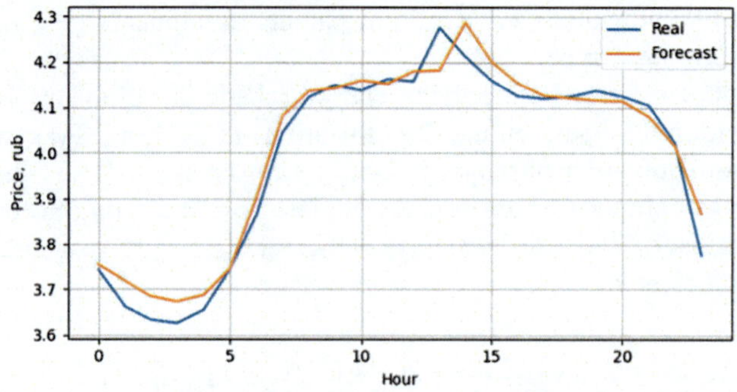

Fig. 2. Actual and forecasted prices for February 6, 2020.

Further, two optimal energy consumption profiles will be considered for each optimization alternative. One for actual price values, and the other – for forecasted prices. This allows estimating the effect of forecasting uncertainty on the optimal energy consumption profile form, as well as on the electricity costs.

To solve the problem (1)–(4), we also need to know the peak load hour forecast for the desired date. This forecast can be obtained using the digital electricity consumer models developed by the authors in [19]. As shown in [19], the forecasting accuracy for the peak load hours is about 50% of cases. The actual data on the peak load hours are available on the Trade System Administrator website [20]. For February 6, 2020, the forecast peak load hour coincided with the real value equal to 10 a.m. Consideration of cases when the peak load hours are forecasted with an error, as well as methods that address the uncertainty associated with the peak load hour forecasting, goes beyond the scope of this study.

4.2 Linear Programming

To solve the problem of determining the optimal daily energy consumption profile (1)–(4) using the LP algorithm, the authors have developed an application using Python and the linprog function of the scipy library. To use this function, data structures were prepared corresponding to Eqs. (1)–(4). For this, the following values of limits were adopted: $X_d = 9275.8$ kW, $x_{il} = 300$ kW, $x_{iu} = 400$ kW, $i = 1,\ldots,24$, $X_{pph} = 200$ kW, $h_{pp} = 10$. The optimal daily energy consumption profile for February 6, 2020 obtained from the solution of the LP problem for actual prices is shown in Fig. 3, and for the forecasted prices – in Fig. 4.

Analysis of Fig. 3 and 4 shows that the optimal energy consumption profile obtained from solving the LP problem significantly differs from the real one. It mainly includes the values of the lower and upper load limits and is a rather radical solution from this point of view. Such an energy consumption profile requires significant restructuring and mobilization of the enterprise's production and energy resources and therefore may be difficult to implement. At the same time, the difference in actual and forecast price values may have a noticeable effect on the optimal energy consumption profile form.

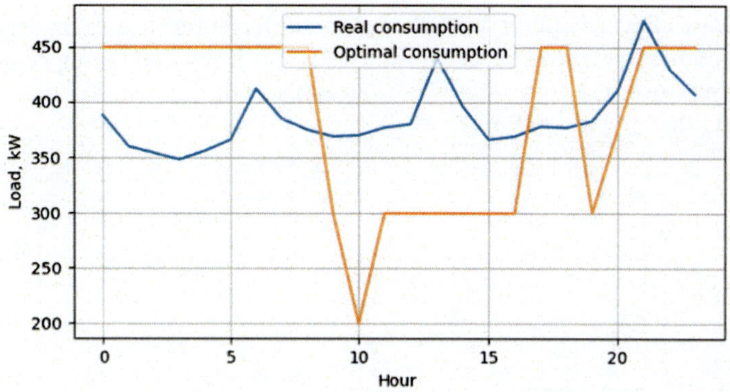

Fig. 3. Optimal energy consumption profile for LP problem and actual prices.

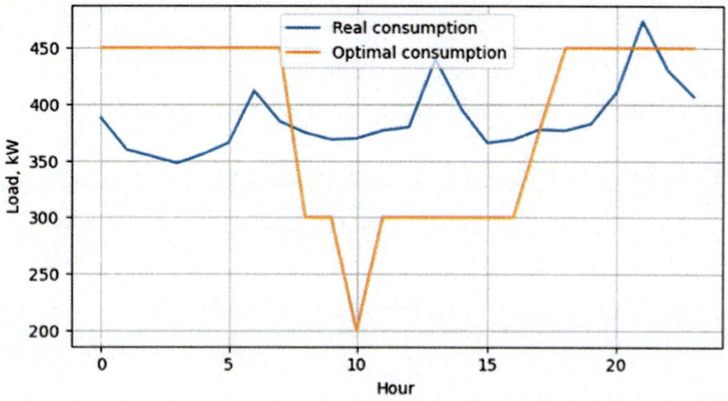

Fig. 4. Optimal energy consumption profile for LP problem and forecasted prices

4.3 Genetic Algorithm

The authors have used Python and geneticalgorithm2 library [21] to develop an application allowing to determine the optimal daily energy consumption profile using the GA algorithm. For this, the following values of GA parameters were used: $N_{ps} = 1000$, $P_m = 0.1$, $P_c = 0.5$, $R_e = 0.01$, $R_p = 0.3$. The roulette method (roulette type) was used as a selection operator. The mutation operator was implemented using uniform distribution (uniform_by_center type). The crossover operator was implemented using a multipoint homogeneous crossover (uniform type). The values of limitations as per conditions (2)–(4) were taken the same as for LP.

The GA algorithm implemented using the geneticalgorithm2 library allows only constraints (3), (4) to be taken into account. Limitation (2) can be taken into account by converting the target function (1) to the form

$$F(x) = f(x) + p(x_1 + x_2 + \ldots + x_{24} - X_d)^2 \qquad (5)$$

where p is the penalty coefficient.

For the case of actual price values, changes of the target function (5) in the process of finding an optimal solution using GA are shown in Fig. 5. The daily energy consumption profile obtained using GA for February 6, 2020, and actual price values is shown in Fig. 6, and for forecasted prices – in Fig. 7.

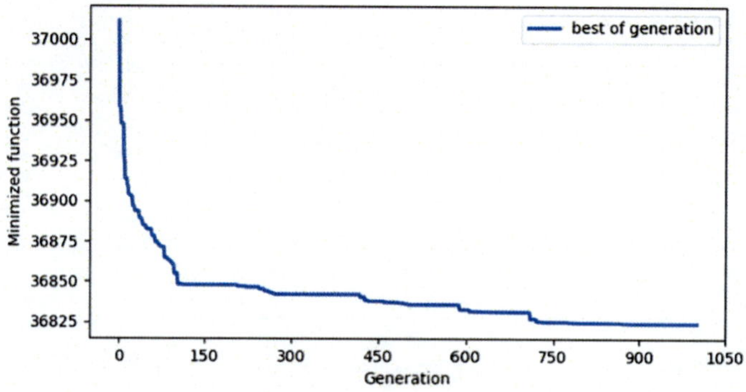

Fig. 5. Changes of the target function during the search process using GA.

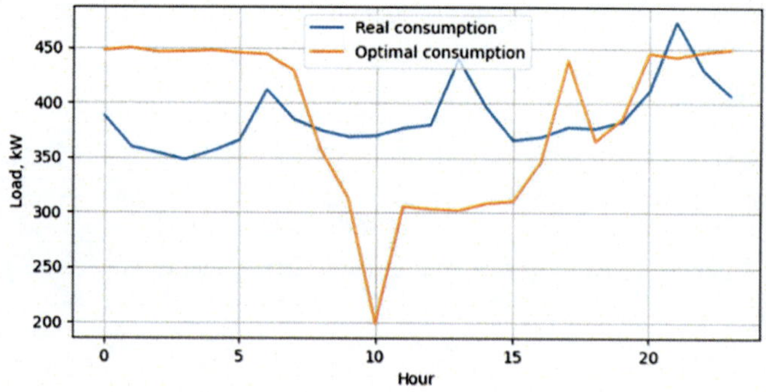

Fig. 6. Optimal energy consumption profile found using GA for actual price values.

Graphs of Fig. 6 and 7 are similar to the graphs shown in Fig. 3 and 4. Therefore, the conclusions made for the LP case are valid for this case as well.

4.4 Particle Swarm Optimization

The authors have used Python and PySwarms library [22] to develop an application to determine the optimal daily energy profile using the PSO algorithm. For this, the following PSO algorithm parameter values were taken: $N_p = 100$, $k_{im} = 0.5$, $k_{cm} = 0.3$, $k_i = 0.9$. Same as GA, the PSO algorithm allows directly considering only the constraints (3), (4), and to take into account constraint (2) requires transforming the target function to the form (5).

For the case of actual price values, changes of the target function in the process of finding an optimal solution using the PSO algorithm are shown in Fig. 8. The daily

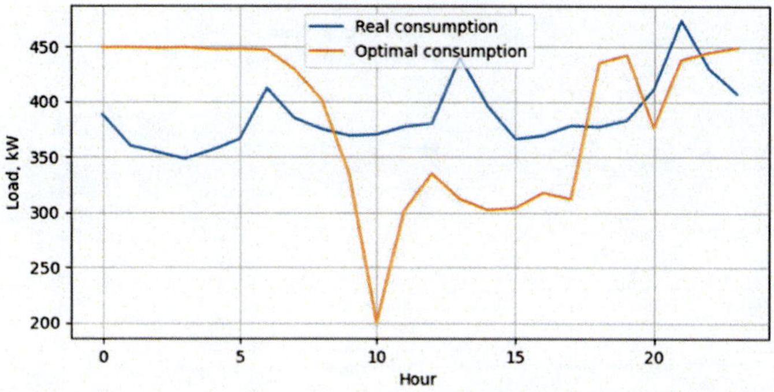

Fig. 7. Optimal energy consumption profile found using GA for forecasted price values.

energy consumption profile obtained using the PSO algorithm for February 6, 2020, and actual price values is shown in Fig. 9, and for forecasted prices – in Fig. 10.

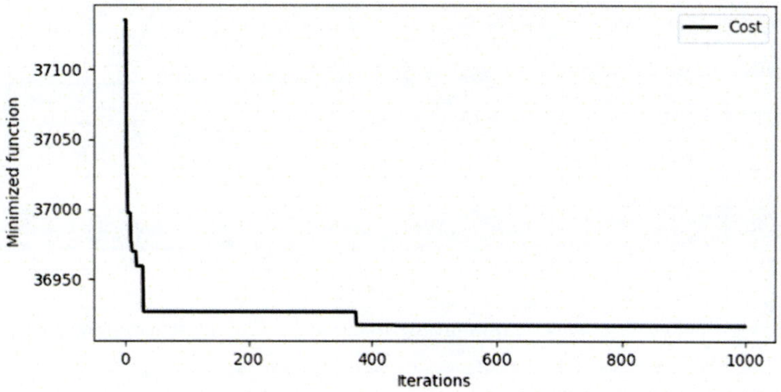

Fig. 8. Changes of the target function during the search process using PSO

The optimal energy consumption profile obtained using the PSO algorithm significantly differs from the profiles obtained using LP and GA methods. This is due to the fact that this algorithm does not find a globally optimal solution in a reasonable time. However, in terms of implementation feasibility, the profile obtained using the PSO algorithm may be more convenient for the enterprise. The difference between the forecasted and actual prices in this case almost does not affect the optimal energy consumption profile form.

4.5 Electricity Cost Estimate

The electricity cost for enterprises being energy consumers 3–6 price categories includes energy cost and the cost of power. The daily total electricity cost determined by the formula (1) for actual hourly market electricity prices and real energy profile of the enterprise on 06.02.2020 is 37,115.19 rubles. This value was used to estimate the savings

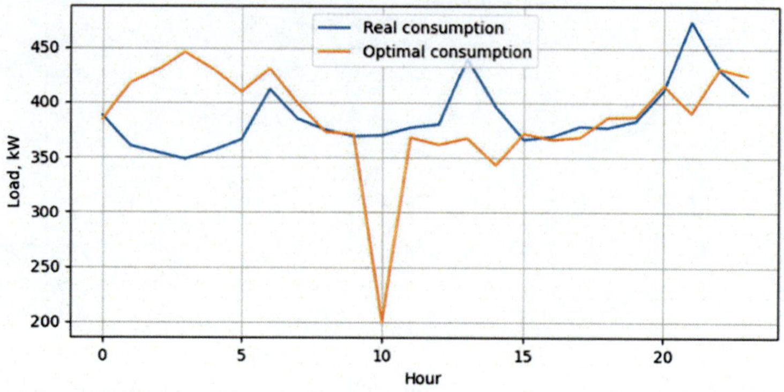

Fig. 9. Optimal energy consumption profile found using PSO for actual price values.

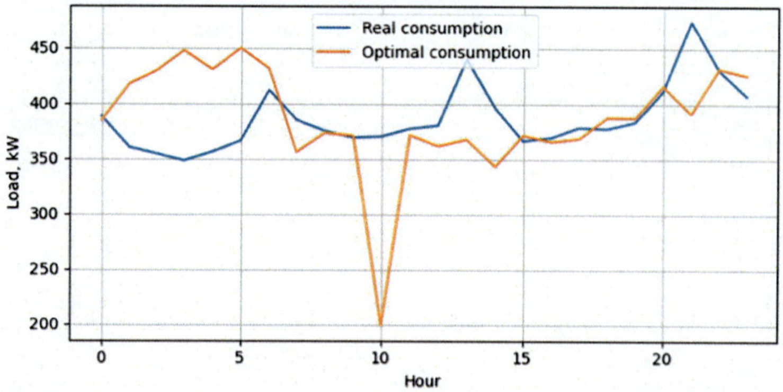

Fig. 10. Optimal energy consumption profile found using PSO for forecasted price values.

on electricity payments when using the optimized energy consumption profile taking into account the hourly market electricity prices.

The cost of power is calculated as a product of the weighted average market price of power and the average energy consumption of the enterprise during the peak load hours for the current month. For the considered case, it amounted to 323,325.5 rubles in February 2020. If the energy consumption of the enterprise in peak load hours was minimal during the month (for example, 200 kW), then the power payment would be 173,502 rubles in this case. Furthermore, savings on electricity bills due to the power payment would be 149,823.5 rubles or 7885.45 rubles per day.

Table 1 shows the daily electricity cost obtained by formula (1) for various methods for determining the optimal energy consumption profile using actual and forecasted prices.

Table 2 shows the daily savings obtained for various methods for determining the optimal energy consumption profile using actual and forecasted prices. In addition, Table 2 contains the values of the total savings obtained by summing up savings due to energy payment and the power payment.

Table 1. Electricity costs for various optimization methods.

Algorithm	Electricity cost (actual prices), rub	Electricity cost (forecasted prices), rub
Linear programming	36801.39	36966.64
Genetic algorithm	36824.29	36983.84
Particle swarm optimization	36925.54	37080.2

Table 2. Savings for various optimization methods.

Algorithm	Savings on electr. (actual prices), rub	Savings on electr. (forecast values), rub	Total savings (actual prices), rub	Total savings (forecast values), rub
Linear programming	313.8	148.55	8199.25	8034.0
Genetic algorithm	290.9	131.35	8176.35	8016.8
Particle swarm optimization	189.65	75.0	8075.1	7960.44

Analysis of the data from Tables 1 and 2 shows that the greatest effect on savings when paying by energy payment can be obtained when optimizing the energy consumption profile using LP. GA provides a rather close result. The least savings due to the energy payment resulted when using the PSO algorithm. It is worth noting that from a practical point of view, the most feasible is the energy consumption profile obtained using the PSO algorithm since it is closest to the real schedule.

The use of forecasted price values, instead of their real values, in this example, reduces the savings by optimizing the energy consumption profile by more than half. The calculations have shown that the positive difference between savings for actual prices and forecasted prices is typical for most days in February 2020. However, this difference may be significantly lower on some days of February, and there are days when the energy savings for the forecasted prices are greater. Generally, this circumstance can stimulate to improve the accuracy of forecasting electricity prices.

5 Discussion

One of the conclusions that can be made from the study is that the daily savings on electricity bills obtained by optimizing the energy consumption profile can be quite minor. This is explained by the fact that the Russian market has low electricity price volatility. The market electricity prices are substantially smoothed taking into account the regulated price component (electricity transmission tariffs and sales surcharges) and may vary within a month in the range of 10–15% of the average electricity price. It should be noted that such stability of market electricity prices is specific to the Russian electricity market. Energy markets of Europe and the United States have a much larger range of price changes during the day and during the season.

To assess the influence of electricity price volatility on the savings obtained when optimizing the energy consumption profile, the following experiment was carried out. Let us define an artificial electricity price (comparable to electricity prices in the European market), which has greater volatility compared to current prices in the Russian Federation and changes during the day as shown in Fig. 11. The standard deviation for this artificial price is 1.55.

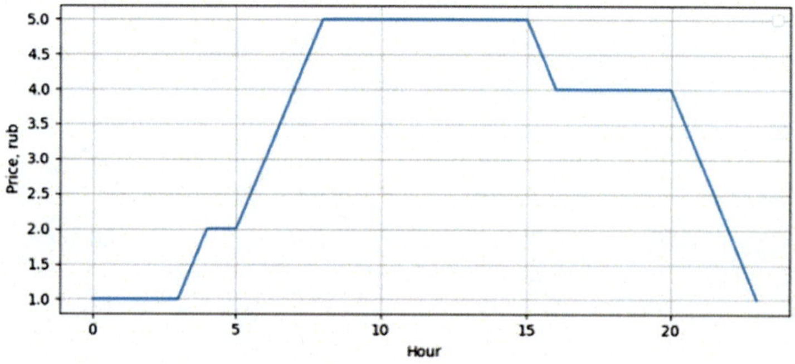

Fig. 11. Changes of the artificial price with great volatility.

Using LP for the same conditions as in the case described above, we shall define the optimal daily energy consumption profile for artificial price. At the same time, the electricity costs calculated by the formula (1) amounted to 29,053.22, while the daily savings – 8061.97 rubles. This experiment shows that with the increase in electricity price volatility, the savings due to energy consumption profile optimization rapidly increase and become comparable to savings due to the power payment. Consequently, if the electricity price volatility is increased to the values specified in the experiment, the application of the optimized energy consumption profile brings significant savings.

This experiment allows concluding that the current two-factor pricing model in the Russian energy and power market can be reduced to a single-factor model. In this case, the daily schedule of electricity price with low volatility is maintained until the peak load hour (for 3 and 5 price category calculations) and the scheduled peak load hours (for 4 and 6 price category calculations), after which the price is again stabilized with a low variability degree. Furthermore, the electricity price may change by tens of times in peak load hour reaching the values of 30–40 rubles per kWh.

Therefore, for the current model of market pricing and current electricity prices, much more significant are savings due to the power payment. In the considered example, the daily savings due to the power payment are at least 25-times bigger than savings achieved due to energy payment. This emphasizes the relevance of forecasting the peak load hours and reducing energy consumption in these hours to the lowest possible values.

6 Conclusions

This study considers the problem of determining the optimal daily energy consumption profile of an enterprise. By solving this problem, a reduction of electricity costs can be achieved by reducing energy consumption during periods of high prices, increasing energy consumption during hours when electricity prices are lower, as well as by cardinally reducing energy consumption during peak load hours. To determine the optimal daily energy consumption profile, three different optimization methods are proposed – linear programming method, genetic algorithm, and particle swarm optimization. The results are provided regarding the solution of the problem using the specified methods and specific input data, data on the energy consumption, and data on actual and forecasted electricity prices. The problem definition of determining the optimal daily energy consumption profile discussed in the study is general, as it does not depend on the specifics of the consumer operation. However, such a general approach can limit the enterprise's possibilities related to reducing electricity costs. The study also includes an estimation of the savings obtained using various optimization methods. The effect of forecasted prices and price volatility on the size of savings is investigated as well.

The methods for determining the optimal daily energy consumption profile can be useful for companies that pay for electricity at market prices. In the case of low electricity price volatility, the greatest effect in terms of savings is provided by the power control during peak load hours. The key factor, in this case, is the possibility of forecasting the peak load hours and reducing energy consumption during these hours to the lowest possible values. This will result in savings up to 15–30% of the total electricity cost for consumers.

References

1. Nguyen, A.T., Reiter, S., Rigo, P.: A review on simulation-based optimization methods applied to building performance analysis. Appl. Energy **113**, 1043–1058 (2014)
2. Rodrigues, G.S., Ferreira, J.C.E., Rocha, C.R.: A novel method for analysis and optimization of electric energy consumption in manufacturing processes. Procedia Manuf. **17**, 1073–1081 (2018)
3. Ilbeigi, M., Ghomeishi, M., Dehghanbanadaki, A.: Prediction and optimization of energy consumption in an office building using artificial neural network and a genetic algorithm. Sustain. Cities Soc. **61**, 1–15 (2020)
4. Shah, A.S., Nasir, H., Fayaz, M., Lajis, A., Shah, A.: A review on energy consumption optimization techniques in IoT based smart building environments. Information **10**, 1–34 (2019)
5. Maryasin, O., Kolodkhina, A.S.: Control of the thermal regime of buildings using predictive models. Bull. SamGTU **1**(53), 122–132 (2017)
6. Afram, A., Janabi-Sharifi, F.: Theory and applications of HVAC control systems – A review of model predictive control (MPC). Build. Environ. **72**, 343–355 (2014)
7. Maharjan, I.K.: Demand Side Management: Load Management, Load Profiling, Load Shifting, Residential and Industrial Consumer, Energy Audit, Reliability, Urban, Semi-Urban and Rural Setting, p. 116. LAP Lambert Academic Publishing, Sunnyvale (2010)
8. Bradac, Z., Kaczmarczyk, V., Fiedler, P.: Optimal scheduling of domestic appliances via MILP. Energies **8**, 217–232 (2014)

9. Zhao, Z., Lee, W.C., Shin, Y., Song, K.: An optimal power scheduling method applied in home energy management system based on demand response. ETRI J. **35**(4), 677–686 (2013)
10. Ahmad, A., et al.: An optimized home energy management system with integrated renewable energy and storage resources. Energies **10**(549), 2–35 (2017)
11. Weron, R.: Electricity price forecasting: a review of the state-of-the-art with a look into the future. Int. J. Forecast. **30**(4), 1030–1081 (2014)
12. Lee, J.Y., Choi, S.G.: Linear programming based hourly peak load shaving method at home area. In: 16th International Conference on Advanced Communication Technology, pp. 310–313 (2014)
13. Awais, M., et al.: An efficient genetic algorithm based demand side management scheme for smart grid. In: 18th International Conference on Network-Based Information Systems, pp. 1–6 (2015)
14. Bharathi, C., Rekha, D., Vijayakumar, V.: Genetic algorithm based demand side management for smart grid. Wirel. Pers. Commun. **93**, 481–502 (2017)
15. Rasheed, M.B., Javaid, N., Ahmad, A., Khan, Z.A., Qasim, U., Alrajeh, N.: An efficient power scheduling scheme for residential load management in smart homes. Appl. Sci. **5**, 1134–1163 (2015)
16. Aslam, S., Iqbal, Z., Javaid, N., Khan, Z.A., Aurangzeb, K., Haider, S.I.: Towards efficient energy management of smart buildings exploiting heuristic optimization with real time and critical peak pricing schemes. Energies **10**(2065), 1–25 (2017)
17. Electricity prices TNS energo Yaroslavl (2020). https://yar.tns-e.ru/legal-entities/prices/
18. Maryasin, O.Y., Lukashov, A.I.: A python application for hourly electricity prices forecasting using neural networks. In: International Russian Automation Conference, pp. 1–6 (2020)
19. Maryasin, O.Y., Lukashov, A.I.: Developing a digital model of an electricity consumer using deep learning. In: 2nd International Conference on Control Systems, Mathematical Modeling, Automation and Energy Efficiency, pp. 1–6 (2020)
20. Peak load hours of ATS (2020). https://www.atsenergo.ru/results/market/calcfacthour
21. GitHub – PasaOpasen/geneticalgorithm2: Supported genetic algorithm package for Python (2020). https://github.com/PasaOpasen/geneticalgorithm2
22. GitHub – ljvmiranda921/pyswarms: A research toolkit for particle swarm optimization in Python (2020). https://github.com/ljvmiranda921/pyswar

A Predictive Nonlinear Regression Model Under Initial Z-Information

O. M. Poleshchuk(✉)

Moscow Bauman State Technical University,
5/1, Vtoraya Baumanskaya Street, Moscow 1105005, Russia

Abstract. The paper developed a nonlinear regression model under initial Z-information. The information received from experts has a certain level of reliability, therefore, to study dependencies and predict such information, is necessary apparatus regression analysis, that takes this reliability into account. However, the development of such apparatus is at an initial stage, therefore, research in this area is relevant and in demand. In the paper, we assume that the components of the initial Z-numbers are the values of full orthogonal semantic spaces. Operations on Z-numbers and distances between them are determined based on their aggregating points, which are the midpoints of aggregating segments. The aggregating segments for Z-numbers are defined as the aggregating segments of the product of both components. The optimization function is defined as the sum of the squares of the distances between the model Z-numbers and the initial Z-numbers and between the first components of the model Z-numbers and the initial Z-numbers. The choice of such an optimization function was made to enhance the influence of the first components of the initial Z-information. To determine the unknown regression coefficients, the minimum of the optimization function is found. The new developed model expands the possibilities of regression analysis and allows you to construct a prediction and investigate dependencies using Z-information, which is shown by the numerical example given in the paper.

Keywords: Z-number · Z-information · Regression model · Semantic space

1 Introduction

Regression analysis plays an essential role in information processing to predict it and identify dependencies between various indicators. The capabilities of classical regression analysis have been significantly expanded with the development of methods of fuzzy regression analysis, which made it possible to take into account not only random uncertainty, but also the fuzziness that arises when using natural language words in the estimation or description procedures [1–8]. However, the developed fuzzy regression models did not make it possible to process fuzzy information with a certain level of its reliability. After the formalized presentation of such information (Z-information) with the help of Z-numbers defined by Professor Lotfi Zadeh [9], the question of developing regression models for prediction Z-information and identifying existing dependencies was especially acute. To develop regression models under initial Z-information, it was

A. A. Radionov and V. R. Gasiyarov (Eds.): RusAutoCon 2021, LNEE 857, pp. 383–392, 2022.
https://doi.org/10.1007/978-3-030-94202-1_36

necessary to develop a mathematical apparatus, including operations on Z-numbers, their ranking, determining the distances between them, and so on. The necessary apparatus has been developed and continues to be developed to this day [10–19].

The first regression model with initial Z-information was developed in 2017 [20] based on operations on fuzzy numbers and probability distributions. In [21], a linear regression model with input and output Z-information and crisp coefficients was developed, as well as approaches to creating a linear regression model with initial Z-information and fuzzy coefficients were given. For the study of dependencies and prediction under Z-information, the few existing models are not enough. Therefore, the paper continues research in this direction and a nonlinear model is being developed in the conditions of input and output Z-numbers, the components of which are values of full orthogonal semantic spaces.

2 Basic Concepts and Definitions

A linguistic variable [22] is called a full orthogonal semantic space, all the membership functions of which have corresponding intervals, for all values of which the values of the functions are equal to one. Functions increase to the left of these intervals and decrease to the right of them. In addition, for each point of the domain of definition of the space, the sum of all functions at this point is equal to one. All the functions are continuous or have at most two discontinuity points of the first kind [23].

Methods for constructing full orthogonal semantic spaces for different input information are developed in detail and presented in [24].

In [6], the aggregating segment $[\beta_1, \beta_2]$ for fuzzy number $\tilde{B} = (b_1, b_2, b_L, b_R)$ has been defined:

$$\beta_1 = b_1 - \frac{1}{6}b_L, \ \beta_2 = b_2 + \frac{1}{6}b_R. \tag{1}$$

In [21], the aggregating point β for fuzzy number $\tilde{B} = (b_1, b_2, b_L, b_R)$ has been defined as the midpoint of the weighted segment (1):

$$\beta = \frac{\beta_1 + \beta_2}{2} = \frac{b_1 + b_2}{2} + \frac{b_R - b_L}{12}. \tag{2}$$

A Z-number $Z = \left(\tilde{B}, \tilde{R}\right)$ is a pair of fuzzy numbers, where $\tilde{B}$ is a fuzzy value of some variable and $\tilde{R}$ is a fuzzy value of reliability $\tilde{B}$ [9].

In [21], the aggregating segment $\left[\beta_Z^1, \beta_Z^2\right]$ for $Z = \left(\tilde{B}, \tilde{R}\right)$ $\left(\tilde{B} = (b_1, b_2, b_L, b_R), \tilde{R} = (r_1, r_2, r_L, r_R)\right)$ has been defined as the aggregating segment of the product of its components $\tilde{B}$, $\tilde{R}$:

$$\beta_Z^1 = r_1\left(b_1 - \frac{1}{6}b_L\right) - r_L\left(\frac{1}{6}b_1 - \frac{1}{12}b_L\right), \beta_Z^2 = r_2\left(b_2 + \frac{1}{6}b_R\right) + r_R\left(\frac{1}{6}b_2 + \frac{1}{12}b_R\right). \tag{3}$$

By analogy with aggregating point (2) of usual fuzzy number, we define for Z - the number $Z = \left(\tilde{B}, \tilde{R}\right) \left(\tilde{B} = \left(b_1, b_2, b_L, b_R\right), \tilde{R} = \left(r_1, r_2, r_L, r_R\right)\right)$ the aggregating point β_Z:

$$\beta_Z = \frac{\beta_Z^1 + \beta_Z^2}{2} = r_1\left(\frac{b_1}{2} - \frac{1}{12}b_L\right) - r_L\left(\frac{1}{12}b_1 - \frac{1}{24}b_L\right) + r_2\left(\frac{b_2}{2} + \frac{1}{12}b_R\right) + r_R\left(\frac{1}{12}b_2 + \frac{1}{24}b_R\right). \tag{4}$$

3 Problem Formulation and Solution

The problem to be solved is to construct a nonlinear regression model with initial input and output Z-information.

Let $Z_k^p = \left(\tilde{A}_k^p, \tilde{R}_k^p\right), k = \overline{1, n}, p = \overline{1, m}$ and $Z^p = \left(\tilde{A}^p, \tilde{R}^p\right), p = \overline{1, m}$ are the input and output Z-numbers accordingly. The first components of initial Z-numbers or fuzzy numbers $\tilde{A}_k^p = \left(a_{1k}^p, a_{2k}^p, a_{Lk}^p, a_{Rk}^p\right), k = \overline{1, n}, p = \overline{1, m}$ are equal to the one of fuzzy numbers $\tilde{C}_{kl}, k = \overline{1, n}, l = \overline{1, L_k}$, which are formalizations of the values of input full orthogonal semantic spaces and $\tilde{A}^p = \left(a_1^p, a_2^p, a_L^p, a_R^p\right), p = \overline{1, m}$ are equal to one of fuzzy numbers $\tilde{C}_l, l = \overline{1, L}$, which are formalizations of the values of output full orthogonal semantic space.

The second components of input and output Z-numbers or fuzzy numbers $\tilde{R}_k^p = \left(r_{1k}^p, r_{2k}^p, r_{Lk}^p, r_{Rk}^p\right), k = \overline{1, n}, p = \overline{1, m}$ and $\tilde{R}^p = \left(r_1^p, r_2^p, r_L^p, r_R^p\right), p = \overline{1, m}$ are equal to one of fuzzy numbers $\tilde{R}_h, h = \overline{1, H}$, which are formalizations of the values of full orthogonal semantic space "Reliability". We assume that all the full orthogonal semantic spaces are constructed on the segment [0, 1]. In this case, the fuzzy numbers $\tilde{A}_k^p, \tilde{A}^p, \tilde{R}_k^p, \tilde{R}^p, k = \overline{1, n}, p = \overline{1, m}$ are non-negative. If for some spaces the universal set is not the segment [0, 1], then this set can be transformed into [0, 1].

The dependence between the input and output Z-numbers will be constructed in the form:

$$Z = a_1 Z_1^2 + \ldots + a_n Z_n^2 + a_{n+1} Z_1 Z_2 + \ldots + a_{\frac{n(n+1)}{2}} Z_{n-1} Z_n + a_{\frac{n^2+n+2}{2}} Z_1 + \ldots + a_{\frac{n(n+3)}{2}} Z_n, \tag{5}$$

where $a_k, k = \overline{1, \frac{n(n+3)}{2}}$ are unknown coefficients of the regression model, which are defined as crisp numbers.

Let us define the aggregating segment $\left[\beta_{Z_k^p}^1, \beta_{Z_k^p}^2\right]$ and the aggregating point $\beta_{Z_k^p}$ for $Z_k^p = \left(\tilde{A}_k^p, \tilde{R}_k^p\right), k = \overline{1, n}, p = \overline{1, m}$ according to respectively (3), (4):

$$\begin{gathered}\beta_{Z_k^p}^1 = r_{1k}^p\left(a_{1k}^p - \tfrac{1}{6}a_{Lk}^p\right) - r_{Lk}^p\left(\tfrac{1}{6}a_{1k}^p - \tfrac{1}{12}a_{Lk}^p\right), \beta_{Z_k^p}^2 = r_{2k}^p\left(a_{2k}^p + \tfrac{1}{6}a_{Rk}^p\right) + r_{Rk}^p\left(\tfrac{1}{6}a_{2k}^p + \tfrac{1}{12}a_{Rk}^p\right), \\ \beta_{Z_k^p} = \tfrac{1}{2}\left(r_{1k}^p\left(a_{1k}^p - \tfrac{1}{6}a_{Lk}^p\right) - r_{Lk}^p\left(\tfrac{1}{6}a_{1k}^p - \tfrac{1}{12}a_{Lk}^p\right)\right) + \tfrac{1}{2}\left(r_{2k}^p\left(a_{2k}^p + \tfrac{1}{6}a_{Rk}^p\right) + r_{Rk}^p\left(\tfrac{1}{6}a_{2k}^p + \tfrac{1}{12}a_{Rk}^p\right)\right), \\ k = \overline{1, n}, p = \overline{1, m}.\end{gathered} \tag{6}$$

We define the aggregating segment $\left[\beta^1_{(Z_k^p)^2}, \beta^2_{(Z_k^p)^2}\right]$ and the aggregating point $\beta_{(Z_k^p)^2}$ for $\left(Z_k^p\right)^2, k=\overline{1,n}, p=\overline{1,m}$ in the following forms:

$$\beta^1_{(Z_k^p)^2} = \left(\beta^1_{Z_k^p}\right)^2 = \left(r^p_{1k}\left(a^p_{1k} - \tfrac{1}{6}a^p_{Lk}\right) - r^p_{Lk}\left(\tfrac{1}{6}a^p_{1k} - \tfrac{1}{12}a^p_{Lk}\right)\right)^2, \beta^2_{(Z_k^p)^2} = \left(\beta^2_{Z_k^p}\right)^2$$
$$= \left(r^p_{2k}\left(a^p_{2k} + \tfrac{1}{6}a^p_{Rk}\right) + r^p_{Rk}\left(\tfrac{1}{6}a^p_{2k} + \tfrac{1}{12}a^p_{Rk}\right)\right)^2, \beta_{(Z_k^p)^2} = \frac{\left(\beta^1_{Z_k^p}\right)^2 + \left(\beta^2_{Z_k^p}\right)^2}{2}, k=\overline{1,n}, p=\overline{1,m}. \tag{7}$$

We define the aggregating segment $\left[\beta^1_{Z_k^pZ_l^p}, \beta^2_{Z_k^pZ_l^p}\right]$ and the aggregating point $\beta_{Z_k^pZ_l^p}$ for $Z_k^pZ_l^p, k=\overline{1,n}, l=\overline{1,n}, p=\overline{1,m}$ in the following forms:

$$\beta^1_{Z_k^p\times Z_l^p} = \left(r^p_{1k}\left(a^p_{1k} - \tfrac{1}{6}a^p_{Lk}\right) - r^p_{Lk}\left(\tfrac{1}{6}a^p_{1k} - \tfrac{1}{12}a^p_{Lk}\right)\right) \times \left(r^p_{1l}\left(a^p_{1l} - \tfrac{1}{6}a^p_{Ll}\right) - r^p_{Ll}\left(\tfrac{1}{6}a^p_{1l} - \tfrac{1}{12}a^p_{Ll}\right)\right),$$
$$\beta^2_{Z_k^p\times Z_l^p} = \left(r^p_{2k}\left(a^p_{2k} + \tfrac{1}{6}a^p_{Rk}\right) + r^p_{Rk}\left(\tfrac{1}{6}a^p_{2k} + \tfrac{1}{12}a^p_{Rk}\right)\right) \times \left(r^p_{2l}\left(a^p_{2l} + \tfrac{1}{6}a^p_{Rl}\right) + r^p_{Rl}\left(\tfrac{1}{6}a^p_{2l} + \tfrac{1}{12}a^p_{Rl}\right)\right),$$
$$\beta_{Z_k^p\times Z_l^p} = \frac{\beta^1_{Z_k^p\times Z_l^p} + \beta^2_{Z_k^p\times Z_l^p}}{2}, k=\overline{1,n}, l=\overline{1,n}, p=\overline{1,m}. \tag{8}$$

Let us define the aggregating point $\beta_{Z^p_{\text{mod}}}$ for

$$Z^p_{\text{mod}} = a_1\left(Z_1^p\right)^2 + \ldots + a_n\left(Z_n^p\right)^2 + a_{n+1}Z_1^pZ_2^p + \ldots + a_{\frac{n(n+1)}{2}} Z^p_{n-1}Z^p_n + a_{\frac{n^2+n+2}{2}} Z_1^p + \ldots + a_{\frac{n(n+3)}{2}} Z_n^p :$$
$$\beta_{Z^p_{\text{mod}}} = \sum_{k=1}^{n} a_k\beta_{Z_k^p} + a_{n+1}\beta_{Z_1^pZ_2^p} + \ldots + +a_{\frac{n(n+1)}{2}}\beta_{Z^p_{n-1}Z^p_n} + \sum_{k=\frac{n^2+n+2}{2}}^{\frac{n(n+3)}{2}} a_k\beta_{Z_k^p}, p=\overline{1,m}. \tag{9}$$

We define the aggregating segment $\left[\beta^1_{A_k^p}, \beta^2_{A_k^p}\right]$ and the aggregating point $\beta_{A_k^p}$ for $\tilde{A}_k^p, k=\overline{1,n}, p=\overline{1,m}$ in the following forms:

$$\beta^1_{A_k^p} = a^p_{1k} - \frac{1}{6}a^p_{Lk}, \beta^2_{A_k^p} = a^p_{2k} + \frac{1}{6}a^p_{Rk}, \beta_{A_k^p} = \frac{a^p_{1k} + a^p_{2k}}{2} + \frac{a^p_{Rk} - a^p_{Lk}}{12}, k=\overline{1,n}, p=\overline{1,m}. \tag{10}$$

We define the aggregating segment $\left[\beta^1_{(A_k^p)^2}, \beta^2_{(A_k^p)^2}\right]$ and the aggregating point $\beta_{(A_k^p)^2}$ for $\left(\tilde{A}_k^p\right)^2, k=\overline{1,n}, p=\overline{1,m}$ in the following forms:

$$\beta^1_{(A_k^p)^2} = \left(a^p_{1k}\right)^2 - \tfrac{1}{3}a^p_{1k}a^p_{Lk} + \tfrac{1}{12}\left(a^p_{Lk}\right)^2, \beta^2_{(A_k^p)^2} = \left(a^p_{2k}\right)^2 + \tfrac{1}{3}a^p_{2k}a^p_{Rk} + \tfrac{1}{12}\left(a^p_{Rk}\right)^2,$$
$$\beta_{(A_k^p)^2} = \frac{\beta^1_{(A_k^p)^2} + \beta^2_{(A_k^p)^2}}{2}, k=\overline{1,n}, p=\overline{1,m}. \tag{11}$$

Let us define the aggregating segment $\left[\beta^1_{A^p_k A^p_l}, \beta^2_{A^p_k A^p_l}\right]$ and the aggregating point $\beta_{A^p_k A^p_l}$ for $\tilde{A}^p_k \tilde{A}^p_l$, $k = \overline{1, n}$, $l = \overline{1, n}$, $p = \overline{1, m}$:

$$\beta^1_{A^p_k A^p_l} = a^p_{1l}(a^p_{1k} - \tfrac{1}{6}a^p_{Lk}) - a^p_{Ll}(\tfrac{1}{6}a^p_{1k} - \tfrac{1}{12}a^p_{Lk}), \beta^2_{A^p_k A^p_l} = a^p_{2l}(a^p_{2k} + \tfrac{1}{6}a^p_{Rk}) + a^p_{Rl}(\tfrac{1}{6}a^p_{2k} + \tfrac{1}{12}a^p_{Rk}),$$
$$\beta_{A^p_k A^p_l} = \frac{\beta^1_{A^p_k A^p_l} + \beta^2_{A^p_k A^p_l}}{2}, k = \overline{1, n}, l = \overline{1, n}, p = \overline{1, m}. \tag{12}$$

Let us define the aggregating point $\beta_{A^p_{\text{mod}}}$ for $\tilde{A}^p_{\text{mod}} = a_1\left(\tilde{A}^p_1\right)^2 + \ldots + a_n\left(\tilde{A}^p_p\right)^2 + a_{n+1}\tilde{A}^p_1\tilde{A}^p_2 + \ldots + +a_{\frac{n(n+1)}{2}}\tilde{A}^p_{n-1}\tilde{A}^p_n + a_{\frac{n^2+n+2}{2}}\tilde{A}^p_1 + \ldots + a_{\frac{n(n+3)}{2}}\tilde{A}^p_n$:

$$\beta_{A^p_{\text{mod}}} = \sum_{k=1}^{n} a_k \beta_{A^p_k} + \sum_{k=n+1}^{\frac{n(n+1)}{2}} a_k \beta_{A^p_{k-n} A^p_{k-n+1}} + \sum_{k=\frac{n^2+n+2}{2}}^{\frac{n(n+3)}{2}} a_k \beta_{A^p_k}, p = \overline{1, m}. \tag{13}$$

We define the aggregating segment $\left[\beta^1_{Z^p}, \beta^2_{Z^p}\right]$ and the aggregating point β_{Z^p} for $Z^p = \left(\tilde{A}^p, \tilde{R}^p\right), p = \overline{1, m}$ in the following forms:

$$\beta^1_{Z^p} = r^p_1(a^p_1 - \tfrac{1}{6}a^p_L) - r^p_L(\tfrac{1}{6}a^p_1 - \tfrac{1}{12}a^p_L), \beta^2_{Z^p} = r^p_2(a^p_2 + \tfrac{1}{6}a^p_R) + r^p_R(\tfrac{1}{6}a^p_2 + \tfrac{1}{12}a^p_R),$$
$$\beta_{Z^p} = \tfrac{1}{2}(r^p_1(a^p_1 - \tfrac{1}{6}a^p_L) - r^p_L(\tfrac{1}{6}a^p_1 - \tfrac{1}{12}a^p_L)) + \tfrac{1}{2}(r^p_2(a^p_2 + \tfrac{1}{6}a^p_R) + r^p_R(\tfrac{1}{6}a^p_2 + \tfrac{1}{12}a^p_R)), p = \overline{1, m}. \tag{14}$$

We define the aggregating segment $\left[\beta^1_{A^p}, \beta^2_{A^p}\right]$ and the aggregating point β_{A^p} for $\tilde{A}^p, p = \overline{1, m}$ in the following forms:

$$\beta^1_{A^p} = a^p_1 - \frac{1}{6}a^p_L, \beta^2_{A^p} = a^p_2 + \frac{1}{6}a^p_R, \beta_{A^p} = \frac{a^p_1 + a^p_2}{2} + \frac{a^p_R - a^p_L}{12}, p = \overline{1, m}. \tag{15}$$

The optimization problem is defined as follows, using (6)–(15): $F\left(a_1, \ldots, a_{\frac{n(n+3)}{2}}\right) = \sum_{p=1}^{m}\left(\left(\beta_{Z^p_{\text{mod}}} - \beta_{Z^p}\right)^2 + \left(\beta_{A^p_{\text{mod}}} - \beta_{A^p}\right)^2\right) \to \min.$

Unknown coefficients $a_k, k = \overline{1, \frac{n(n+3)}{2}}$ are solutions to a system of equations. To obtain this system, the first derivatives with respect to each of the coefficients are found and equated to zero.

Let us describe the prediction procedure for the output model Z-number $Z_{\text{mod}} = \left(\tilde{A}_{\text{mod}}, \tilde{R}_{\text{mod}}\right)$. We designate the aggregating point of the first component $\tilde{A}_{\text{mod}}$ of Z-number $Z_{\text{mod}} = \left(\tilde{A}_{\text{mod}}, \tilde{R}_{\text{mod}}\right)$ by $\beta_{A_{\text{mod}}}$ and the aggregating points of fuzzy numbers $\tilde{C}_l, l = \overline{1, L}$, which are the formalizations of the terms of output full orthogonal semantic space by $\beta_{C_l}, l = \overline{1, L}$.

The first component $\tilde{A}_{\text{mod}}$ of model Z-number $Z_{\text{mod}} = \left(\tilde{A}_{\text{mod}}, \tilde{R}_{\text{mod}}\right)$ is identified to fuzzy number $\tilde{C}_q$, if $\left|\beta_{\text{mod}} - \beta_{C_q}\right| = \min_i \left|\beta_{\text{mod}} - \beta_{C_i}\right|$. In order to predict the reliability of $\tilde{A}_{\text{mod}}$ or the second component of $Z_{\text{mod}} = \left(\tilde{A}_{\text{mod}}, \tilde{R}_{\text{mod}}\right)$, we calculate the aggregating point $\beta_{Z_{\text{mod}}}$ of the model value $Z_{\text{mod}} = a_1(Z_1)^2 + \ldots + a_n(Z_n)^2 + a_{n+1}Z_1Z_2 + \ldots + a_{\frac{n(n+1)}{2}}Z_{n-1}Z_n + a_{\frac{n^2+n+2}{2}}Z_1 + \ldots + a_{\frac{n(n+3)}{2}}Z_n$ and the aggregating points $\beta_l, l = \overline{1, L}$ of the products $\tilde{R}_h \times \tilde{A}_{\text{mod}}$, $h = \overline{1, H}$, where $\tilde{R}_h, h = \overline{1, H}$ are formalizations of the values of full orthogonal semantic space "Reliability". If $\left|\beta_{Z_{\text{mod}}} - \beta_d\right| = \min_h \left|\beta_{Z_{\text{mod}}} - \beta_h\right|$, then the reliability of the value $\tilde{A}_{\text{mod}}$ is fuzzy number $\tilde{R}_d$ and accordingly d-th value of full orthogonal semantic space "Reliability".

Thus, the predicted output value of the model is $Z_{\text{mod}} = \left(\tilde{C}_q, \tilde{R}_d\right)$.

4 Numerical Example

Let $Z^p = \left(\tilde{A}^p, \tilde{R}^p_Z\right), p = \overline{1, 5}$ and $V^p = \left(\tilde{B}^p, \tilde{R}^p_V\right), p = \overline{1, 5}$ are the input and output Z-numbers accordingly. Fuzzy numbers $\tilde{A}^p, \tilde{B}^p, p = \overline{1, m}$ are equal to one of the values $\tilde{C}_1 = (0, 0.2, 0, 0.1)$, $\tilde{C}_2 = (0.3, 0.7, 0.1, 0.2)$, $\tilde{C}_3 = (0.9, 1, 0.2, 0)$ of full orthogonal semantic space. Fuzzy numbers $\tilde{R}^p_Z, \tilde{R}^p_V, p = \overline{1, 5}$ are equal to one of the values $\tilde{R}_1 = (0, 0, 0.25)$, $\tilde{R}_2 = (0.25, 0.25, 0.25)$, $\tilde{R}_3 = (0.5, 0.25, 0.25)$, $\tilde{R}_4 = (0.75, 0.25, 0.25)$, $\tilde{R}_5 = (1, 0.25, 0)$ of full orthogonal semantic space "Reliability". Let $\tilde{A}^1 = (0, 0.2, 0, 0.1)$, $\tilde{R}^1_Z = (0.5, 0.25, 0.25)$, $\tilde{A}^2 = (0.3, 0.7, 0.1, 0.2)$, $\tilde{R}^2_Z = (0.75, 0.25, 0.25)$, $\tilde{A}^3 = (0.9, 1, 0.2, 0)$, $\tilde{R}^3_Z = (0.25, 0.25, 0.25)$, $\tilde{A}^4 = (0.3, 0.7, 0.1, 0.2)$, $\tilde{R}^4_Z = (0, 0, 0.25)$, $\tilde{A}^5 = (0, 0.2, 0, 0.1)$, $\tilde{R}^5_Z = (1, 0.25, 0)$, $\tilde{R}^1_V = (0.75, 0.25, 0.25)$, $\tilde{B}^2 = (0.1, 0.1, 0.1, 0.3)$, $\tilde{B}^1 = (0, 0, 0, 0.1)$, $\tilde{R}^2_V = (0.25, 0.25, 0.25)$, $\tilde{B}^2 = (0.1, 0.1, 0.1, 0.3)$, $\tilde{R}^2_V = (0.25, 0.25, 0.25)$, $\tilde{B}^3 = (0.4, 0.6, 0.3, 0.2)$, $\tilde{R}^3_V = (1, 0.25, 0)$, $\tilde{B}^4 = (0.8, 1, 0.2, 0)$, $\tilde{R}^4_V = (0.5, 0.25, 0.25)$, $\tilde{B}^5 = (0.1, 0.1, 0.1, 0.3)$, $\tilde{R}^5_V = (0, 0, 0.25)$. The dependence will be found in the following form: $V = a_1Z^2 + a_2Z$.

Let us define aggregating segments $\left[\beta^1_{A^p}, \beta^2_{A^p}\right]$, $\left[\beta^1_{R^p_Z}, \beta^2_{R^p_Z}\right]$, $\left[\beta^1_{Z^p}, \beta^2_{Z^p}\right]$, $\left[\beta^1_{B^p}, \beta^2_{B^p}\right]$, $\left[\beta^1_{R^p_V}, \beta^2_{R^p_V}\right]$, $\left[\beta^1_{V^p}, \beta^2_{V^p}\right]$ and aggregating points β_{A^p}, $\beta_{R^p_Z}\beta_{Z^p}$, β_{B^p}, $\beta_{R^p_V}$ β_{V^p} for $\tilde{A}^p, \tilde{R}^p_Z, Z^p = \left(\tilde{A}^p, \tilde{R}^p_Z\right), \tilde{B}^p, \tilde{R}^p_V, V^p = \left(\tilde{B}^p, \tilde{R}^p_V\right), p = \overline{1, 5}$ and enter this data into the Table 1, 2 and 3.

Table 1. Aggregating segments and points of $\tilde{A}^p$, $\tilde{R}_Z^p$.

No	$\left[\beta_{A^p}^1, \beta_{A^p}^2\right]$	β_{A^p}	$\left[\beta_{R_Z^p}^1, \beta_{R_Z^p}^2\right]$	$\beta_{R_Z^p}$
1	[0, 0.216]	0.108	[0.458, 0.542]	0.500
2	[0.283, 0.733]	0.508	[0.708, 0.792]	0.750
3	[0.867, 1]	0.934	[0.208, 0.292]	0.250
4	[0.283, 0.733]	0.508	[0, 0.042]	0.021
5	[0, 0.216]	0.108	[0.958, 1]	0.979

Table 2. Aggregating segments and points of Z^p,$\tilde{B}^p$.

No	$\left[\beta_{Z^p}^1, \beta_{Z^p}^2\right]$	β_{Z^p}	$\left[\beta_{B^p}^1, \beta_{B^p}^2\right]$	β_{B^p}
1	[0, 0.118]	0.059	[0, 0.017]	0.008
2	[0.200, 0.581]	0.390	[0.083, 0.150]	0.233
3	[0.180, 0.292]	0.236	[0.350, 0.633]	0.492
4	[0, 0.031]	0.015	[0.767, 1]	0.883
5	[0, 0.216]	0.108	[0.083, 0.150]	0.233

Table 3. Aggregating segments and points of $\tilde{R}_V^p$, V^p.

No	$\left[\beta_{R_V^p}^1, \beta_{R_V^p}^2\right]$	$\beta_{R_V^p}$	$\left[\beta_{V^p}^1, \beta_{V^p}^2\right]$	β_{V^p}
1	[0.708, 0.792]	0.750	[0, 0.013]	0.007
2	[0.208, 0.292]	0.250	[0.017, 0.044]	0.030
3	[0.958, 1]	0.979	[0.335, 0.633]	0.484
4	[0.458, 0.542]	0.500	[0.351, 0.542]	0.447
5	[0, 0.042]	0.021	[0, 0.006]	0.003

The optimization problem is defined as follows: $F(a_1, a_2) = \sum_{p=1}^{5}\left(\left(\beta_{V^p_{\text{mod}}} - \beta_{V^p}\right)^2 + \left(\beta_{A^p_{\text{mod}}} - \beta_{B^p}\right)^2\right) \to \min$, $F(a_1, a_2) = \sum_{p=1}^{5}\left(\left(a_1\beta_{(Z^p)^2} + a_2\beta_{Z^p} - \beta_{V^p}\right)^2 + \left(a_1\beta_{(A^p)^2} + a_2\beta_{A^p} - \beta_{B^p}\right)^2\right) \to \min.$

$$\begin{cases} \frac{\partial F(a_1 a_2)}{\partial a_1} = 0 \\ \frac{\partial F(a_1 a_2)}{\partial a_2} = 0 \end{cases} \tag{16}$$

We get the system from (16):

$$\begin{cases} a_1\left(\sum_{p=1}^{5}\left(\beta_{(Z^p)^2}\right)^2 + \sum_{p=1}^{5}\left(\beta_{(A^p)^2}\right)^2\right) + a_2\left(\sum_{p=1}^{5}\beta_{(Z^p)^2}\beta_{Z^p} + \sum_{p=1}^{5}\beta_{(A^p)^2}\beta_{A^p}\right) = \sum_{p=1}^{5}\beta_{V^p}\beta_{(Z^p)^2} + \sum_{p=1}^{5}\beta_{B^p}\beta_{(A^p)^2} \\ a_1\left(\sum_{p=1}^{5}\beta_{(Z^p)^2}\beta_{Z^p} + \sum_{p=1}^{5}\beta_{(A^p)^2}\beta_{A^p}\right) + a_2\left(\sum_{p=1}^{5}\beta_{(Z^p)^2} + \sum_{p=1}^{5}\beta_{(A^p)^2}\right) = \sum_{p=1}^{5}\beta_{V^p}\beta_{Z^p} + \sum_{p=1}^{5}\beta_{B^p}\beta_{A^p}. \end{cases}$$

$$\begin{cases} 1.001a_1 + 1.231a_2 = 0.822 \\ 1.231a_1 + 1.917a_2 = 1.187 \end{cases} \Rightarrow a_1 = 0.281, a_2 = 0.439 \Rightarrow V = 0.281Z^2 + 0.439Z.$$

Let input $Z = \left(\tilde{A}, \tilde{R}_Z\right)$, where $\tilde{A} = (0.9, 1, 0.2, 0)$, $\tilde{R}_Z = (0, 0, 0.25)$, then determine output $V = \left(\tilde{B}, \tilde{R}_V\right)$. Let us define $\tilde{B}$. The aggregating point for $\tilde{A}_{\text{mod}} = 0.281\tilde{A}^2 + 0.439\tilde{A}$ is equal to 0.656, aggregating points for $\tilde{C}_1 = (0, 0.2, 0, 0.1)$, $\tilde{C}_2 = (0.3, 0.7, 0.1, 0.2)$, $\tilde{C}_3 = (0.9, 1, 0.2, 0)$ are equal to 0.108, 0.508 and 0.934. As $|0.656 - 0.508| = 0.148 = \min_l\left|\beta_{\text{mod}} - \beta_{C_l}\right|$, $l = \overline{1,3}$, then $\tilde{B} = \tilde{C}_2 = (0.3, 0.7, 0.1, 0.2)$. Let us define $\tilde{R}_V$. The aggregating point for $V = 0.281Z^2 + 0.439Z$ is equal to 0.121, aggregating points of the products $\tilde{R}_h \times \tilde{A}_{\text{mod}}$, $h = \overline{1,5}$ are equal correspondingly to 0.014, 0.164, 0.328, 0.492, 0.642. As $|0.121 - 0.164| = 0.043 = \min_h\left|\beta_{Z_{\text{mod}}} - \beta_h\right|$, $h = \overline{1,5}$, then $\tilde{R}_V = \tilde{R}_2 = (0.25, 0.25, 0.25)$.

Thus, the predicted output Z-numbers is $V = \left(\tilde{C}_2, \tilde{R}_2\right)$, where $\tilde{C}_2 = (0.3, 0.7, 0.1, 0.2)$, $\tilde{R}_2 = (0.25, 0.25, 0.25)$.

5 Conclusion

In the paper, a predictive nonlinear regression model has been developed for the input and output Z-information. The research carried out is relevant and in demand since the regression analysis under Z-information is at the initial stage of development.

It is assumed that the components of input and output Z-numbers are values of full orthogonal semantic spaces. For these components, aggregating segments and aggregating points are determined, based of which, aggregating segments and aggregating points for input and output Z numbers are defined. Based on the aggregating points, the distances between model and initial data are determined. The optimization function is defined as the sum of the squares of the distances between the model Z-numbers and the initial Z-numbers and between the first components of the model Z-numbers and the initial Z-numbers. The unknown coefficients of the regression model are determined from the condition of the minimum optimization function.

The developed model makes it possible to process the reliability of expert information, construct a nonlinear relationship between the initial data, predict the output information and its reliability, which is essential when making decisions.

The paper provides a numerical example that shows how the new model expands the possibilities of regression analysis under initial Z-information.

References

1. Tanaka, H., Ishibuchi, H.: Identification of possibilistic linear models. Fuzzy Sets Syst. **41**, 145–160 (1991)
2. Ishibuchi, H.: Fuzzy regression analysis. Jpn. J. Fuzzy Theory Syst. **4**, 137–148 (1992)
3. Tanaka, H., Ishibuchi, H.: Exponential possibility regression analysis. Fuzzy Sets Syst. **69**, 305–318 (1995)
4. Chang, Y.-H.: Hybrid fuzzy least-squares regression analysis and its reliability measures. Fuzzy Sets Syst. **119**, 225–246 (2001). https://doi.org/10.1016/S0165-0114(99)00092-5
5. Domrachev, V.G., Poleshuk, O.M.: A regression model for fuzzy initial data. Autom. Remote Control **64**(11), 1715–1723 (2003). https://doi.org/10.1023/A:1027322111898
6. Poleshuk, O.M., Komarov, E.G.: Multiple hybrid regression for fuzzy observed data. In: Proceedings of the Annual Conference of the North American Fuzzy Information Processing Society, New York, May 2008, p. 4531224 (2008). https://doi.org/10.1109/NAFIPS.2008.4531224
7. Shavaei, B.N., Kamyad, A.V., Zare, A.: A piecewise type-II fuzzy regression model. Int. J. Comput. Intell. Syst. **10**(1), 734–744 (2017). https://doi.org/10.2291/ijcis.2017.10.1.49
8. Arefi, M.: Quantile fuzzy regression based on fuzzy outputs and fuzzy parameters. Soft Comput. **24**(1), 311–320 (2019). https://doi.org/10.1007/s00500-019-04424-2
9. Zadeh, L.A.: A note on Z-numbers. Inf. Sci. **14**(181), 2923–2932 (2011). https://doi.org/10.1016/j.ins.2011.02.022
10. Kang, B., Wei, D., Li, Y., et al.: Decision making using Z-numbers under uncertain environment. J. Inf. Comput. Sci. **8**(7), 2807–2814 (2012)
11. Kang, B., Wei, D., Li, Y., et al.: A method of converting Z-number to classical fuzzy number. J. Inf. Comput. Sci. **9**(3), 703–709 (2012)
12. Aliev, R.A., Zeinalova, L.M.: Decision making under Z-information. In: Guo, P., Pedrycz, W. (eds.) Human-centric decision-making models for social sciences. SCI, vol. 502, pp. 233–252. Springer, Heidelberg (2014). https://doi.org/10.1007/978-3-642-39307-5_10
13. Gardashova, L.A.: Application of operational approaches to solving decision making problem using Z-numbers. Appl. Math. **5**(9), 1323–1334 (2014)
14. Aliev, R.A., Alizadeh, A.V., Huseynov, O.H.: The arithmetic of discrete Z-numbers. Inf. Sci. **290**(1), 134–155 (2015). https://doi.org/10.1016/j.ins.2014.08.024
15. Aliev, R.K., Huseynov, O.H., Aliyeva, K.R.: Aggregation of an expert group opinion under Z-information. In: Proceedings of the Eighth International Conference on Soft Computing, Computing with Words and Perceptions in System Analysis, Decision and Control, Budva, Montenegro, August 2015, pp. 115–124 (2015)
16. Aliyev, R.R., Mraizid, D.A.T., Huseynov, O.H.: Expected utility-based decision making under Z-information and its application. Comput. Intell. Neurosci. **3**, 364512 (2015). https://doi.org/10.1155/2015/364512

17. Aliyev, R.R.: Similarity based multi-attribute decision making under Z-information. In: Proceedings of the Eighth International Conference on Soft Computing, Computing with Words and Perceptions in System Analysis, Decision and Control, Budva, Montenegro, August 2015, pp. 33–39 (2015)
18. Aliev, R.A., Huseynov, O.H., Zeinalova, L.M.: The arithmetic of continuous Z-numbers. Inf. Sci. **373**, 441–460 (2016). https://doi.org/10.1016/j.ins.2016.08.078
19. Wang, F., Mao, J.: Approach to multicriteria group decision making with Z-numbers based on TOPSIS and power aggregation operators. Math. Probl. Eng. **2**, 1–18 (2019). https://doi.org/10.1155/2019/3014387
20. Zeinalova, L.M., Huseynov, O.H., Sharghi, P.: A Z-number valued regression model and its application. Intell. Autom. Soft Comput. **24**, 187–192 (2017). https://doi.org/10.1080/10798587.2017.1327551
21. Poleshchuk, O.: Fuzzy regression model with input and output Z-numbers. In: IOP Conference Series: Materials Science and Engineering, vol. 919, no. 5, p. 052041 (2020). https://doi.org/10.1088/1757-899X/919/5/052041
22. Zadeh, L.A.: The concept of a linguistic variable and its application to approximate reasoning. Inf. Sci. **8**, 199–249 (1975). https://doi.org/10.1016/0020-0255(75)90036-5
23. Ryjov, A.P.: The concept of a full orthogonal semantic scope and the measuring of semantic uncertainty. In: Proceedings of the Fifth International Conference Information Processing and Management of Uncertainty in Knowledge-Based Systems, Iran, May 1994, pp. 33–34 (1994)
24. Poleshchuk, O.M.: Creation of linguistic scales for expert evaluation of parameters of complex objects based on semantic scopes. In: Proceedings of the 2018 International Russian Automation Conference (RusAutoCon - 2018), Sochi, Russia, September 2018, p. 8501686 (2018)

The Analysis of Student Performance During Face-to-Face and Distance Learning Under Z-Information

S. V. Tumor(✉) and O. M. Poleshchuk

Moscow Bauman State Technical University, 5, 2-nd Baumanskaya, Moscow 105005, Russia
tumor@bmstu.ru

Abstract. In this paper the comparative analysis of student performance during face-to-face and distance learning under Z-information is developed. Importance of similar researches has increased in the context of pandemic, when the quality of education has decreased due to the enforced transition of the majority of universities to online learning. Moreover the reliability of the students' grades has been affected negatively, which cannot be ignored. For this purpose an instrument for the processing of Z-information is essential. Student performance is associated with four Z-numbers, the first component represents the formalization of grades "unsatisfactory", "satisfactory", "good", "excellent". The formalizations of these grades constitute the valuation of a linguistic variable with the completeness and orthogonal properties, which aggregates the linguistic variables by three regulating criteria. The second component of Z-numbers is the fuzzy value of the reliability of formalized grades. Students' performance manifests itself through the total of four Z-numbers for a term of face-to-face learning and for a term of distance learning. To conduct a comparative analysis of the collected data operations with Z-numbers, distances between them and the distance between the populations of Z-numbers are specified. A numerical example is given demonstrating the function and effectiveness of this model.

Keywords: Distance learning · Z-number · Z-information · Weighted segment

1 Introduction

These days, all spheres of human activity are in continuous change and adaptation to current realities. For education this adaptation meant a transition from face-to-face to distance learning. But the student academic performance as well as the professor's degree of confidence about reliability of this information are changed. Since student evaluations are not random variables it is needed to choose another mathematical approach. One of the solutions of this problem can be the concept of Z-number proposed by Professor Lotfi Zadeh.

By definition [1] Z-number is a pair of two fuzzy numbers, where the first fuzzy number represents a fuzzy value of an uncertain real variable. The second fuzzy number is related to the first and it expresses a degree of confidence about the first fuzzy number. The

A. A. Radionov and V. R. Gasiyarov (Eds.): RusAutoCon 2021, LNEE 857, pp. 393–402, 2022.
https://doi.org/10.1007/978-3-030-94202-1_37

data presented in the form of Z-numbers is called Z-information. Z-numbers discovered the possibility to formalize the reliability of information taken from the expert so that this concept is very useful in many areas where the expert opinion plays an essential role. Combining all the advantages of fuzzy sets the concept of Z-number takes soft computing to the next level. The contribution made by Professor Zadeh to the process of modeling the real world is invaluable and many of its possibilities have yet to be discovered.

At present, the theoretical component of Z-numbers is not as developed as the practical, but scientists have made a significant contribution to methods of formalizing information based on Z-numbers [2] operations with them [3–6], and to applied fields of science – decision-making and prediction [7, 8].

In [2], the authors use the concept of t-norm and t-conorm to formalize expert opinion. In [3, 4], arithmetic operations on Z-numbers were developed. In [5], was proposed an approach to operate with Z-numbers based on α - cuts of its components. In [6], the authors integrate both components of Z-number and get one classical fuzzy number.

To compare Z-numbers it is needed to be able to rank them. Nowadays, there is no one unique approach for ranking. However, several options were proposed each with its own advantages and disadvantages. In [9] the authors suggest converting a Z-number to a crisp number by multiplying both of its components, but part of the information is lost with this approach. Authors in [10] use the principle of fuzzy optimality based on utility functions. In [11], the Jaccard similarity measure is used for ranking Z-numbers. Novel approach based on aggregative segments was proposed in the paper [12]. The model of object monitoring under Z-information was developed in [13].

Today, there are no models based on Z-numbers that analyze and compare student performance in face-to-face and distance learning. This paper is an attempt by the authors to fill this gap and also allows using the proposed approaches to operating with Z-numbers in other areas not related to learning.

The paper is organized as follows: Sect. 2 includes the basic concepts and definitions. Section 3 represents the formulation of the problem and its solution. Section 4 proposes a numerical example of the analysis of student performance during face-to-face and distance learning under Z-information. Section 5 gives conclusions.

2 Basic Concepts and Definitions

Let us define [14] a fuzzy set $\tilde{B}$ is a pair $\{x, \mu_B(x)\}, x \in X$, where membership function of fuzzy set $\tilde{B}$ is $\mu_B(x) : X \to [0, 1]$ and X - universal set of $\tilde{B}$.

Let us define [15] linguistic variable $\{X, T(X), U, V, S\}$, where X - is a name of a variable; $T(X) = \left\{X_l, l = \overline{1, m}\right\}$ - a term-set of values of the linguistic variable X (each one is a fuzzy variable on the set U); V - is a syntactical rule for generating names of new values X; S - is a semantic procedure to convert a new name generated by the V procedure into a fuzzy variable (determine the type of the membership function). We will call a semantic space a linguistic variable with a fixed term-set $\{X, T(X), U, S\}$ [15].

The properties of semantic spaces have been studied in [16–19].

Semantic spaces with the following properties 1–4 of terms $X_l, l = \overline{1,m}$ with membership functions $\mu_l(x), l = \overline{1,m}$ have been named Full Orthogonal Semantic Spaces (FOSS) [17]:

- For terms $X_l, l = \overline{1,m}$ the sets $\widehat{U}_l = \{x \in U : \mu_l(x) = 1\}$ are points or intervals.
- The membership functions $\mu_l(x), l = \overline{1,m}$ of terms $X_l, l = \overline{1,m}$ does not increase to the right of $\widehat{U}_l$ and does not decrease to the left of $\widehat{U}_l$.
- $\mu_l(x), l = \overline{1,m}$ have at most two first type discontinuity points.
- $\forall x \in U : \sum_{l=1}^{m} \mu_l(x) = 1.$

Z - number is an ordered pair of fuzzy numbers, $Z = \left(\tilde{A}, \tilde{R}\right)$, where $\tilde{A}$ is a fuzzy number represents fuzzy constraint on a uncertain variable $X \in \mathbb{R}$ with membership function $\mu_A(x) : X \to [0, 1]$ and $\tilde{R}$ with membership function $\mu_R(x) : [0, 1] \to [0, 1]$ is a fuzzy number represents degree of confidence of the first component. In other word $\tilde{R}$ is a fuzzy constraint on the measure of reliability of the first component such as sureness, probability or possibility, strength of belief [1].

Let consider an α - cut of fuzzy number $\tilde{A}$ with membership function $\mu_A(x) = (a, a_L, a_R)$ as the ordinary set A_α such as

$$A_\alpha = \{x \in \mathbb{R} : \mu_A(x) \geq \alpha\} = \left[A_\alpha^1, A_\alpha^2\right] = [a - (1-\alpha)a_L, a + (1-\alpha)a_R], \ \ \alpha \in [0, 1].$$

In [13], for fuzzy number $\tilde{A}$ $(\mu_A(x) = (a_1, a_2, a_L, a_R))$ a weighted segment $[\Theta_1, \Theta_2]$ has been defined in the following form:

$$\begin{aligned}\Theta_1 &= \int_0^1 \frac{2a_1 - (1-\alpha)a_L}{2} 2\alpha d\alpha = a_1 - \frac{1}{6}a_L, \\ \Theta_2 &= \int_0^1 \frac{2a_2 + (1-\alpha)a_R}{2} 2\alpha d\alpha = a_2 + \frac{1}{6}a_R.\end{aligned} \tag{1}$$

To determine the weighted segment, the weighting function for the α - cuts is taken equal to $p(\alpha) = 2\alpha$.

Consider the formulation of the problem and its solution.

3 Problem Formulation and Solution

In this paper is developed the comparative analysis of student performance during face-to-face and distance learning under Z-information.

The approach to the formalization of Z-numbers proposed in this paper is based on the principles that were developed in [12, 13]. Conclusions about the improvement or worsening of student performance are made based on the comparison of formalized Z-numbers with the ideal Z-number.

Let us solve the problem in general form. A numerical example will be represented in the next section.

Let us consider qualitative characteristics K^1 - student performance during face-to-face learning, K^2 - student performance during distance learning where each one has characteristics X_1, X_2, X_3. So that X_1 - «Grades for tests», X_2 - «Grades for home tests», X_3 - «Grades for the exam». These characteristics completely determine the characteristic $K^v, v = \overline{1,2}$.

For characteristics $X_j, j = \overline{1,3}$ let us define the levels of verbal scales as $X_{ij},\ i = \overline{1,4}, j = \overline{1,3}$, accordingly, with corresponding verbal scale: C_1 - "Unsatisfactory", C_2 - "Satisfactory", C_3 - "Good", C_4 - "Excellent". Fuzzy numbers $\widetilde{C}_i,\ i = \overline{1,4}$ have membership functions $\mu_i(x) = (a_{i1}, a_{i2}, a_{iL}, a_{iR}),\ i = \overline{1,4}$.

We construct the FOSSs with names $X_j^v, j = \overline{1,3}, v = \overline{1,2}$ for grades "Unsatisfactory", "Satisfactory", "Good", "Excellent". Term sets for $\tilde{X}_j^v, j = \overline{1,3}, v = \overline{1,2}$ are $\tilde{X}_{ij}^v,\ i = \overline{1,4}, j = \overline{1,3}, v = \overline{1,2}$, accordingly, with membership functions $\mu_{ij}^v(x),\ i = \overline{1,4}, j = \overline{1,3}, v = \overline{1,2}$ [20]. The values for all FOSSs are numbers from universal set [0, 1].

Define reliability of information as FOSS R with linguistic terms: U - «Unlikely», NVL - «Not very likely», L – "Likely", VL – "Very likely", EL – "Extremely likely". Membership functions of these terms are $\mu_U, \mu_{NVL}, \mu_L, \mu_{VL}, \mu_{EL}$ and let us define fuzzy numbers corresponding to its by $\tilde{R}_k,\ k = \overline{1,5}$.

Let us assign estimates of characteristics as Z-numbers $\left(\tilde{X}_{ij}^v, \tilde{R}_k^v\right), i = \overline{1,4}, j = \overline{1,3},\ k = \overline{1,5}, v = \overline{1,2}$ in relation to characteristics $X_j,\ j = \overline{1,3}$. The evaluation of the v-th characteristic we denote by $Z_i^v = \left(\tilde{X}_i^v, \tilde{R}_i^v\right), i = \overline{1,4}, v = \overline{1,2}$ in relation to the levels of characteristics $X_j,\ j = \overline{1,3}$. The levels are represented by $C_i, i = \overline{1,4}$. We determine fuzzy numbers $\tilde{X}_i^v, i = \overline{1,4}, v = \overline{1,2}$ as $\tilde{X}_i^v = \omega_1 \otimes \tilde{X}_{i1}^v \oplus \omega_2 \otimes \tilde{X}_{i2}^v \oplus \omega_3 \otimes \tilde{X}_{i3}^v$ with membership function $\mu_i^v(x) = \left(\sum_{j=1}^{3} \omega_j a_{j1}^v, \sum_{j=1}^{3} \omega_j a_{j2}^v, \sum_{j=1}^{3} \omega_j a_{jL}^v, \sum_{j=1}^{3} \omega_j a_{jR}^v\right), i = \overline{1,4}, v = \overline{1,2}$. To determine fuzzy number $\tilde{R}_i^v, v = \overline{1,2}, i = \overline{1,4}$ with membership function $\mu_i^v(x) = (r_i^v, r_{iL}^v, r_{iR}^v),\ v = \overline{1,2}, i = \overline{1,4}$ we correlate it with one of the fuzzy numbers $\tilde{R}_k, k = \overline{1,5}$ and $r_i^v = \max_j\left(r_j^v\right), r_{iL}^v = \max_j\left(r_{Lj}^v\right), r_{iR}^v = \max_j\left(r_{Rj}^v\right)$, $v = \overline{1,2}, j = \overline{1,3}, i = \overline{1,4}$.

The weights of the characteristics are $\omega_1 = 0.3, \omega_2 = 0.3, \omega_3 = 0.4$ correspondingly for X_1, X_2, X_3.

Fuzzy rating of the v-th characteristic is defined as $Z^v = \left(\widetilde{A}^v, \widetilde{R}^v\right), v = \overline{1,2}$. Z^1 represents student performance during face-to-face learning, Z^2 represents student performance during distance learning. Fuzzy number $\widetilde{A}^v, v = \overline{1,2}$ we determine as $\widetilde{A}^v = \omega_1 \otimes \tilde{X}_1^v \oplus \omega_2 \otimes \tilde{X}_2^v \oplus \omega_3 \otimes \tilde{X}_3^v \oplus \omega_4 \otimes \tilde{X}_4^v$ with membership function $\mu^v(x) = \left(\sum_{i=1}^{4} \omega_i a_{i1}^v, \sum_{i=1}^{4} \omega_i a_{i2}^v, \sum_{i=1}^{4} \omega_i a_{iL}^v, \sum_{i=1}^{4} \omega_i a_{iR}^v\right), v = \overline{1,2}$. The weights of the evaluations $\tilde{X}_i^v, i = \overline{1,4}, v = \overline{1,2}$ are $\omega_i = 0.25, i = \overline{1,4}$. To determine fuzzy number $\tilde{R}^v, v = \overline{1,2}$ with membership function $\mu^v(x) = (r^v, r_L^v, r_R^v),\ v = \overline{1,2}$ we correlate it with one of the fuzzy numbers $\tilde{R}_k, k = \overline{1,5}$ and $r^v = \max_i (r_i^v), r_L^v = \max_i (r_{Li}^v), r_R^v = \max_i (r_{Ri}^v), v = \overline{1,2}, i = \overline{1,4}$.

Based on [13] we define a fn-segment $[\Theta_1, \Theta_2]$ for fuzzy number $\tilde{A}$ with membership function $\mu_A(x) = (a, a_L, a_R)$ as follows:

$$\begin{aligned} \Theta_1 &= \int_0^1 \frac{2a - (1-\alpha)a_L}{2} 3\alpha^2 d\alpha = a - \frac{1}{8} a_L, \\ \Theta_2 &= \int_0^1 \frac{2a + (1-\alpha)a_R}{2} 3\alpha^2 d\alpha = a + \frac{1}{8} a_R. \end{aligned} \tag{2}$$

For fuzzy number $\tilde{A}$ with membership function $\mu_A(x) = (a_1, a_2, a_L, a_R)$ a fn-segment is:

$$\begin{aligned} \Theta_1 &= \int_0^1 \frac{2a_1 - (1-\alpha)a_L}{2} 3\alpha^2 d\alpha = a_1 - \frac{1}{8} a_L, \\ \Theta_2 &= \int_0^1 \frac{2a_2 + (1-\alpha)a_R}{2} 3\alpha^2 d\alpha = a_2 + \frac{1}{8} a_R. \end{aligned} \tag{3}$$

To determine the weighted segment the weighting function for the α- cuts is taken equal to $p(\alpha) = 3\alpha^2$.

$$3\alpha^2 - 2\alpha > 0, \alpha(3\alpha - 2) > 0, \alpha \in [0, 1] \Rightarrow \frac{2}{3} < \alpha \le 1.$$

Taking the weighting function $p(\alpha) = 3\alpha^2$ as opposed to $p(\alpha) = 2\alpha$ in formula (1), we want to give more weight to α - cuts larger than $\frac{2}{3}$, since these α - cuts play the most important role for recognizing fuzzy concepts and definitions, than α - cuts smaller than $\frac{2}{3}$, which in [12] have no priority.

Let us determine the distance between two Z - numbers $Z^1 = (\tilde{A}^1, \tilde{R}^1), Z^2 = (\tilde{A}^2, \tilde{R}^2)$ with fn-segments $\left[\Theta_1^{A^1}, \Theta_2^{A^1}\right], \left[\Theta_1^{R^1}, \Theta_2^{R^1}\right]$ and $\left[\Theta_1^{A^2}, \Theta_2^{A^2}\right], \left[\Theta_1^{R^2}, \Theta_2^{R^2}\right]$:

$$d(Z^1, Z^2) = \sqrt{\begin{array}{l} \left(\Theta_1^{A^1} - \Theta_1^{A^2}\right)^2 + \left(\Theta_2^{A^1} - \Theta_2^{A^2}\right)^2 + \left(\Theta_1^{R^1} - \Theta_1^{R^2}\right)^2 + \left(\Theta_2^{R^1} - \Theta_2^{R^2}\right)^2 \\ + \left(\Theta_1^{A^1}\Theta_1^{R^1} - \Theta_1^{A^2}\Theta^{R^2}\right)^2 + \left(\Theta_2^{A^1}\Theta_2^{R^1} - \Theta_2^{A^2}\Theta_2^{R^2}\right)^2 \end{array}} \tag{4}$$

This formula takes into account both components of the fn-segment and it is more general than the formula based on weighted points which does not allow, for example, to distinguish between two unimodal and two tolerant fuzzy numbers [6].

Determine the left and the right borders of fn-segment $\Theta_1^{A^\nu}$, $\nu = \overline{1,2}$ and $\Theta_2^{A^\nu}$, $\nu = \overline{1,2}$ for the fuzzy number $\widetilde{A}^\nu = \omega_1 \otimes \tilde{X}_1^\nu \oplus \omega_2 \otimes \tilde{X}_2^\nu \oplus \omega_3 \otimes \tilde{X}_3^\nu \oplus \omega_4 \otimes \tilde{X}_4^\nu$ with membership function $\mu^\nu(x) = \left(\sum_{i=1}^{4} \omega_i a_{i1}^\nu, \sum_{i=1}^{4} \omega_i a_{i2}^\nu, \sum_{i=1}^{4} \omega_i a_{iL}^\nu, \sum_{i=1}^{4} \omega_i a_{iR}^\nu\right), \nu = \overline{1,2}$:

$$\Theta_1^{A^\nu} = \sum_{i=1}^{4} \omega_i a_{i1}^\nu - \frac{\sum_{i=1}^{4} \omega_i a_{iL}^\nu}{8}, \ \nu = \overline{1,2};$$

$$\Theta_2^{A^\nu} = \sum_{i=1}^{4} \omega_i a_{i2}^\nu + \frac{\sum_{i=1}^{4} \omega_i a_{iR}^\nu}{8}, \ \nu = \overline{1,2}.$$

Determine the left and the right borders of fn-segment $\Theta_1^{R^\nu}$, $\nu = \overline{1,2}$ and $\Theta_2^{R^\nu}$, $\nu = \overline{1,2}$ for the fuzzy number $\widetilde{R}^\nu$, $\nu = \overline{1,2}$ with membership function $\mu^\nu(x) = \left(r^\nu, r_L^\nu, r_R^\nu\right)$:

$$\Theta_1^{R^\nu} = r^\nu - \frac{r_L^\nu}{8}, \ \nu = \overline{1,2};$$

$$\Theta_2^{R^\nu} = r^\nu + \frac{r_R^\nu}{8}, \ \nu = \overline{1,2}.$$

Let us define the ideal Z-number in the framework of the problem as $Z^{id} = \left(\tilde{A}^{id}, \tilde{R}^{id}\right)$ where $\tilde{A}^{id}$ has membership function $\mu^{A^{id}}(x) = (1, 1, 0, 0)$ and $\tilde{R}^{id}$ has membership function $\mu^{R^{id}}(x) = (1, 0, 0)$. Fn-segments are $\Theta_1^{A^{id}} = 1$, $\Theta_2^{A^{id}} = 1$ and $\Theta_1^{R^{id}} = 1$, $\Theta_2^{R^{id}} = 1$.

Calculate $d(Z^\nu, Z^{id})$, $\nu = \overline{1,2}$. If $d(Z^1, Z^{id}) > d(Z^2, Z^{id})$ then student performance has improved, if $d(Z^1, Z^{id}) < d(Z^2, Z^{id})$ then student performance has become worse, if $d(Z^1, Z^{id}) = d(Z^2, Z^{id})$ then student performance has not changed.

4 Numerical Example

Let the professor evaluates performance of a group of students during face-to-face and distance learning according to the following criteria: «Grades for tests», «Grades for home tests», «Grades for the exam». It is required to evaluate whether performance of the group has improved has become worse or has not changed.

We define «Grades for tests», «Grades for home tests», «Grades for the exam» during face-to-face learning and during distance learning as X_1^1, X_2^1, X_3^1 and X_1^2, X_2^2, X_3^2, accordingly. We will call $\widetilde{X}_1^1, \widetilde{X}_2^1, \widetilde{X}_3^1$ and $\widetilde{X}_1^2, \widetilde{X}_2^2, \widetilde{X}_3^2$ fuzzy numbers corresponding to the formalizations of these characteristics.

The linguistic scale corresponding to these characteristics has the form: C_1 - «Unsatisfactory», C_2 - «Satisfactory», C_3 - «Good», C_4 - «Excellent».

The data obtained are presented in Tables 1 and 2.

We construct membership functions with methods [20].

Table 1. Performance for 100 students during face-to-face learning.

During face-to-face learning	Grades for tests	Grades for home tests	Grades for the exam
Unsatisfactory	10	20	10
Satisfactory	40	40	30
Good	30	30	40
Excellent	20	10	20

Table 2. Performance for 100 students during distance learning.

During distance learning	Grades for tests	Grades for home tests	Grades for the exam
Unsatisfactory	10	10	20
Satisfactory	30	40	40
Good	40	30	30
Excellent	20	20	10

For characteristics $X_j^v, i = \overline{1,3}, v = \overline{1,2}$ membership functions are $\mu_{ij}^v(x),\ i = \overline{1,4}, j = \overline{1,3}, v = \overline{1,2}$. Then.

$\mu_{11}^1 = (0, 0.05, 0, 0.1)$, $\mu_{21}^1 = (0.15, 0.35, 0.1, 0.3)$, $\mu_{31}^1 = (0.65, 0.7, 0.3, 0.2)$, $\mu_{41}^1 = (0.9, 1, 0.2, 0)$;

$\mu_{12}^1 = (0, 0.1, 0, 0.2)$, $\mu_{22}^1 = (0.3, 0.45, 0.2, 0.3)$, $\mu_{32}^1 = (0.75, 0.85, 0.3, 0.1)$, $\mu_{42}^1 = (0.95, 1, 0.1, 0)$;

$\mu_{13}^1 = (0, 0.05, 0, 0.1)$, $\mu_{23}^1 = (0.15, 0.25, 0.1, 0.3)$, $\mu_{33}^1 = (0.55, 0.7, 0.3, 0.2)$, $\mu_{43}^1 = (0.9, 1, 0.2, 0)$;

$\mu_{11}^2 = (0, 0.05, 0, 0.1)$, $\mu_{21}^2 = (0.15, 0.25, 0.1, 0.3)$, $\mu_{31}^2 = (0.55, 0.7, 0.3, 0.2)$, $\mu_{41}^2 = (0.9, 1, 0.2, 0)$;

$\mu_{12}^2 = (0, 0.05, 0, 0.1)$, $\mu_{22}^2 = (0.15, 0.35, 0.1, 0.3)$, $\mu_{32}^2 = (0.65, 0.7, 0.3, 0.2)$, $\mu_{42}^2 = (0.9, 1, 0.2, 0)$;

$\mu_{13}^2 = (0, 0.1, 0, 0.2)$, $\mu_{23}^2 = (0.3, 0.45, 0.2, 0.3)$, $\mu_{33}^2 = (0.75, 0.85, 0.3, 0.1)$, $\mu_{43}^2 = (0.95, 1, 0.1, 0)$.

Define reliability of information as linguistic variable R with linguistic terms: U – "Unlikely", NVL – "Not very likely", L – "Likely", VL – "Very likely", EL – "Extremely likely" with membership functions:

$$\mu_U = (0, 0, 0.25),\ \mu_{NVL} = (0.25, 0.25, 0.25),\ \mu_L = (0.5, 0.25, 0.25),$$
$$\mu_{VL} = (0.75, 0.25, 0.25),\ \mu_{EL} = (1, 0.25, 0).$$

Next step we should aggregate $\tilde{X}_{ij}^v, i = \overline{1,4}, j = \overline{1,3}, v = \overline{1,2}$ to get a fuzzy number $\tilde{X}_i^v, i = \overline{1,4}, v = \overline{1,2}$ for each grade: "Unsatisfactory", "Satisfactory", "Good", "Excellent". The weights of the characteristics are $\omega_1 = 0.3$, $\omega_2 = 0.3$, $\omega_3 = 0.4$ correspondingly for X_1, X_2, X_3. Thus we obtain the following membership functions $\mu_i^v(x), i = \overline{1,4}, v = \overline{1,2}$:

$\mu_1^1 = (0, 0.065, 0, 0.13)$, $\mu_2^1 = (0.195, 0.34, 0.13, 0.3)$, $\mu_3^1 = (0.64, 0.745, 0.3, 0.17)$, $\mu_4^1 = (0.915, 1, 0.17, 0)$;

$\mu_1^2 = (0, 0.07, 0, 0.14)$, $\mu_2^2 = (0.21, 0.36, 0.14, 0.3)$, $\mu_3^2 = (0.66, 0.76, 0.3, 0.16)$, $\mu_4^2 = (0.92, 1, 0.16, 0)$.

Next step we should aggregate reliability $\tilde{R}_k^v$, $k = \overline{1,5}$, $v = \overline{1,2}$ to get a fuzzy number $\tilde{R}_i^v$, $i = \overline{1,4}$, $v = \overline{1,2}$ for each grade: "Unsatisfactory", "Satisfactory", "Good", "Excellent".

The professor expressed degrees of confidence for the characteristics X_1^1, X_2^1, X_3^1: "Likely", "Not very likely", "Likely", correspondingly. For the characteristics X_1^2, X_2^2, X_3^2 the professor's degrees of confidence are: "Not very likely", "Not very likely", "Not very likely", correspondingly.

Thus, we obtain the following membership functions $\mu_i^v(x)$, $i = \overline{1,4}$, $v = \overline{1,2}$:

$\mu_1^1 = (0.5, 0.25, 0.25)$, $\mu_2^1 = (0.5, 0.25, 0.25)$, $\mu_3^1 = (0.5, 0.25, 0.25)$, $\mu_4^1 = (0.5, 0.25, 0.25)$;

$\mu_1^2 = (0.25, 0.25, 0.25)$, $\mu_2^2 = (0.25, 0.25, 0.25)$, $\mu_3^2 = (0.25, 0.25, 0.25)$, $\mu_4^2 = (0.25, 0.25, 0.25)$.

Hence the corresponding Z-numbers take the form $Z_i^v = \left(\tilde{X}_i^v, \tilde{R}_i^v\right)$, $i = \overline{1,4}$, $v = \overline{1,2}$.

Let us calculate fuzzy rating for student performance during face-to-face learning $Z^1 = \left(\tilde{A}^1, \tilde{R}^1\right)$ and during distance learning $Z^2 = \left(\tilde{A}^2, \tilde{R}^2\right)$. Since the weights of the evaluations $\tilde{X}_i^v$, $i = \overline{1,4}$, $v = \overline{1,2}$ are $\omega_i = 0.25$, $i = \overline{1,4}$ we obtain fuzzy number $\tilde{A}^1$ with membership function $\mu^1 = (0.4375, 0.5375, 0.15, 0.15)$ and fuzzy number $\tilde{A}^2$ with membership function $\mu^2 = (0.4475, 0.5475, 0.15, 0.15)$. Fuzzy number represents reliability - $\tilde{R}^1$ has membership function $\mu^1 = (0.5, 0.25, 0.25)$ and fuzzy number represents reliability - $\tilde{R}^2$ has membership function $\mu^1 = (0.25, 0.25, 0.25)$.

As we received both ratings Z^1 and Z^2 then we should calculate fn-segments $\left[\Theta_1^{A^1}, \Theta_2^{A^1}\right], \left[\Theta_1^{R^1}, \Theta_2^{R^1}\right]$ and $\left[\Theta_1^{A^2}, \Theta_2^{A^2}\right], \left[\Theta_1^{R^2}, \Theta_2^{R^2}\right]$ by using formulas (2) and (3):

$\left[\Theta_1^{A^1}, \Theta_2^{A^1}\right] = [0.4188, 0.5563]$, $\left[\Theta_1^{R^1}, \Theta_2^{R^1}\right] = [0.4688, 0.5313]$;

$\left[\Theta_1^{A^2}, \Theta_2^{A^2}\right] = [0.4288, 0.5663]$, $\left[\Theta_1^{R^1}, \Theta_2^{R^1}\right] = [0.2188, 0.2813]$.

Next using the ideal Z-number and applying formula (4) we calculate the distance $d(Z^1, Z^{id})$ and $d(Z^2, Z^{id})$:

$d(Z^1, Z^{id}) = 1.4761$, $d(Z^2, Z^{id}) = 1.7803$.

Since $d(Z^1, Z^{id}) < d(Z^2, Z^{id})$ then student performance during distance learning has become worse than student performance during face-to-face learning.

5 Conclusions

In the paper a model for the comparative analysis of student performance during face-to-face and distance learning under Z-information is developed. The authors determine a new concept of fn-segment for a fuzzy number and a distance between two Z-numbers using fn-segments. This research is relevant because at the present there are no models for comparative analysis of student performance taking into account the degree of reliability of information received from an expert.

Fuzzy rating of student performance is represented by Z-numbers and the first component of Z-numbers is calculated based on four grades «unsatisfactory», «satisfactory», «good», «excellent». Performance of students is assessed according to three criteria «Grades for tests», «Grades for home tests», «Grades for the exam». The second component of Z-numbers expresses the degree of professor's confidence in the given marks according to the corresponding criteria. Both components of Z-numbers are fuzzy numbers which are values of Full Orthogonal Semantic Spaces. Calculating the distance between Z-numbers, both of their components are taken into account. Comparison of student performance during face-to-face and distance learning is made using the ideal Z-number.

The paper provides a numerical example of the proposed approach of the comparative analysis of student performance during face-to-face and distance learning under Z-information.

Developed model, within the framework of Z-information, can be used not only for a comparative analysis of academic performance, but also to determine the state of objects in different periods in comparison with the ideal state.

References

1. Zadeh, L.A.: A note on Z-numbers. Inf. Sci. **181**(14), 2923–2932 (2011)
2. Aliev, R.R., Huseynov, O.H., Aliyeva, K.R.: Z-valued T-norm and T-conorm operators-based aggregation of partially reliable information. In: Proceedings of the 12th International Conference on Application of Fuzzy Systems and Soft Computing (2016)
3. Aliev, R.A., Alizadeh, A.V., Huseynov, O.H.: The arithmetic of discrete Z-numbers. Inf. Sci. **290**, 134–155 (2015)
4. Aliev, R.A., Huseynov, O.H., Zeinalova, L.M.: The arithmetic of continuous Z-numbers. Inf. Sci. **373**, 441–460 (2016)
5. Dutta, P., Boruah, H., Ali, T.: Fuzzy arithmetic with and without - cut method: a comparative study. Int J. Latest Trends Comput. **2**(1), 99–107 (2011)
6. Kang, B., Wei, D., Li, Y., Deng, Y.: A method of converting Z-number to classical fuzzy number. J. Inf. Comput. Sci. **9**(3), 703–709 (2012)
7. Wang, F., Mao, J.: Approach to multicriteria group decision making with Z-numbers based on TOPSIS and power aggregation operators. Math. Probl. Eng. **2019**(3), 1–18 (2019)
8. Yager, R.R.: On Z-valuations using Zadeh's Z-numbers. Int. J. Intell. Syst. **27**(3), 259–278 (2012)
9. Kang, B., Wei, D., Li, Y., Deng, Y.: Decision making using Z-numbers under uncertain environment. J. Inf. Comput. Sci. **8**(7), 2807–2814 (2012)
10. Aliyev, R.R., Mraizid, D.A.T., Huseynov, O.H.: Expected utility based decision making under Z-information and its application. Comput. Intell Neurosci. **2015**(3) (2015). https://doi.org/10.1155/2015/364512
11. Aliyev, R.R.: Similarity based multi-attribute decision making under Z-information. In: 2015 Proceedings of the Eighth International Conference on Soft Computing, Computing with Words and Perceptions in System Analysis, Decision and Control (2015)
12. Poleshchuk, O.M.: Novel approach to multicriteria decision making under Z-information. In: Proceedings of the International Russian Automation Conference (2019). https://doi.org/10.1109/RUSAUTOCON.2019.8867607

13. Poleshchuk, O.M.: Object monitoring under Z-information based on rating points. In: Kahraman, C., Cevik Onar, S., Oztaysi, B., Sari, I.U., Cebi, S., Tolga, A.C. (eds.) INFUS 2020. AISC, vol. 1197, pp. 1191–1198. Springer, Cham (2021). https://doi.org/10.1007/978-3-030-51156-2_139
14. Zadeh, L.A.: Fuzzy sets. Inform. Control **8**(3), 338–353 (1965)
15. Zadeh, L.A.: The concept of a linguistic variable and its application to approximate reasoning. Inf. Sci. **8**(3), 199–249 (1975)
16. Ryjov, A.P.: Description of objects in human-machine information systems. In: 1994 Proceedings of the International Conference on Application of Fuzzy Systems (1994)
17. Ryjov, A.P.: The concept of a full orthogonal semantic scope and the measuring of semantic uncertainty. In: 1994 Proceedings of the Fifth International Conference Information Processing and Management of Uncertainty in Knowledge-Based Systems (1994)
18. Ryjov, A.P.: Fuzzy data bases: description of objects and retrieval of information. In: 1993 Proceedings of the First European Congress on Intelligent Technologies (1993)
19. Lowe, W.: Towards a theory of semantic space. In: 2001 Proceedings of the 23rd Conference of the Cognitive Science Society (2001)
20. Poleshchuk, O.M., Komarov, E.G., Darwish, A.: Comparative analysis of expert criteria on the basis of complete orthogonal semantic spaces. In: 2016 Proceedings of the 19th International Conference on Soft Computing and Measurements (2016)

A Tool to Automate the Assessment of Patient Dynamics in Intensive Care Units, Based on a Specialized Methodology

N. Serzhantova, M. Sidorova, and A. Syomin(✉)

Penza State Technological University, Russia, 1a/11, Ul. Gagarina, Penza 440039, Russia
anton.semin.88@inbox.ru

Abstract. Today the most functional complex control and measurement system (CMS) is one used in intensive care units (ICU). However, the problem of compact and informative work results presentation of the control and measurement resuscitation complexes is still relevant. Problems include the low sensitivity and discriminative ability of existing patient assessment scales. There also are data problems such as limited time for obtaining, processing, analyzing and interpreting. Another problem is the lack of a unified approach to determining the patient's condition dynamics (PCD), and the large amount to be transformed and visualized data. The article is devoted to identifying the problems of automated PCD assessment. The main goal of the study is to develop a methodology for PCD automated assessment. The methodology includes eight stages: the initial stage is the research tasks formulation, the final stage is the results visualization. The developed methodology is the basis for the automated program. This program will help to significantly facilitates and accelerates the diagnosis process, patient dynamics assessment and outcome prediction. The software was tested on the clinical data taken from patients having peritonitis.

Keywords: Methodology · Dynamics assessment · Diagnosis · Prognosis · Resuscitation and intensive care · Outcome · Neural networks

1 Introduction

The widespread and gradual introduction of information technologies into clinical practice is the priority direction of modern medicine and medical technology. Rapid computer technologies development as one of the new requirement in the health policy set the task of creating complex CMS for the diagnosis and diseases monitoring.

Among the integrated CMS, the ICU systems are the most developed and implemented ones. These systems are aimed at optimizing a wide range of tasks. They should minimize the burden of routine operations on the medical staff. This will help to save precious time for the patients' treatment. Another important task is organization and structuring the data flows recorded by CMS, as well as providing doctor intellectual support in decision-making. This helps to track the dynamics of quantitative indicators,

A. A. Radionov and V. R. Gasiyarov (Eds.): RusAutoCon 2021, LNEE 857, pp. 403–413, 2022.
https://doi.org/10.1007/978-3-030-94202-1_38

conduct differential diagnostics and accurately predict the course of the disease and its outcome in lack of time conditions.

The modern ICU is a high-tech structure that allows tracking thousands of body state parameters of each patient 24/7. However, most of this data is not used to understand the patient profile and improve treatment outcomes. For many years, resuscitators have used individual patient data for monitoring and tracking the severity and trajectory of organ failure. They used scoring systems [1], but the current level of CMS development allows implementing powerful assessment systems that can measure vital signs, track dynamic changes and support decision-making. However, automation of a multiple processes at the same time is difficult due to the peculiarities of the measured data. Therefore, the main task of the study is to identify the problems of automating the procedure for assessing the PCD. The main goal is developing a methodology for automated assessment of the PCD, which allows significantly reducing the process of making a diagnosis and predicting the disease outcome.

2 Relevance and Scientific Significance of the Research

There are special tools to make the right decision regarding the assessment of the severity of the patient's condition and the following disease prognosis. It scales (systems) for assessing the disease severity and the outcome predicting. Currently, there are about 1,500 different scales that directly or indirectly reflect the severity of the patient's condition. The probability of their negative effects is represented as a sum of points or other integral assessment. There are universal ones, such as APACHE II, SOFA, SAPS, MODS, and specialized ones for assessing the severity of an injury – ISS, TRISS, RTS. All these quantitative systems for assessing the condition severity make it easier for the doctor to predict the outcome. It also helps to make decisions about the tactics of the patient managing. Another important part of research purposes and controlling the quality of treatment is the capability of comparing the treatment results of different patient groups. The average predictive accuracy of quantitative assessing systems which identifying the severity of the condition is 85%. These systems enable to give a pretty accurate prognosis for patients with low scores (survival) and high scores (death). With intermediate values, the prognostic accuracy is significantly lower, due to the great importance of the timeliness and correctness of treatment [2].

To improve the quality of forecasting, medical information systems based on neural network technologies are currently being widely implemented [3–6]. At the same time, a number of problems remain unresolved due to the accuracy and reliability of automated PCD assessment. All this certainly have great importance for the immediate and effective therapeutic measures adjustment. The precise outcome prognosis depends on the disease diagnosis, as well as body state diagnosis with the identification of its reserves and capabilities at each moment. Another attribute is the correct and qualitative PCD assessment [7]. Therefore, the development of methods and tools for automated patient's condition assessment, determining the condition dynamics and the disease outcome predicting is a very important and urgent task.

3 Problems of Automating the Process of Assessing the patient's Condition Dynamics

Clinicians must have a clear understanding of the patient's medical history, physiological characteristics, and possible response to treatment. All of this will provide safe and effective resuscitation care. Therefore, there is a need to integrate a huge amount of data in real time. Specialists of medical institutions should be able to qualitatively process any information, including physical examination data, information received from colleagues, the results of radiological studies, and mathematically analyze laboratory data. Even a simple clinical decision can be affected by insufficient information or an incorrect assessment and prevent it from being made [2].

Modern systems used in ICU allow to accurately recording a large number of the patient's condition parameters. The software as part of the control and measurement resuscitation complexes is able to simultaneously receive and consolidate the registered clinical indicators from resuscitation monitors. It also helps to form an electronic map of intensive care and a list of prescriptions for medicines and procedures. There is a need in clinical practice to reduce the flow of information, especially digital, coming to the attending physician. This can be done either by using a specially developed mathematical language for describing the problem or visualization methods. Nowadays the problem of compact and informative results presentation of control and measurement resuscitation complexes remains very relevant. Resolving this problem will allow effectively monitoring the PCD and keep in mind all of the necessary and sufficient data volume. All of this builds the basis which gives the user a very quick, accurate and objective assess the severity and the PCD.

There are several aspects that make it difficult to automate the assessing process illustrating the severity of a patient's condition, the dynamics and disease outcome.

The first problem is the imperfection of patient assessment scales and complicated objective diagnosis. Low sensitivity with sufficiently high specificity is a common disadvantage of existing scales. These features of integral scales enable to divide patients into groups for scientific research and report writing, but practically it makes the scales unsuitable for deciding on the patient treatment tactics [8].

The second problem is the shortage of time to make a decision in the ICU. There is also a need to deal with a patient whose nosologic diagnosis may remain unknown for a long time. This problem makes it difficult to optimize the therapeutic and diagnostic process in the ICU. For the anesthesiology and intensive care a syndrome approach to the critical conditions diagnosis may be a promising direction for the unified therapeutic, tactical, organizational and methodological standards creation [9].

Syndrome diagnosis in the ICU is based on the fact that "resuscitation" syndromes are mostly are a variety of nonspecific pathological conditions (nosologic units) with a risk factors range. All this factors have similar manifestations in the case of extreme forms [10].

More than 2.5 thousand syndromes are described in the specialized medical literature. Any syndrome is essentially an integral characteristic of several parameters (symptoms) connected by the unity origin mechanism. At the qualitative level, it is preferable to assess a syndrome, since this approach helps the ICU physician to judge the whole clinical situation more adequately. At that point each physician is guided by his clinical thinking, his experience, his degree of understanding of the disease mechanisms, and the diagnostic techniques available to him. The presence and severity of the syndrome are usually determined approximately ("there is" - "there isn't"), based on the specialist experience. The dynamics of the syndrome ("positive", "negative", and "no dynamics") is very important. However, this approach makes it possible to use the algorithmic diagnostic principle, as opposed to solving diagnostic problems by the heuristic method. The main difference between these two principles is the amount of information. Applying the nosologic approach, the doctor deals with a huge volume of information. The syndrome approach gives to doctor only single elements of information [11].

Automated determination of the PCD and outcome prediction based on the syndrome approach can be included in the functionality of comprehensive diagnostic monitoring and measurement systems. However, the process of automation tools developing is significantly hampered by the lack of conceptual and classification unity. Another problem is the difference in intensive care specialists' opinions, domestic and foreign. The number of leading syndromes and the lack of quantitative measures of syndrome severity are also needed to be considered. All this is the third problem of assessing the PCD and predicting the outcome of the disease.

The fourth important problem is partly correlating with the third. It is the lack of accurate methods for determining the dynamics of the disease.

If a single parameter is analyzed (trend models are mostly used) it is easy to monitor the change dynamics. However, the number of constantly monitored parameters in the ICU is measured in hundreds. Thus the assessing of change severity in one particular parameter makes it very difficult. This problem is also related to the general patient condition and disease development. The values of the linguistic variable characterizing in the textual representation the PCD ("positive", "negative", and "no dynamics") for successful automation of the diagnostic process and outcome prediction must obey a certain semantic rule that determines the algorithmic procedure for calculating the meaning of each value.

In the specialized literature several approaches to determining the patient's state dynamics have been already described. One of them is based on the clinical state index. Another one is based on one of the most significant parameters. The last of them is based on one or more parameters with mandatory consideration of improvement of well-being [7]. Despite this none of these approaches is effective for the ICU.

Currently, the assessment of a resuscitation patient's condition is increasingly performed by using the syndrome approach, which is based on several assumptions. Body parameters deviation degree can be "insignificant", "significant" or "prohibitive" (both increasing and decreasing relatively to the average statistical values). Possible values of any organism parameter are divided into conditional ranges (10 for each parameter). If the parameter value deviation from the average statistical norm is "insignificant", then it falls into the +3S or −3S range (depending on whether the deviation is found to be increasing or decreasing). Similarly, cases where the deviation is "significant" (4S/+4S) or "prohibitive" (−5S/+5S) are also considered. The normal range is further subdivided into 4 subranges (2S, −1S, +1S, +2S). The deviation degrees of any parameters are comparable with each other but their absolute values aren't [11]. The dynamics in this case is the transition of the parameter value between the ranges (or subranges).

Such an approach to the integral assessment of a patient's condition has a number of advantages. For example, a clinician has no need to keep in mind the entire set of defined parameters (a resuscitation patient has more than 100 of them). But only those parameters that deviate from their average statistical values in the case of pathology, established for one of them by expert assessments. However, this method also has one disadvantage. The translation of quantitative parameter values into their "qualitative" characteristics requires the creation of special reference books of correspondence of numerical values intervals of each parameter to its "qualitative" characteristic [12]. The reference books are formed by using expert estimates. For all kinds of pathologies there is its own reference book.

Thus, there are two promising directions in the development of comprehensive intelligent CMS for medical purposes. The first one is the formation of reference books of nosologic units using modern information technology capabilities. The second one is software tool development that allows visualizing the PCD within syndromes for each nosologic unit during the transition from one qualitative characteristic to another.

4 Methodology for Automated Assessment of Patient Dynamics in Intensive Care Units

The methodology of automated patient dynamics assessment in the ICU involves a certain sequence of actions that must be performed in order to successfully applying information technology to find the solution for intensive diagnosis and therapy problems.

The diagnosis process, determining the severity of the patient's condition ("satisfactory", "moderate severity", "severe", etc.) or the dynamics ("positive", "negative", "no change") can be identified with solving a classification problem. The process of predicting the outcome, under certain conditions, can also be referred to as classification problems. The main condition in this case will be the prediction not in the form of a time series, but the linguistic equivalent of the outcome of some development trend.

Various mathematical methods, such as discriminant, correlation, regression analysis, etc., can be used to solve the problems of classification and attributing an object using a certain set of features to one of the known classes. However, the development of automation tools using traditional mathematical methods can be significantly complicated by the nature of the data under study, the complexity of formalizing the dependencies and the large volume of computational operations. The most promising experience for diagnostics and prediction can be considered the applying neural network technologies [13–20] as a part of complex CMS considering all features of the input information.

The methodology for patient dynamics assessing in the ICU is presented in Fig. 1. It includes eight main stages.

The first stage involves a clear formulation of the tasks to be solved. This stage is mandatory, accompanied by a preliminary diagnosis and an approximate assessment of the severity of the patient's condition. The analysis tasks are formulated depending on what kind of neural network apparatus a particular ICU has. For example, a patient has peritonitis. The tasks to be solved can be to determine the outcome of peritonitis (recovery, death in the distant or early postoperative period), to determine the deviation of the set of syndrome parameters (a separate task for each syndrome).

Sometimes, in addition to the task of patient's condition assessing, a differential diagnosis may be required, which is also formulated as a separate task. The list of tasks determines the number of applied neural networks.

At this stage, there must be a simple and intuitive selection of the components of the neural network unit of the system involved in the further PCD assessing in the ICU. This choice is supposed to be implemented using user interface elements.

The second stage is data acquisition. It involves selecting only those body state data that is needed for resolving the tasks formulated in the previous stage. This stage should be performed fully automatically and initiated by the selection of the required tasks using the user interface at the previous stage. For successful automation in the system development it is necessary to clearly specify the correspondence of the selected parameters to the tasks. Another important step is determining the transformation mechanisms of qualitative variables into quantitative ones, and to ensure the transfer of the selected data to the block of the system responsible for the subsequent manipulations with the data.

The third stage is an automated data set formation for the subsequent analysis. It also takes place can be conducted without the user's participation. At this stage the selected data are presented in a form, which is able to be perceived by a neural network unit (it is defined by features of the programming environment, in which neural networks were created). At this stage the transposition of sets and transformation of data types can be carried out. The generated array is fed to the input of all neural networks, which are involved in the solution of the formulated tasks. The degree of user involvement in this process is minimal. The user should simply run the neural network analysis by pressing the button on the interface control.

The fourth stage (the neural network analysis process itself) is also performed automatically. In this case, each neural network involved in the analysis generates an output set, which contains encoded information about the solution of a particular problem. It is supposed to use several neural networks in parallel to solve different types of tasks, because it increases the sensitivity and specificity of the final result.

At the fifth stage all output sets are combined into a common one. Then the results are interpreted. The elements of the row (column) of the final set correspond to one particular solution, and the values presented as numbers represent the code of the general solution. This stage must also be automated as much as possible.

To evaluate and control parameters (dynamics) changes, it is necessary to compare the new results with the data obtained earlier. Such a process is realized at the sixth stage. At this stage user should define the dynamics evaluation period (for example, from the beginning of admission of the patient to the ICU or the previous result of the analysis). Relationship operators can be used in the comparison, but it is more effective to find the difference between the results of each individual solution. In this way it is possible to record not only the fact that the patient's condition has changed, but also to assess the extent to which this condition has changed (e.g., not just worsened, but the deviation from the norm has changed from insignificant to prohibitive). The degree of change in the patient's condition is evaluated at the seventh stage. Then the automatic determination of correspondence between the obtained difference of solutions and the linguistic dynamics expression is implied.

The eighth stage includes an automated procedure for plotting the PCD.

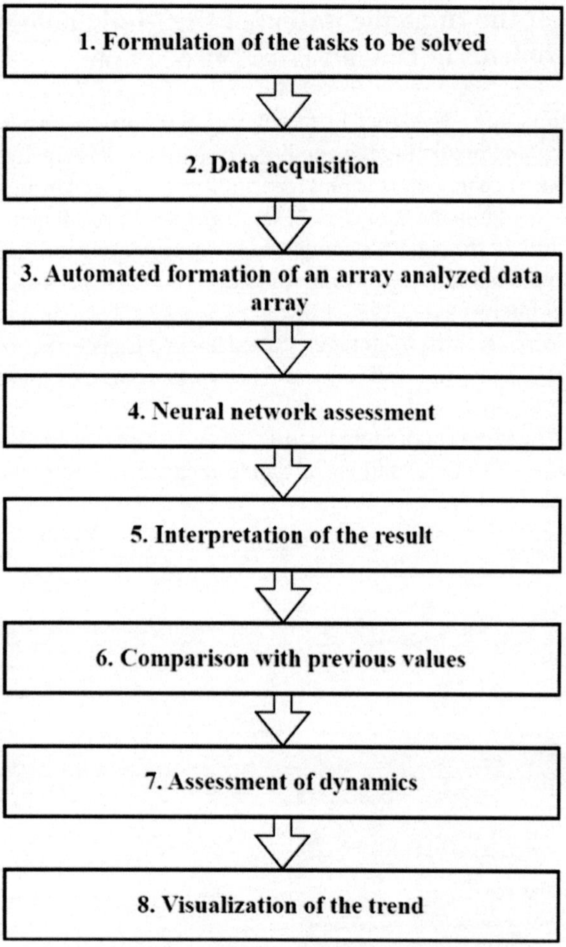

Fig. 1. The methodology for PCD assessing in the ICU

Practical implementation of the technique proposed by the authors of the article will allow implementing of several processes. The first thing is accumulating the measurements results. The second one is analyzing the patient's condition in dynamics according to the selected syndromes. And the third thing is interpretation and visualization of the results of neural network analysis with the required detailing.

5 Examples of the Implementation of the Methodology in Solving Private Problems of Diagnosis and Prediction

Based on the methodology developed by the authors of the article, an automated software tool for diagnosing and predicting the condition of patients was created. It performs the following functions: 1) parameters values measurement characterizing the patient's condition and saving them into the database; 2) set formation that will help with differential diagnosis, evaluation of the patient's condition, PCD, predicting the disease outcome; 3) input sets neural network analysis; 4) interpretation of neural network analysis results; 5) visualization of interpreted results of neural network analysis, including comparison with previous research results; 6) storage, processing and transferring of data sets at any functioning stage; 7) "post-training" of neural networks on new examples; 8) evaluation of neural network unit performance.

For example, the program of automated PCD assessment, created by the authors of the article, allows to visualize markers of syndromes, considering coagulation potential, thrombinemia level, platelet hemostasis, anticoagulation system work (Fig. 2). Color indication of the values is performed according to a developed reference book of nosologic units, a fragment of which is shown in Table 1.

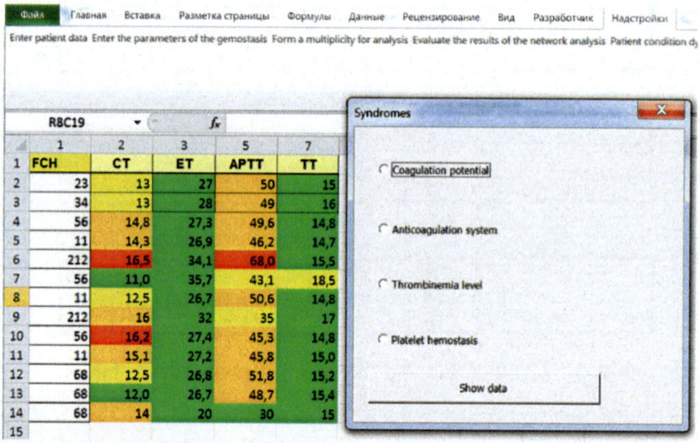

Fig. 2. Visualization of syndrome markers characterizing the coagulation potential.

Currently, the automated program has been tested to solve several ICU tasks. The first one is assessing the condition of a patient with peritonitis. The second one is predicting the outcome of this disease.

Clinical data from four groups of patients were used for testing (Table 2): control group, healthy (0), 53 persons; group 1, patients who recovered (mean hospital stay 30.7 $\pm$ 1.7 bed-days), 40 persons; group 2, patients who died in the long-lasting period (mean hospital stay 20.8 $\pm$ 2.5 bed-days), 47 persons; group 3, patients who died in the early postoperative period (mean hospital stay 1.7 $\pm$ 0.3 bed-days), 57 persons.

The software tool is able to predict the following disease outcomes. Outcome one ("I1") patients stayed alive. Outcome two ("I2") implies that patients died in the remote

Table 1. Handbook of nosologic units

Syndrome	Syndrome Markers	Deviation									
		-5S	*-4S*	*-3S*	*-2S*	*-1S*	*+1S*	*+2S*	*+3S*	*+4S*	*+5S*
Coagulation potential	Activated partial thrombin time (APTT)	Less than 20	From 20	From 22	From 24	From 27	To 31	To 34	To 45	To 55	Over 55
	Echinotoxic time	Less than 13	From 13	From 16	From 9	From 25	To 35	To 41	To 50	To 60	Over 60
	Thrombin time	Less than 10	From 10	From 12	From 14	From 16	To 17	To 18	To 22	To 24	Over 24
	Clotting time	Less than 4	From 4	From 6	From 8	From 9	To 10	To 12	To 14	To 16	Over 16
Anticoagulation system	Antithrombin-3 (AT-3)	Less than 76	From 76	From 78	From 80	From 90	To 100	To 110	To 112	To 114	Over 114

Table 2. Characteristics of patient groups.

Syndrome	Syndrome markers	Patient groups			
		0	*1*	*2*	*3*
Coagulation potential	APTT	31,9 ± 0,8	49,1 ± 1,3	45,5 ± 3,7	60,1 ± 16,5
	Echinotoxic time	25,5 ± 0,5	26,7 ± 0,5	26,8 ± 0,5	31,7 + 2,6
	Thrombin time	14,9 ± 0,3	15,2 ± 0,2	14,7 ± 0,3	20,1 ± 3,6
	Clotting time	11,1 ± 0,9	12,6 ± 0,5	13,5 ± 1,4	17,0 ± 4,4
Anticoagulation system	AT-3	97,4 ± 0,8	90,4 ± 0,5	91,4 ± 1,9	80,6 ± 6,4

period (14 or more days after admission to a hospital). Outcome three ("I3") implies that patients died in the early (including early postoperative) period (2–3 days after admission to a hospital). In the course of resuscitation measures the resuscitator must constantly monitor the changes in the measured and recorded parameters. Also making a large number of calculations is required.

An example of predicting the disease outcome and visualizing the PCD is shown in Fig. 3.

Practical application of the developed software automation tool has shown that its diagnostic efficiency is 92%. Meanwhile the average diagnostic efficiency of resuscitation scales is about 85%.

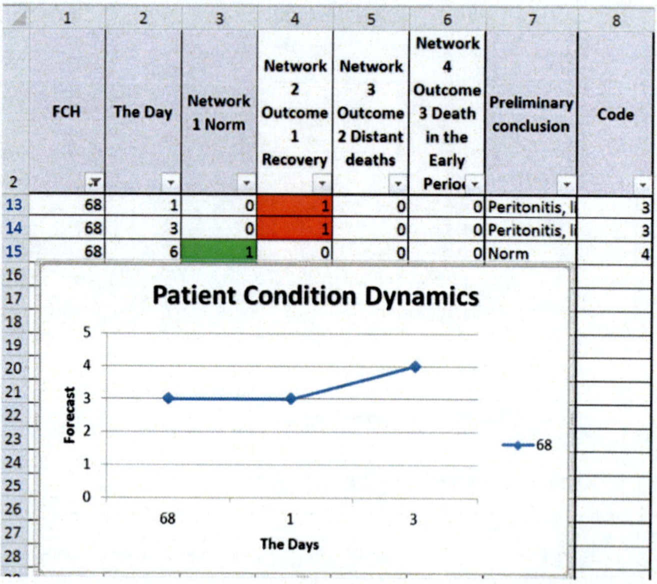

Fig. 3. Example of the disease outcome predicting and visualizing the PCD.

6 Conclusions

As a result of a detailed analysis of the problems that arise when solving the tasks of automating the process of the PCD evaluating and predicting the disease outcome, the methodology for evaluating the PCD has been developed.

When the methodology has been developing, a syndromic approach to parameter assessment was applied, i.e. analysis of the study of changes in the syndrome trend, which combines the dynamics of a limited set of parameters. In this case there was no need to analyze the single integral parameter or the entire set of parameters.

The practical embodiment of the technique was the creation of a software tool and its testing to solving individual tasks of diagnosing and predicting the condition of patients with peritonitis and hemorrhagic fever. Functional tests of the developed program have shown that it has higher sensitivity and specificity in comparison with traditional scales.

Thus scaling up the diagnostic and prognostic system and expanding the tasks range to be solved will provide a universal software automation tool that will help the resuscitator to quickly and accurately determine the PCD and the degree of therapy efficiency.

References

1. da Silva Ramos, F.J., Salluh, J.I.F.: Data-driven management for intensive care units. ICU Manag. Pract. **19**(1), 20–23 (2019)
2. Pickering, B.W., Littell, J.M., Geraskevich, V., Gajic, O.: Clinical review: hospital of the future. Creating a structure for safe and effective treatment of intensive care patients. Med. Technol. Eval. Sel. **3**, 28–39 (2013)

3. Frize, M., Ennett, C.M., Stevenson, M., Trigg, H.C.: Clinical decision support systems for intensive care units: using artificial neural networks. Med. Eng. Phys. **23**(3), 217–225 (2001)
4. Holmgren, G., Andersson, P., Jakobsson, A., Frigyesi, A.: Artificial neural networks improve and simplify intensive care mortality prognostication: a national cohort study of 217,289 first-time intensive care unit admissions. J. Intensive Care **7**, 44 (2019)
5. Hsieh, M.H., Hsieh, M.J., Chen, C., Hsieh, C., Chao, C., Lai, C.: An artificial neural network model for predicting successful extubation in intensive care units. J. Clin. Med. **7**(9), 240 (2018)
6. Klyuchko, O.M.: Application of artificial neural networks method in biotechnology. Biotechnologia Acta **10**(4), 5–14 (2017)
7. Polyakov, G.A.: Trifles in intensive medicine, pp. 7–10. Publishing House "Sovetskaya Kuban", Krasnodar (1998)
8. Alexandrovich, Yu.S., Gordeev, V.I.: Evaluation and prognostic scales in critical condition medicine, guide, pp. 22–43. Publishing House "Sotis", St. Petersburg (2007)
9. Ivanov, R.V., Sadchikov, D.V., Prigorodov, M.V.: Algorithm of diagnostics in patients in critical condition. Fundam. Res. **10**, 501–504 (2011)
10. Sadchikov, D.V., Prigorodov, M.V.: Features of diagnostics in resuscitation patients (review). Saratov Sci. Med. J. **7**(2), 404–409 (2011)
11. Vasilkov, V.G., Safronov, A.I.: Syndromology of critical states in the clinical activity of a practical doctor. Med. Alphabet **2**(9), 56–59 (2015)
12. Safronov, A.I., Istomina, T.V., Minkin, A.V., Lukyanova, A.A.: Visualization of patient condition parameters in critical condition medicine. XXI Century: Results of the Past and Problems of the Present Plus **1**(17), 135–137 (2014)
13. Patil, S.B., Kumaraswamy, Y.S.: Intelligent and effective heart attack prediction system using data mining and artificial neural network. Eur. J. Sci. Res. **31**(4), 642–656 (2009)
14. Andersson, B., Andersson, R., Ohlsson, M.: Prediction of severe acute pancreatitis at admission to hospital using artificial neural networks. Pancreatology **11**, 28–35 (2011)
15. Gorunescu, F., Gorunescu, M., Saftoiu, A.: Competitive/collaborative neural computing system for medical diagnosis in pancreatic cancer detection. Expert Syst. **28**(1), 33–44 (2011)
16. Kurt, I., Ture, M., Kurum, T.A.: Comparing performances of logistic regression, classification and regression tree, and neural networks for predicting coronary artery disease. Expert Syst. Appl. **34**(1), 366–374 (2008)
17. Hirose, H., Takayama, T., Hozawa, S.: Prediction of metabolic syndrome using artificial neural network system based on clinical data including insulin resistance index and serum adiponectin. Comput. Biol. Med. **41**, 1051–1056 (2011)
18. Saraoglu, H., Temurtas, F., Altikat, S.: Quantitative classification of HbAlC and blood glucose level for diabetes diagnosis using neural networks. Australas. Phys. Eng. Sci. Med. **36**(4), 397–403 (2013). https://doi.org/10.1007/s13246-013-0217-x
19. Caliskan, A., Yuksel, M.E.: Classification of coronary artery disease data sets by using a deep neural network. EuroBiotech J. **1**(4), 271–277 (2017)
20. Gusev, A.V., Pliss, M.A.: The basic recommendations for the creation and development of information systemsin health care based on artificial intelligence. Doct. Inf. Technol. **3**, 45–60 (2018)

Researching the Fundamentals of Decision Synthesis into Technical Systems of Intelligent Data Processing

M. Makarov and A. Astafiev(✉)

Vladimir State University, 87, Gorky Street, Vladimir 600000, Russia

Abstract. The work is devoted to the concept of decision synthesis within the components as a part of technical systems of intelligent data processing. A methodology of experimental research aimed at substantiating the scientific and practical significance of this concept has been developed and implemented. As an object of research, we used a computer model of the component of decision-making, which provides a pattern recognition procedure based on data obtained about an external analyzed object. The results of the research represent the response of the object to changes in external conditions that affect the formation of the decision. It was revealed that the incorporation of the principles of decision synthesis into the object of research promoted the emergence of cognitive mechanisms in the process of information processing, which led to an increase in the adaptive abilities of the technical system to change the external conditions of its existence.

Keywords: Artificial intelligence · Artificial neural networks · Decision-making · Pattern recognition · Intelligent data processing

1 Introduction

The transition to innovative technologies for establishing the computational process within systems of intelligent data processing (SIDP) is impossible without eliminating the constraints related to the modern hardware and software implementations of artificial intelligence (AI).

First of all, the fact that the functioning of the discussed systems still has a strict algorithmic form, which is reduced to the use of combinatorial methods, can be considered as a constraining factor. For this reason, system functioning is deterministic at each stage of the process preventing the incorporation of cognitive functions into it, which is important for the autonomous study of the environment and decision-making under changing objectivity conditions.

This aspect is important for SIDP, since the experience of the practical application of these systems for solving various applied problems shows that in most cases, with external conditions of the environment changing, their functioning significantly affects the data processing quality and in some cases makes correct functioning impossible [1].

A. A. Radionov and V. R. Gasiyarov (Eds.): RusAutoCon 2021, LNEE 857, pp. 414–423, 2022.
https://doi.org/10.1007/978-3-030-94202-1_39

2 Current State of Research on the Subject

Analyzing the current state of research on the considered subject, we have studied papers published in leading scientific and technical sources. Today, SIDP can be most effectively implemented by means of technologies that imply using a fundamentally new approach to the synthesis of computational mechanisms in an algorithmic sense. These technologies are associated with a group of methods known as machine learning [2, 3, 4, 5, 6, 7], which grants systems the ability to learn from known examples, which is believed to determine their belonging to the smart type. The experience of the practical application of machine learning shows that it becomes possible to construct mathematical models of dynamic processes without laborious, and often impossible, analytical descriptions of computed functions. Furthermore, a system can operate with fuzzy concepts and, as a result, solve applied problems with a high degree of accuracy, which are reduced to special cases of approximation of mathematical functions, classification and clustering of objects of various nature, pattern recognition in any data types, forecasting, etc.

To assess the SIDP development level related to using cognitive mechanisms to generate solutions for certain tasks, their ability to maintain the specified performance indicators during autonomous functioning under changing objectivity conditions shall be used as the main criterion. The fundamental research carried out so far has contributed to the accumulation of basic theoretical background that reveals the principles underlying the methods of SIDP design and functioning with signs of smart decision-making, but has not provided the final results proving that these technologies are based on the cognitive machine abilities.

An overwhelming number of studies in this area suggest connectionist models based on neural networks to be used as the main tool for building the basic computational architecture of SIDP. Specific examples can be papers devoted to the development of control systems based on neural networks for dynamic processes, image processing in machine vision systems and automated image recognition, information security, big data mining, and decision-making based on expert systems. The considered papers describe experimental studies, which do not always take into account the fact that the data processing quality indicators obtained at the learning stage may not have the same values during functioning in real conditions.

The information found in the considered scientific paper leaves open the question of creating SIDP capable of an autonomous reaction to changes in the external conditions of its environment. As a rule, the applied methods are adapted to solving a practical problem in ideal conditions and do not consider the issues of ensuring that the data processing process stays within the specified accuracy limits when exposed to the external environment.

Certain studies include attempts to provide an autonomous smart behavior for a computing system. In such a case, its functioning is based on methods that are not invariant to the practical application area and are consistent only for solving a single, highly specialized applied problem. Some methods are based on variations of approaches to segmentation and analysis of multivariate time series, which in some cases is not feasible in practice.

The next important criterion for proving the possibility of implementing cognitive mechanisms in SIDP is the use of the potential of modern computational methods by means of innovative approaches at the hardware implementation stage for these systems. The theory and practice of SIDP design prove that these systems shall be implemented in a form of specialized hardware with modern hardware components suitable for the applied algorithms. However, the main widely-used modern approach to the creation of machine learning tools is still based on software emulation using classical computing systems with von Neumann architecture. Thus, the lack of approaches to the creation of a general fundamental concept of AI that takes into account this aspect, as well as the systematization, standardization and unification of the process of engineering design of specialized software and hardware components remains an insurmountable obstacle to the emergence of cognitive mechanisms in SIDP.

Summarizing the abovementioned, we can conclude that only private theoretical and practical developments currently exist that do not sufficiently recreate the complete or partial similarity of cognitive mechanisms in SIDP. The data processing process in the discussed systems is still has a deterministic algorithmic form without the incorporation of functions that are important for the autonomous study of its environment and decision-making under changing objectivity conditions.

Thus, an inevitable stage in AI modernization is conducting research aimed at the formalization of new fundamental laws necessary for the implementation of quasi-cognitive functions providing an autonomous and adaptable behavior of SIDP during functioning under the influence of internal factors and the external environment.

3 Methods

If we consider the human brain as an ideal model of the data transformation and decision-making process organization including the ability to perform cognitive operations, then one of the fundamental differences of this biological cognitive tool compared to the computer technology is the fact that its high effectiveness is due to focusing on the decision synthesis.

Therefore, if we consider an object based on any existing neural network technology related to the field of machine learning as SIDP, then it can be said that such a computational architecture exists as a deterministic finite state machine [8].

In other words, after learning, such a system can adopt a limited number of states, which determines the current value of the output data vector at each stage. Then we can conclude that during functioning, the system has a finite set of basic decisions, and the procedure of mathematical approximation of the function's value is performed if required with specific input parameters. This clearly contradicts the principle of maintaining the quality of the system's functioning when the general rule of data transformation is changed affected by the external destabilizing influences.

The currently known methods of additional learning assume the conditional plasticity of states, which lies in the plane of adjustment of parameter values of computational elements (weight coefficients and biases), which does not imply taking into account changes in the general rule of data transformation. Thus, the approximated function can change its values with variations of the arguments, but not its properties that determine its inherent regularity. Moving away from an algorithmic deterministic behavior requires other methods of establishing plasticity within the considered objects, which enable the

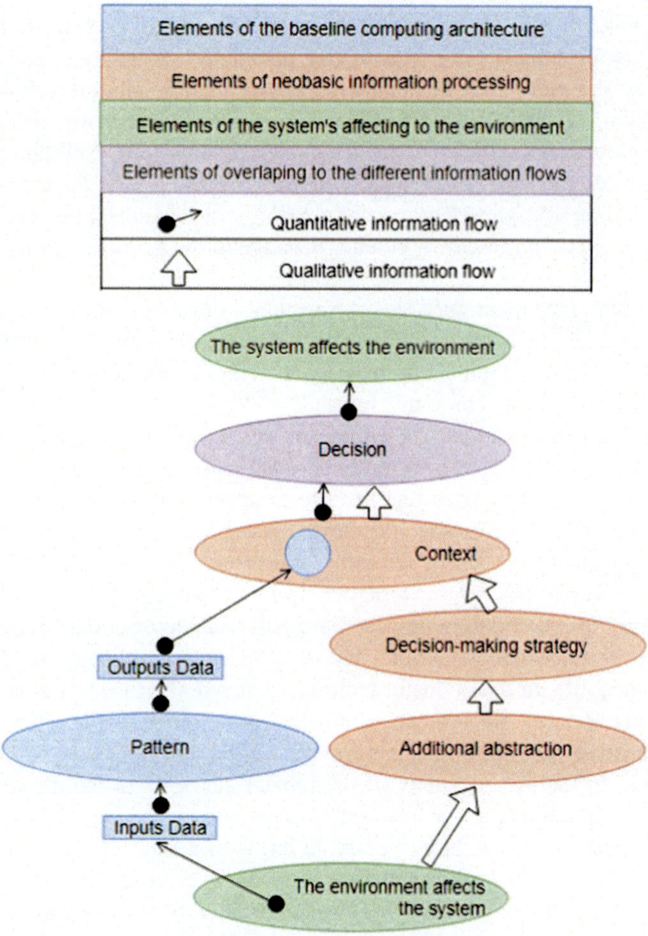

Fig. 1. Illustration of the decision synthesis concept for SIDP.

system to adapt its state to the requirements of the current problem being solved, which is not included in the general set used at the learning stage.

To solve the formulated scientific problem, the decision synthesis concept for SIDP has been proposed (Fig. 1), which is based on four principles described below.

First of all, to reveal the meaning of a schematic representation of the proposed concept shown in Fig. 1, SIDP structural elements shall be determined taking into account the system's interaction with the physical world.

The system's interaction with the physical world is represented by two cases, where the system acts as an object and subject of influence. In the first case, the system is focused on solving a certain applied problem, and this requires obtaining certain data from an external source for its subsequent processing. Let us call the fact of data reception as an influence on the system, prompting it to switch to operation. In another case, the system itself influences the elements of the environment based on the obtained data processing results. In the existing paradigm for SIDP design, each of these two cases of interaction is formalized using quantitative parameters called input and output data, respectively. At the same time, the nature of each of the influences is a physical change in the surrounding environment, while the influence acting both on the system and in the opposite direction is part of the objective reality. It is assumed that excluding the physical measurement from the impact formalization process is the reason for the impossibility of the manifestation of cognitive properties in SIDP, which would contribute to the fault-tolerant system operation in case a potential change in the general data transformation rule.

Based on the abovementioned, let us formulate the first principle of the concept concerning the decision synthesized by the system as a result of its operation: "The output data determines the impact of the system on the external environment, but is not a decision, since it is not aware of this impact."

Now let us introduce the concept of the external environment for SIDP. This term means a certain part of reality where physical interaction of the system with the applied problem being solved takes place. For example, the external environment for a mobile robotic system is the space in which it moves, restrictions imposed on the process of its existence and manifestation of activities, as well as channels for data transmission in both possible directions. The external environment affects the system, and for the classical version of a computing device, the input data vector is the equivalent of this impact. However, under the conditions requiring fault-tolerant operation of the system when the general rule for data transformation is changed, the formalization of the input influence in the form of quantitative parameters is not enough. The impact does not have an exhaustive description by the applied abstraction apparatus via quantitative parameters due to the heterogeneity of the impact elements in the physical sense. It is assumed that the impact of the external environment on the system and systems on the external environment can be expressed in terms of a state of equilibrium, and the developed decision is a tool for establishing this state of equilibrium.

The result of generalization of the presented theoretical information will be the second principle of the concept: "The effects of the system on the external environment can be known by establishing an associative relationship between a cognitive agent and the external environment of its existence, which forms the experience of solving local problems inside the system that are not included in the general set of tasks existing in the object's environment at a given time."

The implementation of the previous principles requires a special tool for organizing the cognitive activity process. For this purpose, the data duality concept is proposed to be incorporated into the system. This principle is based on the idea of the intersection of two fundamental classes of data, mutually influencing each other, but not changing the overall structure. The first class is quantitative data in digital form, which is used by the system to perform the programmed action. The second class of data is the qualitative properties of types of system influence on the external environment, which generate a contradiction within the system and develop its cognitive activity. It is assumed that data duality within the system will act as a new cognitive semantics providing the required effect, which allows to organize the process of combining heuristic solutions with the transfer of known operations that were originally developed under different conditions, but used to achieve current goals.

In this case, the third principle can be represented as follows: "The synthesis of decision is based on the data duality within the system – the division of data flow into two classes, the existence of these classes and their intersection".

As it was specified earlier, the input data received for processing is represented exclusively by quantitative parameters of the properties of the analyzed object and is isolated from a certain integrative entity (which systematizes a given form object, defining it as a part of reality). For instance, the operation of an automated face recognition system is based on biometric data, however, humans recognize other individuals without exact knowledge of these properties and use it in connection with the general picture that generates associative phenomena in the nervous system. Therefore, it can be assumed that each object has a certain integrative essence, which contains additional hidden data that can be used to intellectualize SIDP. In turn, the previously declared strategy of the system's behavior is closely related to the analysis of this integrative entity.

Thus, an additional data class appears from an additional abstraction of the external environment impact on the system: "The strategy is formed within the system based on an additional abstraction of the external environment impact initiating the division of the data flow into two classes, which can be defined as quantitative and qualitative ones."

4 Results

To assess the scientific and practical significance of the proposed concept as an object of the experimental study, we used a computer model of the pattern recognition and decision-making component as part of an abstract SIDP. It is assumed that a certain part of the physical world (the analyzed object) is a source of images. As a result of the image identification, the component must generate the system's response to the external environment. Each image has a strict geometric interpretation, i.e., can be described by a certain two-dimensional figure. To perceive it, multiple parameters are measured, which are supposed to be sufficient to compare the image with one of the four basic states of the analyzed object, four variants of special cases of basic states and other anomalous states that are irrelevant during recognition. The measurement results enter the system as a vector of input data $X = \{x_1, x_2, x_3, x_4, x_5, x_6\}$, where x_1 – object height, x_2 – object width, x_3 – object area, x_4 – object perimeter, x_5 – area of a circle inscribed in the object, x_6 – area of the circumscribed circle. The component transforms the received input information and makes a decision regarding the observed object state. The accepted decision is the output vector $Y = \{y_1, y_2, y_3\}$, where y_1 – image type indicator for the object's base state, y_2 – image indicator for a particular case of the object's base state, y_3 – image indicator for any abnormal condition of the object. A training set containing 5000 examples was used as a training sample. The sample included sample values of geometric parameters of 4 image types of the object's basic states (Fig. 2), 4 image types for particular cases of object's base states (Fig. 2), various image types for anomalous states of the object (multiple polygon varieties), as well as the corresponding set of target values of the output vector Y. Figure 3 shows the principle of general data transformation rule change.

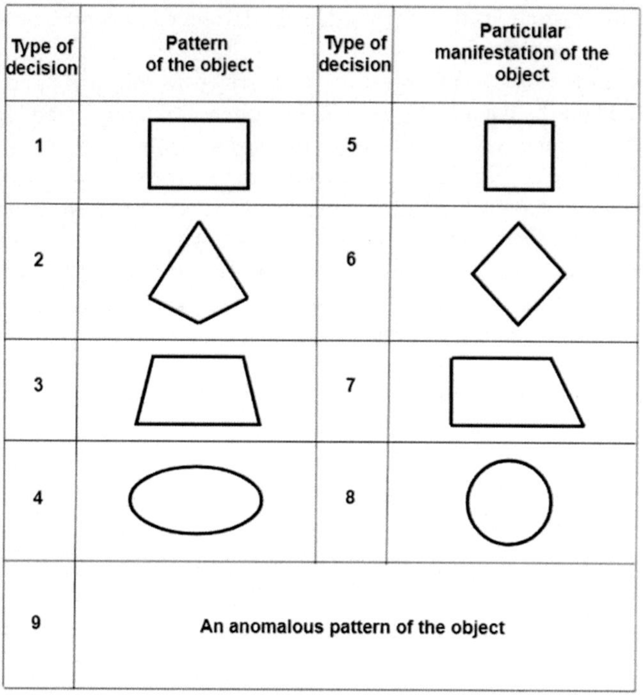

Fig. 2. Geometric interpretation of the types of recognizable images of the analyzed object states, where are the initial learning conditions for the basic computing architecture.

A basic computational architecture model was developed using MATLAB for the investigated component performing pattern recognition and decision-making. The model is an artificial neural network (ANN) capable of performing an approximation of a given function with a high degree of accuracy. It is a three-layer fully-connected feedforward network with 55 neurons in the first layer, 18 in the second, as well as 3 output neurons. The activation function for the first and second layers is tangential, while for the third layer – a threshold function. The Levenberg-Marquardt algorithm with Bayesian regularization (TRAINBR function in MATLAB) was used for training. The training was carried out until the maximum quality of data transformation was obtained, which allowed to obtain the probability of correct pattern recognition of 0.97 for 100 test examples.

The experimental study consisted of several stages. At the first stage, the basic neural network computing architecture has undergone a change in the general data transformation rule. As a factor of such a change was the replacement of each of the four main images with an image resembling the original one, but not completely the same (Fig. 2). This impact significantly affected the probability of correct pattern recognition, which dropped to 0.21 per 100 test cases.

After that, the decision synthesis mechanisms were incorporated based on the proposed principles. Since the proposed study was primarily aimed at obtaining results to

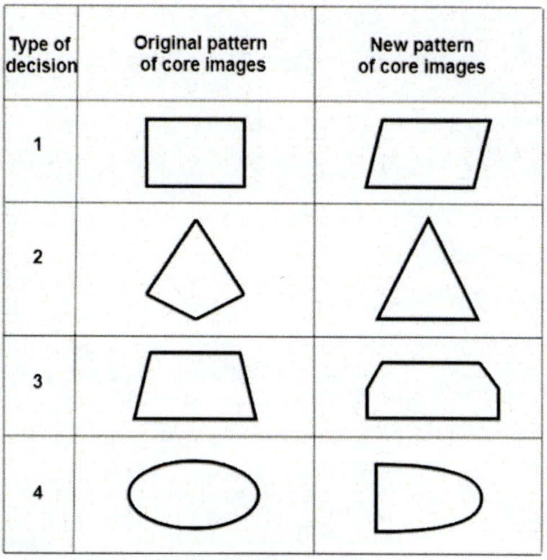

Fig. 3. Geometric interpretation of the types of recognizable images of the analyzed object states, where is the principle of general data transformation rule change.

demonstrate a tendency proving the manifestation of the adaptive abilities of the modeled system, the implementation of the decision synthesis principles in the experimental conditions was carried out programmatically relying on classical methods of data processing. The creation of mechanisms for the emergence, existence and intersection of two different data flows within the neobasic part consists in the representation of the analog aspects of these phenomena through a discrete representation.

To get the ANN's response to the changes, a pattern recognition cycle was performed consisting of 100 epochs. Each iteration contained 100 examples, including modified images of all four basic states with different geometric parameters. Figure 4 shows the values of the experimental results.

The experimental study results presented in Fig. 4 show that due to the incorporation of the decision synthesis mechanisms, the studied system model showed its adaptive abilities and increased the probability of correct recognition of the input image over 100 epochs. The increase in the number of epochs over 100 does not contribute to a significant change in the analyzed indicator. Based on this, we can conclude that during operation, the system strived to adapt to the changed general data transformation rule. However, it should be noted that the maximum probabilities (in the range of 0.59–0.61 for a trapezoid, 0.57–0.58 for a rectangle, 0.49–0.51 for an ellipse, and 0.48–0.49 for a rhombus) did not reach sufficient values to assert that a complete system fault tolerance was achieved.

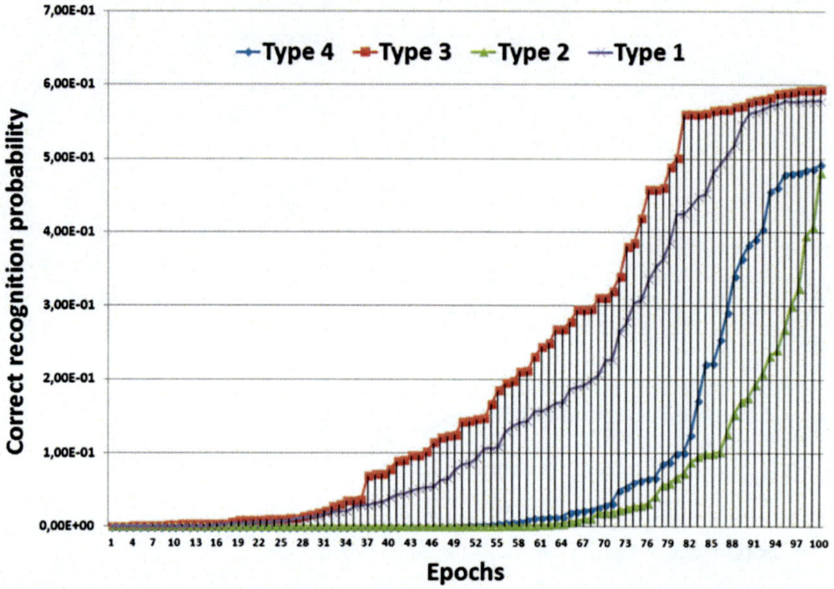

Fig. 4. The results of the experimental study as the dependence of the probability of correct pattern recognition on the system operation epoch after the incorporation of decision synthesis mechanisms – ellipse (blue), rhombus (green), trapezoid (red), rectangle (purple).

5 Conclusions

The obtained results allowed to conclude that new cognitive mechanisms may form that can be used as the base for decision synthesis within the computational components of SIDP. In the considered case, incorporating such innovation into the data processing process leads to an increase in the adaptive abilities of the studied object when the external conditions of its environment change. However, to implement this process with maximum efficiency, natural cognitive semantics shall be developed, as well as an analog and not algorithmically described method shall be developed for the existence of a decision synthesis mechanism capable of finding its hardware implementation when implementing the considered systems as applied devices based on modern electronic components [9, 10].

Acknowledgments. The reported study was funded by RFBR, project number 20-07-00951.

References

1. Danilin, S.N., Makarov, M.V. Shchanikov, S.A.: Design of artificial neural networks with a specified quality of functioning. In: IEEE International Conference Engineering and Telecommunication, pp. 67–71 Russia (2014)
2. Sani, A.: Machine Learning for Decision Making. Université de Lille 1 (2015)
3. Kashyap, P.: Machine Learning for Decision Makers. Apress, Berkeley (2017)

4. Bishop, C.: Pattern Recognition and Machine Learning, p. 738. MIT Press, Cambridge (2018)
5. Chandiok, A., Chaturvedi, D.K.: Machine learning techniques for cognitive decision making. In: IEEE Workshop on Computational Intelligence: Theories, Applications and Future Directions (WCI) (2015). https://doi.org/10.1109/wci.2015.7495529
6. Lu, H.: Artificial Intelligence and Robotics, p. 326. Springer, Berlin (2018)
7. Meyer, G., Adomavicius, G., et al.: A machine learning approach to improving dynamic decision making. Inf. Syst. Res. **25**(2), 239–263 (2014)
8. Minsky, M.: Computation: Finite and Infinite Machines. Prentice-Hall Inc., Hoboken (1967)
9. Makarov, M.V.: Practical analysis of the properties of nanoscale electronic elements aimed at their application when designing parallel architecture computing systems. J. Nano Electron. Phys. **3**(8), 03023-1–03023-4 (2016). https://doi.org/10.21272/jnep.8(3).03023
10. Makarov, M.: Investigating the physical and information parameters of nanoscale electronic elements as part of the computing systems with the neural network architecture. Mater. Phys. Mech. **3**(41), 78–83 (2019). https://doi.org/10.18720/MPM.4112019_12

Computer Vision

A System for Detecting and Detecting Defects in Sheet Metal on Grayscale Images

K. V. Mortin(✉), D. G. Privezentsev, and A. L. Zhiznyakov

Vladimir State University named after Alexander Grigorievich and Nikolai Grigorievich Stoletovs, 87, st. M. Gorky, Murom 600000, Russia
Mortinkv@ya.ru

Abstract. The article discusses the main problems of timely detection of defects in sheet metal by means of technical vision. In the course of the analysis, it was found that artificial neural networks of a typical structure do not allow to reduce the influence of real production factors on digital flaw detection images, and the quality of defect detection will be quite high. Created on the basis of a system of neurons second set and a special structure, and developed specialized algorithms based on the established network. In the course of experimental studies, defects in sheet metal were successfully identified in 89% of the images of the test set.

Keywords: Flaw image defect of metal · Machine vision · Neural networks · The defects detection systems

1 Introduction

The main task of metallurgical enterprises is to produce quality products without defects. In the conditions of production of sheet metal, it is very difficult for a defectoscopist to detect the entire spectrum of defects. Therefore, in production there is an acute issue of the development and application of a process for automating the detection of defects on sheet metal by means of technical vision and ensuring a high reliability indicator. This approach is due to the accumulation of a set of flaw detection images at the production site for detailed research and analysis of segmentation and detection.

Currently, in technical vision [1] at enterprises, non-destructive testing is used using various methods and algorithms for pattern recognition, adapted to a specific non-destructive testing. But it is worth noting that the increasing popularity recruit neural networks, which are composed of convolution algorithms, segmentation, detection and are a relatively new approach to solving the problems of the analysis of flaw images, in particular for the automatic detection and identification of defects [2–6] on sheet rolled metal. This study examines the creation of a system for detecting and detecting defects in sheet metal in production conditions based on a convolutional neural network. The problem of such flaw detection images is noise, uneven pixel contrast [6] and practically the same gradient of the background and defect in the images.

A. A. Radionov and V. R. Gasiyarov (Eds.): RusAutoCon 2021, LNEE 857, pp. 427–435, 2022.
https://doi.org/10.1007/978-3-030-94202-1_40

2 Development of an Ensemble of Filtering and Segmentation Algorithms in a Convolutional Layer

When designing detection systems and detection [4, 5] defects n and sheet metal, must be considered a convolution neural network in terms of its layers individually. This is to ensure that the complex algorithms for filtering, detection and segmentation found camping in different layers of the neural network. The key layer m in the system being developed is the convolutional layer [7]. Based on the structure diagram of the convolutional layer (Fig. 1), the architecture of the input network layer is rectangular input two-dimensional images with a defect in sheet metal.

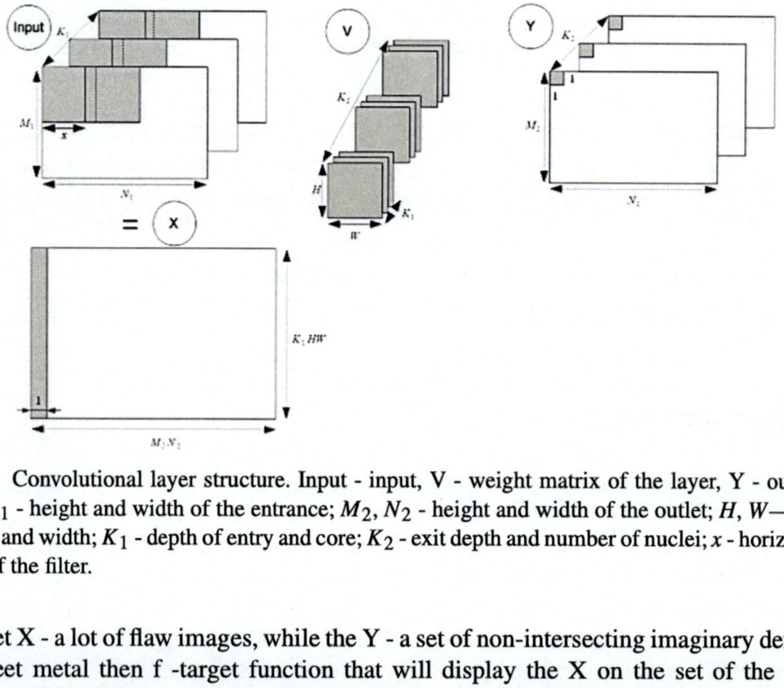

Fig. 1. Convolutional layer structure. Input - input, V - weight matrix of the layer, Y - output; M_1, N_1 - height and width of the entrance; M_2, N_2 - height and width of the outlet; H, W—core height and width; K_1 - depth of entry and core; K_2 - exit depth and number of nuclei; x - horizontal shift of the filter.

Let X - a lot of flaw images, while the Y - a set of non-intersecting imaginary defects of sheet metal then f -target function that will display the X on the set of the Y, f: $X \rightarrow Y$. The values of the objective function Y are known only for a finite set of pairs of precedents (X_i, Y_i)-a training set.

Recording $f(X_k) = Y_k$ will mean that X_k there is a defect on the flaw image Y_k, as shown in Fig. 2.

Based on the described data, the following structure of the alternation of layers of a deep convolutional network is compiled (Fig. 3).

The convolution of the flaw detection image X with is g indicated as $X * g$ and is calculated:

$$(X * g)[m, n] = \sum_{k=-K}^{K} \sum_{l=-L}^{L} f[m-k, n-l] g[k+K, l+L]$$

where g-kernel size $(2K+1) \times (2L+1)$.

Fig. 2. Flaw detection image.

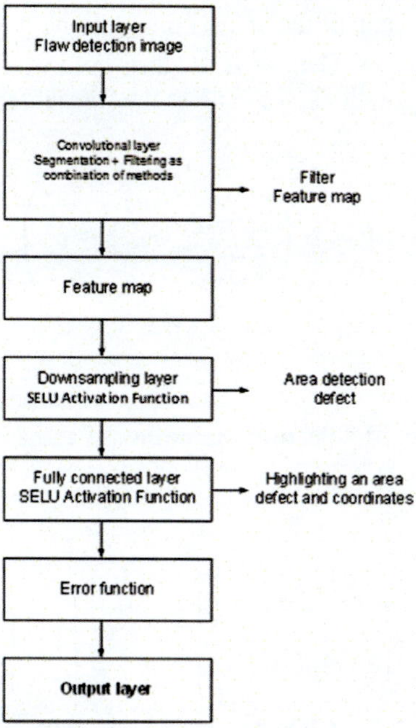

Fig. 3. Block diagram of a deep neural convolutional network with alternating layers

The areas of interest in the flaw detection image are the edges of the defect, namely, a sharp change in the value of the background brightness function to the edge of the defect [8]. Finding such areas can be organized based on the analysis of the derived image.

Algorithm of actions:

1 Divide the feature map of the $n(l-1)$ layer into disjoint blocks of two by two pixels, sum the values of these pixels in each block, and obtain a matrix $Z_n^{l-1} = \{Z_n^{l-1}(i,j)\}$, whose elements are the values of the sums.

The formula for calculating the values of the matrix elements will be as follows:

$$Z_n^{l-1} = y_n^{l-1}(2i-1, 2j-1) + y_n^{l-1}(2i-1, 2j) + y_n^{l-1}(2i, 2j-1) + y_n^{l-1}(2i, 2j)$$

Then the subsampling layer feature map is calculated as:

$$y_n^l = f_l\left(z_n^{l-1} * w_{m,n}^l + b_n^l\right)$$

Thus, it turns out that $H^l * W^l$- the feature maps of the y_n^l subsampling layer l will look like: $H^l = \frac{H^{l-1}}{2}$, $W^l = \frac{w^{l-1}}{2}$.

2 Consider the output layer of a L multilayer convolutional neural network, which consists of single neurons. Let be the N^L number of neurons in the output layer. A filter $w_{m,n}^L$, applied to the feature map of the m last convolutional layer to get the transition to the neuron in the n output layer. Then b_n^L is the threshold value added to the neuron n.
3 Using the introduced designations, we express the formula for calculating the value of the output neuron n:

$$y_n^L = f_L\left(\sum_{m=1}^{N^{L-1}} y_m^{L-1} w_{m,n}^L + b_n^L\right)$$

3 Experimental Part

Figure 4 shows the process of selecting the points of the area (a) and constructing a histogram of the selected defect (b).

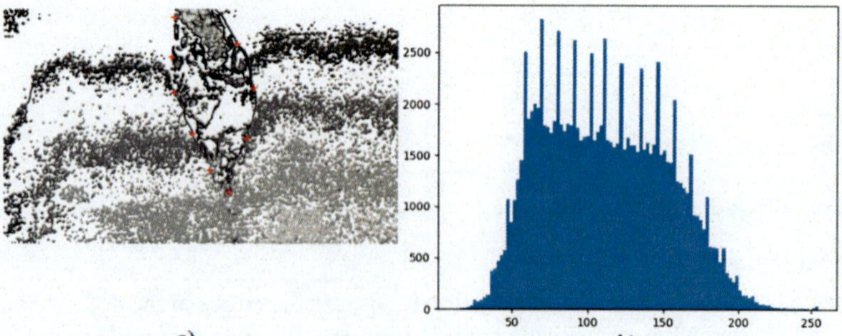

a) b)

Fig. 4. Selecting points of the defect area (a) and constructing a histogram (b).

Thus, the coordinates of the selected contour of the defect are obtained.

koordinata defekta.

= [(188.16330645161287, 242.65080645161285), (185.52016129032256, 200.36048387096773), (188.16330645161287, 162.47540322580642), (209.30846774193543, 118.42298387096773), (227.8104838709677, 78.77580645161291), (248.07459677419354, 55.86854838709678), (267.45766129032256, 114.01774193548385), (274.5060483870967, 167.76169354838706), (256.8850806451613, 214.45725806451608), (241.02620967741933, 242.65080645161285)].

4 Implementation of Algorithm a for Recognition of Flaw Detection Image of Sheet Metal

The developed algorithm is divided into several sequential steps:

1 The original digital flaw image is converted to grayscale and noise is removed.
2 Image segmentation is performed to search for suspected defects [9]. The estimated location of defects in the image is determined for the purpose of their further recognition by a multilayer convolutional neural network.
3 Unsized illumination of the flaw detection image is removed.
4 Detection are and are classified are defects using trained multilayer convolution neural network.
5 Conversion found defect in 3D.

In order for the convolutional neural network to receive a set of original flaw detection images with a size of 1600 × 256 at the input, it is necessary to extract a subwindow from the original image so that the defect occupies the maximum part of the image of the required dimension. In the final convolutional layer, the filtering will double [11]. Thus, an area with a defect is found. The detected defect, if necessary, is scaled, segmented and fed to the input of a multilayer convolutional neural network [10–13].

5 Creation of an Ensemble Method in the System of Detecting and Detecting Defects in Rolled Metal

Figure 5 shows the images after the experiment.

In order for the created neural network to produce a more reliable work result, it must be trained [14, 15].

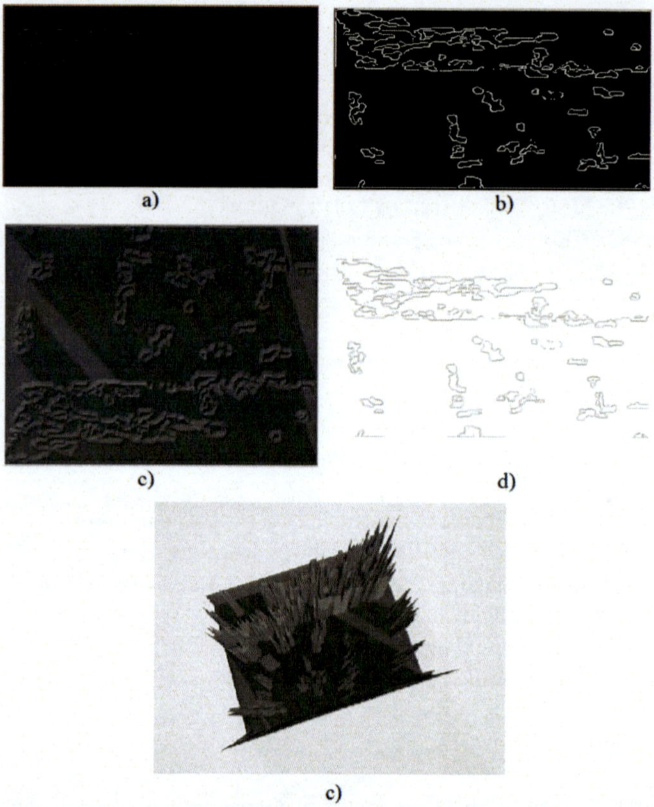

Fig. 5. Experimental result of the system for detecting and detecting defects in sheet metal a) the result of the work after the first step of the algorithm (segmentation) b) the result of the work after the second step of the algorithm (complex filtering) c) the result of the work after the third step of the algorithm (background extraction) d) the result of the work after the fourth step of the algorithm (background removal) e) the result work after the fifth step of the algorithm (transformation of the defect into 3 D)

For more effective training of layers of a deep convolutional neural network, you must first train on a large dataset (training took place on a set of 1,000 0 flaw detection images), and then adjust the layers to the required set of flaw detection images. The last layers are retrained, since the first layers contain more general features, and the latter are responsible for the area of the defect, its outline and determination of coordinates.

The result is a new layer structure of the deep convolutional neural network, as shown in Fig. 6.

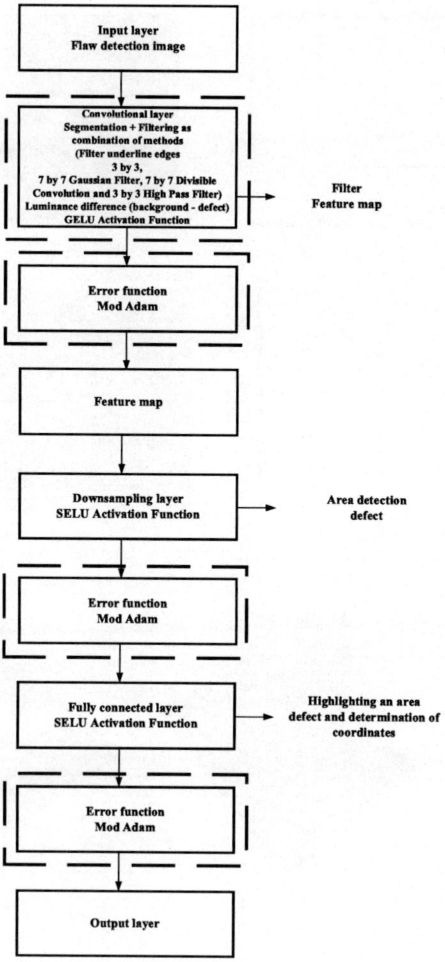

Fig. 6. The resulting structure of layers of a convolutional neural network for a system for detecting and detecting defects in sheet metal

6 Algorithm for Transforming the Background with Its Subsequent Removal from the Flaw Detection Image

Detection of the presumptive location of the defect occurs in several stages. The processing is based on the proposal for cross-correlation of the map of the features of the flaw detection image. First, convolution is performed with a classified known defect, and the resulting convolution determines the centers of local maxima, which will correspond to the assumed locations of the defects, as shown in Fig. 7.

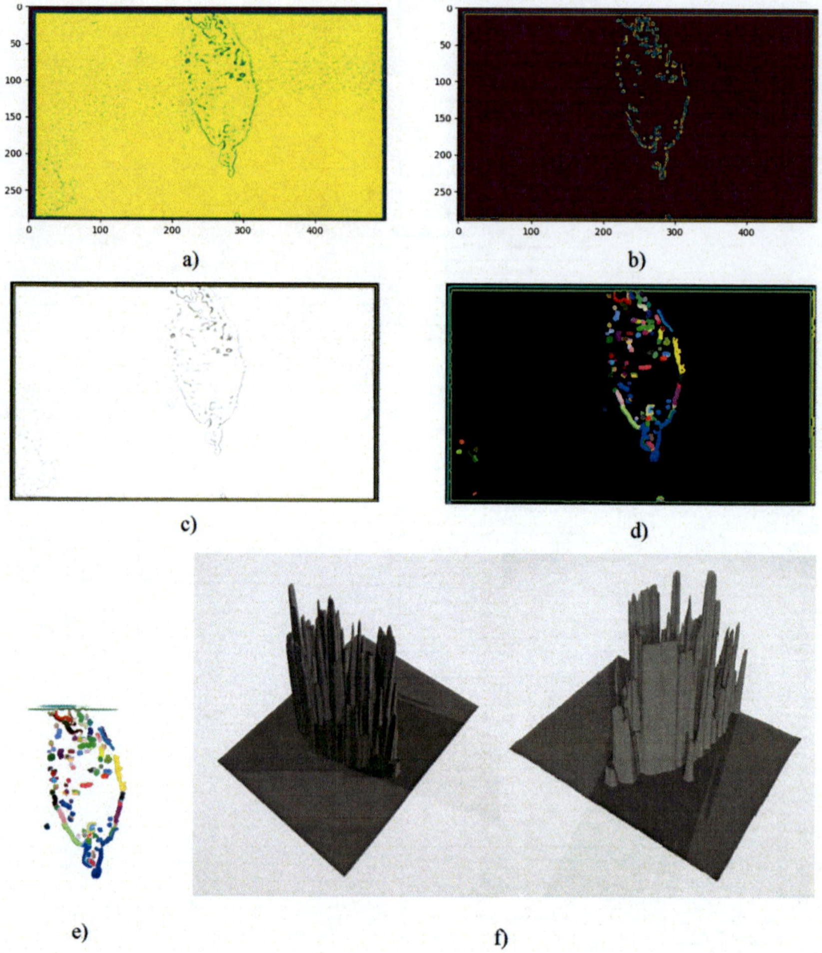

Fig. 7. Operation of the system for detecting and isolating a defect in sheet metal a) defect with the transformed background b) determining the contour of the defect c) a complex filtering convolutional layer d) Semantic Segmentation e) removing background f) conversion of a defect in the 3D

7 Conclusion

The developed system for detecting and detecting defects in sheet metal on rolled products in grayscale images is implemented in the form of a convolutional neural network. The system is aimed at multiple complex filtering in the process of segmentation in the convolutional layer, as well as detection, determination of coordinates and background extraction of a defect in the subsampling layer.

More than 10,000 flaw detection images have been tested in the developed system. As a result, more than 89% of defects in sheet metal products from the set under study were successfully isolated.

The developed system is a prerequisite for a new automatic approach in technical vision to determine the imaginary areas of a defect, its geometry, coordinates. The developed algorithm for converting to 3D is the basis for determining the depth of a defect on sheet metal.

References

1. Zmyzgova, T.R., Kuznetsova, E.M., Karpov, Y.K.: Problems of processing and recognition of digital images in technical vision systems. In: 2019 International Multi-Conference on Industrial Engineering and Modern Technologies (2019)
2. He, H., Yuan, M., Liu, X.: Research on surface defect detection method of metal workpiece based on machine learning. In: 2021 6th International Conference on Intelligent Computing and Signal Processing (2021)
3. Zhou, A., Zheng, H., Li, M., Shao, W.: Defect Inspection Algorithm of Metal Surface Based on Machine Vision. In: 2020 12th International Conference on Measuring Technology and Mechatronics Automation (2020)
4. Sun, J., Li, C., Wu, X.J., Palade, V., Fang, W.: An effective method of weld defect detection and classification based on machine vision. IEEE Trans. Ind. Inf. **15**(12), 6322–6333 (2019)
5. Riaz, F., Kamal, K., Zafar, T., Qayyum, R.: An inspection approach for casting defectsdetection using imagesegmentation. In: 2017 International Conference on Mechanical, System and Control Engineering (2017)
6. Marukatat, S., Kittisuwan, P.: Image enhancement by pixels sorting. In: 15th International Conference on Electrical Engineering/Electronics, Computer, Telecommunications and Information Technology (2018)
7. Privezentsev, D.G., Mortin, K.V., Zhiznyakov, A.L., Titov, D.V.: Development of a convolutional layer of a deep neural network for detecting defects in rolled metal. Inst. Making **64**(3), 202–207 (2021)
8. Akimov, A.V., Sirota, A.A.: Models and algorithms for artificial data multiplication for training face recognition algorithms using the Viola-Jones method. Comput. Opt. **6**, 899–906 (2016)
9. Privezentsev, D.G., Shamshin, M.N., Mortin, K.V., Pugin, E.V.: Fractal digital image model based on fuzzy rank blocks. Telecommunications **11**, 13–21 (2020)
10. Liu, Y.: An improved faster R-CNN for object detection. In: 2018 11th International Symposium on Computational Intelligence and Design (2018)
11. Wang, Y., Wang, C., Zhan, H., Yingbo, G., Wei, S.: Automatic Ship Detection Based on RetinaNet Using Multi-Resolution. Remote Sens. **11**(5), 531 (2019)
12. Silver, D., Huang, A., Maddison, C.: Mastering the game of go with deep neural networks and tree search. Nature **529**(7587), 484–489 (2016)
13. Wainberg, M., Alipanahi, B., Frey, B.J.: Are random forests truly the best classifiers? J. Mach. Learn. Res. **17**(110), 1–5 (2016)
14. Zhang, Z.: Improved Adam Optimizer for Deep Neural Networks. In: IEEE/ACM 26th International Symposium on Quality of Service (2018)
15. Sun, H., Gu, L., Sun, B.: Adaptive Gradient Method Based on Estimates of Third-Order Moments. In: IEEE Fourth International Conference on Data Science in Cyberspace (2019)

Industrial Automation Systems Cybersecurity

Software Package for Training Users to Respond to Information Security Incidents in Industrial Automated Systems

M. Tumbinskaya[1], A. Abzalov[1], and I. Davydova[2](✉)

[1] Kazan National Research Technical University named after A.N. Tupolev, 10, K.Marx Street, Kazan 420111, Russia
[2] Kazan Innovative University named after V. G. Timiryasov, 42, Moskovskaya Street, Kazan 420111, Russia
davydova@ieml.ru

Abstract. The article considers the issue of increasing the level of security of industrial automated systems by automating the training of users to respond to information security incidents. A software package for training users to respond to information security incidents in industrial automated systems is proposed. The article presents an overview of the information security state in industrial automated systems: the damage from cyber threats, types of attacks and objects of attacks in industrial automated systems, confidential information of the attackers' interest. The study examined the impact of the users' information security knowledge level on improving the quality of users' response to information security incidents to increase the level of industrial automated systems security and prevent information security threats. 4 classes of attacks were investigated: 1) attacks using malicious software, 2) DoS-type attacks, 3) attacks using social engineering methods, 4) credential matching attacks. The results of the study and experimental data confirm the effectiveness of the proposed software package, which made it possible to increase the industrial automated system security level by an average of 28%.

Keywords: Information security incidents · Software package · User · Industrial automated system · Attack · Attacker · Cyber threat

1 Introduction

The tasks of incident management and countering attacks in industrial automated systems have been sufficiently studied, but they remain relevant. With the global growth of cybercrime and the constant improvement of cyberattacks, it is necessary to increase the security level of information systems, web resources, industrial automated systems, etc. Increasing the level of security is possible by solving the problem of training users to respond to information security incidents in the information systems operation. Usually, the training of industrial automated systems users is carried out through advanced training courses, which require on-the-job training, time, and material resources. A review of

A. A. Radionov and V. R. Gasiyarov (Eds.): RusAutoCon 2021, LNEE 857, pp. 439–451, 2022.
https://doi.org/10.1007/978-3-030-94202-1_41

the literature [1–3] showed that currently there are software solutions for training users to protect information. They are aimed at increasing the users' awareness about cyberattacks, but the use of practical simulation tasks is not presented in them. The works [4–7] published the results of research on the development of software that allows users to reduce information security incidents through configuration settings, both in the physical and in the virtual environment of the information system. The authors [8–14] describe tools for mass education aimed at various categories of users (students, schoolchildren, programmers, professional users of information systems of various fields of activity, etc.) who are interested in information security, but are characterized by a limited set of sections and topics, practical tasks. The works [15–18] describe interactive and practical courses on information security with limited access. The article proposes a software package for training users to respond to information security incidents in industrial automated systems. Its main task is to provide the industrial automated system users with practical skills for an adequate response to incidents. The package will increase the knowledge level of industrial automated systems users, thereby increasing the security level of these systems.

2 Review of the Information Security State of Industrial Automated Systems

Analysis of works [19–21] on damage from crimes in the digital space shows that potential losses from cybercrimes are estimated at $ 10.2 billion. Thanks to the prompt investigation of incidents, it was possible to prevent about 70–80% of cybercrimes in the world. Such statistics should motivate an increase in the information security of any enterprise. Table 1 shows the total losses from various types of cyber threats [22].

Table 1. Total losses from cyber threats in 2019.

Type of cyber threat	Damage, $
Spoofing	300 478 433
Identity theft	160 305 789
Personal data breach	120 102 501
Phishing/Vishing/Smishing/Pharming	57 836 379
Tech support	54 041 053
Corporate data breach	53 398 278
Harassment/Threats of violence	19 866 654
Misrepresentation	12 371 573
Ransomware	8 965 847
DenialofService/TDoS	7 598 198
Malware/Scareware/Virus	2 009 119
Hacktivist	129 000

Tables 2, 3 show the distribution of attacks types and attacks objects in industrial automated systems in 2019 [23–25].

Table 2. Percentage distribution of attacks types on industrial automated systems in 2019.

Class of attacks	1 quarter	2 quarter	3 quarter	4 quarter
Malicious software	79	96	92	91
Social engineering	71	78	95	88
Hacking	21	4	5	3
Exploiting web vulnerabilities	0	7	5	3
Selection of credentials	0	4	0	6
Other	0	0	3	9

Table 3. Percentage distribution of attack targets in industrial automated systems in 2019.

Type of attack object	1 quarter	2 quarter	3 quarter	4 quarter
Infrastructure	96	96	92	94
Web-ресурсы	0	4	3	6
Users/employees	4	0	5	0

Analysis of statistical data showed that, on average, in 89.5% of cases, attackers use malicious software to gain access to information unauthorized, and in 83% of cases, they use social engineering methods. The infrastructure of industrial automated systems is the main target of attacks. Figure 1 shows the distribution of confidential information in industrial automated systems of enterprises that was stolen in 2019 and is of interest to the attacker. The average for each of these types of confidential information is 25%.

To gain access to confidential information of industrial automated systems, attackers all over the world act according to typical scenarios. The most common scenarios are:

1. hacking corporate e-mail to gain access to financial and management contacts of the company, to understand internal processes and correspondence;
2. exploitation of the users trust in the technical support service of the enterprise (the damage is $ 54 million);
3. ransomware viruses that get into industrial automated systems through phishing mailings.

Except for material damage, there may be other types of damage associated with the downtime of an industrial automated system, reduced efficiency, incident response costs, and others. That makes developing special software to increase the security level of industrial automated systems relevant.

The developed software package is aimed at automated users training to respond to information security incidents in industrial automated systems. It can reduce costs to ensure information security at industrial enterprises (material, time resources, etc.).

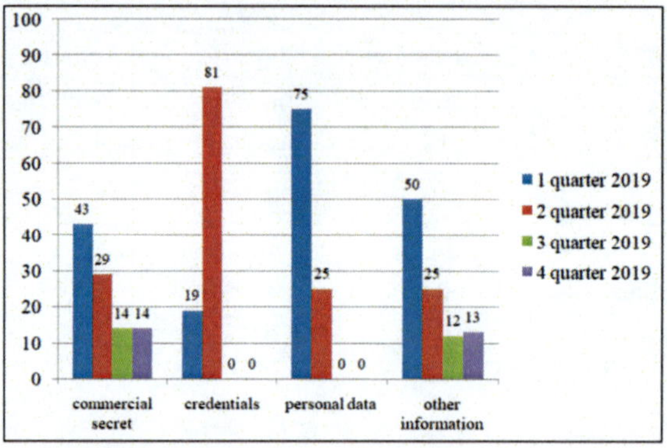

Fig. 1. Histogram of the distribution of confidential information in industrial automated systems of enterprises stolen in 2019.

3 Software Package for Training Users to Respond to Information Security Incidents in Industrial Automated Systems

The software for training users to respond to information security incidents in industrial automated systems consists of three main modules: a user authorization and accounting module, a training module (theoretical part) and a contingency modeling module (virtual simulator). Figure 2 shows a schematic diagram of the software package.

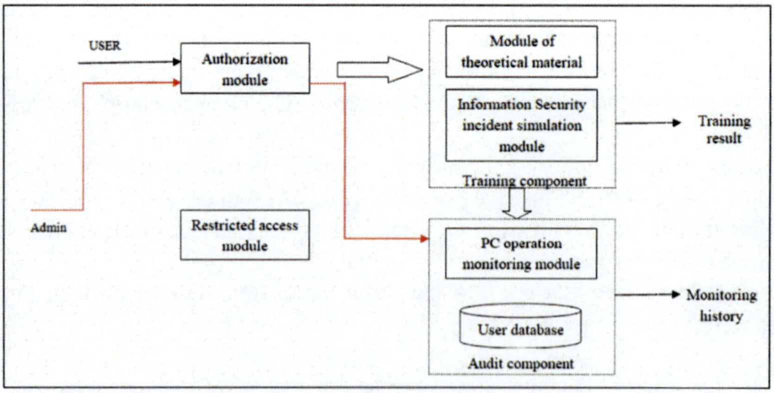

Fig. 2. Diagram of a software package for training users to respond to information security incidents.

3.1 User Authorization and Accounting Module

Successful learning requires controlling the assimilation of the received information. Therefore, the beginning of the training is preceded by the user authorization procedure. In a separate window, the user is prompted to enter the username and password previously issued by the administrator (it is assumed that the login and password for the software matches the user's credentials in the operating system). An account is created for each user. Their ID and training progress data are stored in that account. This data includes a list of chapters studied and the results of answers to control questions after each chapter. All data is stored on the server and is available only to system administrators. At the same time, each user sees their own progress in the learning process. Figure 3 shows the authorization dialog box for the software package users.

3.2 Training Module

The training module consists of a means of displaying text and graphic material and a testing mechanism. The training material is presented in hypertext format, for easy navigation between chapters (Fig. 4).

Fig. 3. User authorization window of the software package.

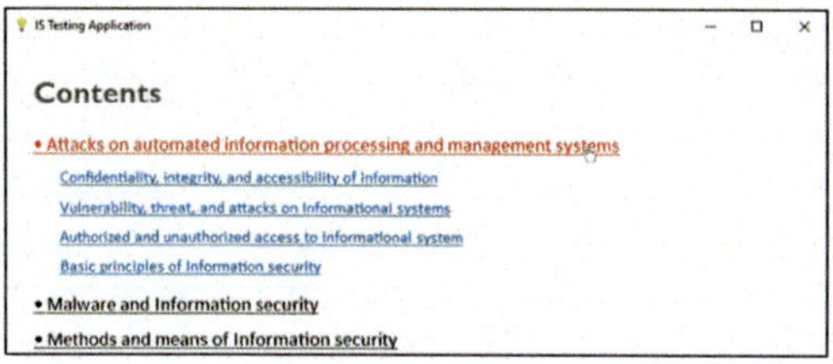

Fig. 4. Window of the educational module of the software packagel.

The study of theoretical material consists of two main sections: general principles and basic concepts of information security, practical aspects of automated process control systems information security ensuring. The first section covers basic information security concepts such as confidentiality, integrity and availability, vulnerability, threat and attack; authorized and unauthorized access and others. To help the user understand information security problems in a short time, the training material reveals the meaning of the information security basic principles: consistency, complexity, continuity of protection, reasonable sufficiency, flexibility in the management and application of information security measures, openness of algorithms and protection mechanisms, ease of use of protective means and tools. Figure 5 shows a block diagram of the user's work algorithm with the software package.

After studying each chapter, the user is asked to take a control test. You cannot move on to the next chapter without passing the test successfully. The ability to re-pass testing based on the studied chapter materials opens after a certain period, set by the administrator. (The default is 30 min.) This gives the users time to review the studied material, to pay attention to their mistakes, and also to minimize the likelihood of finding the correct answers. For the same purpose, the test questions and answer options are displayed in a different order for each attempt.

3.3 Modeling of Contingency

Modeling of contingency is based on the following idea. As different diseases can have the same symptoms and only by the totality of all the symptoms can the doctor make the correct diagnosis, so in computer systems, only by the totality of signs can one try to determine what kind of emergency has arisen. The simulator randomly generates a set of features and displays them on the screen. After that, the user must determine the essence of the contingency by a combination of different signs and respond adequately.

Below is a list of signs of abnormal situations (accidental failures, malicious programs running in the automated control system, or other malicious effects on the system: 1) "deadlock" of the operating system, 2) software "deadlock", 3) slowdown of the operating system, 4) software slowdown, 5) completing some software functions faster than usual, 6) error messages or program crashes appearing, 7) inability to access network resources, 8) inability to open previously saved files, 9) data disappearing from previously saved files, 10) programs or files disappearing from disk, 11) the appearance of unknown programs or files on the disk, 12) disappearing shortcuts on the desktop, 13) the appearance of unknown shortcuts on the desktop, 14) display disappears, 15) complete shutdown of the computer etc.

It is obvious that the concepts of "contingency" and "sign of a contingency" are not equivalent, and in this work, a contingency is considered to be the manifestation of a non-empty set of signs. An example of a combination of signs and their corresponding abnormal situations is given in Table 4. Table 4 is given as an example, the complete list of contingency and combinations of their signs is much more extensive and cannot be contained in this article. For each contingency there is a script of user actions. Scripts help to prevent malicious impact on the system or minimize damage from such impact.

A script is a non-empty set of the following simple actions:

- do not take any action.
- close the running applications and start again.
- close the running applications and run a full scan of the computer with an antivirus.
- check the antivirus for individual suspicious files.
- restart the system.

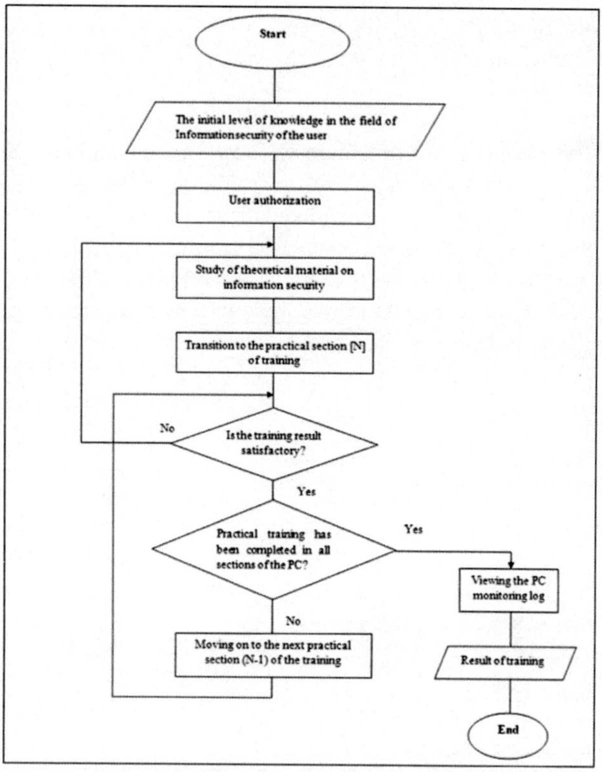

Fig. 5. Block diagram of the algorithm for the user with the software package.

Table 4. Contingencies and their signs.

Contingencies and their signs	"Deadlock" of the operating system	Software "deadlock"	Slowing down the operating system	Slow down the software	Data disappearing from previously saved files	Error messages or program crashes appear	Inability to access network resources	Complete shutdown of the computer
Emergency power off	0	0	0	0	0	0	0	1
Network problems. No network connection	0	0	0	0	0	0	1	0
The system is affected by DoS attack	0	0	1	0	0	0	1	0
Malicious software is running on the system	0	1	1	1	1	0	0	0

- disable the computer's access to the network.
- inform the administrator about what is happening.
- turn off the computer, etc.

Figure 6 shows the simulator operation window. In the simulator window, the user sees an image of the elements of his workplace: a computer display, a button for turning on the system unit, a telephone, etc., provided with appropriate inscriptions. By clicking on such shortcuts, the user simulates one action from the list above. For example, by clicking on the phone image, the user imitates the action "inform the administrator." The user of the simulator learns about the signs of abnormal situations from the text messages that appear at the top of the window.

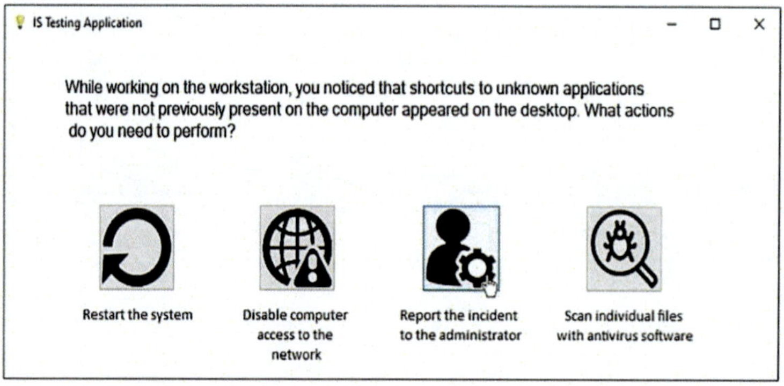

Fig. 6. Simulator window of the software package.

To respond to various situations, the user is given a certain amount of time from a few seconds to a minute. The results of the work on the simulator are displayed on the screen: the number of simulated situations and how many of them were worked out correctly. According to the results, the user can be offered to re-examine individual chapters from the theoretical part.

4 Experimental Results

The study was conducted during 15 months of 2019–2020. The results of the experiment were processed in the statistical package Statistica 10.0. All participants of the experiment gave their written consent and voluntarily agreed to participate in the study.

The study examined the impact of the users' knowledge level in information security on improving the quality of user response to information security incidents to increase the security level of industrial automated systems and prevent information security threats. An industrial automated system of a large machine - building industry enterprise [26–28] of the Republic of Tatarstan was selected for the experiment. The study analyzes information security incidents [29, 30], which revealed 4 most common classes of attacks: 1) attacks using malicious software, 2) DoS-type attacks, 3) attacks using social engineering methods, 4) credential matching attacks.

The experiment was conducted in 3 stages:

1. Collection of statistical data on the number of information security incidents (1st-2nd quarters of 2019).
2. Training of users to respond to information security incidents (3rd-4th quarters of 2019).
3. Analytical study of information security incidents of industrial automated systems (1st quarter of 2020).

During the period of self-isolation due to the global COVID-19 pandemic, the experiment was not conducted (2–4 quarters of 2020). After achieving a favorable sanitary and epidemiological situation, the experimental study will resume.

As a result of the analysis of the experimental data of stage 1, it can be concluded (Fig. 7) that in industrial automated systems, attacks of two types are predominantly carried out: social engineering attacks (average value 95%), attacks using malicious software (average value 90%).

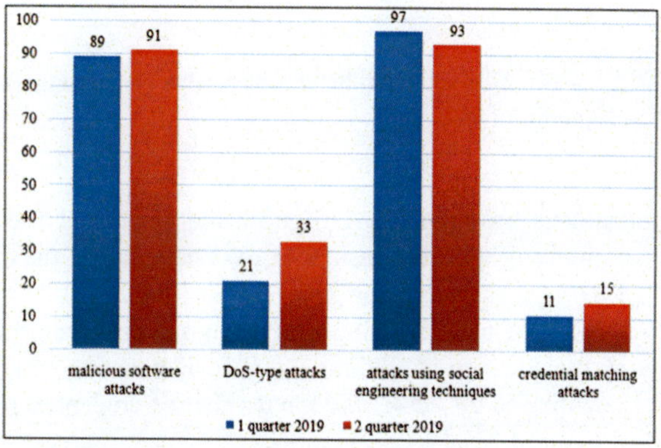

Fig. 7. Histogram of the distribution of attacks in the industrial automated system of the enterprise (1st-2nd quarters of 2019).

At the 2nd stage of the experiment, 1,500 users of industrial automated systems were trained for 6 months. During this period of the experiment, information security incidents were also monitored. Over the analyzed period of time, the number of attacks on an industrial automated system (Fig. 8) decreased by an average of 11.87%, the number of attacks implemented using social engineering methods decreased by 16.5%, the number of attacks using malicious software decreased by 17.5%, the number of DoS–type attacks decreased by 8.5%, and the number of credential matching attacks decreased by 5%.

At the 3rd stage of the analytical study, based on the results of user training, an analysis of information security incidents of an industrial automated system was carried out. The analysis showed (Fig. 8) that the number of attacks on an industrial automated system in comparison with the initial results of the study (Stage 1) decreased by an average of 33.75%, which proves the effectiveness of the proposed software package. The number of attacks implemented using social engineering methods decreased by 60%, the number of attacks using malicious software decreased by 49%, the number of DoS–type attacks decreased by 16%, the number of credential matching attacks – by 10%.

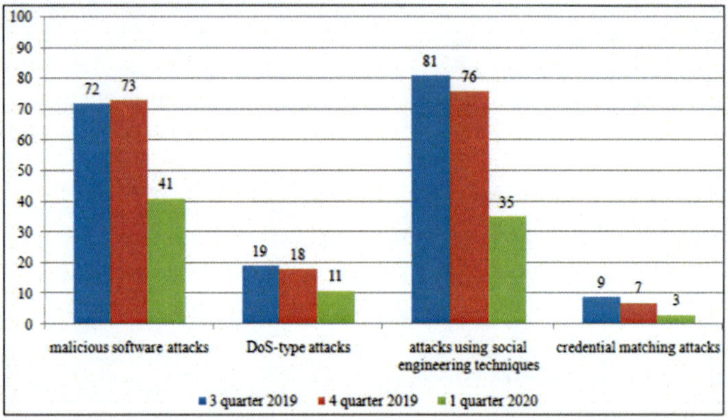

Fig. 8. Histogram of the distribution of attacks in the industrial automated system of the enterprise (3-4 quarters of 2019, 1 quarter of 2020).

The number of attacks on an industrial automated system at the end of the 3rd stage of the study, comparing with the results of the 2nd stage of the study, decreased by an average of 21.87%, which also confirms the positive effect of the proposed set of programs. The number of attacks using social engineering methods decreased by 43.5%, the number of attacks using malicious software decreased by 31.5%, the number of DoS–type attacks decreased by 7.5%, and the number of credential matching attacks decreased by 5%.

The analysis showed that as a result of training, users respond to emerging information security incidents more correctly and adequately because most situations were considered and analyzed while training using the software package. On average, the number of attacks at the end of 2019 compared to the 1st quarter of 2020 as a whole decreased by 28%: the number of attacks implemented using social engineering methods decreased by 51.75%, the number of attacks using malicious software decreased by 40.25%, the number of DoS–type attacks decreased by 11.75%, the number of credential matching attacks decreased by 7.5%.

The results of the study show that the level and quality of user knowledge on information security positively affect the quality of user response to information security incidents, which allows to increase the level of security of an industrial automated system by 28%.

5 Conclusion

The article presents an overview of the state of information security in industrial automated systems. Confidential information that is of value to an attacker has been identified in industrial automated systems. The damage from cyber threats, types of attacks, objects of attacks in industrial automated systems is analyzed. A software package for automated training of users to respond to information security incidents in industrial automated systems is proposed. The software package for training users to respond to information security incidents in industrial automated systems consists of three components: authorization and accounting of users, training, and simulation of emergency situations.

The influence of the users' knowledge level in information security on improving the quality of user response to information security incidents has been investigated. The study was conducted with the aim of increasing the level of security of industrial automated systems and preventing threats to information security. The research results and experimental data confirm the effectiveness of the proposed software package, which can increase the security level of an industrial automated system by an average of 28%. In the future, we plan to improve the software package in terms of modeling information security incidents by expanding the set of situations that simulate the implementation of attacks, as well as expanding the types of attacks. It is planned to refine and automate the algorithm of the software package to be able to work in a large-scale network using virtual machines. The results of the study allow to apply a new level of training approach to the information security incidents study in the industrial enterprises information systems, obtaining interesting and visual results.

References

1. Wang, X., Shuai, N.: The multilevel comprehensive grey evaluation model for enterprise training effect. In: International Workshop on Education Technology and Training & 2008 International Workshop on Geoscience and Remote Sensing, pp. 51–54 (2009)
2. Mokhtar, I.A., Majid, S., Foo, S.: Using information technology to improve health information literacy in Singapore-an exploratory study. In: IEEE International Conference on Information & Communications Technology, pp. 1–2 (2007)
3. Komalasari, N., Murad, D.F., Agustine, D., Irsan, M., Budiman, J., Fernando, E.: Effect of education, performance, position and information technology competency of information systems to performance of information system. In: IEEE International Seminar on Research of Information Technology and Intelligent Systems, pp. 221–226 (2019)
4. Lee, J., Kim, Y.S., Kim, J.H., Kim, I.K.: Toward the SIEM architecture for cloud-based security services. In: IEEE Conference on Communications and Network Security, pp. 398–399 (2017)
5. Okubo, N., Nara, K., Takemura, S., Ueda, Y.: Applying an instructional design process to development of an independent verification and validation training program. In: IEEE 29th International Conference on Software Engineering Education and Training, pp. 237–240 (2016)
6. Kumar, S., Arekar, K., Jain, R.: The impact of effectiveness of the simulator training program on different factors of needs and interest of the training. In: IEEE International Conference on Next Generation Computing Technologies, pp. 485–489 (2017)

7. Belikova, T., Lekakh, A., Dovbenko, O., Dodukh, O.: Method of increasing the capacity of information threat detection filters in modern information and communication systems. In: IEEE International Conference on Advanced Information and Communications Technologies, pp. 426–429 (2019)
8. Hershey, P., Silio, C.B.: Systems engineering approach for event monitoring and analysis in high speed enterprise communications systems. In: Annual IEEE Systems Conference, pp. 344–349 (2009)
9. Rathnayake, N., Meedeniya, D., Perera, I., Welivita, A.: A framework for adaptive user interface generation based on user behavioural patterns. In: IEEE Moratuwa Engineering Research Conference, pp. 698–703 (2019)
10. Tomilin, A., Tumbinskaya, M., Tregubov, V., Smolevitskaya, M.: The BESM-6 virtualization project. In: IEEE International Conference on Computer Technology in Russia and in the Former Soviet Union, pp. 241–245 (2017)
11. Gizatullin, Z., Gizatullin, R., Drozdikov, V.: Research of noise immunity of computer equipment of control systems under action of pulsed magnetic field. In: IEEE International Russian Automation Conference, pp. 65–69 (2019)
12. Gizatullin, Z., Shkinderov, M., Arkhipov, A.: Research of resonant effects in interconnects of multilayer PCB of computing equipment. In: IEEE International Conference of Russian Young Researchers in Electrical and Electronic Engineering, pp. 116–119 (2020)
13. Gizatullin, Z., Konstantinov, E.: Technique for research spurious electromagnetic emission from electronic means. In: IEEE International Russian Automation Conference, pp. 380–384 (2020)
14. Akanmu, S.A., Jamaludin, Z.: A user-centered design methodology for students' data-focused InfoVis. In: IEEE International Conference on User Science and Engineering, pp. 115–118 (2015)
15. Ahma, I., Jaafar, A.: Games design and integration with user's emotion. In: IEEE International Conference on User Science and Engineering, pp. 69–72 (2012)
16. Ying, J., Gračanin, D.: Poster: an approach to development of adaptive 3D user interfaces. In: IEEE Symposium on 3D User Interfaces, pp. 169–170 (2012)
17. Hasim, W., Wibirama, S., Nugroho, H.A.: Redesign of E-participation using user-centered design approach for improving user experience. In: IEEE International Conference on Information and Communications Technology, pp. 857–861 (2019)
18. Li, M., Li, L., Jiao, R., Xiao, H.: Virtrul reality and artificial intelligence support future training development. In: IEEE Chinese Automation Congress, pp. 416–419 (2018)
19. Gibadullin, R.F., Baimukhametova, G.A., Perukhin, M.Yu.: Service-oriented distributed energy data management using big data technologies. In: IEEE International Conference on Industrial Engineering, Applications and Manufacturing, pp. 45–49 (2019)
20. Sharipov, R., Tumbinskaya, M., Abzalov, A.: Analysis of users' keyboard handwriting based on Gaussian reference signals. In: IEEE International Russian Automation Conference, pp. 28–32 (2019)
21. Garae, J., Ko, R.K.L., Apperley, M.: A full-scale security visualization effectiveness measurement and presentation approach. In: IEEE International Conference on Trust, Security and Privacy in Computing and Communications, pp. 639–650 (2018)
22. The total losses from cyber incidents in the world are about $ 3.5 billion. https://www.securitylab.ru/blog/company/AngaraTech/347774.php
23. Industrial companies: attack vectors. https://www.ptsecurity.com/ru-ru/research/analytics/ics-attacks-2018/
24. Do not extinguish the light: how to protect industrial control systems from attacks from Industroyer and similar malware. https://www.ptsecurity.com/ru-ru/research/analytics/ics-attacks-2018/

25. Gibadullin, R.F., Vershinin, I.S., Minyazev, R.Sh.: Development of load balancer and parallel database management module. In: IEEE International Conference on Industrial Engineering, Applications and Manufacturing, pp. 33–39 (2018)
26. Yakimov, I., Tregubov, V., Tumbinskaya, M.: Creation of a unified industrial basis for the development and supply of applied research program packages at the Kazan computer factory. In: IEEE International Conference on Computer Technology in Russia and in the Former Soviet Union, pp. 126–130 (2017)
27. Petrovsky, V., Tumbinskaya, M.: The history and prospects of information security at Russian enterprises. In: IEEE International Conference on Computer Technology in Russia and in the Former Soviet Union, pp. 150–153 (2014)
28. Gumerov, V., et al.: Quality functions modeling of industrial enterprises products. Int. Rev. Manage. Market. **6**(1), 165–169 (2016)
29. Tumbinskaya, M.V.: Process of distribution of undesirable information in social networks. Bus. Inform. **3**, 65–76 (2017)
30. Tumbinskaya, M.V., Bayanov, B.I., Rakhimov, R., Kormiltcev, N.V., Uvarov, A.D.: Analysis and forecast of undesirable cloud services traffic. Bus. Inform. **13**(1), 71–81 (2019)

Application TRIKE Methodology When Modeling Threats to APCs Information Security

D. Chernov(✉)

Tula State University, 92, Lenina Avenue, Tula 300012, Russia

Abstract. In the modern world automation technologies of industrial processes have found wide application. Automated process control systems have become an important part of enterprises that operate in different economic fields and life support facilities all over the world. Nevertheless, the high growth of automation means raises the acute problem of providing APCs information security from external and internal threats. There are systematic reports in the media about new critical vulnerabilities in industrial equipment and attacks based on the exploitation of such vulnerabilities. When designing such systems it is necessary to estimate possible information security threats that already exist in the system or that are predicted to appear. That is why threat modeling is an important part of providing information security at industrial facilities using automation. In this work the author investigates the methodology of estimating threats to software security TRIKE. The research is aimed at finding general approaches to determine threat sources, tactics, and techniques for making implementation scenarios of threats to software information security, for applying them to provide cyber security of automated process control systems. In the frameworks of the subject mainstreaming, the research gives a description of cyber-attacks at big industrial facilities, and their basic vulnerabilities are emphasized. In the article the analysis of TRIKE methodology has been done. It is aimed at determining basic threat modeling stages applicable to industrial automation systems. The approach of TRIKE methodology to generating a list of threats to information security is formalized. To achieve the aims the author constructs DFD data flow diagrams with the decomposition of peculiar elements of the algorithm for modeling threats to information security of automated process control systems using TRIKE methodology.

Keywords: Automated process control systems · Information security · Threat · Threat model · TRIKE · Data flow diagrams · Attack library

1 Introduction

In recent years cyber-attacks at industrial systems often appear in security bulletins of the leading companies that develop information security means. For example, in 2019 there were several large-scale information security attacks at automated process control systems (APCs), which had consequences for residents of entire cities and even countries. Attacks at Venezuela power facilities such as the automated process control system of the hydroelectric power plant and others led to massive blackouts all over the

A. A. Radionov and V. R. Gasiyarov (Eds.): RusAutoCon 2021, LNEE 857, pp. 452–461, 2022.
https://doi.org/10.1007/978-3-030-94202-1_42

country, including state strategic facilities [1]. Malicious actions of a cryptographer to the company City Power that provides electricity to Johannesburg, South Africa, left its citizens without power till the company restored information systems [2]. Hackers often attack industrial facilities web-sites. Intruders placed Trojan Emotet on the Uniden web-site. The web-site of the oil and gas company Petrobangla was hacked twice on the same day. In the latter case the intruder insisted that he showed the web-site owners the security problems [3]. Dragos, the company specializing in industrial cyber security, presented the results of a study of vulnerabilities in industrial systems (ICS-CERT) excluding related equipment (network, system, etc.). According to the results 77% of the estimated vulnerabilities demand substantial access to operation control network and are considered to be "deeply interior" ones; 26% of the known vulnerabilities had no mending (patches) at that moment; 46% of vulnerabilities refer to workstations and user interface software that require an Internet connection to operate, which is often unacceptable in the industry.

By vulnerabilities types the leaders are [4]:

- Lack of monitoring of user input;
- Buffer overflow;
- Incorrect processing of input in web-interfaces (cross-site scripting);
- Using hard-coded credentials;
- Lack of monitoring the use of system resources (memory leaks).

In order to minimize the consequences of the threats using vulnerabilities information security specialists model information security threats. At the present time there is a lot of threat modeling methodologies, though only a few of them meet the needs in the field of information security of automated process control systems.

Threat modeling is the process by which potential threats, such as structural vulnerabilities or the lack of appropriate security measures, can be identified, listed, and measures to address them can be prioritized. The goal of threat modeling is to provide defenders with a systematic analysis of what controls or defenses need to be enabled, given the nature of the system, the profile of the likely attacker, the most likely attack vectors, and the assets most sought after by the attacker. Threat modeling answers questions like "Where am I most vulnerable to attacks?", "What threats are most relevant?" and "What do I need to do to protect myself from these threats?". When developing the threat model, the features of the territorial location of remote objects of automated process control systems with the distributed nature of the functioning of automated control systems, their social and economic status, operating modes, equipment used, personnel, and other factors that determine the possibility of threats to information security and their predicted consequences are taken into account [5].

In 2005 the methodology TRYKE was published and it has found a wide application by specialists when modeling threats to APCs software [6]. However, this technique can also be applied at the hardware level of industrial systems.

The trike is based on a risk-based approach to building information security and is designed for conducting information security audits and building threat models.

The distinctive features of this methodology are:

- its initial focus was on using specialized software to build threat models;
- using "attack trees" to describe security threats;
- using libraries of typical attacks.

2 Methodologies

The methodology TRIKE is based on the structure of using information security threat models as a risk managing tool. In the frameworks of the structure, information security threat models are used to satisfy the security audit process. The threat models are based on the ratio of "requirements models" and "system implementation model", built on the basis of DFD - data flow diagrams. The requirements model sets an "acceptable" risk level determined by the expert method for each of the asset classes obtained from the results of the inventory. Based on the requirements model an information security threat model is formed where threats are listed and appropriate risk values are assigned. The complete information security threat model is used for building a risk model based on assets, roles, activities and computed risk exposure. The given description fully complies with the possible application of the considered methodology for modeling information security threats to APCs information security.

Let us show graphical notation IDEF0 of the process of assessing security threats to APCS in accordance with TRIKE methodology in Fig. 1 [7–9].

An expert group participates in the process of estimating APCs information security threat. It is formed of the subject area specialists who use respective software and hardware in their work. Information security threat estimation is carried out on the basis of operational documentation, ongoing technological processes, as well as local regulatory documents of the organization that operates APCs. The threat model formed by the results of the estimation is built on the basis of information assets, operations, attacks, rules and possible threats.

All the threats of the considered methodology are divided into two categories:

- Service denial;
- Privilege elevation.

A service denial threat appears if the access subject fails to perform a legitimate, planned operation in an APCs, acting in accordance with the access differentiation rules. A privilege elevation threat occurs in one of the threes situations:

- An access subject fulfills the operation that none of the subjects should fulfill with the asset (a completely prohibited action);
- An access subject fulfills an operation with the asset despite the access differentiation rules for this operation (in particular a prohibited action);
- An access subject uses APCs to fulfill operations with another system asset (a threat to "social responsibility").

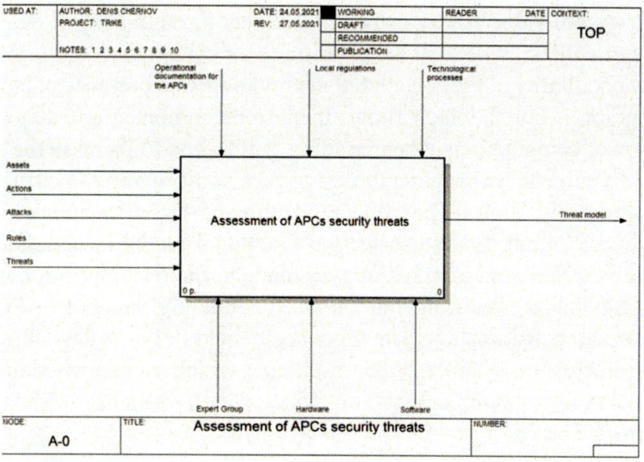

Fig. 1. IDEF0 – diagram of APCs security threat assessment

It is noteworthy that the latter case of "social responsibility" threats includes, for example, an open SMTP relay server in the local APCs network, which can be used by potential information security intruders [10].

The list of threats to APCs information security is generated on the basis of the obtained "requirements model" as follows: one "service denial" threat is made for each assumed operation. The next step is to invert the set of planned operations in order to form a set of prohibited operations. Doing so, a "privilege elevation" threat is created for each set of prohibited operations. Then for each planned operation "privilege elevation" threats to completely or partly prohibited actions are generated. After that a "social responsibility" threat is added to the obtained threat set (a threat that a subject will use the system to take measures against another system). Thus TRIKE methodology covers all possible threats to APCs information security. The obtained by the given algorithm information security threat model can be formalized in accordance with the expression (1).

$$U_{di} = \sum_{d=1}^{n} \left(U_d(a) + \sum_{i=1}^{3} U_{\overline{d}}(b_i) \right) \tag{1}$$

where U_d - threats to operations $d = \overline{1,n}$; $U_d(a)$ - the function that characterizes a "service denial" threat in relation to operations d; $U_{\overline{d}}(b_i)$ - function describes "privilege elevation" threats b_i, $i = \overline{1,3}$, for the inverted operation $\overline{d}$ under the following conditions: b_1 - a threat of a completely prohibited action; b_2 - a threat of a partly prohibited action; b_3 - a "social responsibility" threat.

3 Discussion

TRIKE methodology allows determining APCs vulnerabilities and uses particular measures to eliminate them. It is important to note that in accordance with the considered methodology, system vulnerability is a complete way through a threat tree from one or several leaves to a threat, where all the conditions for every attack are fulfilled.

As the key peculiarity of the considered methodology the presence of repetitive paths of the attack graph is noted, which allows them to be combined into attack libraries in order to minimize computations when building graphs and to increase the efficiency of using graphs when identifying actual threats to APCs information security.

Based on the analysis of the possible application of TRIKE methodology for modeling threats to APCs information security a functional model for assessing threats to APCs security was built and a transition was made to the decomposition of functional blocks characterizing the main operations when modeling threats to APCs using the methodology under consideration. The model presented in Fig. 2 describes in detail all the stages of information security threat modeling, which include obtaining a requirements model, obtaining an implementation model, generating a list of actual threats and creating an attack library.

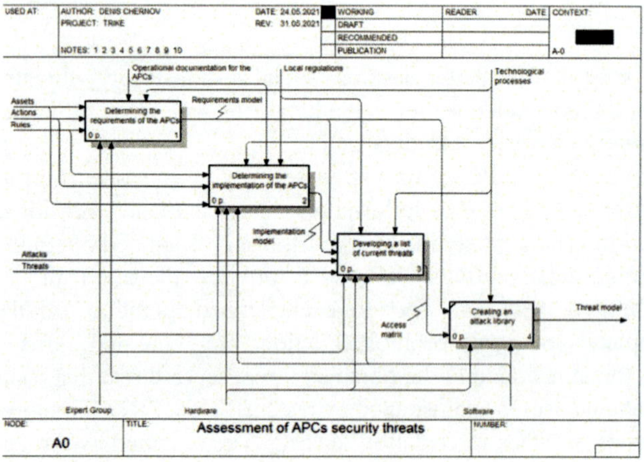

Fig. 2. APCs security threat assessment model.

The decomposition of the functional block Defining the requirements of the automated control system, which includes operations for identifying the assets of the automated control system and developing an access matrix for the list of assets, is shown in Fig. 3. The result of the execution of this functional block is a model of the requirements of the automated control system. At the stage of asset identification, the expert group conducts interviews with the personnel of the automated control system in order to identify the assets used. The assets of the automated process control system are a component or part of the overall system in which the organization operating the system directly invests and which, accordingly, require protection from the organization. When identifying assets, it should be borne in mind that the automated process control system includes not only hardware, but also software [11–13]. The description of information assets is carried out by constructing binary statements. The following types of information assets can exist:

Information (files containing information about the technological process).

- Hardware (PLC, SCADA, computers).
- Software, including application programs.
- Local network.
- Software and hardware (electronic media).

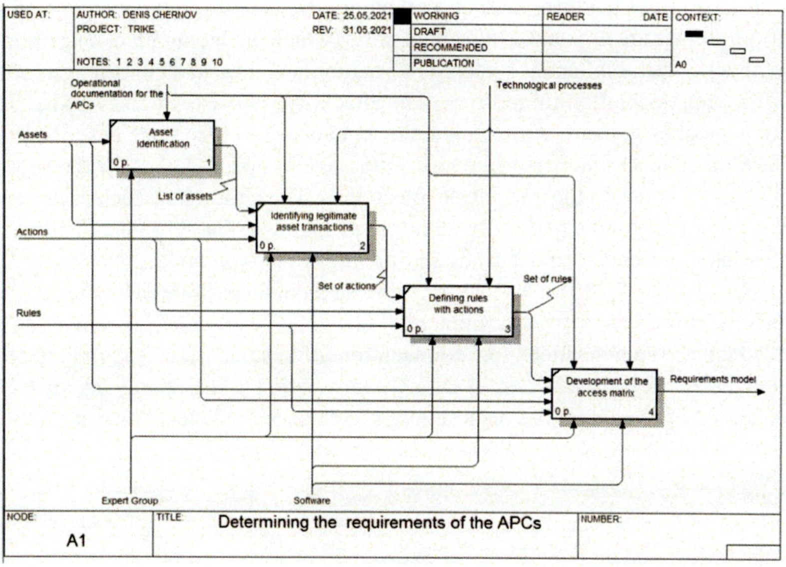

Fig. 3. Decomposition of the functional block determining APCs requirements.

Decomposition of the functional block Determining APCs implementation including the operations identifying possible operations and forming DFD data flow diagram, is presented in Fig. 4. The result of this functional block execution is an APCs implementation model. Each automated control system has steps that the user performs in the system that are not included in the set of planned actions. These operations indicate how the user interacts with the rules of the automated control system, which affect how and when he can take actions and what requirements he must meet to do so. In other words, these auxiliary operations affect the state of the automated process control system itself, as opposed to assets. Logging in is a prime example of such a supportive operation—it puts the user in a new role, which is a prerequisite for other, scheduled actions in the system. Defining a set of auxiliary operations is similar to defining a set of intended actions. To find the set of possible operations, you must first find the set of possible objects of these actions [6]. After determining the set of possible operations, the TRIKE methodology involves the transition to building a data flow diagram in the automated process control system. DFD diagrams provide a logical description of the system implementation and show the large-scale architecture of the system. They show which entities exist in the implementation of the system and in which ways these entities exchange information [14–16].

An important stage in the process of modeling threat to APCs information security is developing a list of actual threats to the system, as it includes basic functions of a threat model, such as threat generation, identification of vulnerable links and vulnerabilities [17, 18].

Based on the results of the formation of the list of threats to the security of information of the automated control system, described by the expression (1), the TRIKE methodology implies the formation of a graph of attacks. Based on the results of graph generation, attacks are organized into attack trees, which are hierarchical descriptions of how an attacker can implement a specific threat to the automated control system using the system implementation in question. The attack tree consists of tasks and subtasks. The root node of each tree is a threat, and the child nodes of each node describe in more detail how an attacker can perform a task in the parent node. Except for the root node of the threat, all nodes in the attack tree are attacks. The children of each node are sub-targets for the node, and together the children of the node should indicate all possible ways in which that node can occur. In addition to target-type nodes, attack graphs can contain logical links. Some nodes may require that all of their children be executed independently, while others may require that only one node be executed. Usually, you don't need to create a complete attack tree for each threat. In such cases, it is only necessary to expand the attack tree until there is enough information to reasonably decide whether the risk caused by the threat has been reduced to an acceptable level of risk or not [6].

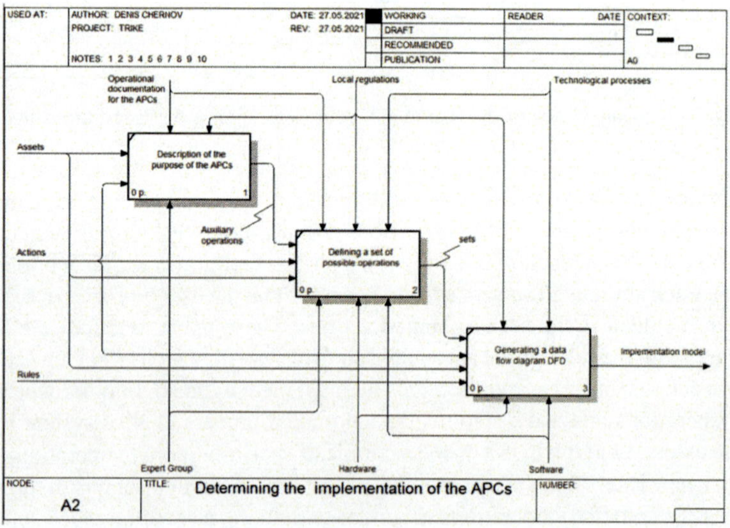

Fig. 4. Decomposition of the functional block determining APCs implementation.

A vulnerable link in the system can be a software, hardware, or hardware-software tool, including information security tools, in respect of which security threats can be implemented. Vulnerable links are identified at each level of the automated control system [19, 20].

Vulnerability is the properties inherent in the object of informatization that lead to a violation of the security of information on a particular object and are caused by the peculiarities of the process of functioning of the object of informatization, the properties of the architecture of the automated system, the exchange protocols and interfaces used by the software and hardware platform, as well as the operating conditions [21].

Decomposition of the functional block generating a list of actual threats to APCs is presented in Fig. 5. The result of this functional block execution is an APCs threat graph.

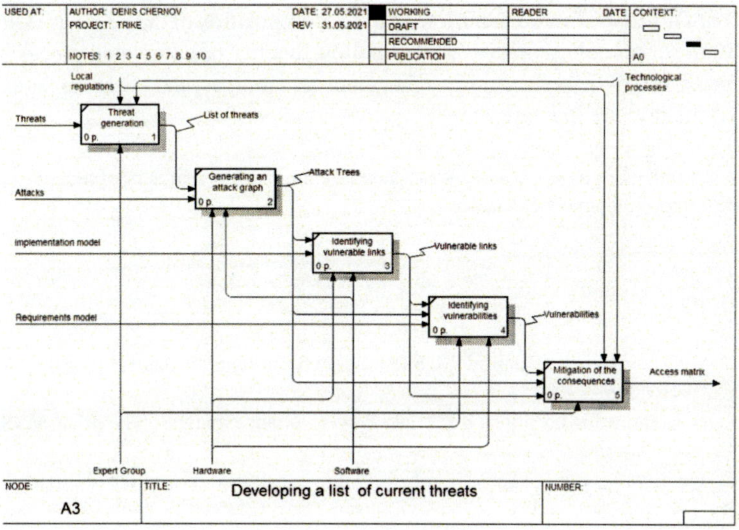

Fig. 5. Decomposition of the functional block generating a list of actual threats to APCS.

Mitigation is a guarantee that reduces or eliminates the risk associated with a particular weakness. The goal of mitigation is to reduce the risk to an acceptable level determined by the stakeholders. Mitigation can either reduce the likelihood of a successful attack, or reduce the impact of a successful attack.

Developing a functional model allows to display visually and effectively the entire mechanism of modeling threats to APCS information security. The conducted analysis of TRIKE methodology allowed to structure information, required for supporting APCs threat modeling functions, emphasize basic stages of threat modeling, trace the dynamics of functions, information and resources.

As a difference from other existing approaches to threat modeling let us emphasize the focus on modeling threats to information security from the point of view of an information protection system and not a potential intruder. This difference is an essential dignity of the TRIKE methodology.

The main lack of the considered methodology, which is especially evident when modeling threats to APCs information security, is the complexity of its implementation for all system components situated at different levels of APCs [22]. This lack justifies the assumption that some threats will be excluded from consideration in a threat model as insignificant. Minimizing the consequences from the possible exclusion of actual threat from consideration due to the impossibility of using TRIKE methodology for modeling threats to APCs information security is the subject of further research.

4 Conclusion

Based on the results of TRIKE methodology analysis aimed at determining basic stages of threat modeling, applicable to industrial automation systems, the approach of the considered method to generating a list of information security threats was formalized. Diagrams of DFD data flow were built with the decomposition of the peculiar elements of the algorithm for modeling threats to information security of automated process control systems using TRIKE methodology. The main advantages and disadvantages of the considered method are indicated.

Acknowledgment. The reported study was funded by the Russian Ministry of Science (information security), project number 15/2020.

References

1. Aktual'nye kiberugrozy. 2 kvartal 2018 (Actual cyber threats. II quarter of 2018), Moscow, Positive Technologies (2018)
2. Panettieri, J.: Ransomware attack Rock City Power. Johannesburg, South Africa, MSSP Allert (2019)
3. Aktual'nye kiberugrozy. 2 kvartal 2019 (Actual cyber threats. II quarter of 2019). Moscow, Positive Technologies (2019)
4. Aktual'nye kiberugrozy: itogi 2019 goda (Cybersecurity threatscape-2019), Moscow, Positive Technologies (2020)
5. Chernov, D.V., Sychugov, A.A.: Development of a mathematical model of threat to information security of automated process control systems. In: 2019 International Russian Automation Conference (RusAutoCon), Sochi, Russia, pp. 1–5 (2019). https://doi.org/10.1109/RUSAUTOCON.2019.8867708
6. Eddington, M., Larcom, B., Saitta, E.: TRIKE v1 Methodology Document, USA. Octotrike (2005)
7. Khanji, S., Jabir, R., Ahmad, L.: Evaluation of Linux SMTP server security aspects? a case study. In: 7th International Conference on Information and Communication Systems (ICICS), pp. 252–257 (2016). https://doi.org/10.1109/IACS.2016.7476120
8. Fu, M., Wang, D.: Modeling method of operational task combined with IDEF and UML. In: IEEE 3rd Advanced Information Technology Electronic and Automation Control Conference, pp. 1443–1447 (2018). https://doi.org/10.1109/IAEAC.2018.8577660
9. Yu, H., Wu, D.: Enterprise modeling based on IDEF and UML. In: 4th International Conference on Advanced Information Technology and Sensor Application (AITS), pp. 59–62 (2015). https://doi.org/10.1109/AITS.2015.22
10. Manenti, G., Ebrahimiarjestan, M.: Functional modelling and IDEF0 to enhance and support process tailoring in systems engineering. In: International Symposium on Systems Engineering (ISSE), pp. 1–8 (2019). https://doi.org/10.1109/ISSE46696.2019.8984539

11. Pletnev, P.V., Belov, V.M.: Metodika ocenki riskov informacionnoj bezopasnosti na predpriyatiyah malogo i srednego biznesa (Methodology for assessing information security risks in small and medium-sized businesses). Doklady Tomskogo gosudarstvennogo universiteta sistem upravleniya i radioelektroniki **25**(1–2), 83–86 (2012)
12. Semenov, D.A., Chichikin, G.Y.: Zashchita informacionnyh aktivov (Protection of information assets). Academy **57**(6), 31–33 (2020)
13. Rozdestvenskaya, T.E., Guznov, A.: Cifrovye finansovye aktivy: problemy i perspektivy pravovogo regulirovaniya (Digital financial assets: problems and prospects of legal regulation). Aktual'nye problemy rossijskogo prava **115**(6), 43–54 (2020)
14. Tasvaeva, A.N.: Diagrammy potokov dannyh i variantov ispol'zovaniya kak instrumenty proektirovaniya informacionnyh sistem (Data flow diagrams and use cases as information system design tools). Modeli, sistemy, seti v ekonomike, tekhnike, prirode i obshchestve **3**(2), 143–146 (2012)
15. Kalyanov, G.N.: Konceptual'naya model' DFD-tekhnologii (The conceptual model of DFD technology). Otkrytoe obrazovanie **21**(4), 21–26 (2017)
16. Zimovetc, O.A., Matorin, S.I.: Predstavlenie diagramm v notaciyah DFD, IDEF0 i BPMN s pomoshch'yu sistemno-ob"ektnyh modelej "Uzel-funkciya-ob"ekt" (Representation of diagrams in DFD, IDEF0, and BPMN notations using Node-Function-Object system-object models). Ekonomika. Informatika **20**(19–1), 134–144 (2011)
17. Chernov, D.V., Sychugov, A.A.: Mathematical modeling of information security threats of automated process control systems. In: International Conference on Electrotechnical Complexes and Systems (ICOECS) (2019). https://doi.org/10.1109/ICOECS46375.2019.8950023
18. Verigo, A.A., Csapko, G.P., Katashev, A.S.: Ocenka uyazvimostej avtomatizirovannyh sistem upravleniya tekhnologicheskimi processami (Vulnerability assessment of automated process control systems). Mezhdunarodnyj nauchno-issledovatel'skij zhurnal **53**(11–4), 47–49 (2016)
19. Chernov, D.V., Sychugov, A.A.: Analysis of modern requirements and problems of information security of automated process control systems. Neurocomputers **8**, 38–46 (2018). https://doi.org/10.18127/j19998554-201808-04
20. Konovalenko, S.A., Korolyov, I.D.: Vyyavlenie uyazvimostej informacionnyh sistem (Identifying information system vulnerabilities). Innovacii v nauke **58**(9), 12–20 (2016)
21. Muhanova, A., Revnivikh, A.V., Fedotov, A.M.: Klassifikaciya ugroz i uyazvimostej informacionnoj bezopasnosti v korporativnyh sistemah (Classification of threats and vulnerabilities of information security in corporate systems). Vestnik Novosibirskogo gosudarstvennogo universiteta. Seriya: Informacionnye tekhnologii **11**(2), 55–72 (2013)
22. Chernov, D.V., Sychugov, A.A.: Application of the method of determining the degree of danger of destructive actions to solve the problem of information security of APCs. In: 2020 International Conference on Electrotechnical Complexes and Systems (ICOECS), Ufa, pp. 1–4 (2020). https://doi.org/10.1109/ICOECS50468.2020.9278479

Industrial Control System Cybersecurity Assessment Handling Delay Estimation

A. A. Baybulatov(✉) and V. G. Promyslov

V. A. Trapeznikov Institute of Control Sciences of Russian Academy of Sciences, 65, Profsoyuznaya Street, Moscow 117997, Russia
bajbulatov@mail.ru

Abstract. Nowadays, industrial control systems are becoming more digital, more complex, and more interconnected causing growing anxiety about their safety, security, and especially cybersecurity. For dealing with all security problems including cybersecurity assessment, security programs are utilized where the properties of confidentiality and integrity are characterized in detail. But the availability attribute often suffers due to a lack of attention, which makes the assessment of availability grow into one of the thorniest issues. The article investigates cybersecurity in the industrial control systems context, clarifies the great value of availability, and explains a reasonable shift between cybersecurity and availability assessment problems. A delay of the signal transmission is discovered to be a suitable measure of the quantitative availability assessment, and a theory of deterministic queuing systems Network calculus is advocated to be a relevant tool for the delay estimation and availability modelling. A reference model for the availability assessment and also an appropriate metric based on delay and system dependency are proposed. The results of the verification of the applicability of Network calculus to solving the delay estimation and cybersecurity assessment problems are presented.

Keywords: Cybersecurity · Availability · Industrial control system · Delay · Measure · Metric · Network calculus

1 Introduction

Since the Information Age began, a great amount of data containing personal, business, and industrial information have been processed daily. Many consequent problems have been unsurprisingly raised. As the major part of information needs to be protected, the problem of security of information proves to be the burning one. Because today's information is stored mostly in digital format and operated by digital means, aka computers, this problem is likely to be better referred to as a cybersecurity problem. Indeed, the term cybersecurity is probably the most popular among all the other notions concerning the security of information.

As for the precise meaning, the concept of cybersecurity usually implies "the protection of computer systems and networks from information disclosure, theft of or damage

A. A. Radionov and V. R. Gasiyarov (Eds.): RusAutoCon 2021, LNEE 857, pp. 462–473, 2022.
https://doi.org/10.1007/978-3-030-94202-1_43

to their hardware, software, or electronic data, as well as from the disruption or misdirection of the services they provide [1].” Taking into account this definition and comparing cybersecurity of digital systems with the security of traditional engineering systems in relation to restricted access, preventing attacks, user errors, and complexity of construction, it is easy to realize that cybersecurity is significantly different and more complex [2]. From this, the issues related to cybersecurity happen to be very serious.

The first important issue is that historically the cybersecurity problem has been fully recognized for classic information technology applications like business, government, and academic systems, but cybersecurity for industrial control has not been a popular realm of investigation for many years.

Indeed, cybersecurity for classic information technology is widely debated, many international conferences organized every year have special information- and cybersecurity sessions, many conferences and events are fully devoted to cybersecurity. As a result of these activities, five laws of cybersecurity focused on vulnerabilities, innovations and exploitations, as well as human's trust and doubts were advised for a broad range of software, developers and consumers [3].

However, cybersecurity in the industrial control context had not gained much attention until the second decade of this century when in 2010 the Stuxnet attack happened. Since then, a concern about industrial control systems cybersecurity is becoming more considerable and deeper [4]. As for accepted approaches that can help to ensure industrial cybersecurity, the process hazard analysis (PHA) method [5] can be distinguished.

The second issue is that cybersecurity is widely assumed to be all about protecting the Internet and online computerized systems [6]. Up to now, local area networks have not been considered seriously. The most dangerous cyberattacks are expected to be based on Internet vulnerabilities. Such online threats as social engineering, ransomware, distributed denial-of-service (DDoS) attacks, third party software, and cloud computing vulnerabilities as well as solutions on how to protect against them are thoroughly discussed.

This issue is mostly regarded to classic information technology, as many modern appliances are Internet-connected now forming the Internet of Things. As for the industrial world, the list of threats and solutions is different herein [7] but also related to the Internet and Industrial Internet of Things, which is becoming a trendy innovation these days.

Even investigating Supervisory Control and Data Acquisition (SCADA) systems, open access networks are considered as a primary threat [8]. To solve this problem, unidirectional gateways have been advised instead of standard firewalls [7].

The third issue is the reality that a vast majority of cybersecurity studies and discussions are hyper-focused on threats and methods to prevent them. The problem of how to assess cybersecurity is not raised often: very little number of investigations deal with cybersecurity assessment. Furthermore, some researchers point out to behavioral, not technological, nature of cybersecurity [9]. Others claim that one of the critical attributes of cybersecurity, availability, can be either 0 which means a lack of availability or 1 that corresponds to any usable value of availability [10], which leads to the conclusion that availability and so cybersecurity is not needed to be quantitatively assessed at all. However, the truth is that the problem of proper cybersecurity assessment with accurate quantitative measures and metrics happens to be very serious [11].

The presented study is devoted to the cybersecurity issues related to industrial control systems and their local area networks. The article starts with a review of cybersecurity modelling and an in-depth explanation of the great value of availability, especially for industrial control. Then the problem of availability assessment is considered: the scope of application is elucidated, the techniques based on a theory of deterministic queuing systems Network calculus are explained, and a quantitative metric is proposed. In the last but one section, the results of the verification of the applicability of Network calculus to the cybersecurity assessment is presented. The article concludes with a discussion of the contributions made in the research.

2 Cybersecurity and Availability Modeling for Industrial Control

When considering cybersecurity issues, the majority keep in mind a set of attributes: confidentiality, integrity, and availability called the CIA triad. The attributes mean the following. Confidentiality: only authorized users or processes should be able to access data. Integrity: nobody should be able to improperly modify data. Availability: authorized users should be always able to access data. According to the CIA triad, cybersecurity assurance implies maintaining all of these three attributes. There are other more complicated alternative models, e.g., the Parkerian hexad, but they usually rely on these three basic concepts, moreover, they are not so popular [12].

Historically, cybersecurity as well as information security began with confidentiality. This is because most of the commonly used information is business, corporate, or personal, which must be obviously confidential. Also, correctness, or integrity, is a property that traditionally must be maintained for this type of information. At the same time, availability is usually being forgotten. As a result, confidentiality plays a major role in the triad, integrity, the second, and availability, the last.

Considering formally the classic CIA triad, we must admit that it has the equilateral shape with each of the attributes on its own side, treating all three concepts equally valuable. But in reality, there is a dependence of confidentiality and integrity on availability [10]. The dependence can be explained as follows. For the information to be available, i.e., for the existence of availability, the properties of confidentiality and integrity are not needed to be maintained. But if the information is not available, i.e., availability does not exist, the properties of confidentiality and integrity can never be achieved. Following the idea of the prevalence of availability, Fig. 1 presents the more realistic shape of the CIA triad.

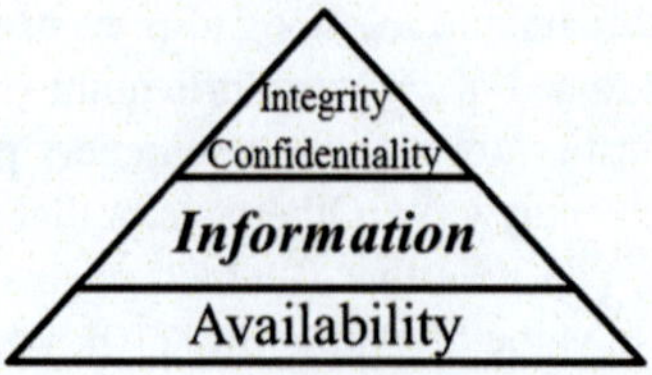

Fig. 1. The realistic CIA triad.

If we consider the CIA triad with respect to industrial control, the priority of availability becomes much more clearly evident. Indeed, a control system must be always in operation, i.e., it must be constantly available. Even in the case of stealing and changing information, which means disrupting confidentiality and integrity, operability, or availability, must be maintained.

Nevertheless, in special documents devoted to the cybersecurity assurance and assessment called security policies, the properties of confidentiality and integrity are usually described in detail, but it is not the case for the availability attribute, which suffers due to a lack of attention. Considering the industrial cybersecurity model of foundational requirements [13], it is interesting to note that it is just the last seventh requirement that corresponds directly to availability.

In security programs, for the cybersecurity assessment to be mathematically clearly described, formal models are utilized [14]. The most commonly used models: the Bell-LaPadula model and the Biba model focus on the properties of confidentiality and integrity respectively. In short, the Bell-LaPadula model states "write up, read down," i.e., data of higher security level cannot be read by the subject of the lower level. The Biba model proposes "read up, write down," i.e., data cannot be written to a higher integrity level by the subject of the lower level. There are other popular models, e.g., the take-grant model, but none of them corresponds directly to availability.

Besides, formal security models are not absolutely ideal: they have theoretical limits, they can lead to unusable systems, following the model is time-consuming and costly, based on assumptions, they do not establish security, moreover, provable security is not necessarily beneficent [15].

As we can see from the above discussion, the availability attribute happens to be the most important, especially for control systems, but we must also admit that it is the least researched one. That is why the shift from the problem of cybersecurity assessment towards the problem of availability assessment seems to be reasonable.

In order to make a useful and accurate availability assessment, a suitable description of availability is needed. There are a vast variety of availability interpretations presented in different international standards; the most commonly used of them revolve around the statement "the fraction of time for which a system is capable of fulfilling its intended purpose". Referring to a real value of time and therefore making availability quantitatively assessable, such definitions look advantageous, but they are closely connected to classic reliability and do not suit properly to availability of industrial control [16].

We propose to use another quantifiable definition, which designed especially for industrial control, the IEC 62443 one: "the property of ensuring timely and reliable access to and use of control system information and functionality [17]." Following this definition, we can select a time of the signal transmission from a source to a receiver, or delay, as a measure of the availability assessment.

The major advantage of delay is its natural correspondence to the operation of industrial control systems because it is the delay, not "the fraction of time", that describes the behavior of a system during and after a cyberattack [18]. In addition, it should be mentioned that "real time network traffic flow statistics," a measure similar to delay, is also usually encountered in today's industrial control systems investigations [19].

3 The Problem of Availability Assessment

Having clarified a delay to be a suitable measure of the quantitative availability assessment, one can realize that all the availability problems are closely connected to the concept of delay.

3.1 General Techniques for Availability Assessment

When considering the availability assessment, one should distinguish two different general scopes where this assessment is commonly applied: normal operation and the prediction for the future.

The appropriate techniques utilized for the assessment can also be classified into two general classes: 1) direct delay (time) measurement, further for short referred to "measurement," 2) indirect delay (time) measurement, further referred to "estimation."

The first technique, measurement, is preferable in most cases because it provides results with high confidence. Besides, it is easy to attain since the measurement of delays in industrial control systems is a common practice for the operational personnel: a special diagnostic function delivers measured values of delays during normal operation [20].

Unfortunately, in many practical cases, measurements are not possible. The problem might be related to the absence of a suitable test configuration of the system due to the high reconfiguration costs, or to the fact that some modes of the system operation are not available at the moment.

The second technique, estimation, is potentially more attractive but requires an extensive verification of the model or other testing environment, which provides a link between estimated results and real characteristics of the system.

For both techniques, the results are usually subjected to subsequent statistical processing, but for digital industrial control systems, the statistical interpretation of the estimation might be extremely challenging. The matter is that in many practical cases, statistical distributions of the time parameters tend to be non-gaussian heavy-tailed, which require sophisticated processing [21].

That is why, resolving this issue, for more convenient delay estimation, we advise using a deterministic approach of Network calculus.

3.2 Network Calculus Basic Idea

Being a competitor to classic queueing theory, a theory of deterministic queuing systems Network calculus is specially designed for calculating deterministic worst-case bounds, which are naturally not needed to be subjected to the statistical processing and especially useful for critical important plants with high operation risk [22]. The basic idea of Network calculus can be outlined as follows.

In Network calculus, a lossless system (single buffer, communication node, etc.) is considered as a black box with arrival and departure cumulative functions $F(t)$ and $F^*(t)$. It is easy to see that a delay at any time in this model equals to a horizontal deviation between these functions.

However, a delay is proved to be calculated using bounds, not actual cumulative functions [23]:

$$d(t) \leq \sup_{s \geq 0}\{\inf\{\tau \geq 0 : \alpha(s) \leq \beta(s+\tau)\}\} \tag{1}$$

where $\alpha(t)$ is the minimum arrival curve, a bound of $F(t)$,

$$\alpha(t) = \sup_{s \geq 0}\{F(t+s) - F(s)\} = F \oslash F \tag{2}$$

and $\beta(t)$ is the service curve, a bound of $F^*(t)$. Herein, two different types of service curves can be used. The minimum service curve $\beta_{\min}(t)$ defined as

$$F^*(t) \geq \inf_{0 \leq \tau \leq t}\{F(\tau) + \beta_{min}(t-\tau)\} = F \otimes \beta_{min} \tag{3}$$

describes the minimal performance of the system. The maximum service curve $\beta_{\max}(t)$ defined as

$$F^*(t) \leq \inf_{0 \leq \tau \leq t}\{F(\tau) + \beta_{max}(t-\tau)\} = F \otimes \beta_{max} \tag{4}$$

characterizes the system performance under normal conditions. For all service curves

$$\beta_{min}(0) = \beta_{max}(0) = 0 \tag{5}$$

Symbols $\otimes$ and $\oslash$ means min-plus convolution and de-convolution respectively [24].

As for the values of delays calculated with the help of these service curves, the minimum service curve (2) gives the maximum, or worst, delay. The maximum service curve (3) gives the operational delay under normal conditions.

3.3 Network Calculus Techniques for Delay Estimation

A couple of different techniques for delay estimation can be derived from the basic Network calculus idea discussed above.

The first technique, classic delay calculation, consists of the immediate usage of (1) in the case when the arrival process and service are known. Unfortunately, straightforward calculations with arbitrary arrival and service curves can lead to insuperable problems. For convenience, the arrival and service curves must be expressed by linear functions. Two linear models of the arrival curve are commonly used: a "leaky bucket" specified by the affine function and a traffic specification that can be described by a couple of "leaky buckets". The most broadly applicable model for the service curve is determined by the rate-latency function. For linear arrival and service curves, the calculations according to (1) were proved to be simple and easy to handle [23].

The second technique, delay from backlog bound calculation, is applicable when the arrival process is unknown, but there is relevant information about a backlog, i.e., a number of items (bits, processes, etc.) kept inside the system, and both types (a framework) of service curves (3) and (4). The technique consists of the calculation of the delay with the help of the maximum value of a backlog and a framework of service curves.

The solution for the linear relationship between service curves, particularly defined by rate-latency functions, was proved to be easy to calculate [25].

Taking a glance at the applications where considered techniques can be used, we should note the following. The first technique, classic delay calculation, is useful for the modelling of the availability dependence on potential cyberattacks [16]. The second technique, the approach of utilizing a backlog bound and a framework of service curves, is useful for the modelling of process scheduling [25].

Table 1 summarizes considered techniques for handling delays and the scopes of their application.

Table 1. Delay techniques and scopes of application.

Technique	Scope of application
Measurement	Normal operation
Diagnostic function	Availability diagnostic
Estimation	Future
Classic delay calculation	Modelling of the availability dependence on potential cyberattacks
Delay from backlog bound calculation	Modelling of process scheduling

3.4 Availability Metric Modeling

Having explained the techniques for the delay estimation, we need an availability model utilizing the delay measure and providing a metric for the availability assessment. Construct now this model.

Taking into consideration a set of organizational levels that is usually regarded during availability investigations [10], we should note that we consider just the software level and omit other levels, such as physical site, personnel, hardware, etc.

For the availability assessment, we propose using the reference model presented in Table 2. The six concepts that constitute the model mean the following. Availability metric is an abstract attribute based on related measures, which facilitates the quantitative availability assessment. The function is one of the functions for which implementation the system is designed. The system is system architecture. The platform is a set of assets that are needed to be protected. Delay is a measured or estimated value of delay considered with respect to system architecture. Time dependent parameters are parameters for which the delay is measured for.

In order to make a metric according this reference model, firstly, we should consider the function domain. For simplicity, we can select just one function, e.g., the monitoring function, for the assessment. Then the availability metric we are vouching for can be specified as follows

$$A = \begin{cases} 1 - \frac{d}{d_{max}}, d < d_{max} \\ 0, d \geq d_{max} \end{cases} \tag{6}$$

Table 2. Availability reference model.

Availability	Availability metric
	Function
	System
	Platform
	Delay
	Time dependent parameters

where d is a measured or estimated value of delay according to the specified time dependent parameters, d_{max} is the preassigned maximum value of delay, for example, according to the technical assignment.

To make (6) more accurate, we should consider the system concept and understand that in reality delay d depends on system architecture, i.e., it is a complex variable containing some time slots.

Following the idea that availability is known to be an attribute of dependability [26], we can apply the dependability notion for utilizing the system concept and making metric (6) more general. To do this, consider a control system to be composed of a number of software components.

Regard a dependency matrix for representing the dependency between system components with the elements $c_{ij} = 1$ if i-component depends on j-component, and $c_{ij} = 0$ otherwise [27]. Then the dependency of the whole control system can be specified as follows

$$\sum_{i=1}^{n}\sum_{j=1}^{n} c_{ij} \tag{7}$$

where n is the number of software components.

With the use of (7), the delay can be defined

$$d = \sum_{i=1}^{n}\sum_{j=1}^{n} c_{ij} d_{ij} \tag{8}$$

where d_{ij} means the time of the signal transmission from i-component to j-component if $i \neq j$, and time of the execution of i-component if $i = j$.

From (6) and (8) the availability metric

$$A = \begin{cases} 1 - \dfrac{\sum_{i=1}^{n}\sum_{j=1}^{n} c_{ij} d_{ij}}{d_{max}}, \sum\limits_{i=1}^{n}\sum\limits_{j=1}^{n} c_{ij} d_{ij} < d_{max} \\ 0, \sum\limits_{i=1}^{n}\sum\limits_{j=1}^{n} c_{ij} d_{ij} < d_{max} \end{cases} \tag{9}$$

Equation (9) allows calculation of the availability metric based on the knowledge of the system structure with the use of the measured time slots of the signal propagation and the specified maximum value of delay.

4 Verification of Network Calculus Modeling

As already mentioned, we advocate Network calculus to be a suitable tool for the implementation of techniques for the delay estimation and availability modelling (see Table 1). Moreover, Network calculus happens to be useful for solving a number of other problems related to industrial control systems [18].

Now our wish is to verify the applicability of Network calculus to solving problems related to delays, in particular, how accurate the metric derived with Network calculus happens to be for the availability assessment. In other words, the idea is to analyse how close the calculated values of delay to the actual ones. To do this, some tests have been carried out. The arrival flow was simulated. For this, different distributions were used: normal distribution, Rayleigh distribution, generalized Pareto distribution, and uniform distribution. The volume of samples was about 1000 elements.

Two different types of simulation tests were performed. The first type of tests was implemented via straightforward computer simulation in a numeric computing environment, namely, in Wolfram Mathematica. For the second type of tests, special software was developed; these tests were much closer to real signal transmission in industrial control systems.

Each test for a particular arrival flow distribution was organized as follows. Arrival flow was simulated. Actual values of delay were measured, from which the maximum value was obtained. A delay was also calculated according (1) with the minimum (2) and maximum (3) service curves. Then the maximum value from measured values of delay was compared to the calculated ones.

Table 3 presents the results of the tests. The designations of the rows stand for: N(0,5;0,3) and $N^*(0{,}5;0{,}3)$ are normal distributions with the mean $\mu = 0{,}5$ and the variance $\sigma^2 = 0{,}3$; $N^*(0;0{,}3)$ is a normal distribution with the mean $\mu = 0$ and the variance $\sigma^2 = 0{,}3$; $R^*(0{,}5)$ is the Rayleigh distribution with the scale parameter $\sigma = 0{,}5$; $GP^*(0{,}1;0{,}01;1)$ is the generalized Pareto distribution with location $\mu = 0{,}1$, scale $\sigma = 0{,}01$, and shape $k = 1$; U (0;0,3) is the continuous uniform distribution with the bounds (0;0,3).

An asterisk * indicates that a test was carried out with real software execution. The lack of asterisk means that results are gained by simulation in a numeric computing environment.

The columns are denoted as follows: d_x is the maximum among measured values of delay; d_1 is the calculated with the minimum service curve value of delay; d_2 is the calculated with the minimal of the maximum service curve value of delay; P(d) (where d can be either d_x, d_1, or d_2) is the probability that the actual delay is smaller than d.

From Table 3, it is easy to see that the delay calculated with the minimum service curve (d_1) is always greater than the maximum among measured values of delay (d_x), i.e., $d_1/d_x > 1$ for all d_1 and d_x. It means that delay d_1 as a measure and the appropriate metric can be used for characterizing the worst behavior of the system.

On the other hand, the delay calculated with the minimal of the maximum service curve (d_2) is in most cases smaller than the maximum among measured values of delay (d_x), i.e., $d_2/d_x < 1$ for all d_2 and d_x except for the uniform distribution. It means that delay d_2 as a measure and the appropriate metric can be used for characterizing the normal operation of the system.

Table 3. Simulation tests results.

Distribution	d_x	P(d_x)	d_1	P(d_1)	d_1/d_x	d_2	P(d_2)	d_2/d_x
N(0,5;0,3)	1,61	0,99	26,78	1	16,63	1,10	0,99	0,68
N*(0,5;0,3)	1,58	0,99	34,33	1	21,73	1,27	0,99	0,80
N*(0;0,3)	1,08	0,99	12,41	1	11,49	0,94	0,99	0,87
R*(0,5)	1,86	0,99	24,84	1	13,35	1,61	0,99	0,87
GP*(0,1;0,01;1)	19,23	1	35,37	1	1,84	17,67	1	0,92
U(0;0,3)	0,30	1	6,86	1	22,86	0,41	1	1,37

5 Conclusion

Industrial control systems are faced with many challenges related to cybersecurity; one of the most serious and least researched of them is the cybersecurity assessment. Because the availability attribute is much more critical in the CIA triad for industrial control, and the confidentiality and integrity attributes are usually carefully investigated and accurately described in security policies, the problem of availability assessment turns to be the thorniest one. Keeping this in mind allows shifting the problem of cybersecurity assessment towards the problem of availability assessment.

For the availability assessment, the IEC 62443 availability representation happens to be the most relevant because it is quantitatively applicable and allows direct calculation. Following this description, we selected a delay of the signal transmission to be a suitable measure of the availability assessment. For the delay estimation, a theory of deterministic queuing systems Network calculus was used.

A reference model for the availability assessment and the appropriate metric based on the delay measure and the system dependency were proposed. Two general scopes of application related to the availability assessment where the metric is applicable were distinguished: normal operation and the prediction for the future.

A couple of Network calculus techniques are proposed to be applicable for the estimation. The first, the classic delay calculation technique, is helpful for solving the problem of availability dependence on potential cyberattacks. The second, delay from backlog bound calculation technique, is useful during process scheduling.

In order to verify the applicability of Network calculus to solving problems related to delays, special tests were carried out. As a result, Network calculus was revealed to be applicable to solving the problems of delay estimation and availability assessment.

The contributions made in this research are applicable to any industrial control system. Future works can be directed towards applying other Network calculus delay estimation techniques to the industrial control system availability and cybersecurity assessment.

Acknowledgments. The reported study (Sect. 3) was partially funded by RFBR, project number 19-29-06044.

References

1. Schatz, D., Bashroush, R., Wall, J.: Towards a more representative definition of cyber security. J. Digit. Forensics Secur. Law **12**(2), 53–74 (2017)
2. Denning, P.J., Denning, D.E.: Cybersecurity is harder than building bridges. Am. Sci. **104**,154–157 (2016)
3. Espinosa, N.: The five laws of cybersecurity (2018). TEDxFondduLac. http://www.ted.com/talks/nick_espinosa_the_five_laws_of_cybersecurity
4. Karnouskos, S.: Stuxnet worm impact on industrial cyber-physical system security. In: IECON 2011–37th Annual Conference of the IEEE Industrial Electronics Society, pp. 4490–4494 (2011)
5. Marszal, E.M., McGlone, J.: Security PHA Review for Consequence-Based Cybersecurity. NC 27709, ISA, USA (2019)
6. Stevens, T.: Global cybersecurity: new directions in theory and methods. Polit. Govern. **6**(2), 1–4 (2018)
7. Ginter, A.: Secure Operations Technology. Abterra Technologies Inc., Calgary (2018)
8. Ghosh, S., Sampalli, S.A.: Survey of security in SCADA networks: current issues and future challenges. IEEE Access **7,** 135812–135831 (2019)
9. Anderson, A.: Built to Survive: A Business Owner's Guide on how to Survive a Cyber Attack. CreateSpace Independent Publishing Platform, Scotts Valley (2018)
10. Qadir, S., Quadri, S.M.K.: Information availability: an Insight into the most important attribute of information security. J. Inf. Secur. 7, 185–194 (2016)
11. Black, P.E., Scarfone, K., Souppaya, M.: Cyber Security Metrics and Measures. Wiley Handbook of Science and Technology for Homeland Security, pp. 1–8. John Wiley & Sons, New York (2008)
12. Andress, J.: The Basics of Information Security: Understanding the Fundamentals of InfoSec in Theory and Practice. Syngress, Oxford (2014)
13. IEC 62443-1-1: Industrial communication networks - Network and system security - Part 1-1: Terminology, concepts and models. IEC, Geneva (2009)
14. Bishop, M.: Computer Security: Art and Science. Addison-Wesley Professional, Boston (2003)
15. Denning, D.E.: The Limits of Formal Security Models. National Computer Systems Security Award Acceptance Speech (1999)
16. Baybulatov, A.A., Promyslov, V.G., Control system availability assessment via maximum delay calculation. In: Proceedings of the 2019 International Conference on Industrial Engineering, Applications and Manufacturing (ICIEAM) (2019). http://ieeexplore.ieee.org/document/8743012
17. IEC 62443-3-3: Industrial communication networks - Network and system security - Part 3-3: System security requirements and security levels. IEC, Geneva (2013)
18. Baybulatov, A.A., Promyslov, V.G.: On a deterministic approach to solving industrial control system problems. In: Proceedings of the 2020 International Russian Automation Conference (RusAutoCon), pp. 115–120 (2020). http://ieeexplore.ieee.org/document/9208149

19. Gunter, D.G., Medoff, M.D., O'Brien, P.C.: Implementing IEC 62443 - A Pragmatic Approach to Cybersecurity. Sellersville, PA, 18960, exida.com LLC (2018)
20. Promyslov, V.G., Masolkin, S.I.: NPP APCS diagnostics implementation as a routine task of APCS. IFAC Proc. Vol. **42**(2), 221–225 (2009)
21. Promyslov, V., Semenkov, K.: Non-statistical method for validation the time characteristics of digital control systems with a cyclic processing algorithm. submitted to review in Mathematics. Special Issue "Distributed Computer and Communication Networks"
22. Baybulatov, A.A.: Towards network calculus. A review of theories for dealing with facilities of increased danger. In: Proceedings of the 2018 11th International Conference "Management of large-scale system development" (MLSD 2018), pp. 1–5 (2018). https://ieeexplore.ieee.org/document/8551896
23. Bouillard, A., Boyer, M., Le Corronc, E.: Deterministic Network Calculus: From Theory to Practical Implementation. Wiley-ISTE, London (2018)
24. Litvinov, G.L.: Dequantization of mathematical structures and tropical/idempotent mathematics. An introductory lecture. In: Litvinov, G.L., Maslov, V.P., Kushner, A.G., Sergeev, S.N. (eds.) Tropical and Idempotent Mathematics, pp. 5–21, Moscow (2012)
25. Baybulatov, A.A., Promyslov, V.G.: Cybersecurity assessment using delay from backlog bound calculation. In: Proceedings of the 14th IEEE International Conference on Application of Information and Communication Technologies (AICT2020) (2020). http://ieeexplore.ieee.org/document/9368731
26. Avizienis, A., Laprie, J., Randell, B.: Fundamental concepts of dependability. In: Proceedings of the 3rd IEEE Information Survivability Workshop (ISW-2000), pp. 7–12 (2000)
27. Qadir, S.M., Quadri, S.M.K.: Metric for evaluating availability of an information system: a quantitative approach based on component dependency. Int. J. Netw. Secur. Appl. (IJNSA) **9**(2), 1–11 (2017)

Efficient Application of the Residue Number System in Elliptic Cryptography

M. Babenko[1(✉)], A. Redvanov[1], and A. Djurabaev[2]

[1] North-Caucasus Federal University, 1, Pushkin Street, Stavropol 355017, Russia
mgbabenko@ncfu.ru
[2] ITMO University, bldg. A, 49, Kronverksky Pr., St. Petersburg 197101, Russia

Abstract. The article is devoted to the study of the efficiency of arithmetic operations with points of an elliptic curve using the residue number system. Based on the obtained data, it was found that the use of the residue number system with moduli of a special type from the operations with an elliptic curve from NIST FIPS 186 allows gaining 7.72% for the operation of addition and 7.50% for the operation of doubling points of an elliptic curve on average.

Keywords: Residue number system · Elliptic curve · Arithmetic operations · Doubling points · Weighted number system · Cryptography

1 Introduction

In the modern world, information security is one of the most important problems. If earlier it was quite easy to hide information due to the poor communication, now, with more developed communication systems, such as the Internet, and with the world transforming into an information community, we cannot be sure that every bit of information is safe. Therefore, simply hiding information is no longer enough. Now even if the information is lost or stolen, it should remain inaccessible to those for whom it was not intended.

Today, elliptic curves application in cryptography is used to deal with the difficulty of solving the discrete logarithm problem. The main purpose of building cryptographic systems has always been to protect information during its transmission and storage. This problem remains relevant to this day, but the development of computing systems has given it a new quality: the question is no longer just about data protection, but also about the speed of encryption and decryption algorithms execution. In this regard, we propose to apply a residue number system to obtain the maximum speed of performing arithmetic operations with the points of an elliptic curve given over a simple field. This work aims to efficiently implement cryptographic algorithms on elliptic curves using the residue number system and the C++ programming language.

2 Elliptic Curves

Elliptic curves can be found in many different areas of mathematics such as algebra, geometry, number theory, complex analysis, etc. Factoring and cryptography were the

A. A. Radionov and V. R. Gasiyarov (Eds.): RusAutoCon 2021, LNEE 857, pp. 474–486, 2022.
https://doi.org/10.1007/978-3-030-94202-1_44

latest of the most significant applications of elliptic curves. Elliptic curves are groups defined over fields. Elliptic curve groups allow only one binary operation (referred to as the low addition operation of the addition group), which arises from arithmetic operations over a finite field. The representation of elliptic curves can be performed in several ways, such as the Legendre equation, cubic equations, fourth-order equations, and the intersection of two quadratic surfaces [1]. This can also be expressed in terms of the Weierstrass equation.

Elliptic curve $E(\mathbb{F})$ is determined by the Weierstrass equation [2, 3]:

$$E(\mathbb{F}) : y^2 + a_1xy + a_3y = x^3 + a_2x^2 + a_4x + a_6 \tag{1}$$

where $a1, a2, a3, a4, a6 \in \mathbb{F}$ and $\Delta \neq 0$.

Here Δ is the discriminant of $E(\mathbb{F})$. Equation (1) is called the general Weierstrass equation for elliptic curves. The possibility of solving the discrete logarithm problem in a public-key cryptosystem by a group of rational points on the elliptic curve $E(\mathbb{F})$ was shown by Miller and Koblitz. Besides all $(x, y) \in \mathbb{F}$ solutions of the above equation, there is an additional point, which is called a point at infinity and denoted as O, that cannot be determined using the affine equation but is necessary to complete the definition of the group.

If characteristic of the field $\mathbb{F}$ $char\,\mathbb{F} \notin \{2, 3\}$, then $E(\mathbb{F})$ can be transformed into [2]

$$E(\mathbb{F}_p) : y^2 = 4x^3 + b_2x^2 + 2b_4x + b_6 \tag{2}$$

and then we apply coordinate change $(x, y) \rightarrow \left(\frac{x-3b_2}{36}, \frac{y}{108}\right)$, which leads to the following simplified equation [2]

$$(F_p) : y^2 = x^3 + ax + b \tag{3}$$

where $a, b \in \mathbb{F}_p$.

If $p = 3$ then the canonical form of the equation has the form:

$$y^2 = x^3 + a_2x^2 + a_4x + a_6 \tag{4}$$

If $p = 2$, then the equation is reduced to one of the forms:

$$y^2 + y = x^3 + ax + b \tag{5}$$

or

$$y^2 + xy = x^3 + ax^2 + b \tag{6}$$

Equation (5) corresponds to super singular curves, and Eq. (6) to non-super singular curves.

Equation (3) is called the short Weierstrass equation for elliptic curves. In elliptic cryptography, the equations of the Weierstrass form are used. To prove that (2) has three different roots the condition $4a^3 + 27b^2 \neq 0$ is necessary and sufficient. An elliptic curve with different roots is called a nonsingular curve and forms an Abelian group with respect to the binary operation.

If *char* $\mathbb{F} = 2$ then an acceptable substitution of variables transforms $E(\mathbb{F})$ into a curve of the equation [2]

$$E(\mathbb{F}_{2^m}) : y^2 + xy = x^3 + ax^2 + b \tag{7}$$

where $a, b \in \mathbb{F}_{2^m}$. Such a curve is called non-singular and has a discriminant $\Delta = b$ [2].

The set of points $\{(x, y) \in E(\mathbb{F}_p)\} \cup \{\mathcal{O}\}$ by the operation rule of the additive group, i.e., $\otimes$, which forms an additive Abelian group, the sum is the gain of a point on the same curve. The group operation on elliptic curves is defined by the operations of adding points (ADD) and doubling points (DBL).

3 Group Operations in Affine Coordinates

The calculations of the binary group addition operation in affine coordinates over simple fields are summarized in the Algorithm 1. As we can see from the algorithm, it requires a division operation. Since all the coordinates of the elliptic curve points are represented as finite elements of the field, the proposed division operation is implemented as an expensive and complex inversion operation in the finite field.

Algorithm 1. The rule of addition for an elliptic curve E over $\mathbb{F}_p$ in affine coordinates [2]

Input: $P_1 = (x_1, y_1)$, $P_2 = (x_2, y_2)$, $\mathcal{O} \in \mathbb{F}_p$.

Output: $P_3 = (x_3, y_3) = P_1 \oplus P_2$.

1: If $P_1 = \mathcal{O}$ Then Return P_2 ;

2: Else If $P_2 = 0$ Then Return P_1;

3: Else If $P_2 = -P_1$ Then Return O /* $x_1 = x_2$ and $y_1 = -y_2$ */

4: Else If $P_2 = P_1$ Then /* Perform a DBL operation */

4.1: $\lambda = \dfrac{3x_1^2 + a}{2y_1}$ *mod p*;

5: Else If $P_2 \neq \pm P_1$ Then /* Perform an ADD operation */

5.1: $\lambda = \dfrac{y_2 - y_1}{x_2 - x_1}$ *mod p*;

6: $x_3 = (\lambda^2 - x_1 - x_2) \bmod p; y_3 = (\lambda(x_1 - x_3) - y_1) \bmod p = (\lambda(x_2 - x_3) - y_2) \bmod p;$

7: Return (x_3, y_3).

Algorithm 1 shows that for affine representation of the points, calculations with elliptic curve points involve arithmetic addition/subtraction over the finite field, multiplication, squaring, and an expensive inversion operation. Since the operation of arithmetic inversion over the field is relatively expensive compared to the operations of arithmetic multiplication and squaring, and the calculation of the modular inversion in the residue number system is quite problematic, it is necessary to represent the points of the elliptic curve in another coordinate systems [3]. The transition to projective coordinate systems makes it possible to exclude the operation of modular inversion, which enables efficient application of the residue number systems.

3.1 Group Operations in Projective Coordinate System

The general way of defining a set of points in a projective space for curves defined over $\mathbb{F}_p$, that is (2), is to homogenize an elliptic curve: to replace $x = X/Z$ and $y = Y/Z$ and multiply it by Z^3 to remove the denominators, which gives

$$E\varphi(\mathbb{F}_p) : Y^2Z = X^3 + aXZ^2 + bZ^3 \tag{8}$$

Then, the projective coordinates (X_P, Y_P, Z_P) can be used to replace the affine coordinates $\left(x_p, y_p\right)$. These substitutions $x_p = X_P/Z_P, y_p = Y_P/Z_P$, when $Z_P \neq 0$, are the simplest and commonly used way to get the projective coordinates but are not limited to this substitution choice. In general, the projection remains the same, when projections obtained by substitutions of the form $x = X/Z^i$ and $y = Y/Z^i$.

If we have an affine representation of a point $\left(x_p, y_p\right)$, to convert it to projective representation, it is necessary to set the Z coordinate 1, i.e., $\left(x_p, y_p, 1\right)$. Using projective coordinates has several advantages, the most important of which is that it eliminates the need to perform arithmetic inversion in low-value addition algorithms. On the other hand, it increases the number of multiplications and squares required for each bit of the scalar. Moreover, the projective coordinates are usually used for internal computations, and the result must be converted to the affine representation before transmission. This can be achieved by using the modular exponentiation given by Fermat's little theorem, according to which the inverse of $A \in \mathbb{F}_{2^m}$ is $A^{-1} = A^{p-2} \, mod \, p$ if the $\gcd(A, p) = 1$. We can also implement modular inversion using the extended Euclidian algorithm and the Montgomery inversion algorithm [2].

The operation of adding the points $(P + Q)$ in projective coordinates has the form:

$$\begin{aligned} X_R &= A \cdot D \\ Y_R &= B \cdot \left(X_P \cdot Z_Q \cdot A^2 - D\right) - A^3 \cdot Y_P \cdot Z_Q \\ Z_R &= A \cdot Z_P \cdot Z_Q \end{aligned}$$

where $A = X_Q \cdot Z_P - X_P \cdot Z_Q$, $B = Y_Q \cdot Z_P - Y_P \cdot Z_Q$, $C = X_Q \cdot Z_P + X_P \cdot Z_Q$ and $D = B^2 \cdot Z_P \cdot Z_Q - A^2 \cdot C$.

The point doubling operation (2P) is specified as follows:

$$\begin{aligned} X_R &= 2B \cdot D \\ Y_R &= A \cdot (4C - D) - 8Y_P^2 \cdot B^2 \\ Z_R &= 8B^3, \end{aligned}$$

where $A = 3X_P^2 + aZ_P^2$, $B = Y_P \cdot Z_P$, $C = X_P \cdot Y_P \cdot B$ and $o = A^2 - 8C$.

3.2 Jacobian Projective Coordinates

Jacobian projective coordinate system allows effective representation of the elliptic curves defined over $\mathbb{F}_p$. There, the point represented in projective coordinates as (X, Y, Z), $Z \neq 0$, corresponds to the point represented in affine coordinates as $X/Z^2, Y/Z^3$. The corresponding Weierstrass equation of an elliptic curve represented in the Jacobian coordinates is [2]:

$$E_J(\mathbb{F}_p) : Y^2 = X^3 + aXZ^4 + bZ^6 \quad (9)$$

The point $\mathcal{O}$ corresponds to $(1, 1, 0)$ and the negative value of the point (X, Y, Z) is $(X, -Y, Z)$. By replacing the point $(X/Z^2, Y/Z^3)$ in the equation of the affine curve, that is, in Algorithm 1, we can get the operations of doubling and adding points. The point $Q = (X_Q, Y_Q, Z_Q)$, resulting from doubling the point $P = (X_P, Y_P, Z_P)$ with $P \neq -P$, that is, $Y_P \neq 0$, can be written as follows [2]:

$$\begin{aligned} X_Q &\leftarrow \left(3X_P^2 + a \cdot Z_P^4\right)^2 - 8X_P \cdot Y_P^2, \\ Y_Q &\leftarrow \left(3X_P^2 + a \cdot Z_P^4\right)\left(4X_P \cdot Y_P^2 - X_Q\right) - 8Y_P^4, \\ Z_Q &\leftarrow 2Y_P \cdot Z_P. \end{aligned} \quad (10)$$

If the time values are stored in registers from A to C, then the coordinates (X_Q, Y_Q, Z_Q) of the doubling points requires the following operations: 3 arithmetic multiplications (M), 1 arithmetic multiplication by a constant (D), 6 arithmetic squares (S), and 11 arithmetic additions (A) [4]:

$$\begin{aligned} &A \leftarrow 2Y_P^2, B \leftarrow 2X_P \cdot A, C \leftarrow 3X_P^2 + a \cdot Z_P^4, \\ &X_Q \leftarrow C^2 - 2B, Y_Q \leftarrow C \cdot (B - X_Q) - 2A^2, Z_Q \leftarrow 2Y_P \cdot Z_P \end{aligned} \quad (11)$$

When fast squaring can be used, the doubling operation costs $1M + 11S + 1D$. When the curve parameter $a = -3$, we can perform a quick doubling of the point by storing two arithmetic squares in (10) using [4]:

$$C \leftarrow 3\left(X_P + Z_P^2\right)\left(X_P - Z_P^2\right) \quad (12)$$

Field operations require up to $4M + 5S + 12A$ arithmetic operations to implement fast doubling. Point $R = (X_R, Y_R, Z_R)$, resulting from the addition of point $P = (X_P, Y_P, Z_P)$ and point $Q = (X_Q, Y_Q, Z_Q)$ with $P \neq \pm Q$ and $Z_P, Z_Q \neq 0$, can be expressed as follows [4]:

$$\begin{aligned} X_R &\leftarrow F^2 - E^3 - 2A \cdot E^2, \\ Y_R &\leftarrow F\left(A \cdot E^2 - X_R\right) - C \cdot E^3, \\ Z_R &\leftarrow Z_P \cdot Z_Q \cdot E. \end{aligned} \quad (13)$$

where

$$A \leftarrow X_P \cdot Z_Q^2, B \leftarrow X_Q \cdot Z_P^2, C \leftarrow Y_P \cdot Z_Q^3,$$

$$D \leftarrow Y_Q \cdot Z_P^3, E \leftarrow B - A, F \leftarrow D - C.$$

Field operations require up to $14M + 9S + 7A$ arithmetic operations for general addition. During performing addition for projective Jacobi coordinates one square and four multiplications can be stored in (12) if any of the points P or Q are represented in affine coordinate system, which gives $8M + 3S + 7A$ arithmetic operations [2, 4].

3.3 Jacobian-Chudnovsky Projective Coordinates

The Jacobian-Chudnovsky projective coordinate system is based on the Jacobian coordinate system, and therefore the equation of the elliptic curve remains the same as in the Jacobian coordinate system. However, here the point of the elliptic curve is given by the five coordinates $P = \left(X_P : Y_P : Z_P : Z_P^2 : Z_P^3\right)$. If $Z_P = 0$, then $P = \mathcal{O}$ is a point at infinity.

The adding and doubling the points of an elliptic curve formulas remain unchanged. But instead of the operation Z_P^2 and Z_P^3, the coordinates of this point are used [4].

3.4 Modified Jacobian Coordinates

The modified Jacobian coordinate system is also based on the Jacobian coordinate system, due to which the Weierstrass equation remains the same as in the Jacobi coordinate system. But the point of an elliptic curve in the given coordinate system is given by four coordinates $P = \left(X_P : Y_P : Z_P : aZ_P^4\right)$.

The formulas for adding and doubling points on an elliptic curve remain unchanged. However, instead of the operation aZ_P^4, the coordinate of the origin point [4] is used.

3.5 Comparative Analysis of Computational Complexity

In this section we perform a comparative analysis of the arithmetic operations with points of an elliptic curve given over $\mathbb{F}_p$ computational complexity in the coordinate systems described earlier. We use the following notations:

I – modular inversion computed over the field $\mathbb{F}_p$.
M – multiplication in $\mathbb{F}_p$.
S – squaring in $\mathbb{F}_p$.

According to Table 1, it follows that:

- The computational complexity of the number multiplication in a finite simple field is greater than the squaring computational complexity for the Jacobian-Chudnovsky coordinate system. In this regard, the Jacobian-Chudnovsky coordinate system is most effective for implementing the operation of adding points.
- The modified Jacobian coordinate system is an effective coordinate system for implementing the arithmetic operation of doubling point.

Table 1. Calculating the time of addition and doubling of elliptic curve points.

Coordinate systems	$P + Q$		$2P$	
	$Z \neq 1$	$Z = 1$	$a \neq -3$	$a = -3$
Affine	–	$I + 2M + S$	$I + 2M + 2S$	
Projective	$15M + 3S$	$9M + 2S$	$7M + 5S$	$7M + 6S$
Jacobian	$14M + 9S$	$8M + 3S$	$4M + 6S$	$4M + 5S$
Jacobian-Chudnovsky	$11M + 3S$	$8M + 3S$	$5M + 6S$	$5M + 4S$
Modified Jacobian	$13M + 6S$	$9M + 5S$	$4M + 4S$	

3.6 Standard Elliptic Curves and Their Parameters

It is necessary to be careful choosing elliptic curve parameters, because the wrong choice can lead to an unsafe system, and other specific parameters can increase the safety and optimization of the implementation. The two main NIST standards in FIPS 186-3 [5] and GOST 34.10-2018 have recommendations about setting curve parameters for each finite field. These curves provide the greatest cryptographic strength and allow efficient implementation of algorithms. NIST recommends the use of five trailing fields in binary fields, i.e. $\mathbb{F}_{2^{163}}$, $\mathbb{F}_{2^{233}}$, $\mathbb{F}_{2^{283}}$, $\mathbb{F}_{2^{409}}$ and $\mathbb{F}_{2^{571}}$ for use in ECDSA [5]. In simple fields, NIST and GOST recommend using five trailing fields, i.e. $\mathbb{F}_{2^{192}}$, $\mathbb{F}_{2^{224}}$, $\mathbb{F}_{2^{256}}$, $\mathbb{F}_{2^{384}}$ and $\mathbb{F}_{2^{521}}$ to use in ECDSA [5]. These fields and their corresponding elliptic curves with parameters are listed in Table 2.

Table 2. NIST FISP 186 Recommended finite fields and their corresponding elliptic curve parameters.

	Elliptic curve			
	Field size m	Reduction polynomial $p(x)$	a	b
1	192	$2^{192} - 2^{64} - 1$	−3	2455155546008943817740293915197451784769108058161191238065
2	224	$2^{224} - 2^{96} + 1$	−3	1895828628556660800040866854449392641550468096867932107578723467 2564
3	256	$2^{256} - 2^{224} + 2^{192} + 2^{96} - 1$	−3	4105836372515214212932612978004726840911444101599372555483525631403946740 1291
4	384	$2^{384} - 2^{128} - 2^{96} + 2^{32} - 1$	−3	27580193559959705877849011840389048093056905856361568521428707301988689241309860865136260764883745107765439761230575
5	521	$2^{521} - 1$	−3	1093849038073734274511112390766805569936207598951683748994586394495953116150735016013708737573759623248592132296706313309438452531591012912142327488478985984

4 Residue Number System

The residue number system (RNS) was proposed by Svoboda and Wallach in 1955. It was also independently proposed by Garner in 1959. It uses the base of relatively prime moduli $\{m_1, m_2, ..., m_N\}$ to split an integer X into smaller integers $\{x_1, x_2, ..., x_N\}$, where x_i is the residue of X divided by m_i: $x_i = X \; mod \; m_i$, or just $x_i = |X|_{m_i}$.

Converting a number to RNS is a simple task. The inverse converting is complex. It requires the use of the Chinese Remainder Theorem (CRT) [6–9]. Modular operations, such as addition, subtraction and multiplication can be performed in RNS with residues extremely efficient. The peculiarities of RNS allows parallel carry-free implementation of these operations. This and other properties of RNS accelerate computations and enable a high level of modularity and scalability.

RNS applications in cryptography became one of the most promising developments [6]. Some cryptographic algorithms that require large word lengths, such as the RSA (Rivest-Shamir-Adleman) algorithm, were implemented using RNS [9]. RNS is also an attractive technique for elliptic curve cryptography, where sizes range from 160 to 521 bits.

The residue number system allows not only reducing the digit capacity of numbers and performing arithmetic operations component-wise, but also providing such advantages, as [10, 11]:

- low-bit residue;
- maximum parallelism;
- security;
- high accuracy;
- enabling tabular arithmetic;
- ability of the system to self-correct;
- implementation of the pipeline information processing principle.

Modular arithmetic operations with integers $A \overset{RNS}{\to} (a_1, a_2, \ldots, a_N)$ and $B \overset{RNS}{\to} (b_1, b_2, \ldots, b_N)$ in the RNS are performed component-wise, if the result falls in $[0, M - 1]$, where $M = \prod_{i=1}^{N} m_i$:

$$A \pm B \overset{RNS}{\to} \left(|a_1 \pm b_1|_{m_1}, |a_2 \pm b_2|_{m_2}, \ldots, |a_N \pm b_N|_{m_N}\right),$$
$$A \cdot B \overset{RNS}{\to} \left(|a_1 \cdot b_1|_{m_1}, |a_2 \cdot b_2|_{m_2}, \ldots, |a_N \cdot b_N|_{m_N}\right).$$

However, the division operation is computationally complex. This operation is performed only under the condition that $B|A$, and then division is performed similarly to the multiplication operation using the operation of calculating the multiplicatively inverse element modulo (modular inversion):

$$\frac{A}{B} \overset{RNS}{\to} \left(|a_1 \cdot b_1^{-1}|_{m_1}, |a_2 \cdot b_2^{-1}|_{m_2}, \ldots, |a_n \cdot b_n^{-1}|_{m_n}\right)$$

4.1 Residue Number System Parameters

As moduli of the residue number system, four moduli RNS of a special type can be used:

$$\{2^n - 3, 2^n - 1, 2^n + 1, 2^n + 3\}.$$

The use of these moduli in the development of algorithms makes it possible to efficiently implement arithmetic operations by using ideas from the work of Chervyakov et al., 2016. In this regard, the use of the above-described moduli of a special type is perfect for the cryptography on elliptic curves implementation in the RNS [11–20].

5 Simulation

To implement the software, we use standard elliptic curves SoCs with moduli of a special type described above, as well as the C++ programming language. NTL is selected as the long arithmetic library.

During the development of the software for the study, the RNS that represents a software implementation of the algorithms described above was created. For the computational experiment, a PC with the following characteristics was used: an Intel Core i5 processor with a frequency of 2.7 GHz, 8 GB 1867 MHz DDR3, an operating system macOS High Sierra ver. 10.13.6.

5.1 Simulation Results

Table 3 and Fig. 1 represent the execution time of 1000 operations of points of elliptic curves addition from NIST FIPS in the above-mentioned coordinate systems, respectively.

Table 3. Time (in seconds) of addition operation with the points of an elliptic curve.

Coordinate systems	P-192		P-224		P-256	
	without RNS	with RNS	without RNS	with RNS	without RNS	with RNS
Affine	432	361	480	453	523	465
Projective	352	320	430	413	425	415
Jacobian	421	395	480	428	485	422
Jacobian- Chudnovsky	505	478	583	525	578	521
Modified Jacobian	490	433	554	520	565	487

Coordinate systems	P-384		P-521	
	without RNS	with RNS	without RNS	with RNS
Affine	730	708	994	940
Projective	620	592	1002	960
Jacobian	634	560	922	912
Jacobian- Chudnovsky	778	750	1194	1122
Modified Jacobian	763	670	1120	1052

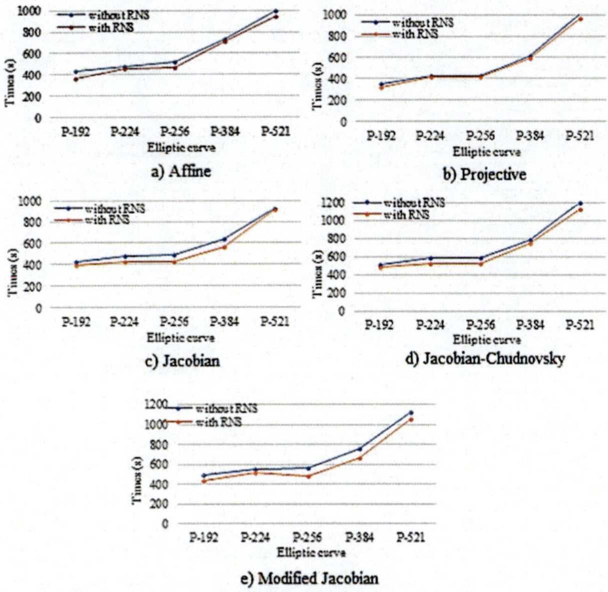

Fig. 1. Execution time in seconds for addition operation with points of elliptic curves: P-192, P-224, P-256, P-384, P-521, using coordinate systems: a) Affine; b) Projective; c) Jacobian; d) Jacobian-Chudnovsky; e) Modified Jacobian.

Based on Tables 3 and 4, a conclusion can be made that the RNS with special moduli enables obtaining an average gain for affine coordinates from 3.01% to 16.44%, for projective coordinates from 2.35% to 9.09%, for Jacobian coordinates in 1.08%–12.99%, for Jacobian-Chudnovsky – 3.60%–9.95%, and for Modified Jacobian – from 6.07% to 13.81%.

Table 4. Comparative analysis of the execution time of the addition operation with the points of an elliptic curve (%).

Coordinate systems	|with RNS/without RNS - 1|				
	P-192	P-224	P-256	P-384	P-521
Affine	16,44	5,63	11,09	3,01	5,43
Projective	9,09	3,95	2,35	4,52	4,19
Jacobian	6,18	10,83	12,99	11,67	1,08
Jacobian-Chudnovsky	5,35	9,95	9,86	3,60	6,03
Modified Jacobian	11,63	6,14	13,81	12,19	6,07

Figure 2 and Table 5 represent the execution time of 1000 operations of doubling points of elliptic curves from NIST FIPS in the above-described coordinate systems, respectively.

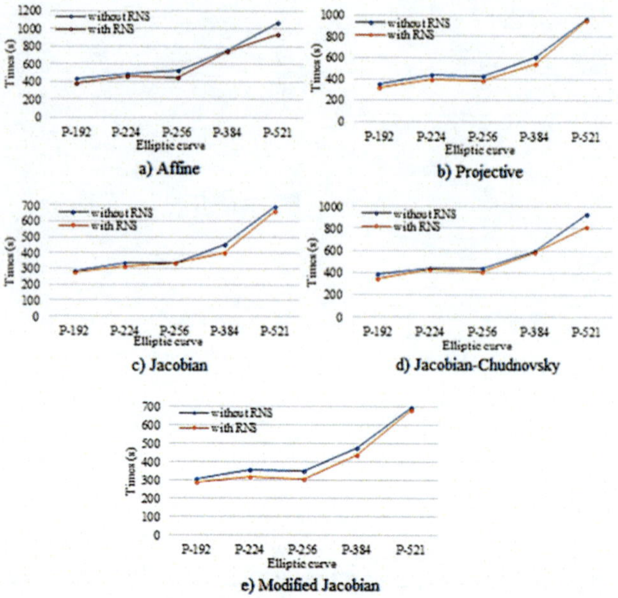

Fig. 2. Execution time in seconds for doubling points of elliptic curves: P-192, P-224, P-256, P-384, P-521, using the coordinate systems: a) Affine; b) Projective; c) Jacobian; d) Jacobian-Chudnovsky; e) Modified Jacobian.

Table 5. Time (in seconds) to complete the operation of doubling the points of an elliptic curve.

Coordinate systems	P-192		P-224		P-256	
	without RNS	with RNS	without RNS	with RNS	without RNS	with RNS
Affine	432	386	490	460	521	448
Projective	350	321	440	395	427	383
Jacobian	289	280	340	312	338	334
Jacobian- Chudnovsky	388	345	440	429	439	410
Modified Jacobian	310	290	360	319	355	307

Coordinate systems	P-384		P-521	
	without RNS	with RNS	without RNS	with RNS
Affine	759	742	1063	945
Projective	612	549	967	951
Jacobian	455	405	696	668
Jacobian- Chudnovsky	600	586	926	818
Modified Jacobian	478	441	698	685

Based on the data from Tables 5 and 6, we can conclude that the use of RNS with moduli of a special type makes it possible to obtain on average a gain for the operation of doubling points of affine coordinates from 2.44% to 14.01%, for projective coordinates 1.65%–10.30%, for Jacobian coordinates 1.18%–10.99%, for Jacobian-Chudnovsky 2.33%–11.66% and for Modified Jacobian – from 1.86% to 13.52%.

Also, from Figs. 1 and 2, we can see that not only the use of RNS affects the speed of algorithms, but also the choice of the size of the final field. If the use of RNS allows getting advantage, then the choice of a finite field with a large size gives a significant loss.

Table 6. Comparative analysis of the execution time of the doubling operation with the points of an elliptic curve (%)

Coordinate systems	|with RNS/without RNS - 1|				
	P-192	P-224	P-256	P-384	P-521
Affine	10,65	6,12	14,01	2,24	11,10
Projective	8,29	10,23	10,30	10,29	1,65
Jacobian	3,11	8,24	1,18	10,99	4,02
Jacobian-Chudnovsky	11,08	2,50	6,61	2,33	11,66
Modified Jacobian	6,45	11,39	13,52	7,74	1,86

6 Conclusion

In general, these main results are consistent with the study of arithmetic operations with points of an elliptic curve in various coordinate systems efficient implementation, showing that the use of RNS in elliptic cryptography provides not only acceleration in calculations, but also the ability to parallelize algorithms to obtain maximum speed. The simulation results presented in Tables 3, 4, 5 and 6 indicate that the residue number system with special moduli from NIST FIPS 186 allows obtaining gain for the operation of adding the points of the elliptic curve on average 7.72% and for the operation of doubling the points of the elliptic curve 7.50%.

Further research could fruitfully continue to consider this issue, using a large number of moduli of the residue number system to improve data processing.

Acknowledgment. The reported study was funded by RFBR, project number 20-37-70023.

References

1. Washington, L.C.: Elliptic Curves: Number Theory and Cryptography. Discrete Mathematics and Its Applications. Chapman & Hall/CRC, New York (2003)
2. Hankerson, D., Vanstone, S., Menezes, A.: Guide to Elliptic Curve Cryptography. Springer, New York (2004). https://doi.org/10.1007/b97644
3. Avanzi, R., et al.: Handbook of Elliptic and Hyperelliptic Curve Cryptography. CRC Press, New York (2005)
4. Verneuil, V.: Elliptic curve cryptography and security of embedded devices. Cryptography and Security, Université de Bordeaux (2012)

5. Mishra, P., Sarkar, P.: Application of Montgomery's trick to scalar multiplication for elliptic and hyperelliptic curves using a fixed base point. In: Bao, F., Deng, R., Zhou, J. (eds.) PKC 2004. LNCS, vol. 2947, pp. 41–54. Springer, Heidelberg (2004). https://doi.org/10.1007/978-3-540-24632-9_4
6. Mohan, P.V.A.: Residue Number Systems: Theory and Applications. Springer, New York (2016). https://doi.org/10.1007/978-3-319-41385-3
7. Garner, H.L.: The residue number system. Papers presented at the March 3–5 Western Joint Computer Conference, pp. 146–153 (1959)
8. Schinianakis, D.M., Kakarountas, A.P., Stouraitis, T.: A new approach to elliptic curve cryptography: an RNS architecture. In: IEEE Mediterranean Electrotechnical Conference, pp. 1241–1245 (2006)
9. Chervyakov, N., Babenko, M., Tchernykh, A., Kucherov, N., Miranda-López, V., Cortés-Mendoza, J.M.: AR-RRNS: configurable reliable distributed data storage systems for Internet of Things to ensure security. Future Gener. Comput. Syst. **92**, 1080–1092 (2019)
10. Chervyakov, N.I., et al.: An approximate method for comparing modular numbers and its application to the division of numbers in residue number systems. Cybern. Syst. Anal. **50**(6), 977–984 (2014)
11. Chervyakov, N.I., Molahosseini, A.S., Lyakhov, P.A., Babenko, M.G., Deryabin, M.A.: Residue-to-binary conversion for general moduli sets based on approximate Chinese remainder theorem. Int. J. Comput. Math. **94**(9), 1833–1849 (2017)
12. Esmaeildoust, M., Schinianakis, D., Javashi, H., Stouraitis, T., Navi, K.: Efficient RNS implementation of elliptic curve point multiplication over GF(p). IEEE Trans. Very Large Scale Integr. (VLSI) Syst. **21**(8), 1545–1549 (2013)
13. Asif, S., Kong, Y.: Highly parallel modular multiplier for elliptic curve cryptography in residue number system. Circuits Syst. Signal Process. **36**(3), 1027–1051 (2017)
14. Babenko, M., et al.: RNS number comparator based on a modified diagonal function. Electronics **9**, 1784 (2020)
15. Vershkov, N., Babenko, M., Kuchukov, V., Kuchukova, N.: Search for the global extremum using the correlation indicator for neural networks supervised learning. Program. Comput. Softw. **46**, 609–618 (2020)
16. Tchernykh, A., et al.: Scalable data storage design for nonstationary IoT environment with adaptive security and reliability. IEEE Internet Things J. **7**, 10171–10188 (2020)
17. Tchernykh, A., et al.: Performance evaluation of secret sharing schemes with data recovery in secured and reliable heterogeneous multi-cloud storage. Cluster Comput. **22**(4), 1173–1185 (2019). https://doi.org/10.1007/s10586-018-02896-9
18. Chervyakov, N., Babenko, M., Tchernykh, A., Kucherov, N., Miranda-López, V., Cortés-Mendoza, J.M.: AR-RRNS: configurable reliable distributed data storage systems for Internet of Things to ensure security. Future Gener. Comput. Syst. **92**, 1080–1092 (2019)
19. Tchernykh, A., Schwiegelsohn, U., Talbi, E., Babenko, M.: Towards understanding uncertainty in cloud computing with risks of confidentiality, integrity, and availability. J. Comput. Sci. **36**, 100581 (2019)
20. Tchernykh, A., et al.: AC-RRNS: anti-collusion secured data sharing scheme for cloud storage. Int. J. Approx. Reason. **102**, 60–73 (2018)

Diagnostics and Reliability of Automatic Control Systems

Features of Software Development for Data Mining of Storage System State

A. Zarubin[1], V. Moshkin[2](✉), and A. Koval[1]

[1] The Bonch-Bruevich Saint - Petersburg State University of Telecommunication, 22/1, Bol'shevikov, St. Petersburg 193232, Russia
[2] Second Ulyanovsk State Technical University, 32, Severny Venets Street, Ulyanovsk 432027, Russia
v.moshkin@ulstu.ru

Abstract. This article proposes the architecture of a diagnostic data mining software system for data storage systems. Analysis of logs consists of several stages: extraction of metrics from big data storage diagnostic information using artificial neural networks, analysis of time series anomalies for diagnostic data metrics using machine learning methods, generalization of the results of the analysis of anomalies and patterns of change and existence of storage performance metrics based on the characteristics of the subject area using a subject ontology, the inference of the recommendation to the administrator for correcting the storage system state through the composition of the results of the analysis of anomalies and subject ontology. In the article, an effective solution (software or algorithm) is found for the implementation of each stage. In the future, it is planned to develop this system and evaluate its effectiveness during experiments.

Keywords: Metrics · Forecasting · Log analysis · Neural network · SWRL · Ontology

1 Introduction

One of the ways to improve the performance of the storage system is monitoring the state of the storage system and timely response to failures and collisions.

The solution to this problem includes the development of models and methods for intelligent monitoring and analysis of storage system diagnostic information in real time. The software platform should implement these models and generate human-readable recommendations based on expert knowledge in natural language.

Online monitoring of storage system diagnostic information using an intelligent software platform solves the following tasks:

- Reduction of equipment downtime.
- Improving storage performance.
- Reducing the cost of maintaining storage systems.

It is necessary to solve the following tasks to achieve the goal:

A. A. Radionov and V. R. Gasiyarov (Eds.): RusAutoCon 2021, LNEE 857, pp. 489–499, 2022.
https://doi.org/10.1007/978-3-030-94202-1_45

- Selection of an effective technology and software for working with big data of diagnostic information in the form of corpuses of semi-structured information (logs).
- Automatic extraction of storage metrics from big data diagnostic information.
- Analysis of time series anomalies of characteristics in diagnostic data.
- Generalization of the results of the analysis of anomalies and descriptions of storage metrics based on the characteristics of the subject area.
- Training the static and dynamic components of the knowledge base of the storage model to update the interpretation of its characteristic collisions.
- The logical conclusion of the recommendation in natural language for correcting the state of the supported storage system.

Consider the currently used basic approaches to solving these problems.

2 Automatic Log Analysis Software

Currently, there is a set of software products that automate the analysis of logs of complex technical systems. Let's consider the most popular ones.

2.1 Papertrail

Papertrail is software that collects diagnostic data from syslog, text files, Apache, MySQL DBMS, Heroku applications, Windows event log, routers and other objects [1].

In addition, Papertrail can search online using the http protocol, command line and API (Fig. 1).

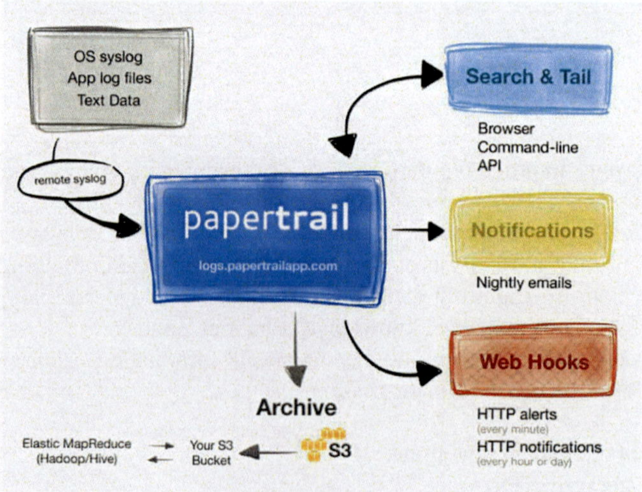

Fig. 1. Papertrail flow chart.

2.2 Loggly

Loggly is one of the most widely used log analysis applications [2]. Loggly can receive logs of the following software systems:

- Web servers Apache, Nginx, interpreters PHP, JavaScript, NET, Java, etc.
- OS family Windows and Unix-like OS.

Loggly can collect data for analysis and create dynamic dashboards to monitor specific performance metrics (Fig. 2).

Loggly can search for data in text and structured databases and archives.

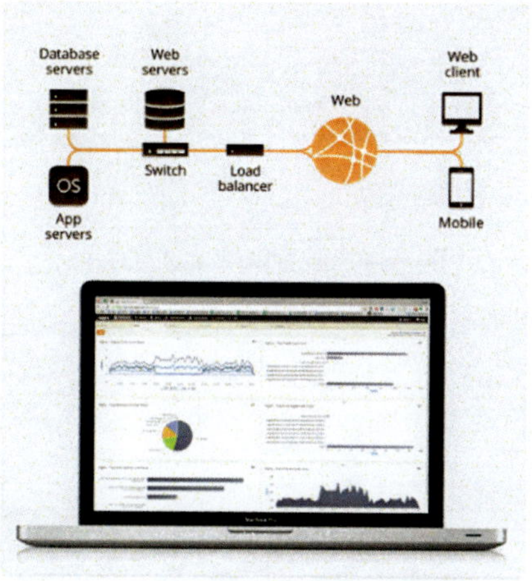

Fig. 2. Loggly log analyzer.

2.3 Sumo Logic

Sumo Logic has a software collector and parser of diagnostic data [3]. Sumo Logic is a service for monitoring and analyzing application and hardware logs. The service can translate and interpret any types of logs online to monitor the status of specific software and hardware (Fig. 3).

Sumo Logic includes built-in algorithms for analyzing diagnostic information for the following applications: mongoDB, AWS Lambda, AWS, Salesforce, Trend Micro, Docker, Linux, Nginx, Apache, IIS, MySQL.

The system primarily serves to automatically analyze and fix performance problems, analyze traffic, proactively monitor security, and create advanced analytics.

2.4 Splunk

Splunk Cloud is a service for collecting, indexing, analyzing and visualizing information of any structured level (Fig. 4).

Splunk can monitor and visualize data and set alert thresholds [4].

Fig. 3. Sumo logic log analyzer.

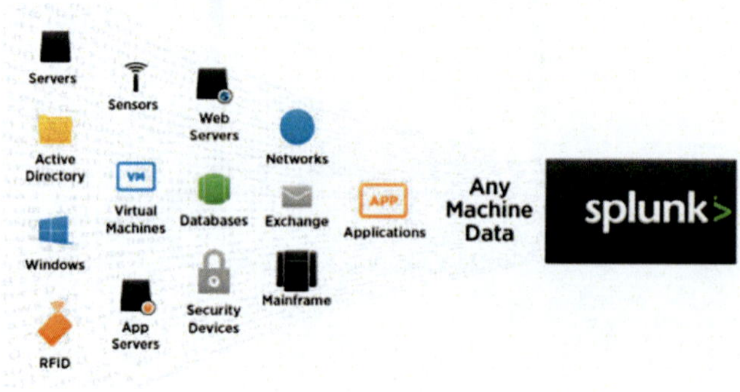

Fig. 4. Splunk log analyzer.

2.5 Logz.Io

Logz.io service hosts ELK (Elastic Search, Logstash, Kibana) as a service [5]. ELK is a free structured and semi-structured data analysis platform.

Logz.io implements the function of data indexing of most modern technologies used in popular information systems (Apache HTTP/Tomcat, Nginx, Hadoop, Heroku, Node.js, mongoDB, postgreSQL, Windows, Linux, AWS), and also provides analysis with the ability to visualize analysis results.

2.6 Timber

The Timber system solves the problems of online monitoring of applications, provides a modern filtering system, fast search [6]. In addition, the Timber service can be integrated with a specific application or platform. Libraries are available for Node, Ruby and Elixir and support the following software platforms: Logstash, Zeit, AWS Lambda, Docker, Linux, Heroku, etc.

2.7 Logsene

Logsene is a cloud-based ELK Stack for managing, automatically collecting, managing and analyzing logs [7]. Logsene supports security protocols and allows you to send logs over encrypted channels from any source, including Syslog.

Logsene runs on AWS and supports SOC, SSAE, FISMA, DIACAP, HIPPA, and more.

It is required to develop a semantic training methodology for analyzing semi-structured resources with the ability to search for anomalies and generate a human-oriented set of recommendations and its implementation in the form of an intelligent software platform, despite a large number of software systems that provide flexible and fast analysis of diagnostic information.

3 The Architecture of the Software for the Intelligent Analysis of Diagnostic Information

3.1 Formal Model of Diagnostic Information Logs

Formally storage diagnostic information logs are:

$$L = \{P, V_P, T_V\},$$

where P is a set of technical characteristics (metrics) of the storage system; V is the set of values of the corresponding elements of P; T is the timestamp of recording an element of the set P for the metric $p \epsilon P$. An example of a fragment of the storage status log looks like this (Table 1):

Table 1. Fragment of the storage log

2021-03-19T12:21:17.000Z,"researchvisor01","researchvisor01 ceph-osd: 2021-03-19 15:21:17.126930 7f259b8eb700 1 -- 172.17.1.111:6802/886215 --> 172.17.1.112:0/1396165398 -- osd_op_reply (387847 rbd_data.11f56b8b4567.000000000000041b [read 2191360~40960] v0'0 uv67 ondisk = 0) v8 -- 0x562fb74b42c0 con 0"
2021-03-19T12:21:45.000Z,"researchvisor01","researchvisor01 ceph-osd: 2021-03-19 15:21:45.234788 7ff95ea32700 1 -- 172.17.1.111:6804/886327 <== client.4557 172.17.1.112:0/1396165398 26931 ==== osd_op (client.4557.0:415098 1.bc 1.dfe99ebc (undecoded) ondisk+read+known_if_redirected e138) v8 ==== 247+0+0 (2572540661 0 0) 0x56402dd99080 con 0x56402deb0800"
2021-03-19T12:20:44.000Z,"researchvisor01","researchvisor01 ceph-mon: 2021-03-19 15:20:44.572710 7f6666ce0700 20 is_capable service=mon command= exec on cap allow *"

Metrics are various characteristics of the analyzed object (software or hardware environment). Most often, metric data is a time series. Several thousand metrics (from each node) can be taken from a working storage system, forming a space of metrics (multidimensional time series).

One of the main tasks to be solved in achieving the goal of increasing the performance of storage systems is the search, detection, analysis and prediction of anomalies in multidimensional time series of metrics taken from the storage system.

Anomaly is some deviation of a technical process from its usual course, such as an increase in the duration of a certain stage of the process or a change in its metrics, etc.

3.2 Anomaly Analysis

Distance-based models use some metric as anomaly rating at [8–10]. Normal data is assumed to be closer together than abnormal data. For example, a distance based on the nearest neighbors method:

$$a_t = \sum_{x_i \in kNN(x_t)} D(x_i, x_t)$$

where D is some metric, $kNN(x_t)$ - k nearest neighbors of x_t.

The model will not be able to give adequate estimates on the data on which it was not trained if it is a weak extrapolator. Almost all machine learning algorithms have these properties. The use of recurrent neural networks solves this problem:

- Recurrent networks naturally support time series. In order to take into account the dependence on time in the data, for such a model it is only necessary to supply the vectors xt in the required order without using complex construction of features.
- At the moment, neural networks give the highest quality of regression in such problems, when the data on the test sample is similar to the data in training, but at the same time they are greatly mistaken in the opposite case. This strong contrast for detecting rare events is a key success factor.

In deep learning, the most effective were LSTM [11] and GRU [12]. These architectures have several advantages:

- The ability to "memorize" very long sequences.
- Fast convergence rate.

The general scheme for detecting anomalies using recurrent neural networks is shown in Fig. 5.

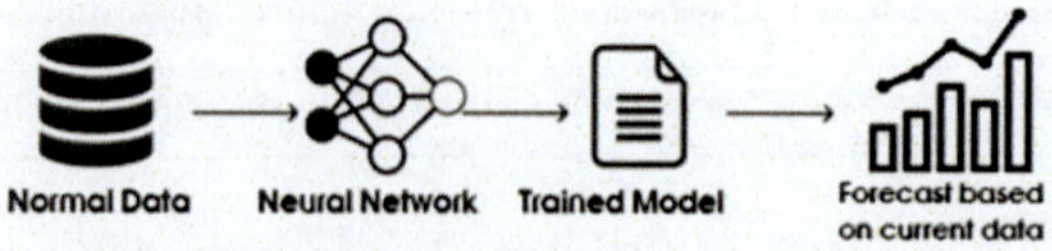

Fig. 5. Search for anomalies using a neural network.

Thus, when using recurrent neural networks, the following values are predicted based on the trained model, and if the real value of the metric diverges from the predicted value, this case can be considered anomalous. In this case, for the model to work effectively, it will be necessary to solve a number of problems that arise.

- The training dataset should not contain abnormal data;
- At any moment of time it is necessary to have an up-to-date neural network model for correct anomaly detection. To do this, it is necessary to determine in advance the "lifetime" of each version of the model and conduct reinforcement learning.

3.3 Choice of Knowledge Base Model

1) Ontology

Ontology is an extended interpretation of semantic networks based on descriptive logic.

In general, any ontology can be viewed as a collection of "triplets": object attribute value (RDF standard). On the other hand, an ontology can contain the following specific elements:

- Classes
- Individuals (class objects)
- Relations:

  A. Datatype Properties;
  B. Object Properties;
  C. Annotation Properties.

- Axioms or functions of interpretation.

It is the possibility of defining axioms that distinguishes ontologies from other semantic structures (thesauri, dictionaries, applied semantic networks). Axioms define the dependencies and constraints imposed on the class, relations and objects of the subject area, represented in the ontology.

The principles of building ontologies are schematically presented in Fig. 6.

Ontologies have become widespread when used as a knowledge base in many expert systems due to:

- Mathematical basis of construction;
- Simplicity of presentation and perception by a person;
- Availability of a large number of presentation formats necessary for efficient processing and analysis of ontologies with appropriate software (RDF/RDFS, Turtle, OWL, OWL 2, OWL Lite, OWL Full, OWL DL, FuzzyOWL, etc.) [13].

2) SWRL

Logical rules include a set of prerequisites (conditions) and consequences. In general, any rule can be represented as follows:

$$a_1 \wedge a_2 \wedge \ldots \wedge a_n \rightarrow b,$$

where $a_1 \ldots a_n$ are atoms of the antecedent; b is the consequent of the rule. The consequent can also consist of several atoms.

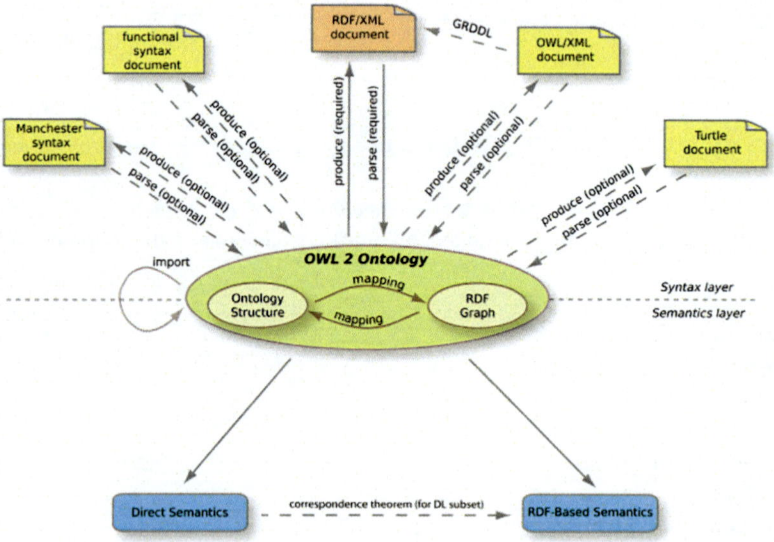

Fig. 6. Principles of constructing subject ontologies.

Rules are a form of knowledge on the basis of which a logical conclusion is made in most expert systems. Elements of the rules must be present in the intelligent platform for semantic analysis of storage diagnostic information. An important advantage of this form of knowledge representation is the presence of a standard for representing rules that is integrated into the ontological form - SWRL (Semantic Web Rule Language).

SWRL is a methodology for describing a mechanism for managing objects and patterns of a problem area. The main advantage of SWRL is the ability to generate new facts from existing knowledge and statements [14].

The simplest SWRL constructs implement production building technology and consist of a set of conditions and consequences written in accordance with the W3C standard. Formally, SWRL rules can be represented in the form of the diagram shown in Fig. 7.

Basic advantages of representing knowledge in the form of SWRL rules:

- Rule atoms include references to domain objects, not the objects themselves. Therefore, each rule can be used for a wide class of objects;
- SWRL notation can be integrated into OWL ontology;
- The format for presenting rules is unified and described in the W3C standard.

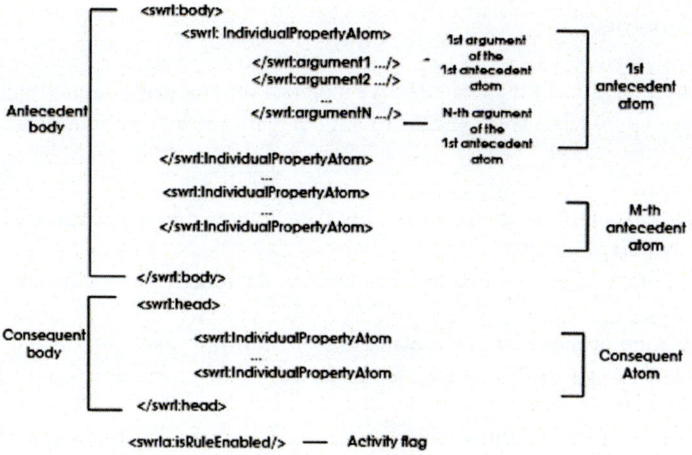

Fig. 7. Scheme of representation of SWRL-rules.

The most effective option for the formation of a knowledge base is the integration of various forms of storage and processing of knowledge with the possibility of its training, by expanding the static (terminological) and dynamic (based on rules and precedents) components of the expert knowledge base.

One of the most effective options for building a knowledge base is the semantic integration of the ontological representation of the subject area [15].

A preliminary diagram of the operation of an intelligent storage system condition monitoring platform is shown in Fig. 8.

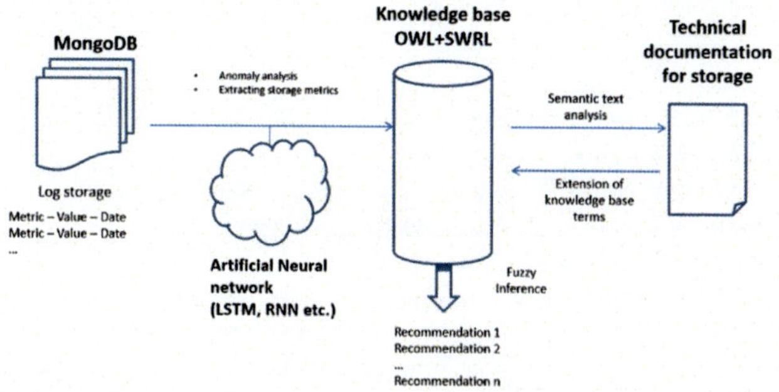

Fig. 8. The principle of operation of an intelligent platform for analyzing storage diagnostic data.

Thus, when constructing an expert knowledge base and logical inference of a recommendation based on the results of the analysis of anomalies extracted from the diagnostic information of the storage logs, it is planned to implement a hybrid knowledge representation model. This model will be obtained by integrating a subject ontology and a set of rules with the implementation of logical inference to generate recommendations and analysis results in a form natural to humans.

4 Conclusions

Based on the analysis of the most effective methods for analyzing semi-structured data, the following approaches were chosen to build a platform for intelligent monitoring of the storage system state:

- Automatic extraction of metrics from big data storage diagnostic information using artificial neural networks.
- Analysis of time series anomalies for diagnostic data metrics using machine learning methods.
- Generalization of the results of the analysis of anomalies and patterns of change and existence of storage performance metrics based on the characteristics of the subject area using a subject ontology.
- The inference of the recommendation to the administrator for correcting the storage system state through the composition of the results of the analysis of anomalies and subject ontology.
- Extension of the static and dynamic components of the storage ontological model through semantic analysis of semi-structured information (extraction and classification of short texts and terms).

Acknowledgments. This work was carried out within the framework of the state contract No. P33-1-26/5 dated February 26, 2021 for the implementation of research work on the topic "Development of methods, models and tools for increasing the productivity and fault tolerance of software-defined storage systems using automated control and analysis of results based on a neural network".

References

1. Papertrail. https://www.papertrail.com. Accessed 15 Sep 2021
2. Loggly. https://www.loggly.com. Accessed 15 Sep 2021
3. Sumo Logic. https://www.sumologic.com. Accessed 15 Sep 2021
4. Splunk. https://www.splunk.com. Accessed 15 Sep 2021
5. Logz.io. https://logz.io. Accessed 15 Sep 2021
6. Timber. https://github.com/JakeWharton/timber. Accessed 15 Sep 2021
7. Logsene. https://sematext.com/logsene. Accessed 15 Sep 2021
8. Ramaswamy, S., Rastogi, R., Shim, K.: Efficient algorithms for mining outliers from large data sets. ACM SIGMOD Rec. **29**, 427–438 (2000)
9. Angiulli F., Pizzuti C.: Fast outlier detection in high dimensional spaces. In: Elomaa T., Mannila H., Toivonen H. (eds.) Principles of Data Mining and Knowledge Discovery. PKDD 2002. Lecture Notes in Computer Science (Lecture Notes in Artificial Intelligence), vol. 2431. Springer, Berlin (2002)
10. Bay, S.D., Schwabacher, M.: Mining distance-based outliers in near linear time with randomization and a simple pruning rule. In: Proceedings of the ninth ACM SIGKDD International Conference on Knowledge Discovery and Data Mining, pp. 29–38 (2003)
11. Hochreiter, S., Schmidhuber, J.: Long short-term memory. Neural Comput. **9**(8), 1735–1780 (1997). https://doi.org/10.1162/neco.1997.9.8.1735

12. Cho, K., et al.: Learning Phrase Representations using RNN Encoder-Decoder for Statistical Machine Translation. arXiv:1406.1078 (2014)
13. Yarushkina, N., Moshkin, V., Filippov, A.: Development of a knowledge base based on context analysis of external information resources. In: Proceedings of the International Conference Information Technology and Nanotechnology Session Data Science, pp. 328–337 (2018)
14. O'Connor, M., et al.: Supporting rule system interoperability on the semantic web with SWRL. In: Gil, Y., Motta, E., Benjamins, V.R., Musen, M.A. (eds.) ISWC 2005. LNCS, vol. 3729, pp. 974–986. Springer, Heidelberg (2005). https://doi.org/10.1007/11574620_69
15. Yarushkina, N., Moshkin, V., Filippov, A., Guskov, G.: Developing a fuzzy knowledge base and filling it with knowledge extracted from various documents. In: Rutkowski, L., Scherer, R., Korytkowski, M., Pedrycz, W., Tadeusiewicz, R., Zurada, J.M. (eds.) ICAISC 2018. LNCS (LNAI), vol. 10842, pp. 799–810. Springer, Cham (2018). https://doi.org/10.1007/978-3-319-91262-2_70

Automatic Diagnostics System for Well Tubing Wax Cleaning Devices

S. N. Fedorov, A. N. Krasnov, and M. Yu. Prakhova(✉)

Ufa State Petroleum Technological University, 1, Kosmonavtov Street, Ufa 450064, Russia

Abstract. Oil extraction by the artificial lift method leads to the formation of asphalt, resin, and paraffin wax deposits (ARPD) in well tubing. Narrowing of the useful pipe cross-section decreases the well flow rate and increases the power consumption of the electric centrifugal pump. Therefore, cleaning of well tubing from paraffin wax is a routine procedure, which is most often performed mechanically using special cleaning devices—scrapers. This method has many advantages, though sometimes scrapers might jam in the well, or the scraper wire might slacken or break off. To ensure the fully automatic operation of such units, they shall have an automatic diagnostic system, which evaluates certain indirect signs to identify any problems, and performs the necessary actions as per a preset algorithm. This study proposes a diagnostic model of the paraffin wax cleaning device based on the results of an active experiment at the operating oil well, as well as the diagnostic algorithm allowing to increase the reliability and control of the dewaxing unit.

Keywords: Well tubing · Paraffin wax · Scraper · Diagnostic model · Diagnostic algorithm

1 Introduction

During oil deposit development, the quality of the produced oil inevitably deteriorates, which obstructs oil extraction, increases the overheads, and leads to various operating problems. For instance, the oil becomes more saturated with the so-called ARPD—asphalt, resin, and paraffin wax deposits. These deposits represent a natural composite material consisting of 10–15 organic and mineral substances and compounds. This material is a cream-like suspension or an emulsion having high adhesion to various surfaces [1]. The ARPD composition to a certain extent depends on the properties and composition of the initial oil, as well as on the deposit location along the path of oil movement. The ARPD composition may change significantly within a single oil-producing region and even a single field. Oil of most fields may contain traces of ARPD up to 30% or more. The amount of dissolved paraffin wax varies from below 1% to 2%.

The deposition of ARPD significantly affects such factors as pressure decrease at the well bottom and the associated disruption of the hydrodynamic equilibrium in the gas-liquid system, hydrocarbon composition of oil phases, temperature decrease in the reservoir and the wellbore, intensive gas emission, gas-liquid mixture speed change, pipe surface condition [2–4]. ARPD deposition intensity depends on the predominance

A. A. Radionov and V. R. Gasiyarov (Eds.): RusAutoCon 2021, LNEE 857, pp. 500–512, 2022.
https://doi.org/10.1007/978-3-030-94202-1_46

of one or more factors that may vary in time and depth, therefore the volume and nature of deposits are not constant as well.

The main objects susceptible to the formation of paraffin wax deposits are submersible pumps, well tubing, flow lines, tanks of commercial oil gathering stations [5]. The most intense paraffin wax deposition happens on the inner surface of the rising pipes.

Field studies [5] have shown that the paraffin wax deposit distribution patterns in pipes of various diameters are approximately the same. The deposit thickness gradually increases from the depth of 500–900 m where their formation starts and reaches its maximum at a depth of 50–200 m from the wellhead, and then decreases to a thickness of 1–2 mm in the wellhead area.

ARPD causes many problems during oil extraction. ARPD reduces the useful cross-section of the well tubing, which significantly reduces the fluid phase flow rate, and increases the power consumption during oil extraction [6–10]. In extreme cases, ARPD blocks well operation due to the complete construction of the well tubing cross-section. For this reason, various methods are used to prevent the ARPD deposition and further removal. Smooth protective coatings are used as preventive measures including varnishes, glass and enamels, as well as physical methods based on the effects of mechanical and ultrasonic oscillations or electrical, magnetic and electromagnetic fields on the extracted and transported products. Special paraffin wax inhibitors are used as well.

Despite the abovementioned measures used to prevent ARPD deposition, certain thermobaric conditions may lead to paraffin wax being released from oil and crystallized on the inner surface of pipes with a sufficiently strong cohesion with pipe surface that excludes the possibility of deposits' falling off under the pressure of the gas-liquid mixture or oil at the corresponding operating conditions. In such cases, paraffin wax is removed using other methods—thermal, chemical or mechanical.

2 Relevance and Scientific Significance

2.1 Relevance of Research

An analysis of the existing ways to combat ARPD deposition shows that one of the most common, simple and affordable methods is mechanical cleaning. The mechanical cleaning of ARPD in oil wells using scrapers (pigs) is a reliable and time-proven method. This method is flexible, simple, has low cost, and a wide range of existing scrapers that meet the most diverse requirements. Nevertheless, many companies are constantly working on the further improvement of paraffin wax removal units, which confirms the scientific and practical significance of the studies.

Mechanical methods imply the removal of the already formed ARPD in the well tubing [11]. For this purpose, a wide range of scrapers of various designs has been developed. The so-called "flying" scrapers are the most widespread and are outfitted with blades, which unfold when moving upwards, providing the additional lift.

The scraper is descended on the wire or a thin steel cable, scraping the deposited paraffin wax from pipe walls as its travels. Scrapers travel down affected by the gravity and attached weights (up to 10 kg). While hoists are used to bring them back up (Fig. 1).

Scraper application frequency for cleaning well tubing from ARPD varies depending on the well flow rate and the “tendency” of the well fluid to deposit paraffin wax. The cleaning procedure typically lasts from 3 to 5 days [6]. Timely cleaning from paraffin wax allows to increase the well flow rate, as well as to decrease the frequency of overhauls.

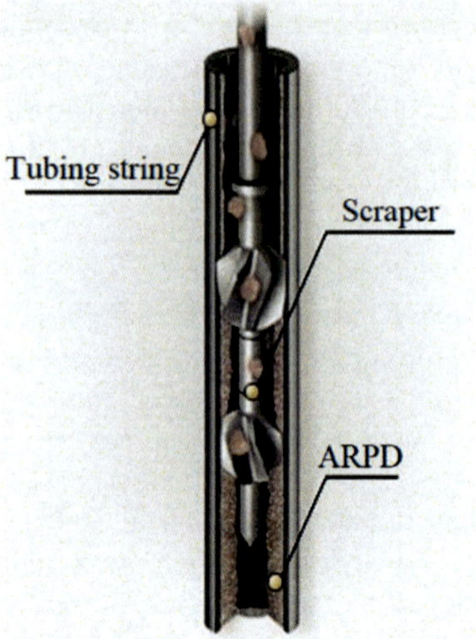

Fig. 1. Cleaning the tubing string from ARPD.

A large number of paraffin wax removal devices are available on the market, which can be controlled manually or automatically, as well as using special control stations. The examples are an automatic paraffin wax removal unit ADU-3, semi-automatic paraffin wax removal units PADU and UDS, installed at a distance of 25–30 m from the wellhead. In these units, the scraper descent is performed manually controlled by the operator using the brake lever, while lifting is automatic. Well paraffin wax removal (WPR) mechanism is designed to clean the inner surface of the well tubing using puncher scrapers in wells with electric submersible pumps in manual and automatic mode. This fully automatic unit is installed directly in the wellhead with the wellhead control equipment.

Despite the large variety of paraffin wax removal units, they all have a number of problems associated with the essence of the mechanical scraping method. The main problems are scraper jamming in the well tubing, scraper wire break, formation of wire loops due to wire slackening, low performance due to well blockage, existing deposits serving as points of additional paraffin wax deposition.

2.2 Problem Definition

Oil producers are interested in the paraffin wax removal units that can operate in fully automatic mode without operator supervision maintaining an optimal oil flow rate, while registering the operating parameters into the non-volatile memory.

To ensure reliable operation of such units, constant state monitoring of the scraper descent and lifting in the well is required for any measured parameters. If the attached weight is insufficient, the scraper may jam. Certain problems may also arise associated with jamming and abrupt upward motion of the scraper in the well, which may break the wire. Monitoring allows diagnosing the problem and preventing the transition of an abnormal situation into an emergency.

The objective of this study is to develop a diagnostic model and an algorithm for diagnosing the condition of well tubing cleaning device, which will increase the service life of the mechanisms, reduce the number of unscheduled repairs and downtime, improve the operation reliability and efficiency of the well tubing cleaning device.

3 Materials and Methods

The creation of diagnostic systems involves solving such problems as the analysis of diagnosed objects, the choice of diagnostic methods, and the development of diagnostic algorithms [12]. The theoretical foundation for engineering diagnostics is the pattern recognition theory and the theory of diagnosability. The pattern recognition in its turn is based on the decision rules. The pattern recognition theory implies the construction of diagnostic models and recognition algorithms—describing the action sequences during diagnostics [13]. Using diagnostic models when studying the object, the pattern recognition theory allows determining the decision rules for recognizing the current state and type of malfunction. The optimal recognition algorithm (sequence) can be developed if any malfunction properties are known. The theory of diagnosability solves problems of determining the rational sequence of finding failed or faulty elements, as well as controlling the object state. Decisions are based on the diagnostic information characterizing the object state.

Thus, the successful operation of a diagnostic system primarily depends on the completeness and adequacy of the diagnostic model, which is used as the base. The diagnostic model (DM) implies a formalized description of an object requiring technical diagnostics, which is necessary to solve diagnostic problems. The description may include analytical, tabular, vector, and graphic forms [14]. The model is required to connect the so-called diagnostic parameters between each other, i.e. those signs of an object that numerically or qualitatively characterize the object's technical condition. Diagnostic parameters are typically divided into four categories—normal, maximum permissible, extreme and emergency values. Diagnostic models can be based on the results of generalization of production experience, logical analysis or theoretical and experimental studies. In this study, diagnostic models were based on the identification from the results of the active experiment during the normal functioning of the object. This method allows addressing the physical nature of the diagnosed object in the fullest scope.

Generally, the essence of engineering diagnostics is to evaluate and forecast the technical condition of the diagnosed object by the results of direct or indirect measurements of the object's state parameters or diagnostic parameters, as well as their deviations from normal values (Salisbury principle) [15]. At the same time, the state parameter or diagnostic parameter values do not provide any estimates of the object's technical condition.

The diagnostic symptom necessary for the operation of any diagnostic algorithm is the difference in the actual and reference values of the diagnostic parameter. Therefore, the diagnostic symptom can be calculated using not only the actual values of the parameters, but also the corresponding reference values obtained during the normal operation of the object.

3.1 Constructing the Diagnostic Algorithm

The diagnostic algorithm is typically developed in several stages [16].

At the first stage, the design specifics are determined, as well as the nature of loading of mechanism elements. The failures and operating conditions are analyzed.

At the second stage, the state dictionary is created, which is a list of possible states and their properties. These data are used to select diagnostic parameters, as well as perform object's condition estimates.

At the third stage, diagnostic parameters and measurement tools are selected. The measurement program is developed.

At the fourth stage, a diagnostic table (matrix) is prepared, containing states and decision rules. In the recognition system, the data required for diagnosis is represented by the rules and logical conclusions such as "if…, then…".

At the fifth stage, the model of a diagnosed object is developed, which is the description of an object in an analytical, graphical, tabular or other forms.

3.2 Active Experiment

The active experiment with the creation of a diagnostic model was carried out on an operating well outfitted with an electric centrifugal pump (ECP). The UOK-NKT device was mounted in the well for cleaning the well tubing string together with the SULS-16 control station remotely controlled via the RS-485 interface.

The experimental studies were conducted to solve the following objectives:

- establish the possibility of developing a diagnostic model based on the available sensors of the tubing string cleaning device;
- determine the dependencies of sensor readings on the tubing string cleaning device condition;
- develop the diagnostic algorithms.

The experimental installation layout is shown in Fig. 2. The scraper is pulled upwards by winding the wire over the drum driven by the electric motor. Weight sensor monitors wire tension and scraper position. The rotational speed converter ensures smooth lifting and lowering of the scraper, measuring of the power consumed by the electric motor, measuring of the power supply voltage, automatic adjustment of the electric motor rotation speed, and stopping the drum rotation as per the weight sensor signal. RPM sensors are based on magnetic reed switches. The sequential triggering of magnetic reed switches allows calculating the roller rotation speed and determining the scraper depth. The direct sequence of triggering indicates scraper descent (increase in the roller rotation) and vice versa. The total number of roller rotations is calculated by the controller. The temperature sensor measures the temperature inside the control station. The current sensor is intended for remote indication of the electrical equipment state (off/on). The indicator's output circuit closes if the current of the controlled circuit is above the setpoint. In case of a stop, the indicator's output circuit opens and the signal is sent to the controller to stop the UOK-NKT unit.

The experiment lasted several months and involved modeling various situations using the control station, while the UOK-NKT unit worked in the well and performed cleaning in normal mode. Numeric data were registered using a special software.

4 Results and Discussion

The results of the experiments allowed to develop a diagnostic algorithm for monitoring the paraffin wax removal unit state.

As already noted, the development of the diagnostic algorithm is carried out in several stages.

1. Analyzing the diagnosed object. For this, the paraffin wax removal unit (PRU) was represented by individual interrelated elements, its possible states were considered (possible combinations of element failures), as well as the technical capabilities of the system of pattern recognition were analyzed characterizing the diagnosed object state. Statistical data were also collected and processed to determine the distribution of possible object state probabilities, as well as the occurrence of failures of its individual elements.

   This stage allowed to create the structure of the well cleaning process model (Fig. 3).
2. Preparation of the object state dictionary involved determining the most informative indirect diagnostic signs of possible PRU state changes, as well as possible defects, which represent the greatest danger to PRU functioning and should be detected during the diagnostic process. These data are used to select diagnostic parameters, as well as perform object's condition estimates. Some of the typical conditions of the well cleaning device and the form of their occurrence are shown in Table 1.
3. Selecting the diagnostic parameters and measurement tools. The diagnostic process usually consists of a series of individual experiments (elementary checks), during which the object is affected in a certain way, and a certain set of control points is used to measure the object's response (reaction) to this effect. The elementary check result is the object response, that is, a set of signals received from the control points. In this case, a formal description of the diagnosis process (diagnostic algorithm) is an unconditional or conditional sequence of elementary checks and rules for analyzing the results of the latter. This stage was performed during the active experiment.

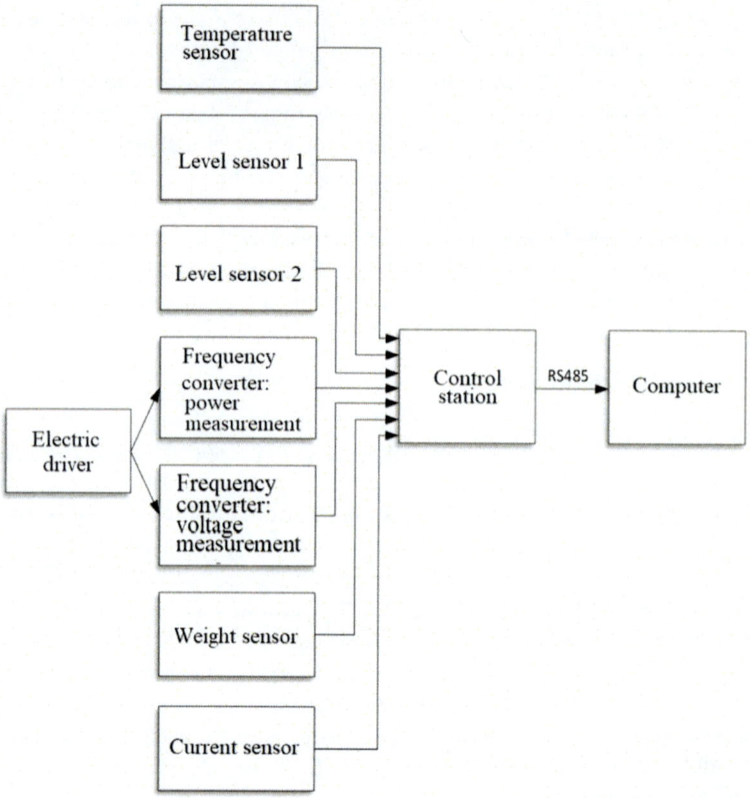

Fig. 2. Schematic diagram of experimental installation.

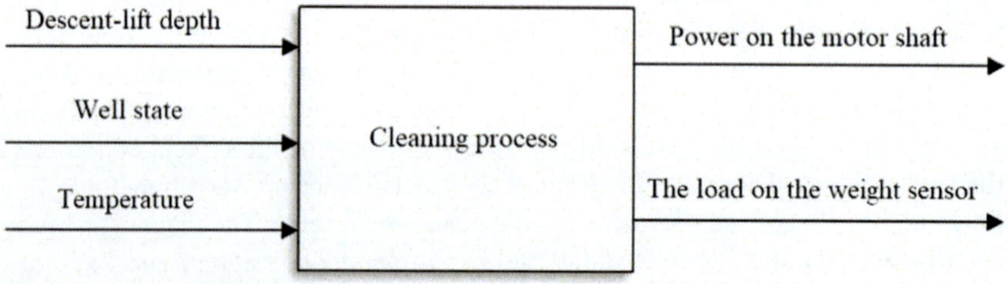

Fig. 3. Structure of the well cleaning process model.

Figure 4 shows the start of the scraper descent from the parking position. The graph shows that when the scraper descent starts, the power consumed by the engine increases and then stabilizes at a certain level. When the scraper passes the obstacle during the descent (Fig. 5), the scraper is repeatedly pulled up and down until the obstacle is passed. During lifting, the power consumed by the drive increases. Figure 6 shows the end of the scraper's descent from the top position to the bottom

Table 1. Dictionary of UOK-NKT unit states.

State property	Cause of occurrence
Reduction of the drive power consumption and the weight sensor readings below the setpoints	Downwards obstacle
Increase in the drive power consumption and the weight sensor readings above the setpoints	Upwards obstacle
Reduction of the weight sensor readings below the setpoint when lifting	Scraper break off
Scraper reaches lifting height setpoint	Top position
Scraper reaches descent depth setpoint	Bottom position
Power consumption by the engine above setpoint	Engine overload failure
Power supply voltage below the setpoint	Low power supply voltage
Power supply voltage above the setpoint	High power supply voltage
Incorrect movement direction—incorrect sequence of level sensor responses	Incorrectly connected level sensor
The specified scraper top level has not been detected during search in the specified top position range	Well top not found in the specified limits
The control station temperature above the setpoint	High control station temperature
Current indicator output circuit open	ECP shutdown
Discrepancy between the values rotation count and the depth of descent at СУЛС control station	Scraper level not confirmed

position and the beginning of lifting. The graph shows that at the beginning of lifting, the power consumed by the engine increases and then stabilizes at a certain level, which is higher compared to the descent. When the scraper is fully lifted into the parking position (Fig. 7), the power consumed by the engine becomes zero.

4. Preparing a diagnostic table (matrix) of states with decision rules. This table is necessary for the optimal arrangement of the diagnostic processes. The controlled object is represented in the diagnostic matrix by the object state function table, which columns correspond to diagnostic symptoms, while the rows represent the causes of failures [17]. Intersections of rows and columns contain symbols that indicate the presence or absence of certain signs for this diagnosis. In fact, this matrix is a recognition system, in which the data required for diagnosis is represented by the rules and logical conclusions such as "if…, then…" (Table 2).
5. Developing the model of the diagnosed object and diagnosing. This stage is the most important. Actually, this stage is the ultimate goal of building a system of automatic PRU state diagnostics. Complex systems and mechanisms require the diagnostic algorithm to be based on a diagnostic matrix (see Table 2). This matrix is a logical model describing the relationship between the sets of diagnostic parameters S and the possible object states. Analyzing this table allows formulating the properties of a set of checks required to solve technical diagnostic problems.

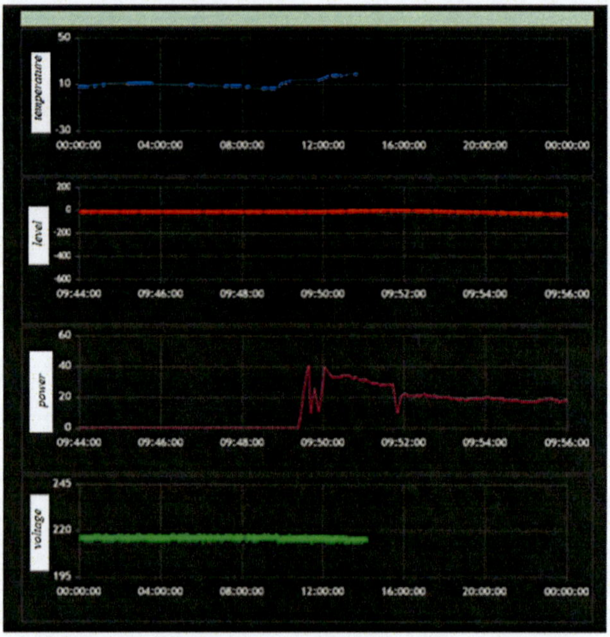

Fig. 4. The start of scraper descent from the parking position.

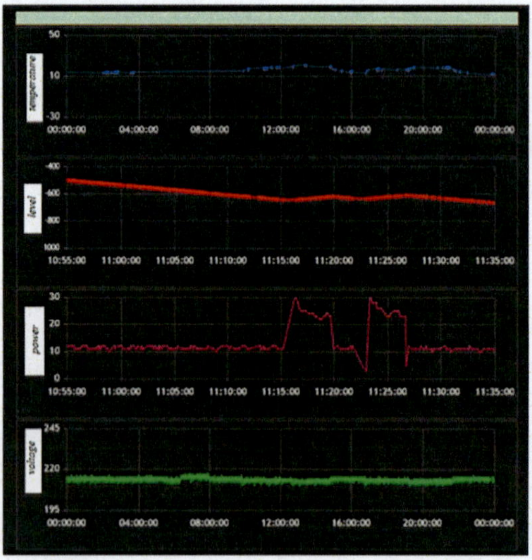

Fig. 5. Scraper descent from the top position to the bottom position passing through the obstacle.

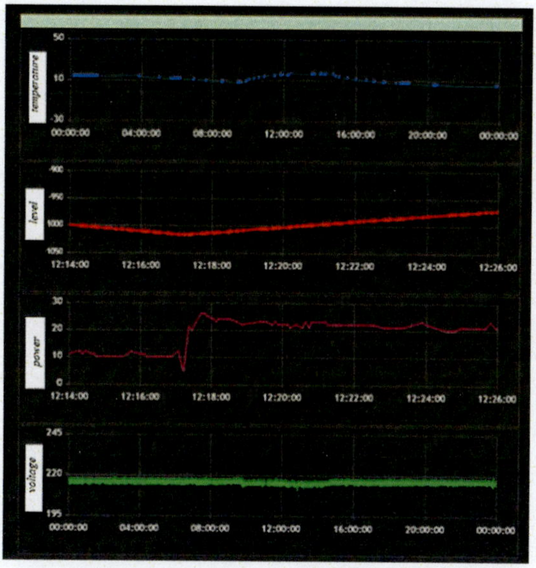

Fig. 6. Scraper descent from the top position to the bottom position and beginning of lifting.

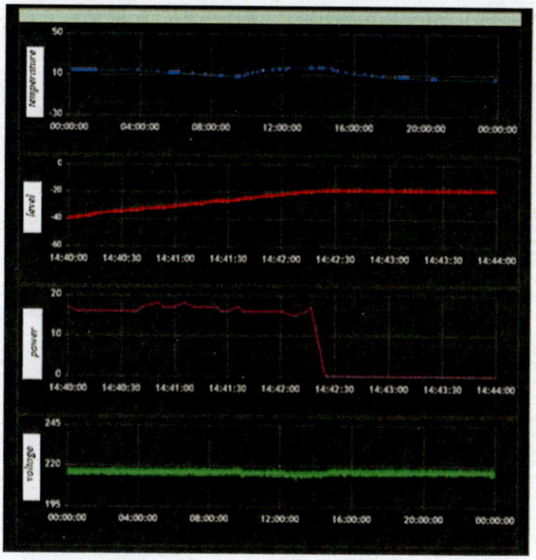

Fig. 7. Scraper is fully lifted into the parking position.

Table 2. States and decision rules.

Diagnosis	Diagnostic parameters								Decision rules
	Temperature, °C (S1)	*Power, W (S2)*	*Voltage, V (S3)*	*Weight, kg (S4)*	*ECP (S5)*	*Level*			
						Rotations, m (S6)	*Direction 1 (S7)*	*Direction 2 (S8)*	
Downwards obstacle (D1)	0	$S2 \leq 75$	0	$S4 \leq 50$	0	0	0	0	$S2 \leq 75$ & $S4 \leq 50 \rightarrow D1$
Upwards obstacle (D2)	0	$S2 \geq 150$	0	$S4 \geq 150$	0	0	0	0	$S2 \geq 150$ & $S4 \geq 150 \rightarrow D2$
Scraper break off (D3)	0	0	0	$S4 \leq 50$	0	0	0	0	$S4 \leq 50 \rightarrow D3$
Top position (D4)	0	$S2 \geq 200$	0	0	0	$S6 \geq -10$	0	0	$S6 \geq -10$ & $S2 \geq 200 \rightarrow D4$
Bottom position (D5)	0	0	0	0	0	$S6 = -1600$	0	0	$S6 = -1600 \rightarrow D5$
Low voltage (D6)	0	0	$S3 \leq 140$	0	0	0	0	0	$S3 \leq 140 \rightarrow D6$
High voltage (D7)	0	0	$S3 \geq 300$	0	0	0	0	0	$S3 \geq 300 \rightarrow D7$
High temperature (D8)	$S1 \geq 80$	0	0	0	0	0	0	0	$S1 \geq 80 \rightarrow D8$
Overload (D9)	0	$S2 \geq 300$	0	0	0	0	0	0	$S2 \geq 200 \rightarrow D9$
ECP shutdown (D10)	0	0	0	0	true	0	0	0	S5 = true → D10
Scraper level not confirmed (D11)	0	0	0	0	0	$S6 < -1600$			$S6 < -1600 \rightarrow D11$
Invalid direction during descent (D12)	0	0	0	0	0	0	false	true	S7 = false & S8 = true → D12
Invalid direction during lifting (D13)	0	0	0	0	0	0	true	false	S7 = tue & S8 = false → D13
Top side not found (D14)	0	0	0	0	0	$S6 > 0$	0	0	$S6 > 0 \rightarrow D14$

The diagnostic matrix allows constructing a diagnosed object model, which represents the diagnostic algorithm (Fig. 8).

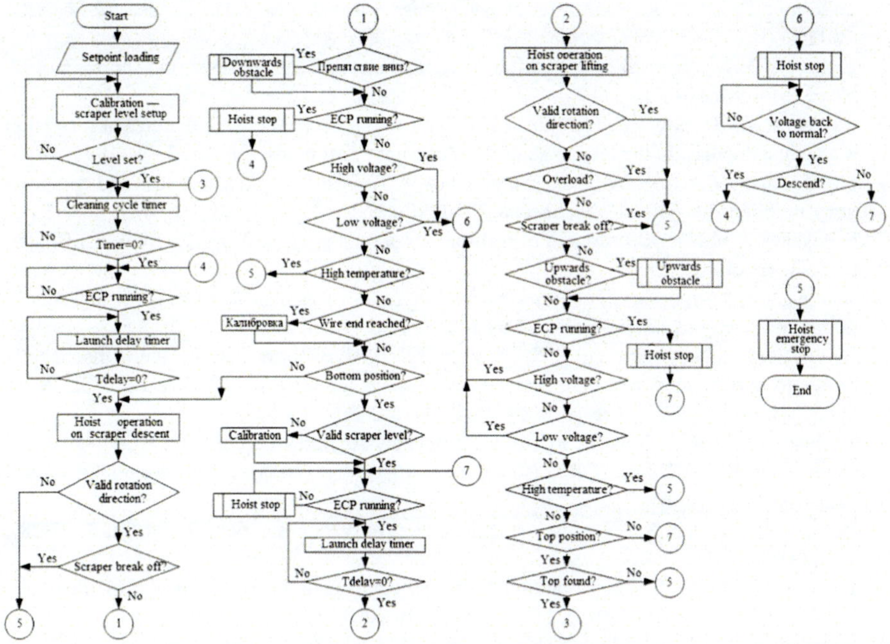

Fig. 8. Diagnostics algorithm.

5 Conclusions

The main results of theoretical, practical and experimental studies are described below.

- The analysis of the existing ways to combat ARPD deposition has shown that one of the most common, simple and affordable methods is mechanical cleaning. A wide range of the existing paraffin wax removal units allows addressing the specifics of operational wells. The disadvantage of the method is the possibility of a scraper wire break off and the formation of loops due to wire slackening. Therefore, the paraffin wax removal units are of particular interest operating in fully automatic mode without operator supervision and connected to an automatic system of scrapper device diagnostics based on certain indirect signs.
- The diagnostic model describing scraper descent and lifting in the well tubing based on the results of an active experiment at the operating oil well allowed to develop the algorithm for diagnosing the state of the paraffin wax removal device. The proposed algorithm allows to facilitate the process of adopting technological solutions in various conditions, increase the reliability and efficiency of the paraffin wax removal unit control.

References

1. Ivanova, L.V., Burov, E.A., Koshelev, V.N.: Asphaltene-resin-paraffin deposits in the processes of oil production, transportation and storage. Electron. Sci. J. Oil Gas Bus. **1**, 268–284 (2011)
2. Tronov, V.P.: Mechanism of formation of resin-wax deposits and their prevention, p. 192. Nedra, Moscow (1970)
3. Kayumov, M., Tronov, V.P., Gus'kov, I.A., Lipaev, A.A.: Account of the features of formation asphaltene deposits in the late stage of development of oil fields. Oil Ind. **3**, 48–49 (2006)
4. Sharifullin, A.V., Baibekova, L.R., Suleimanov, A.T.: Features of the structure and composition of oil deposits. Oil Gas Technol. **6**, 19–24 (2006)
5. Persiyantsev, M.N.: Oil production in difficult conditions, p. 653. Nedra-Business Center, Moscow (2000)
6. Belkina, S.A., Nagaeva, S.N.: Reasons for formation of asphalt resin paraffin deposits in tubing. Bull. Yugorsk State Univ. **3**(42), 7–11 (2016)
7. Minnivaleev, A.N., Bakhtizin, R.N., Kuzeev, I.R., Gabdrahimov, M.S.: Hydro-mechanical device for cleanning pipes inner surface from asphaltene-resin-paraffin deposits. OIJ **2020**, 65–67 (2020). https://doi.org/10.24887/0028-2448-2020-2-65-67
8. Turbakov, M.S., Riabokon, E.P.: Cleaning efficiency upgrade of oil pipeline from wax deposition. Bull. PNRPU Geol. Oil Gas Eng. Min. **17**, 54–62 (2015). https://doi.org/10.15593/2224-9923/2015.17.6
9. Herman, J., Ivanhoe, K.: Paraffin, asphaltene control practices surveyed. Oil Gas J. **97**(28), 61–63 (1999)
10. Bai, J., Jin, X., Wu, J.-T.: Multifunctional anti-wax coatings for paraffin control in oil pipelines. Pet. Sci. **16**(3), 619–631 (2019). https://doi.org/10.1007/s12182-019-0309-7
11. Polyakov, D.B., Shaimardanov, R.F., Aminev, M.Kh., Chudnovsky, A.A., Yauk, V.D.: A method of cleaning a tubing string during oil production by a mechanized method. Russian Federation Patent 2157447. https://findpatent.ru/patent/215/2157447.html
12. Birger, I.A.: Technical diagnostics, p. 234. Mechanical Engineering, Moscow (1978)
13. Lyubimov, I.V., Meshkov, S.A., Ushakov, A.P., Roan, R.V.: Methods and tools for diagnosing technical systems, p. 95. Publishing House of BSTU, SPb (2012)
14. Mekkel, A.M.: Diagnostic model of possible states of an object. T-Comm **11**(7), 31–37 (2017)
15. Pronyakin, V.I.: Diagnostic features in the assessment of the technical condition of machines and mechanisms. J. High. Educ. Inst. Eng. **10**(679), 64–72 (2016)
16. Dubrovin, V.I.: Automated system of technical diagnostics. Radio Electron. Comput. Sci. Control **2**, 111–116 (2005)
17. Khaliullin, F., Akhmetzyanov, I.R.: Peculiarities of composition of diagnostic matrix of Bayes in cleanless diagnostics of internal combustion engines. Int. Res. J. **5**(47), 205–209 (2016). https://doi.org/10.18454/IRJ.2016.47.270

Predicting the Decrease in the Metrological Reliability of Ultrasonic Flow Meters in Conditions of Wax Deposition

A. N. Krasnov, M. Yu. Prakhova, and Yu.V. Novikova(✉)

Ufa State Petroleum Technological University, 1, Kosmonavtov Street, Ufa 450064, Russia
nuv@npaufa.ru

Abstract. Commercial accounting is carried out at all stages of delivery of the produced and prepared oil to the final consumer. The most common are dynamic methods, direct (mass) and indirect (volume-mass). As a rule, accounting is carried out by systems for measuring the quantity and quality of oil (SIQO). In the process of accounting, the work of SIQO is influenced by a large number of external factors that cause the appearance of additional errors. When accounting for crude oil, wax deposition on the inner surfaces of meter lines and flow meters becomes significant. The influence of this factor in practice is not taken into account due to the immeasurability of the thickness of the paraffin layer. The article discusses the effect of paraffin on the metrological reliability of an ultrasonic flow meter. It is proposed to automatically estimate the current value of the thickness of the paraffin layer by the model and determine the value of the additional error. This allows, when approaching the limits of metrological reliability, to timely carry out maintenance of the flow meter and extraordinary control of its metrological characteristics and, in general, to increase the accuracy of accounting.

Keywords: Oil commercial accounting · Metering system · Ultrasonic flowmeter · Paraffin deposition · Predictive diagnostics · Metrological reliability

1 Introduction

After being extracted from the reservoir, oil is subjected to various technological influences, the purpose of which is to prepare it for long-distance transport and bring it to standard conditions. When oil is transferred from production companies to companies that pump it through trunk pipelines, its fiscal accounting is carried out. It should be noted that such accounting is one of the most common operations in the process of delivering produced oil to the final consumer.

According to [1], the mass of oil can be determined by the direct static, direct dynamic, indirect volume-mass dynamic, indirect volume-weight static, and indirect hydrostatic measurement techniques.

Among all the listed techniques, the dynamic ones are the most widespread in the fiscal metering of oil. The operation of systems for measuring the quantity and quality of both commercial (SIQO) and crude (SIQOC) oil is based on these techniques.

A. A. Radionov and V. R. Gasiyarov (Eds.): RusAutoCon 2021, LNEE 857, pp. 513–521, 2022.
https://doi.org/10.1007/978-3-030-94202-1_47

The SIQO comprises a filter unit, a measuring line unit (MLU), the oil quality measuring unit (QMU), a verifier unit (fixed mechanical displacement prover (MDP) or compact prover), as well as the required shut-off and control valves and piping (Fig. 1). All the calculations required are performed by the data processing system (DPS) located in the SIQO's control room [2].

Both SIQO and SIQOC are a measuring system designed for direct or indirect dynamic measurements of mass and quality indicators of oil and oil products. Therefore, each system has a Pattern Approval Certificate of Measuring Instrument [3]. Both systems are almost identical. The SIQOC difference is an additional measurement of the amount of associated petroleum gas along with the measurement of the quantity and quality of the oil produced.

The main measuring instrument in the MLU is a flow meter, volumetric or mass.

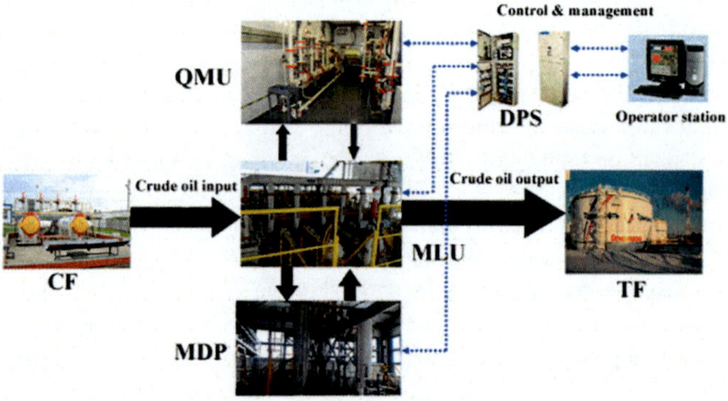

Fig. 1. The SIQO process flow sheet. CF-coarse filter, MDP-mechanical displacement prover, MLU-measuring line unit, QMU-oil quality measuring unit, DPS-data processing system, TF-tank farm.

2 Research Essentials

2.1 Relevance and Overview of the Existing Solutions

In the Russian Federation, Coriolis and ultrasonic flowmeters are the most common ones, which are expanding their presence due to the gradual downturn of turbine meters (Fig. 2).

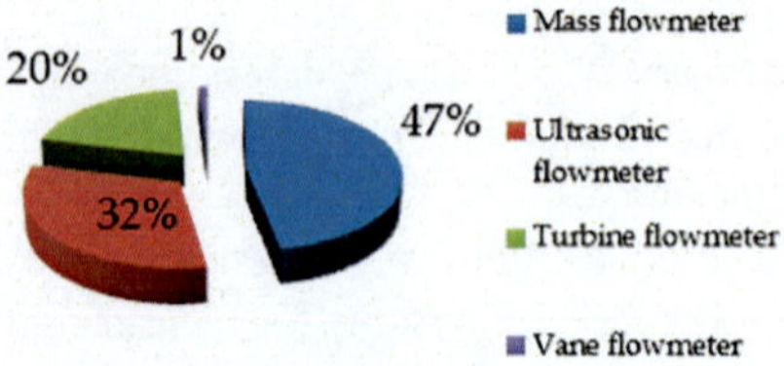

Fig. 2. Percentage of flowmeters used in the SIQO.

Flow measurement is generally a more complex problem than measuring parameters such as temperature or pressure. In the case of measuring the flow rate of crude oil, the flowmeter is subjected to a large number of external impacts, that randomly affect each other.

These are the properties of the medium being measured, such as density, viscosity, the content of paraffins, water, mechanical impurities, and dissolved gases. Further, these are the pipeline operation process parameters (temperature, pressure) and the rheological characteristics of the flow. And these are random external disturbing factors generated by the conditions of the process equipment and measuring instruments (the conditions of the inner pipeline surface, e.g., paraffin deposits both in the pipes and on the sensitive elements of flow meters such as turbine blades or a Coriolis spool, etc.).

The influence of wax deposits on the condition of the measuring lines and the operation of the Coriolis flowmeter is discussed in detail in [4–7]. However, as follows from Fig. 2, about a third of the SIQO (SIQOC) use an ultrasonic flow meter, in which wax deposits also lead to additional error.

So, a major problem of the crude oil pipeline transportation is also paraffin deposition, which occurs under the effect of many factors [8–17]. It is very important to note that in fact, these deposits are uncontrollable since that their thickness cannot be measured directly in the deposition place. This leads to unpredictable changes in the operating conditions of both equipment and measuring instruments, and, as a consequence, affects the efficiency of their maintenance and metrological service.

Let us make a short digression. Maintenance has also passed through several stages in its development. Initially, it was performed based on a 'Run-to-Failure' strategy; then the concept of preventive and predictive maintenance (PPM) was introduced. Despite the discernible advantages of this system, such as control over the TBO duration and regulating the equipment downtime for repairs, it also has obvious drawbacks. In particular, depending on the actual conditions at a particular SIQO, either the PPM is not yet needed as of the moment of its scheduled performance or an emergency arises earlier, requiring unscheduled repair.

With the technological development, primarily due to the emergence of the microprocessor technology, a third strategy has emerged, i.e., repairs based on monitoring of conditions. These three strategies in various combinations are currently used in most enterprises for any type of equipment [18].

Today's APCS systems have a certain redundancy for the number of parameters monitored and the MIs used. Thereat, despite the long measurement data storage period, this information is not used in maintenance or repair, possibly because the existing APCSs ensure control over parameters in very wide ranges without considering their relationship (correlation). Herewith, the analysis of the parameter trends may predict, e.g., the imminent onset of metrological failures of the MIs, which is far from always manifested explicitly.

All these reasons have led to developing predictive diagnostics systems, or predictive analytics, which allow building an equipment operation model (a set of interrelated parameters), train this model based on historical data corresponding to normal operating modes, and then use it in a real-time environment to early predict the equipment or MI failures. The new technology allows passing from the reactive and calendar approaches to the proactive ones (i.e., recognizing signs of approaching failures and preventing them) [19].

Such systems are highly effective since they eliminate sudden failures and provide the possibility of flexible management of risks and production losses due to expanding the planning horizon when making decisions.

Therefore, all studies that allow qualitatively and quantitatively estimating any factors affecting the accuracy of measurements and metrological reliability using the appropriate parameters being measured to promptly intervene in the SIQO operation remain relevant.

As for the paraffin deposits, the urgency of this problem can hardly be overestimated. In many studies, e.g., [20], it is called The Cholesterol of the Oil Industry. According to global estimates [21], the paraffinic oil volume is about 20% of the world's reserves. Its production and transportation require additional measures to control and eliminate paraffin deposits. The solution to this problem is complicated by the still understudied mechanism of paraffin deposition [22]. Herewith, it strongly affects the economic performance of oil industry enterprises since underestimating the deposition intensity and untimely cleaning pipes from paraffin may cause their complete clogging, and overestimating may lead to too frequent cleaning and increasing the production cost. Therefore, predictive diagnostics in this area is very relevant.

So, all studies that allow qualitatively and quantitatively estimating any factors affecting the measurement accuracy using the related parameters being measured to promptly intervene in the SIQO operation remain relevant.

2.2 Statement of Problem

The article examines the influence of wax deposits on the metrological reliability of ultrasonic flow meters. The purpose of the research is to create additional functionality for SIQOC for assessing the current state of an ultrasonic flow meter to predict the moment of its withdrawal for maintenance due to the loss of metrological reliability.

2.3 Theory

According to Fig. 2, ultrasonic flowmeters (USMs) are the second most popular in the SIQOs.

Ultrasonic flowmeters are those measuring the characteristics of ultrasonic waves propagating in a gas or liquid flow by analyzing one or another acoustic effect. The most widespread ones are flowmeters operating on the principle of measuring the difference in the upwind and downwind transit times of sound oscillations in the measured medium stream (the so-called time-pulse flowmeters) (Fig. 3). When sound travels downwind or upwind in the medium studied, the flow rate is added up with or subtracted from the sound velocity, respectively. Then, the measured medium flow rate can be determined by the time the acoustic signal passes the same segment downwind and upwind. Thus, the time difference in the signal travel through the medium, which is directly proportional to the average flow rate, is converted into a volumetric flow output.

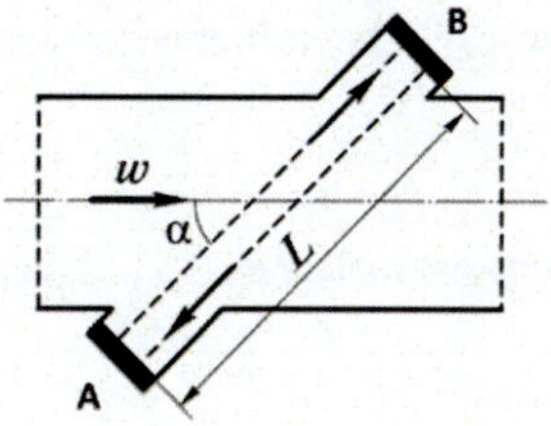

Fig. 3. Principle of ultrasonic flowmeter operation.

A, B are piezoelectric transducers (PETs); L_A is the measuring beam active part length; v is the average oil flow rate in the pipeline; α is the PET to the pipeline axis inclination angle; D is the inner pipeline diameter in the PET installation area.

To pass acoustic oscillations through the flow and read them at the exit, ultrasonic emitters and receivers are used, which are piezoelectric transducers made of monocrystals of zirconate, and barium and lead titanates. They operate in the so-called dual mode, in which each PET alternately acts as a receiver or emitter.

The time of downwind τ_1 and upwind τ_2 ultrasound propagation is determined by the equations

$$\tau_1 = \frac{L}{c + V\cos\alpha}, \quad \tau_2 = \frac{L}{c - V\cos\alpha}. \tag{1}$$

The time difference $\Delta\tau$ is determined by the equation

$$\Delta\tau = \frac{2LV\cos\alpha}{c^2 - V^2\cos^2\alpha}, \tag{2}$$

where L_A is the ultrasonic beam path; V is the flow rate; c is the ultrasound speed in a stationary flow of the measured medium; α is the PET to the pipeline axis inclination angle.

The volumetric flow rate Q is determined by the equation [23]

$$Q = \frac{\pi D^2}{4}\frac{\Delta\tau C_0^2}{2L_A cos\alpha}K \tag{3}$$

where K is the correction factor calculated depending on the flow pattern and the design parameters of the pipeline.

USM is a flowmeter measuring volumetric flow on the area-velocity principle, i.e., the flow rate is determined by the local velocities in the flow section and the area of this section. Therefore, any changes in this area due to, e.g., paraffin deposition on the inner pipe surface and the reduced useful area entail additional errors. Table 1 shows the results of calculating the additional error arising from the reduced pipe section. All velocity meters will have this error since it is caused by the flow contraction and, as a result, an increase in the local velocity. Therefore, at any contamination deposited on the inner pipe surface, all flowmeters of this group begin to overestimate the readings and generate an additional error. An additional error has been calculated for several nominal pipeline diameters for the paraffin layer thickness obtained as a result of simulating its deposition in the measuring line [6]. The values after a month and a year of operation have been taken as an example.

Table 1. Additional error of velocity meters due to a change in the pipeline useful area.

Nominal diameter DN	Rated section area, mm^2	Section area, mm^2, in the presence of a paraffin layer with a thickness, mm		Additional error, %, for a paraffin layer thickness, mm	
		0.082	0.007	0.082	0.007
50	1962.5	1949.65	1 961.40	0.65	0.06
100	7850	7824.27	7 847.80	0.33	0.03
150	17662.5	17623.90	17 659.20	0.22	0.02
200	31400	31348.53	31 395.60	0.16	0.01

The USM principle of measurements determines their two more drawbacks [23].

- Dependence of the ultrasonic velocity on the pipeline temperature and pressure and the physicochemical properties of the measured medium.
- The flow velocity is averaged along the ultrasonic signal path and not over the pipe section.

The second drawback is easily eliminated in multi-beam USMs, and the first one is a stable source of additional error. And while a change in the flow rate pattern due to a change in the pipeline temperature or pressure may introduce an error of up to a few percent (approximately 1.2% per 10 °C), then the impact of deposits may cause an overestimation of readings by tens of percent [24].

2.4 Experimental Results and Practical Significance

As for paraffin deposits, when the ultrasonic beam propagates through them, its speed differs from that laid down in the formula for calculating the flow rate. In oils from Russian fields at normal temperatures, the ultrasound speed is 1,335–1,379 m/s, and in paraffin at the same temperatures –1,460 m/s [25]. However, calculations show that this difference does not virtually affect the oscillation travel time.

Since the USM is actually the same pipeline as the measurement line, in which it is installed, to estimate the additional error caused by the paraffin deposition in its measurement channel, the results of simulating this process in the measurement lines themselves in the Aspen HYSYS environment can be used (Fig. 4).

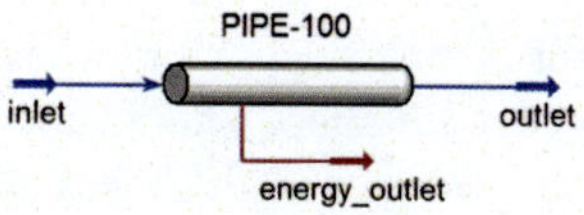

Fig. 4. Measuring section of an ultrasonic flow meter in Aspen HYSYS.

The calculation results are shown in Fig. 5 for a paraffin layer thickness of 0.007; 0.049 and 0.082 mm. These values correspond to a time period of one, six and 12 months. As can be seen, the error value depends on the nominal diameter and after a year it becomes unacceptably large even for relatively large diameters.

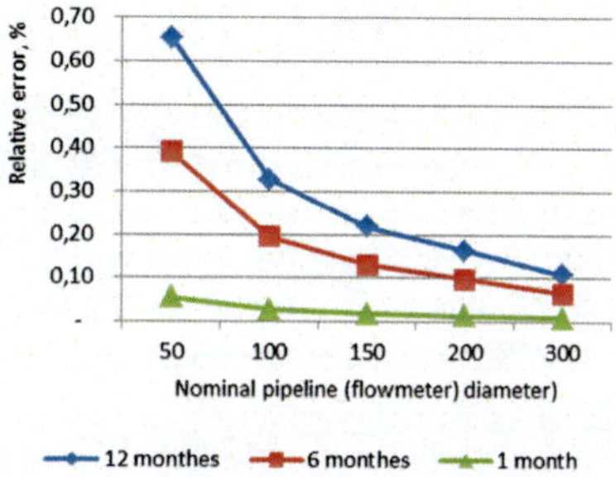

Fig. 5. Relative USM error for a paraffin layer of 0.007; 0.049 and 0.082 mm.

The data obtained allow real-time evaluation of the additional error arising in the ultrasonic flow meter due to wax deposits. This, in turn, makes it possible to control the metrological characteristics (CMC) according to the real state of the flow meter, thereby eliminating an unacceptable decrease in metrological reliability.

In general, the system for diagnosing paraffin deposits in the measuring elements of the SIQOC operates according to the following algorithm. On the basis of the experimentally obtained data on the change in pressure in the ML at the same flow rate, regression equations are constructed that determine the dependence P (Q), their adequacy and significance of individual coefficients are checked [26–28]. Then, in real time, for all MLs participating in the accounting, a comparative analysis of the current regression equations with the original ones is carried out, and the actual value of the thickness of the paraffin layer at a given time is determined. Then, according to the model presented in the article, the additional error of the ultrasonic flowmeter is determined and the proximity to the limit of metrological reliability is estimated. If necessary, a decision is made to hold an extraordinary CMC. All necessary calculations are carried out in DPS, and the results of calculations are displayed upon request to the operator's workstation.

3 Conclusions

The study results allow drawing the following conclusions.

In the case of commercial accounting of oil, it is necessary to exclude or take into account additional errors arising at specific metering units. When it comes to crude oil, one of the sources of this error is wax deposition on all internal surfaces, including where the ultrasonic flow meter is installed. This factor is essentially uncontrollable. The proposed model for assessing the current thickness of the paraffin layer by indirect indications (changes in pressure in the measuring line) makes it possible to timely predict the approach of the flow meter to the limits of metrological reliability and to carry out the necessary maintenance. As a consequence, the accuracy of custody transfer of crude oil is improved.

References

1. GOST R 8.595-2004. State system for ensuring the uniformity of measurements. Mass of petroleum and petroleum products. General requirements for procedures of measurements. https://docs.cntd.ru/document/1200038229
2. Godnev, A., Zorya, E., Nesgovorov, D., Davydov, N.: Custody Transfer of Commodity Flows of Petroleum Products by Automated Systems, pp. 160–170. Max Press, Moscow (2008)
3. MI 2693-2001. The procedure for conducting commercial accounting of crude oil at oil production enterprises. Basic provisions. http://skbpa.ru/publish/mi_2693_2001.pdf
4. Mymrin, I.N., Krasnov, A.N., Prakhova, M.Yu.: Improving the crude oil metering accuracy. In: RusAutoCon: International Russian Automation Conference, pp. 318–324 (2020). https://doi.org/10.1109/RusAutoCon49822.2020.9208099
5. ATabet N.K., Fetisov, V.S.: Information and measuring system with predictive function for determining the thickness of wax deposits in oil pipelines. https://www.elibrary.ru/download/elibrary_35253198_39536416.pdf
6. Krasnov, A.N., Khoroshavina, E.A., Prakhova, M.Yu.: Preventing paraffination of pumping equipment of oil wells. In: AER-Advances in Engineering Research: Proceedings of the International Conference "Actual Issues of Mechanical Engineering" vol. 133, pp. 370–375 (2017)
7. Krasnov, A.N., Lyalin, V.E.: Modeling of flow of two-phase mixture in curved channel pipeline. Vibroeng. Procedia 17534(2016)
8. Theyab, M.A.: Experimental methodology followed to evaluate wax deposition process. J. Pet. Environ. Biotechnol. **9**, 357 (2018). https://doi.org/10.4172/2157-7463.1000357
9. Theyab, M.A.: Wax deposition process: mechanisms, affecting factors and mitigation methods. Open Access J. Sci. **2**(2), 109–115 (2018). https://doi.org/10.15406/oajs.2018.02.00054
10. Botne, K.K.: Modelling wax thickness in single-phase turbulent flow. MSc thesis, Norwegian University of Science and Technology, Department of Petroleum Engineering and Applied Geophysics (2012)
11. Theyab, M.A., Diaz, P.: An experimental and simulation study of wax deposition in hydrocarbon pipeline. Glob. J. Eng. Sci. Res. **4**(7), 27–40 (2017)
12. Pham, S.T., Truong, M.H., Pham, B.T.: Flow assurance in subsea pipeline design for transportation of petroleum products. Open J. Civil Eng. **7**, 311–323 (2017). https://doi.org/10.4236/ojce.2017.72021
13. Leontaritis, K.J., Geroulis, E.: Wax deposition correlation-application in multiphase wax deposition models. Asph Wax, Inc: Offshore Technology Conference, Texas, USA (2011)
14. Solaimany, N.A.R., Dabir, B., Islam, M.R.: Experimental and mathematical modeling of wax deposition and propagation in pipes transporting crude oil. Energy Sour. **27**(1–2), 185–207 (2005)
15. Kasumu, A.S.: An Investigation of solids deposition from two-phase wax–solvent–water mixtures. PhD Thesis, Galgary University (2014)

16. Noville, I., Naveira, L.: Comparison between real field data and the results of wax deposition simulation. SPE 152575 presented at SPE Latin American and Caribbean Petroleum Engineering Conference, Mexico (2012)
17. Zheng, S.: Wax deposition from single-phase oil flows and water-oil two-phase flows in oil transportation pipelines. PhD thesis (Chemical Engineering), The University of Michigan (2017)
18. Bobrovitsky, V.I., Sidorov, V.A.: Mechanical Equipment: Maintenance and Repair, p. 238. Donetsk, Ukraine (2011)
19. Vlasov, A.I., Grigoryev, P.V., Krivoshein, A.I.: Model of predictive equipment maintenance with application of wireless touch networks. Reliab. Qual. Complex Syst. **2**(22) (2018). https://doi.org/10.21685/2307-4205-2018-2-4
20. Sousa, A.: Numerical simulation of wax deposition in pipelines and wells. In: Proceedings of the Annual Meeting Master of Petroleum Engineering, Instituto Superior Técnico (2016). https://fenix.tecnico.ulisboa.pt/downloadFile/282093452021100/06.AnaSousa_AM16.pdf
21. StubsjØen, M.: Analytical and Numerical Modeling of paraffin Wax in Pipelines. Master thesis. Norwegian University of Science and Technology (2013)
22. Rosvold, K.: Wax deposition models. Master thesis. Norwegian University of Science and Technology (2008)
23. Gershman, E.M., Pruglo, S.D., Fafurin, V.A., Borisov, A.A., Sabirzyanov, A.N.: Two-phase flow metering by ultrasonic flow meters. Trans. Acad. **1**, 30–41 (2017)
24. Vatin, N.I., Kurkin, A.G.: Consideration of the effect of deposits on the pipe wall on the readings of a traditional ultrasonic flow meter. https://elib.spbstu.ru/dl/016.pdf/download/016.pdf
25. Utkin, A.V., Sosikov, V.A., Zubareva, A.N.: Experimental study of shock-wave processes in solid and liquid paraffins. J. Tech. Phys. **84**(12), 65–72 (2014)
26. Falovsky, V.I., Khoroshev, A.S., Shakhov, V.G.: Modern approach to modeling the phase transformations of hydrocarbon systems using the peng-robinson equation of state. Bull. Samara Sci. Center Russian Acad. Sci. **4,** 120–125 (2011)
27. Peng, D.Y.: A new two–constant equation of state. Ind. Eng. Chem. Fundam **15**, 59–64 (1976)
28. Timofeeva, D.A., Mymrin, I.N.: Modeling the process of wax deposition in the measuring line SIKNS. In: In: Proceedings of the VIII All-Russian Scientific and Technical Conference "Prospects for the Automation Of Technological Processes of Oil And Gas Production, Transportation And Processing", Ufa, Russian Federation, pp. 52–54 (2019)

A Model of a Distributed Information System with the Possibility of Dynamic Distribution of the Functions Performed

A. A. Sychugov(✉)

Tula State University, 2, Lenin Avenue, Tula 300012, Russia
xru2003@list.ru

Abstract. This article proposes a model of a distributed information system with the possibility of dynamically distributing the functions performed across the physical nodes of the system by changing its structure. The model is based on a matrix representation of the physical and logical structures of the system and the procedure for establishing correspondence between them. Restrictions are introduced, the observance of which is necessary for the formation of a set of equivalent transformations. The implementation of the proposed model is based on the theory of multi-agent systems. The main provisions of this theory, the types of agents used, and their structure are described. The properties of mobile agents as the main tool for implementing the model are described. A fuzzy relational model is proposed as the basis of the implementing agent. It is proposed to use a vertically organized architecture of a multi-agent system. The application of the proposed model in the construction of distributed information systems for various purposes, in particular, systems for monitoring the state of industrial facilities, will increase their reliability and safety.

Keywords: Information systems · Multi-agent systems · Monitoring systems · Dynamic structure

1 Introduction

Operational and reliable collection of information by monitoring systems does not lose its relevance despite constant work in this direction [1–3]. New structures of such systems and technologies for data collection and transmission are being developed. In particular, the joint use of stationary and movable sensors makes it possible to create monitoring systems that are resistant both to external influences (IW) [4–6] and to the conditions of data collection and transmission.

The distribution and dynamism of the monitoring processes of modern technical systems motivates the construction of monitoring systems in the form of distributed information systems with the possibility of dynamic changes in the structure, which undoubtedly entails the need to use mechanisms for automatic distribution of functions performed by individual nodes (distribution of intelligence). The construction of such systems is a difficult task [7], associated with a complex hierarchy, the inclusion of a

A. A. Radionov and V. R. Gasiyarov (Eds.): RusAutoCon 2021, LNEE 857, pp. 522–529, 2022.
https://doi.org/10.1007/978-3-030-94202-1_48

large number of technical elements in it, the heterogeneity of data transmission channels between the nodes of the system.

2 Materials and Methods

The construction of a distributed information system with the possibility of dynamic distribution of the functions performed when its structure changes (DIS DD) is associated with a number of difficulties [7] associated with the complex hierarchy of the system, the inclusion of a large number of technical elements in it, the heterogeneity of data transmission channels between the nodes of the system.

DIS DD has logical and physical structures [8–10]. The logical structure is understood as the entire volume of monitoring and analytical functions performed by the system, the physical structure is a set of computing devices and physical communication channels that implement the logical structure.

The logical structure can be represented as a graph [11] of the logical structure, which is given by the set:

$$\Lambda = \{V, E\} \tag{1}$$

where $V = \{v_1, v_2, ..., v_{N_V}\}$ – is the set of vertices of the graph, each of which is a separate node of the logical structure of the system, which makes sense to be called a logical node; $E = \{e_1, e_2, ..., e_{N_V}\}$ – is the set of arcs of the graph, meaning the transfer of data between individual logical nodes of the system.

The nature of the transmission flows and information processing procedures in the system is reflected by setting the vertices and arcs of the graph of weights. Each vertex $v_i \in V$ is assigned a weight r_i, which denotes the number of elementary computational operations processed in the node per unit of time. Each arc $e_i \in E$ is assigned a weight d_{ij}, which denotes the amount of data transmitted per unit of time by a logical node $v_i \in V$ to a logical node $v_j \in V$.

The given characteristics are more convenient to denote in the form of a processing matrix **R** and a transmission matrix **D**:

$$R = \begin{pmatrix} r_1 \\ r_2 \\ ... \\ r_{N_V} \end{pmatrix} \tag{2}$$

$$D = \begin{pmatrix} d_{11} & d_{12} & ... & d_{1N_V} \\ d_{21} & d_{22} & ... & d_{2N_V} \\ ... & ... & ... & ... \\ d_{N_V1} & d_{N_V2} & ... & d_{N_Vn} \end{pmatrix} \tag{3}$$

The physical structure of DIS DD can also be represented as a graph reflecting a set of physical nodes of DIS DD and a set of connections between them [12]:

$$\Theta = (U, \Xi) \tag{4}$$

$U = \{u_1, u_2, ..., u_{N_U}\}$ – a set of graph vertices, each of which represents a physical node (computing device) of the monitoring system; $\Xi = \{\zeta_1, \zeta_2, \ldots, \zeta_{N_U}\}$ – the set of edges of the graph, meaning a physical digital communication channel between nodes.

The technical characteristics of the physical nodes and communication channels in the system are reflected by setting the vertices and arcs of the graph of weights. Each vertex $u_i \in U$ is assigned to a weight β_i indicating the performance of the physical node (the maximum number of operations that a node can perform per unit of time). Each edge $\zeta_i \in \Xi$ is assigned to a weight s_{ij}, which indicates the maximum amount of data that can be transmitted per unit of time [13]. Here it is assumed that the graph of the physical structure is fully connected, that is, there is a physical possibility of data transfer between any pairs of physical nodes.

The task of implementing DIS DD is to establish a correspondence between the logical and physical nodes of the system [7], which can be set by a matrix of assignments Φ, each element $\phi_{ij} \in \Phi$ denotes the assignment of a logical node $v_i \in V$ to a physical node $u_j \in U$, where the rows are logical nodes and the columns are physical ones:

$$\Phi = \begin{pmatrix} \phi_{11} & \phi_{12} & \ldots & \phi_{1N_U} \\ \phi_{21} & \phi_{22} & \ldots & \phi_{2N_U} \\ \ldots & \ldots & \ldots & \ldots \\ \phi_{N_V 1} & \phi_{N_V 2} & \ldots & \phi_{N_V N_U} \end{pmatrix}. \tag{5}$$

Here the element φ_{ij} is determined in the following way:

$$\varphi_{ij} = \begin{cases} 1, & v_i \longrightarrow u_j \\ 0, & \text{otherwise.} \end{cases}, \tag{6}$$

The design process of DIS DD is characterized by the presence of a certain set of possible equivalent transformations of this system – various implementation options:

$$\{\Phi_1, \Phi_2, \ldots\}. \tag{7}$$

The presence of this set indicates the fundamental presence of the adaptivity property of DIS DD.

For all Φ_i, we can define some criterion $C(\Phi_i)$ that is actually an objective function and has the property that if an option Φ_1 is preferable to an option Φ_2, then $C(\Phi_1) > C(\Phi_2)$ and vice versa. The introduction of this criterion allows us to cut off obviously suboptimal transformations when implementing the adaptivity property of the system.

At the same time, it can be assumed that any equivalent transformation of the system [14] leads to unambiguously known consequences and the given criterion $C(F_i)$ numerically expresses an estimate of these consequences, then an acceptable variant of the transformation F^* is the one that satisfies the condition:

$$C_{\min}(F) < C(F^*) < C_{\max}(F) \tag{8}$$

It should be noted that the set of equivalent transformations should be formed under the following restrictions.

The physical node cannot be overloaded, that is, the condition of matching the node load and its performance must be met:

$$\sum_{i=1}^{N_V} r_i\phi_{i1} \leq \beta_1, \ \sum_{i=1}^{N_V} r_i\phi_{i2} \leq \beta_2, \ldots \ \sum_{i=1}^{N_V} r_i\phi_{iN_U} \leq \beta_{N_U} \tag{9}$$

The physical communication channel cannot be overloaded, that is:

$$\sum_{i=1}^{N_V} d_{i1}\phi_{i1} \leq s_{i1}, \ \sum_{i=1}^{N_V} d_{i2}\phi_{i2} \leq s_{i2}, \ldots \ \sum_{i=1}^{N_V} d_{i3}\phi_{iN_U} \leq s_{iN_U} \tag{10}$$

Several logical nodes can be assigned to one physical node, and all logical nodes must be implemented:

$$\sum_{j=1}^{N_U} \phi_{1j} > 0, \sum_{j=1}^{N_U} \phi_{2j} > 0, \ldots \ \sum_{j=1}^{N_U} \phi_{N_V j} > 0 \tag{11}$$

Thus, we can talk about the problem of finding a set of permissible equivalent transformations satisfying the condition (8), under given constraints (8), under given constraints.

The minimum necessary requirement for the possibility of forming a set () is the fundamental possibility of performing the entire set of operations (implementing a logical structure) of DIS DD on the existing physical structure. This requirement can be expressed by the following relation:

$$\sum_{i=1}^{N_U} r_i \leq \sum_{i=1}^{N_V} \beta_i \tag{12}$$

At the same time, it should be noted that a similar condition for transmission channels is not mandatory, since, in general, there may be such an option that there will be no data transmission over physical channels, in a particular case, when all logical nodes are implemented on one physical node:

At the same time, it is obvious that in the presence of condition (11), the only option for the objective function will be the amount of data transmitted over physical communication channels:

$$C(F) = \sum_{j=1}^{N_U} \sum_{i=1}^{N_V} \left(1 - \phi_{ij}\right) d_{ij} \tag{13}$$

Next, we need to determine the minimum and maximum values of the objective function.

The minimum value of the objective function $C_{\min}(F) = 0$ in the case of implementing a logical structure on a single physical node.

The maximum value of the target function:

$$C_{\max}(F) = \sum_{j=1}^{N_U} \sum_{i=1}^{N_V} d_{ij} \tag{14}$$

In this case, the following condition must be met:

$$\sum_{j=1}^{N_V} \sum_{i=1}^{N_V} d_{ij} \le \sum_{j=1}^{N_U} \sum_{i=1}^{N_U} s_{ij} \tag{15}$$

To determine the assignment matrix (5), it is proposed to use the mechanisms of multi-agent systems [15, 16] based on the concept of a software agent. A software agent is an independent process that has the characteristics of artificial intelligence, works independently or jointly with other agents, able to respond in a timely manner to changes in its environment and initiate actions that affect its environment [16]. It should be noted that the theory of software agents and multi-agent systems based on them is well developed and described in the special literature [15–21]. The author managed to find the most qualitative review on this issue in [15].

The generalized model of software agents includes four levels:

- The software agent
- Agent management component: Tracks agents on a specific computing device, and also provides a mechanism for creating and destroying agents
- Directory Service: The agent can use it to find out about the existence of other agents
- The communication channel between the agents.

Of the currently known four main types of software agents [16]: 1) cooperative; 2) mobile; 3) interface and 4) information, mobile agents that can move from one node to another are of the greatest interest.

A mobile agent (MA) can move from one location to another until approval is received from the delivery location. At the same time, the places of dispatch and delivery can be both on the same or on different computing devices. In any case, the mobile agent initiates the transfer by executing a command that has the address or name of the delivery point as an argument. The following command is executed by the MA at the place of delivery, spaced away from the place of dispatch. Thus, information technology based on MA reduces the work of a distributed automated system for accessing program instructions.

MA can programmatically interact with the environment it visits, as well as with other agents it meets in these places, with their consent. The MA usually travels in order to receive the service offered to it in a remote place, independently implementing the method, time and place of fulfilling the final goal set by the user.

The main advantages of mobile agents are: productivity; the ability to automate processing processes; ease of software installation [16].

To solve the problem, the following types of mobile agents are offered:

- Agent Manager (central authority) conducts a decomposition of the initial problem into separate subproblems, which are distributed among the Executive Agents, that is, in fact forms the logical structure of the system (1), and monitors meeting the conditions (9), (10), (11).

- Executing Agent solves a particular task assigned to it by the Agent Manager;
- Client Agent constantly functions on each physical node $u_i \in U$ in order to control the overall load, processor load and memory usage, as well as ensuring the safe functioning of the mobile Executing Agent on this node, protecting the node from potentially destructive actions of the agent.

Separately, we should focus on the Agent Manager as a mobile agent. It is assumed that this type of agent, as a key element of the system, can also move through individual nodes of the information system and, in principle, function on any available computing device in order to prevent possible negative impacts on this type of agents.

Taking into account the specifics of the tasks solved by the information system, in the process of work, implementing agents will process different types of data, for this reason, it is proposed to use a fuzzy relational model as the basis of the implementing agent.

$$\mathrm{R: A \to B}, \quad R = \begin{pmatrix} r_{11} & r_{12} & \cdots & r_{1p} \\ r_{21} & r_{22} & \cdots & r_{2p} \\ \cdots & \cdots & r_{jk} & \cdots \\ r_{m1} & r_{m2} & \cdots & r_{mp} \end{pmatrix} \tag{16}$$

where $A = \{A_1, A_2, ..., A_m\}$ - a set of linguistic terms defined on X with membership functions $\mu_{\mathrm{A_j}}(x) \in [0, 1]$ for j = 1,…, m; $x = (x_1, x_2, ..., x_m)$ - a vector of fuzzy current parameters of the network functioning; B $= \{B_1, B_2, ..., B_m\}$ - a set of linguistic terms defined on Y with membership functions f $\mu_{\mathrm{B_k}}(\mathrm{y}) \in [0, 1]$ for k = 1,…, p; y – a fuzzy output variable which value from 0 to 1 is an estimate of the presence of an anomaly; $r_{jk} \in [0, 1]$, j = 1,…, m, k = 1,…, p.

When obtaining the values of terms $A' = \{A'_1, A'_2, \ldots, A'_m\}$ - a fuzzy set reflecting a vector of values of different types, including symbolic, variables with the values of membership functions $\mu_{\mathrm{A}'_1}(x) \in [0, 1]$, it allows us to get an estimate of the anomaly as a result of fuzzy inference

$$\mathrm{B}' = \mathrm{A}' \circ (\mathrm{A} \to \mathrm{B})$$

The architecture of a multi-agent system is determined by the conceptual model of the agent and the mechanism of interaction of agents in joint functioning [16].

Based on the analysis of the advantages and disadvantages, as well as the features of the functioning of each of the currently known architectures (horizontally organized and vertically organized) [10], this work uses a vertically organized architecture in which only one of the levels has access to the level of perception and actions, and each of the other levels communicates only with a pair of directly adjacent levels.

One of the advantages of such architecture is that the problem of managing the interaction of levels is quite simple, since the output information of each of the levels always has an addressee.

The disadvantage of a vertically organized architecture is the possibility of overloading the execution (actions) level. It should be noted that this disadvantage is not critical and therefore does not have a tangible impact on the operation of the entire information system as a whole.

3 Discussion

The application of the model proposed in this article in the construction of systems for monitoring the state of industrial facilities will increase their reliability and safety due to the dynamic redistribution of the performed logical functions for individual nodes of the system in case of failures of individual nodes or communication channels as a result of the impact of destructive factors of man-made, anthropogenic or natural character.

4 Results

The main result of the work is a model of a distributed information system with the ability to dynamically distribute the functions performed over the physical nodes of the system. The model is based on a matrix representation of physical and logical structures. The formation of a set of equivalent transformations is performed subject to a number of restrictions. The theory of multi-agent systems is used to implement the model. The properties of the used mobile agents are described. A vertically organized architecture of a multi-agent system was used.

Acknowledgments. The research was carried out with the financial support of the Russian Foundation for Basic Research within the framework of the scientific project No. 19-07-01107\19.

References

1. Carminati, M., Kanoun, O., Ullo, S.L., Marcuccio, S.: Prospects of distributed wireless sensor networks for urban environmental monitoring. IEEE Aerosp. Electron. Syst. Mag. **34**(6), 44–52 (2019). https://doi.org/10.1109/MAES.2019.2916294
2. Zhang, S., Wang, H., He, S., Zhang, C., Liu, J.: An autonomous air-ground cooperative field surveillance system with quadrotor UAV and unmanned ATV Robots. In: IEEE 8th Annual International Conference on CYBER Technology in Automation, Control, and Intelligent Systems (CYBER), pp. 1527–1532 (2018). https://doi.org/10.1109/CYBER.2018.8688331
3. Popescu, D., Vlasceanu, E., Dima, M., Stoican, F., Ichim, L.: Hybrid sensor network for monitoring environmental parameters. In: 28th Mediterranean Conference on Control and Automation (MED), pp. 933–938 (2020). https://doi.org/10.1109/MED48518.2020.9183165
4. Andreeva, O.N., Kournasova, E.V.: Fuzzy cognitive model for identification and analysis of destabilizing factors and technogenic situations. Bull. Mech. Eng. **2**, 81–88 (2019)
5. Dmitriev, O.N., Novikov, S.V.: Preventing faults in machine tools for critical cooperative and distributed industrial productions. Russ. Engin. Res. **39**, 55–59 (2019). https://doi.org/10.3103/S1068798X19010027
6. Evdokimenkov, V.N., Kim, R.V., Popov, S.S.: Risk management by trend analysis of flight information. Russ. Engin. Res. **40**, 160–163 (2020). https://doi.org/10.3103/S1068798X20020136
7. Buslenko, N.: Modeling of Complex Systems. Nauka, Moscow (1986)

8. Danilkin, F., Novikov, A., Sedelnikov, Y., Sychugov, A.: Application of integer quadratic programming in the problems of designing monitoring systems. Izvestiya TSU. Tech. Sci. **3**, 288–295 (2012)
9. Sedelnikov, Yu.: Methods and means of constructing information and measuring systems for ensuring integrated safety of industrial facilities, Dissertation, Tula State University (2012)
10. Sovetov, B.: Modeling of Systems. Higher School, Moscow (2005)
11. Park, J., McKay, S., Wright, E.: Data transmission in control and management systems. IDT Group (2007)
12. Swami, M., Thulaliraman, K.: Graphs, Networks and Algorithms. Mir, Moscow (1984)
13. Sedelnikov, Y., Kiselev, V., Martynenko, A.: Methods and algorithms for solving problems of integer quadratic programming based on the duality theory. In: Scientific works of the 28th Scientific session dedicated to the Radio Day, pp. 50–56 (2010)
14. Tanenbaum, E., Van Steen, M.: Distributed systems. Principles and paradigms. Peter, St. Petersburg (2003)
15. Gorodetsky, V., Grushinsky, M., Khabalov, A.: Multi-agent systems (review). Artificial intelligence news. TSNIEIugol, Moscow (1998)
16. Emerson, E., Halpern, J.: Sometimes and not never revisited: on branching time versus linear time temporal logic. J. ACM **33**(1), 218–233 (1986)
17. FIPA (Federation of Intelligent Physical Agents – Home Page). http://www.cselt.stet.it/fipa/fipa_rationale.htm
18. Hintikka, J.: Knowledge and Belief. Cornel University Press Ithaca, New York (1962)
19. Maes, P.: Agent that reduce work and information overload. Commun. ACM **37**(7), 30–40 (1994)
20. Wooldridge, M., Jennings, N.R.: Agent theories, architectures, and languages: a survey. In: Wooldridge, M.J., Jennings, N.R. (eds.) ATAL 1994. LNCS, vol. 890, pp. 1–39. Springer, Heidelberg (1995). https://doi.org/10.1007/3-540-58855-8_1
21. Müller, J.P., Pischel, M., Thiel, M.: Modeling reactive behaviour in vertically layered agent architectures. In: Wooldridge, M.J., Jennings, N.R. (eds.) ATAL 1994. LNCS, vol. 890, pp. 261–276. Springer, Heidelberg (1995). https://doi.org/10.1007/3-540-58855-8_17
22. Glaschenko, A., Ivaschenko, A., Rzevski, G., Skobelev, P.: Multi-agent real time scheduling system for taxi companies. In: Proceedings of 8th International Conference on Autonomous Agents and Multiagent Systems (2009). http://5fan.ru/wievjob.php?id=31838

Author Index

A. A. Radionov and V. R. Gasiyarov (Eds.): RusAutoCon 2021, LNEE 857, pp. 531–532, 2022.
https://doi.org/10.1007/978-3-030-94202-1